CONNECTIVITY IN GEOMORPHOLOGY

This edited work provides the first comprehensive account of how connectivity concepts and methods are applied in geomorphology. Addressing both qualitative and quantitative aspects, this volume demonstrates how the powerful conceptual framework of connectivity can be used to effectively describe material transfer between geomorphic system components.

The book begins by introducing the key elements of connectivity science, drawing from a broad range of disciplines. The latest research on connectivity is then presented for each major process domain, including hillslopes, rivers and glaciers. Methods of quantification and measurement are described, providing an overview of methodologies and indices that can be used to assess connectivity as a property of soils and landscapes, and approaches for modelling connectivity are reviewed. The book concludes with an examination of applications of connectivity thinking in environmental management.

Accessible and self-contained, this text is a key resource for practitioners, researchers and graduate students in geomorphology.

RONALD E. PÖPPL is Senior Researcher at the BOKU University Vienna and a research group leader of the HI-CONN Working Group (Human Impact and Connectivity). His research interests include water and sediment connectivity in river (catchment) systems, as well as fluvial and hillslope geomorphology, with a focus on human impact.

ANTHONY J. PARSONS is Emeritus Professor at the University of Sheffield. His research interests include hillslopes, dryland geomorphology and soil erosion. He has chaired Working Group 1 of EU COST Action ES1306 (Connecting European Connectivity Research). He is a former chair of the International Geographical Union Commission on Geomorphic Challenges for the Twenty-First Century and a former Chair of the British Geomorphological Research Group (now the British Society for Geomorphology).

SASKIA D. KEESSTRA is Senior Researcher at EIT Climate-KIC and Wageningen University & Research. Her research interests include soils as part of a larger system, system dynamics, regional transformations and the consequences of natural and human drivers on connectivity, with a focus on methodology development and the dissemination of science. She was the chair of EU COST Action ES1306 (Connecting European Connectivity Research) and promoted connectivity research during her time as President of the Soil System Science Division of the European Geosciences Union. Her current work focusses mainly on policy advice related to sustainable soil and water management at the EU level funded under the Mission A Soil Deal for Europe and Horizon Europe.

CONNECTIVITY
IN GEOMORPHOLOGY

Edited by

RONALD E. PÖPPL
BOKU University Vienna

ANTHONY J. PARSONS
University of Sheffield

SASKIA D. KEESSTRA
Wageningen University & Research

CAMBRIDGE
UNIVERSITY PRESS

Shaftesbury Road, Cambridge CB2 8EA, United Kingdom

One Liberty Plaza, 20th Floor, New York, NY 10006, USA

477 Williamstown Road, Port Melbourne, VIC 3207, Australia

314–321, 3rd Floor, Plot 3, Splendor Forum, Jasola District Centre, New Delhi – 110025, India

103 Penang Road, #05–06/07, Visioncrest Commercial, Singapore 238467

Cambridge University Press is part of Cambridge University Press & Assessment,
a department of the University of Cambridge.

We share the University's mission to contribute to society through the pursuit of
education, learning and research at the highest international levels of excellence.

www.cambridge.org
Information on this title: www.cambridge.org/9781108842402

DOI: 10.1017/9781108903196

© Cambridge University Press & Assessment 2025

First published 2025

Printed in the United Kingdom by CPI Group Ltd, Croydon CR0 4YY

A catalogue record for this publication is available from the British Library

Library of Congress Cataloging-in-Publication Data
Names: Pöppl, Ronald E. 1981– editor. | Parsons, Anthony J., editor. | Keesstra, Saskia D., author.
Title: Connectivity in geomorphology / edited by Ronald E. Pöppl, BOKU University Vienna;
Anthony J. Parsons, University of Sheffield; Saskia D. Keesstra,
Wageningen University & Research
Description: New York : Cambridge University Press, [2025] | Includes
bibliographical references and index.
Identifiers: LCCN 2024019467 (print) | LCCN 2024019468 (ebook) |
ISBN 9781108842402 (hardback) | ISBN 9781108903196 (ebook)
Subjects: LCSH: Geomorphology.
Classification: LCC GB401.5 .C66 2025 (print) | LCC GB401.5 (ebook) |
DDC 551.41–dc23/eng/20240729
LC record available at https://lccn.loc.gov/2024019467
LC ebook record available at https://lccn.loc.gov/2024019468

ISBN 978-1-108-84240-2 Hardback

Contents

The plates will be found between pages 212 and 213

Contributors

Mathilde Bayens, Institute of Earth Surface Dynamics, University of Lausanne, Switzerland

Richard E. Brazier, Department of Geography, University of Exeter, UK

Gary Brierley, School of Environment, University of Auckland, New Zealand

Artemi Cerdà, Soil Erosion and Degradation Research Group, Department of Geography, University of Valencia, Spain

Kirstie Fryirs, School of Natural Sciences, Macquarie University, Australia

Ian Fuller, Physical Geography Group, School of Agriculture & Environment, Massey University, New Zealand

Tobias Heckmann, Physical Geography, Catholic University of Eichstaett-Ingolstadt, Germany

Janet Hooke, School of Environmental Sciences, University of Liverpool, Chatham Street, Liverpool, UK

Saskia D. Keesstra, Team Soil Water and Land Use, Wageningen University & Research, The Netherlands and Resilient and Climate Neutral Regions cluster, EIT Climate-KIC, The Netherlands

Peter G. Knight, School of Geography, Geology and the Environment, Keele University, UK

Stuart N. Lane, Institute of Earth Surface Dynamics, University of Lausanne, Switzerland

Yi-Fan Liu, State Key Laboratory of Soil Erosion and Dryland Farming on the Loess Plateau, Institute of Soil and Water Conservation, Northwest A&F University, Yangling, China

Manuel López-Vicente, Research group 'Aquaterra', University of A Coruña, La Coruña, Spain

David Mohrig, Jackson School of Geosciences, University of Texas at Austin, USA

Mariano Moreno-de las Heras, Department of Geography, Faculty of Geography and History, University of Barcelona, Spain

Saeed Najafi, Department of Range and Watershed Management, Faculty of Agriculture and Natural Resources, Urmia University, Iran

Gregory S. Okin, Department of Geography, University of California, Los Angeles, USA

Anthony J. Parsons, Department of Geography, University of Sheffield, UK

Paola Passalacqua, Department of Civil, Architectural, and Environmental Engineering, The University of Texas at Austin, USA

Ronald E. Pöppl, Institute of Hydrobiology and Aquatic Ecosystem Management, BOKU University Vienna

Juan Quijano-Baron, School of Engineering and Centre for Water Security and Environmental Sustainability, The University of Newcastle, Australia

Tim Ralph, Department of Earth and Environmental Sciences, Macquarie University, Australia

Jesús Rodrigo-Comino, Department of Regional Geographic Analysis and Physical Geography, University of Granada, Spain

Jose Rodriguez, School of Engineering and Centre for Water Security and Environmental Sustainability, The University of Newcastle, Australia

Steven Sandi Rojas, School of Engineering, Deakin University, Australia

Patricia Saco, School of Civil and Environmental Engineering, University of Technology Sydney, Australia

Robert Twilley, Department of Oceanography and Coastal Sciences, Louisiana State University, Baton Rouge, USA

John Wainwright, Department of Geography, Science Laboratories, Durham University, UK

Gao-Lin Wu, State Key Laboratory of Soil Erosion and Dryland Farming on the Loess Plateau, Institute of Water and Soil Conservation, Northwest A&F University, China

Preface

Connectivity has emerged, in recent years, as a significant conceptual framework for understanding transfers of materials through the landscape. There have been successful sessions on the topic at the General Assembly of the European Geosciences Union every year since 2012, the European Cooperation in Science and Technology (COST) Action (ES1306 – Connecteur) on the topic that ran from 2014 to 2018, a Binghampton Symposium in Fort Collins/Colorado in 2016, connectivity sessions at the TERRAenVISION conferences 2018, 2019, 2022 and 2024 the International Association of Geomorphologists conferences in New Delhi, India, and Coimbra, Portugal, as well as numerous papers and special issues of journals on the topic. Moreover, an IAG working group on connectivity in geomorphology was launched in 2018, and a Marie Curie International Training Network on interdisciplinary connectivity (including geomorphology) was granted in 2019. To the best of the authors' knowledge, this book brings together in one volume a first comprehensive account of the topic. The book draws upon authorities in their fields to provide state-of-the art surveys across the discipline of geomorphology, both thematically and methodologically, and in some principal areas where the concept has been utilised in the application of geomorphology.

Acknowledgements

We would like to thank the following publishers, organisations and individuals for permission to reproduce the following figures.

American Geophysical Union: Figure 4.8
Copernicus Publications: Figure 10.6
CRC Press: Figure 15.3
Elsevier B.V.: Figures 4.3, 7.1, 7.2, 10.1, 10.3, 10.4, 10.7, 11.2, 11.4, 11.5, 11.6 and 11.7
John Wiley & Sons: Figures 4.1, 4.2, 4.5, 4.6. 4.7, 10.5, 10.8 and 11.3
C. Lambiel: Figure 7.6
Springer: Figures 10.2 and 10.9
D. Tongway: Figure 14.3

Part I

Introduction

1

Connectivity Concepts

ANTHONY J. PARSONS, RONALD E. PÖPPL AND SASKIA D. KEESSTRA

1.1 The Development of 'Connectivity Science'

The notion of connectivity can be dated back to Euler's modelling of the bridges of Königsberg in 1736. However, the use of the term connectivity itself is a later development. Publications using the term can be traced back at least to the 1930s (e.g., Whyburn, 1931). Initially a topic within mathematics, the field grew to encompass a wide range of disciplines. The concept appeared in several disciplines in the 1950s and 1960s but did not enter geomorphology until the 1980s. By 2020, a search on Web of Science under the topic 'connectivity' yielded over 237,000 publications encompassing a wide range of disciplines (Figure 1.1) of which engineering neurosciences and computer sciences had the most publications (around 70,000 each), but geomorphology had only 1,102. As early as the 1950s, ideas on connectivity were crossing into new disciplines across unlikely discipline boundaries. For example, Prihar (1956) working in the field of telecommunications cited the work of Luce and Perry (1949) from the study of social groups. Notwithstanding this evidence of cross-fertilisation in the initiation of connectivity concepts, once connectivity is established in a discipline, subsequent studies proceed largely independently of developments in other disciplines within which the concept has been applied (Turnbull et al., 2018). For example, Kool et al. (2013) in a review of connectivity concepts in population dynamics that specifically claims to 'highlight potential linkages with other fields of research' (p. 165) cites 167 sources but, other than methodological studies, only 7 from outside the broad area of ecology. Consequently, many of the ideas that have developed in connectivity science within a particular discipline have failed to have the widest impact. Even so, it is fair to say that in many of the disciplines in which the concept of connectivity has been applied, it has led to profound insights into the behaviour of the system being studied and has had significant applications in management of some of these systems (see, e.g., Hulme, 2009; Cerdeira et al., 2010; Iori & Mantegna, 2018; Poeppl et al., 2020).

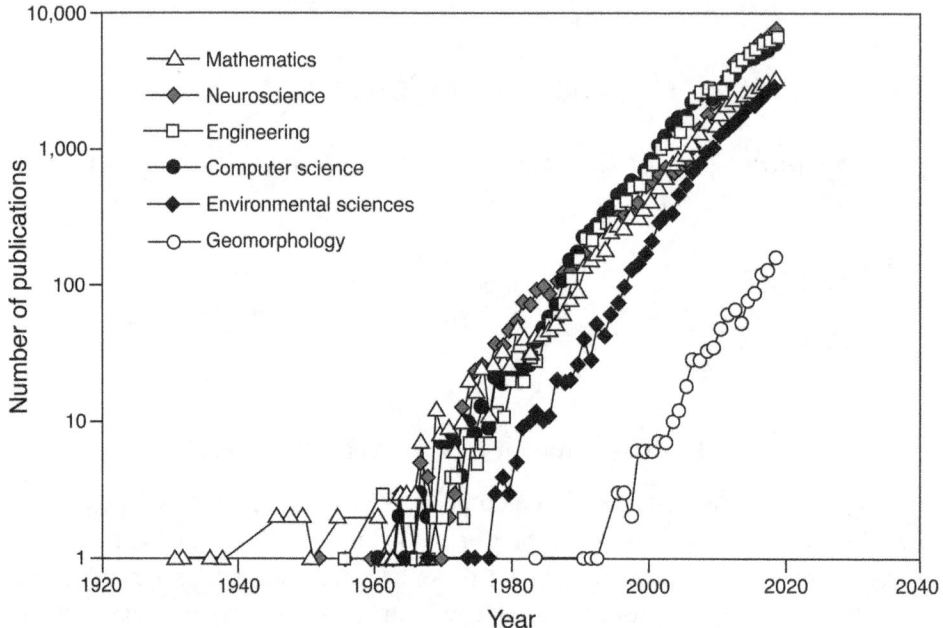

Figure 1.1   The growth of connectivity studies in selected disciplines.

Why has connectivity had such a profound impact on so many disciplines? Fundamentally, all disciplines study phenomena, and those phenomena are typically part of a system of such phenomena, be they galaxies or social groups. In order to understand phenomena, two things are necessary. First, the phenomena must be classified. Without classification, every object is unique. Their study is limited to a description of each object. Put simply, at this level, each discipline might be summarised by the statement 'Things are thus.' Classification moves the discipline forward such that one might say, 'These things are thus, but those things are not thus.' The process of classification is not straightforward. Knowing which attributes of objects A and B enable them to be characterised as 'thus' and of objects C and D that make them 'not thus' has been the subject of much dispute. We may all recognise a chair when we see one, but defining one is far from easy. As Wells (1908, p. 16) put it, 'I would undertake to defeat any definition of chair or chairishness that you gave me.' One difficulty of classification lies in deciding the extent to which the relationships among the phenomena should be included in their classification and, indeed, whether those relationships can be known a priori in any case or are secondary properties to be inferred from morphological similarities. Classification often dominates early phases of a discipline. For example, the voyages of discovery in the sixteenth century led to the discovery of new species and a strong desire to names these new species and to relate them, through classification, to known species. As a discipline evolves, however, relationships among groups and individuals become a greater focus of study. As Mabbutt (1968, p. 27) commented, 'With progressive development it is the links rather than the breaks … with which we become increasingly concerned.' Connectivity is, in effect, a means to study these links separately from issues surrounding classification. The growth of connectivity science might also be seen as both complementary to and a consequence of the growth of an interest in systems stemming from the seminal work by von Bertalanffy (1950) on general systems theory. Von Bertalanffy defined a system (p. 143) as 'a complex of interacting elements'. Connectivity science focuses on the interactions of the elements of a system. The field has gained further significance with the identification of complex systems, the characteristics of which cannot be easily determined and which display such features as emergence and non-linear behaviour. Connectivity science has played a significant role in understanding the processes behind such features (Comin et al., 2020). For example, in the field of climate science, the analysis of teleconnections has thrown light on understanding the stability of the climate system (Tsonis et al., 2008).

1.2 Definitions and Terminology

Connectivity may be defined as a structured set of relationships between spatially and/or temporally distinct entities (Kool et al., 2013) or as a degree to which a system facilitates [or impedes] the movement of matter and energy through itself (CONNECTEUR). The former definition focuses on the structure of a system and the latter on the functioning of it. The two definitions may thus be seen as complementary rather than alternatives. They give rise to the separate concepts of structural and functional connectivity. The former may be defined as the configuration or arrangement of the system elements, whereas the latter describes dynamical processes operating within a structurally connected system (Turnbull et al., 2018). Although the terms structural connectivity and functional connectivity are well embedded in many disciples, variants do exist. Bracken et al. (2013) prefer the term process-based connectivity over functional connectivity for studies of hydrological connectivity, arguing that the term functional has many uses/interpretations in hydrology already, and Wohl et al. (2019) advocates system configuration instead of structural connectivity. Notwithstanding these arguments, we will adhere in this book to the terms structural and functional connectivity because of their widespread acceptance across many disciplines.

A fundamental difference between structural and functional connectivity lies in the fact that, whereas the former can be relatively easily measured, and a variety of tools exist to do so (see Chapters 9 and 10), the latter tends to be inferred from system behaviour, so that measurement is somewhat indirect. Inferring functional connectivity from system behaviour raises two important issues.

The first is equifinality. Different functional relationships may lead to the same behaviour. Thus, there may not be a one-to-one link between system behaviour and a set of functional relationships between elements of a system. Equifinality is particularly a problem in cases where measurements of system behaviour are restricted. In the field of hydrology, it is common to measure system behaviour as a run-off hydrograph at the outlet of a catchment or base of a hillslope. But as Grayson and Moore (1992) demonstrated, many forms of system behaviour can lead to similar hydrographs. Secondly, is the issue of the consistency of system behaviour and the timescales over which consistency of behaviour, and hence a specific functional connectivity, may exist. System behaviour may be inconsistent for one of two reasons. Either the system itself is changing over time, so that its response to two identical external stimuli occurring at different times is not the same, or its behaviour may vary, both qualitatively and quantitatively, in response to differing external stimuli. Consequently, convenient as it may be, to separate structural and functional connectivity of a system, in reality, the two are interdependent. Most obviously, functional connectivity depends on structural connectivity. Dynamical processes are governed by system architecture. For example, a road network structure determines traffic flows. However, unlike road networks, many systems change their architecture in response to events. That is to say that system architecture, or structural connectivity, changes itself in response to the functioning of the system. Commonly used parts of the system may become better developed, whereas those seldom used become moribund. The system may 'learn' to adjust its behaviour in response to events, or it may be evolving under the influence of some other external drivers. Most obviously, the brain's neural network adapts in response to learning, leading to changes in system architecture and, hence, the relationship between structural and functional connectivity. Conversely, in a system with fixed architecture (such as a road network, in the short term), functional behaviour may adapt to maximise the use of the system's connections. Drivers take longer routes because traffic density on them results in shorter travel times.

1.3 Why Does Connectivity Matter?

There are probably two answers to the question why anything matters. First, there is the human desire for understanding. The discussion in Section 1.2 indicates how connectivity science aids understanding of systems. Second, there is a more utilitarian viewpoint. Does connectivity science tell us anything that is of practical value? Again, examples given in Section 1.1 demonstrate that this is the case. It can, therefore, be concluded that investigating connectivity has both theoretical and practical benefits.

1.4 The Challenges of Connectivity Science

Notwithstanding the compelling arguments in favour of studying connectivity, the ability to apply the ideas of connectivity science in any discipline requires a number of challenges to be addressed. In this section, we will investigate these challenges and some of the attempts to solve them.

The first challenge arises from the definition of a system as a complex of interacting elements. Before interactions can be studied, elements themselves have to be identified. This identification may be far from straightforward. It will depend on the scale(s), both temporal and spatial, at which interactions can be meaningfully measured. The definition of such scales may be conceptual or driven by the practicalities of techniques of measurement. In addition, though it might be self-evident that these elements would be physical entities, they need not be. For example, in social network science, elements may be ideas and behaviours, as well as individuals or social groups.

The second challenge is the nature of the interaction that needs to be measured for connectivity science to be meaningfully applied. Interactions may be directional (A affects B, but B does not affect A), non-directional (A and B are mutually interactive), qualitative (an interaction exists or it does not) or quantitative (the strengths of interactions are measured on some scale). Non-directional, qualitative interactions are more amenable to a wider range of analytical techniques than quantitative directional ones, but the benefits of applying such techniques may be outweighed by the loss of vital information about the interactions being investigated.

The third challenge is to be able to address the relationships between structural and functional connectivity and also to address the issues associated with the evolution of the system. This challenge is associated with the first insofar as these relationships are likely to be dependent on the scales of measurement.

All of the issues discussed in this chapter affect the application of connectivity concepts to geomorphology as much as they do to any discipline. The particular characteristics of these issues that may be specific to the discipline of geomorphology are the focus of Chapter 2.

References

Bertalanffy, L. von 1950. 'An outline of general system theory.' *British Journal for the Philosophy of Science*, 1, 114–129.

Bracken, L. J., Wainwright, J., Ali, G. A., Tetzlaff, D., Smith, M. W., Reaney, S. M., & Roy, A. G. 2013. Concepts of hydrological connectivity: research approaches, pathways and future agendas. *Earth Science Review*, 119, 17–34.

Cerdeira, J. O., Pinto, L. S., Cabeza, M., & Gaston, K. J. 2010. Species specific connectivity in reserve-network design using graphs. *Biological Conservation*, 143, 408–415.

Comin, C. H., Peron, T., Silva, F. N., Amancio, D. R., Rodrigues, F. A., & Costa, L. da F. 2020. Complex systems: features, similarity and connectivity. *Physics Reports*, 861, 1–41. doi: 10.1016/j.phys.rep.2020.03.002

Grayson, R. B. & Moore, I. D. 1992. Effect of land-surface configuration on catchment hydrology. In A. J. Parsons & A. D. Abrahams (eds.), *Overland Flow*. London: UCL Press, 147–175.

Hulme, P. E. 2009. Trade, transport and trouble: managing invasive species pathways in an era of globalization. *Journal of Applied Ecology*, 46, 10–18.

Iori, G. & Mantegna, R. N. 2018. Empirical analyses of networks in finance. In C. Hommes & B. LeBaron (eds.), *Handbook of Computational Economics*, Vol. 4. New York: Elsevier, 637–685. doi: 10.1016/bs.hescom.2018.02.005

Kool, J. T., Moilanen, A., & Treml, E. A. 2013. Population connectivity: recent advances and new perspectives, *Landscape Ecology*, 28, 165–185. doi: 10.1007/s10980-012-9819-z

Luce, R. D. & Perry, A. D. 1949. A method of matrix analysis of group structure. *Psychometrika*, 14, 95–116. doi: 10.1007/BF02289146

Mabbutt, J. A. 1968. Review of concepts of land classification. In G. A. Stewart (ed.), *Land Evaluation*. Melbourne: Macmillan of Australia, 11–28.

Poeppl, R. E., Fryirs, K. A., Tunnicliffe, J., & Brierley, G. J. 2020. Managing sediment (dis) connectivity in fluvial systems. *Science of the Total Environment*, 736. doi: 10.1016/j .scitotenv.2020.1.139627

Prihar, Z. 1956. Topological properties of communication networks. *Proceedings of the Institute of Radio Engineers*, 44, 927–933.

Tsonis, A. A., Swanson, K. L., & Wang, G. 2008. On the role of atmospheric teleconnections in climate. *Journal of Climate*, 21, 2990–3001. doi: 10.1175/2007JCLI1907.1

Turnbull, L., Hütt, M.-T., Ioannides, A. A., Kininmonth, S., Poeppl, R., Tockner, K., Bracken, L. B., Keesstra, S., Liu, L., Masselink, R., & Parsons, A. J. 2018. Connectivity and complex systems: learning from a multi-disciplinary perspective. *Applied Network Science*, 3, 11. doi: 10.1007/s41109-018-0067-2

Wells, H. G. 1908. *First and Last Things*. London: Robinson, 307 pp.

Whyburn, G. T. 1931. The cyclic and higher connectivity of locally connected spaces. *American Journal of Mathematics*, 52, 427–442. doi: 10.2307/2370795

Wohl, E., Brierley, G., Cadol, D., Coulthard, T. J., Covino, T., Fryirs, K. A., Grant, G., Hilton, R. G., Lane, S. N., Magilligan, F. J., Meitzen, K. M., Passalacqua, P., Poeppl, R. E., Rathburn, S. L., & Sklar, L. S. 2019. Connectivity as an emergent property of geomorphic systems. *Earth Surface Processes and Landforms*, 44, 4–26.

2

Connectivity in Geomorphology

RONALD E. PÖPPL, ANTHONY J. PARSONS AND SASKIA D. KEESSTRA

This chapter describes the history and development of the concept of connectivity in geomorphology (including definitions; Section 2.1), provides an overview of connectivity terminology and the underlying concepts (Section 2.2), and identifies the benefits of connectivity for geomorphological research and applications (Section 2.3). In the final Section (2.4), general key challenges in using connectivity to understand complex geomorphic systems will be addressed.

2.1 History and Development of the Concept of Connectivity in Geomorphology (Including Definitions)

According to Gregory and Lewin (2015), who reviewed the nature, prevalence and function of concepts in geomorphology, many 'concepts' are taken from other sciences and explicitly applied to or imported into geomorphology. These include concepts such as uniformitarianism, equilibrium (Bertalanffy, 1976) and complexity (e.g., Lorenz, 1963; cf. Gregory & Lewin, 2015). Connectivity is no exception (see later; see also Chapter 1).

Studying potential sediment pathways (connections) and transport processes has always been one of the core tasks in geomorphology and thus 'connectivity thinking' may be thought to have a long history in geomorphology (cf. Pöppl et al., 2017; Turnbull et al., 2018). To our knowledge, the earliest documented mention of 'connectivity' in a geomorphological context can be found in Chorley and Kennedy's 'Physical geography: a systems approach' (1971), where 'connectivity' is defined as the transfer of energy and matter between two landscape compartments or within a system as a whole. Since the beginning of the twenty-first century, geomorphologists started to develop a stronger interest in connectivity concepts to better understand the complexity of geomorphic systems and system response to change by incorporating connectivity concepts from other disciplines, especially ecology and hydrology. In ecology, the concept of connectivity has already been used to explain the persistence of spatially structured populations (Metzger & Décamps, 1997), where 'landscape connectivity' has been broadly defined as being the degree to which a landscape facilitates or impedes the movement of individuals along resource patches (Taylor et al., 1993). Important landscape characteristics influencing landscape connectivity include shape, size and location of different features in the landscape (Brooks, 2003). Some ecologists applied the concept of landscape connectivity to river ecosystems (Ward, 1989, 1997), resulting in concepts of hydrological connectivity – defined as being the water-mediated transfer of matter, energy, and/or organisms within or between elements of the hydrologic cycle (Pringle, 2001). The concept of hydrological connectivity further formed the basis for connectivity concepts in hydrological sciences for understanding runoff and run on (e.g., Bracken & Croke, 2007; Ali & Roy, 2009). Later on, these concepts became applied, and developed a specific geomorphic interpretation, in geomorphology. Hydrological connectivity concepts have been expanded by adding sediment-related aspects resulting in concepts on 'sediment connectivity' in geomorphology, where sediment connectivity is generally being defined as the potential for sediments to move through geomorphic systems (cf. Hooke, 2003).

Based on Ward's (1989) conceptual model on the four-dimensional nature of lotic ecosystems, Brierley et al. (2006) developed a connectivity concept in which they defined different forms of landscape connectivity, differentiating between longitudinal (in-stream), lateral (hillslope/floodplain-channel) and vertical (surface–surface) linkages/relationships within catchment systems to provide 'a platform to interpret the operation of geomorphic processes in any given system' (Brierley et al., 2006; p. 166). They identified key processes and control variables for these different types of linkage dependent on scale (i.e., within landscape compartment, between landscape compartment, subcatchment, catchment), further proposing measures to be used to assess the strength of these linkages to explain landscape sensitivity and geomorphic change. Moreover, they indicated that these linkages may be connected (coupled) or disconnected (decoupled; cf. Brunsden & Thornes, 1979; Harvey, 2002) over differing timescales, as various landforms may disrupt or enhance the strength of longitudinal, lateral and vertical linkages. This framework has been further elaborated by Fryirs et al. (2007), who developed a conceptual model on (dis)connectivity of catchment sediment cascades. In this model, catchment disconnectivity is defined by the degree to which different types of landforms, called buffers, barriers and blankets operate as a series of switches which turn on/off sediment transfer processes (Fryirs et al., 2007). Buffers are landforms that prevent sediment from entering the channel network (e.g., alluvial pockets in floodplains), barriers can disrupt sediment moving along the channel (e.g., sediment slugs), while blankets are features that smother other landforms protecting them from reworking (e.g., bed armouring; Fryirs et al., 2007). In contrast, features such as gorges may enhance connectivity, acting as boosters to sediment propagation (Fryirs et al., 2007). These linkages between landscape compartments operate with different degrees of efficiency, in response to events of different magnitude and frequency, further impacting landscape sensitivity and (offsite) geomorphic responses (cf. Brunsden, 2001; Fryirs et al., 2007; Fryirs, 2017).

Bracken and Croke (2007) developed a conceptual model on hydrological connectivity in which they identified five major components of catchment connectivity affecting the conditions for run-off generation and transmission, further influencing sediment connectivity in run-off-dominated geomorphic systems, that is, climate, hillslope runoff potential, landscape position, delivery pathway and lateral buffering. Moreover, they distinguished between static (i.e., static spatial patterns such as hydrological run-off units (HRU)) and dynamic representations of hydrological connectivity (e.g., rainfall variability or longer-term landscape development), further identifying the combination of static and dynamic process responses as a key aspect of connectivity. Based on the work of Bracken and Croke (2007), Lexartza-Artza and Wainwright (2009) elaborated on structural and dynamic aspects of hydrological connectivity by conceptualizing structure–function relationships and feedbacks and their role in influencing pathways and transfer processes in catchment systems, and developed a framework (incl. guidelines) to study structural and functional aspects of connectivity in the field.

Building on advances in understanding hydrological and sediment connectivity, Bracken et al. (2015) developed a concept in which they integrated the concept of sediment-transport distances within a sediment connectivity framework consisting of three key elements and considerations of the characteristics of the relationships among them. These three key elements are: frequency-magnitude distributions of sediment detachment, transport and deposition processes; spatial and temporal feedbacks between sediment detachment and transport processes; mechanisms of sediment detachment and transport. Based on studies of sediment-travel distance suggesting that all parts of the landscape are potentially involved in sediment production, transfer and storage (Wainwright et al., 2001; Parsons et al., 2006), Bracken et al. (2015) conceptualized and defined sediment connectivity as 'integrated transfer of sediment across all possible sources to all potential sinks in a system over the continuum of detachment, transport and deposition, which is controlled by how the sediment moves between all geomorphic zones' (Bracken et al., 2015; p. 179).

Pöppl et al. (2017) developed a connectivity concept to account for the causes and trajectories of geomorphic change in human-impacted fluvial systems, consisting of three key elements: (i) systematic outline of human impacts on rivers and catchments and their effects on longitudinal, lateral and vertical connectivity (incl. feedback mechanisms); (ii) relationship between connectivity, sensitivity and resilience and the role of system-intrinsic self-regulatory processes in governing geomorphic response to anthropogenic disturbance; (iii) coupling of social and geomorphic systems (coupled human-geomorphic systems) and the role of connectivity relationships in governing the systems' sensitivity and resilience to change (incl. implications for river and catchment management; see also Chapter 13 'Rivers and wetland systems' as well as Pöppl et al. (2020)). The impact of different types of management practices on sediment connectivity has also been conceptualized by Keesstra et al. (2018), who developed a connectivity framework for a systems approach in sediment and water transfer, in which system properties called phases and fluxes (representing static/structural and dynamic/functional connectivity) are differentiated to enable quantification of connectivity (see also Part III 'Quantifying Connectivity in Geomorphology'). Similarly, notions of static and dynamic (called 'fixed' and 'variable') system properties and the role of connectivity as a means to describe nature of and controls on specific linkages and the evolving state of geomorphic systems (with a focus on drainage basins) can be found in a conceptualization of connectivity by Wohl et al. (2019).

2.2 Terminology and Underlying Concepts

As presented in Section 2.1 several concepts (incl. definitions) of connectivity have been developed in different contexts of geomorphology (for an overview on definitions see also Wohl, 2017; Wohl et al., 2019; Singh et al., 2021). In this section we will have a closer look at the terminology used in these concepts, that is, by listing key terms, further examining their conceptual roots/underpinnings (Table 2.1).

As demonstrated by Table 2.1, connectivity concepts in geomorphology have their roots in other disciplines (see also Section 2.1), and/or can be seen as developments of earlier concepts in geomorphology. Concepts from other disciplines include the 'Variable source area concept' (Hewlett & Hibbert, 1967), 'Dynamic contributing areas' (Barling, 1992; Beven, 1997), 'Effective hillslope length' (Aryal et al., 2003), 'Active contributing areas' (Ambroise, 2004), and 'Hydrological response units (HRU)' (Flügel, 1995) from hydrology, and 'Complexity of economic, ecological, and social systems' (Holling, 2001) from social sciences. Earlier concepts in geomorphology which have been incorporated into connectivity concepts in geomorphology are the 'coupling/sensitivity concept' (Brunsden & Thornes, 1979), systems approaches (Chorley & Kennedy, 1971), feedback relationships in geomorphology (e.g., King, 1970; cf. Chorley, 1962), the 'magnitude–frequency' concept (Wolman & Miller, 1960), or the 'geomorphic threshold' concept (Schumm, 1973), as well as notions of sediment-transport distance, sediment budgets/cascades/sediment delivery.

Table 2.1 *Key terms used in connectivity concepts in geomorphology and their conceptual roots/underpinnings*

Researchers (reference)	Key terms	Conceptual roots/underpinnings (incl. references)
Brierley et al., 2006/ Fryirs et al., 2007	Forms of (landscape) (dis) connectivity (buffers, barriers, blankets), landscape sensitivity and (off-site) geomorphic responses	'Coupling/sensitivity' concept (Brunsden & Thornes, 1979)
	Forms of (landscape) connectivity (longitudinal, lateral, vertical)	'Four-dimensional nature of lotic ecosystems' concept (Ward, 1989)
	Effective catchment area; effective timescales Frequency and magnitude of disturbance events Breaching capacity (i.e., recurrence interval of events required to breach disconnecting landforms)	'Variable source area concept' (Hewlett & Hibbert, 1967); 'Effective timescales of coupling within fluvial systems' (Harvey, 2002); 'Magnitude–frequency of forces in geomorphic processes' (Wolman & Miller, 1960) 'Coupling/sensitivity' concept (Brunsden, 1993) and 'geomorphic threshold' concept (Schumm, 1973)
	Sediment budgets; Sediment cascades/ sources and sinks/ conveyance; Sediment delivery ratio (SDR)/problem	'Sediment budget' (e.g., Trimble, 1983); 'Sources, sinks and storage of river sediment' (cf. Meade, 1982); 'Sediment delivery problem' (Walling, 1983)
Bracken and Croke, 2007	(Dynamic) contributing areas (run-off generation)	'Dynamic contributing areas' concept (Barling, 1992; Beven, 1997); 'Effective hillslope length' concept (Aryal et al., 2003); 'Active contributing areas' concept (Ambroise, 2004)
	'Hydrological' run-off units (HRU)	'Hydrological response units (HRU)' concept (Flügel, 1995); 'Hydrological similar surfaces (HYSS)' concept (Kirkby et al., 2002)
	Volume to breakthrough (i.e., accumulated run-off volume per unit width to be applied at a point before flow appears at a downslope point)	'HYSS' concept (threshold values; Kirkby et al., 2002)

Table 2.1   (*cont.*)

Researchers (reference)	Key terms	Conceptual roots/underpinnings (incl. references)
Lexartza-Artza and Wainwright, 2009	Forms of (hydrological and sediment) connectivity (structural, functional) Structure-function 'feedbacks'	'Structural and functional connectivity' concept/terminology in (landscape) ecology (Baudry & Merriam, 1988) Structure-function interactions in hydrological connectivity (Ambroise, 2004; Bracken & Croke, 2007)
Bracken et al., 2015	Sediment-travel distance	Concepts of sediment-transport distance (e.g., Einstein, 1950; Wainwright et al., 2001; Parsons et al., 2004; Furbish et al., 2012)
	Sediment detachment and transport processes: (i) Frequency-magnitude distributions (ii) Feedbacks	'Magnitude–frequency of forces in geomorphic processes' (Wolman & Miller, 1960) 'Feedback relationships in geomorphology' (King, 1970; cf. Chorley, 1962); 'process-form relationships' (cf. Richards, 1999)
Pöppl et al., 2017	Effects of ramped and pulsed type human impacts on the different spatial dimensions of connectivity in fluvial systems	'Coupling/sensitivity' concept (Brunsden & Thornes, 1979); 'Four-dimensional nature of lotic ecosystems' concept (Ward, 1989)
	Relationship between connectivity, sensitivity and resilience in geomorphic systems	'Coupling/sensitivity' concept (Brunsden & Thornes, 1979)
	Connectivity and complex geomorphic response to human disturbance: role of system-intrinsic feedback processes (self-regulation)	'Feedback relationships in geomorphology' (King, 1970; cf. Chorley, 1962); 'complex response' (Schumm, 1973); 'Coupling/ sensitivity' concept (Brunsden & Thornes, 1979)
	Connectivity in coupled human-geomorphic systems	Complexity of economic, ecological and social systems (Holling, 2001); Feedbacks in Human–Landscape Systems (Chin et al., 2014)
Keesstra et al., 2018	Connectivity and system state, phases and fluxes (incl. self-organization)	'Physical Geography: A Systems Approach' (Chorley and Kennedy, 1971)

Table 2.1   (*cont.*)

Researchers (reference)	Key terms	Conceptual roots/underpinnings (incl. references)
	'Co-evolution' of system structures, processes and functions	'Ecogeomorphic coevolution' (Saco and Moreno-de las Heras, 2013); Structure-function interactions in hydrological/ sediment connectivity (Ambroise, 2004; Bracken & Croke, 2007; Lexartza-Artza & Wainwright, 2009)
Wohl et al., 2019	Connectivity and system properties (static or 'fixed' and dynamic or 'variable') / state	'Physical Geography: A Systems Approach' (Chorley & Kennedy, 1971); 'Time, space, and causality in geomorphology' (Schumm & Lichty, 1965)

In all connectivity concepts mentioned above the term 'complexity' or 'complex system' can be found, implying complexity as an overarching paradigm in geomorphological connectivity concepts. For example, according to Bracken and Croke (2007) connectivity can be considered an emergent property of the landscape system, which results from the complex interaction of different connectivity factors, that is, vegetation, soils, topography and landforms (cf. Wainwright et al., 2011; Keesstra et al., 2018). According to Fryirs et al. (2007, p. 65) 'Unravelling [...] complexities (!) represents a significant challenge to geomorphologists. A common element linking these phenomena, and at least part of the reason for their complexity, is the role of landscape (dis)connectivity'. Some concepts even explicitly claim to use a 'complex systems approach' to study connectivity (cf. Bracken et al., 2015) or 'complex system response' to human disturbance (Pöppl et al., 2017).

2.3 Benefits of Connectivity for Geomorphological Research and Applications: Connectivity as a Key Concept in Geomorphology?

In the literature, different benefits of connectivity for geomorphological research and applications are highlighted. According to Wohl et al. (2019), connectivity is seen as a useful conceptual framing for investigating the spatial and temporal variability of fluxes and (complex) system behaviour, since it focusses on (i) interactions among geomorphic system components, (ii) the response of geomorphic systems to varying inputs, (iii) specific features of geomorphic systems that govern connectivity, (iv) how human alterations of geomorphic systems influence system behaviour. For each of these 'benefits', some examples from different branches and in different contexts of geomorphology are presented.

2.3.1 Interactions among System Components

Heckmann and Schwanghart (2013) studied geomorphic coupling and sediment connectivity in an *Alpine catchment* by delineating interactions among geomorphic systems components using *graph theoretical methods*, showing the importance of mass movements and their interaction with debris flow processes in governing sediment cascades. Based on neural *network studies*, Passalacqua (2017) used an adjacency matrix to quantify connectivity and interactions between elements of *deltas*, exhibiting that delta systems emerge as complex 'leaky networks' that continuously and dynamically exchange fluxes of matter, energy and information with their surroundings. Moreover, with a specific emphasis on *hillslope (incl. aeolian) processes* in *semi-arid environments,* numerous studies have shown the importance of complex eco-geomorphic interactions in determining connectivity (e.g., Turnbull et al., 2008; Okin et al., 2009; Calvo-Cases et al., 2021), including co-evolution of vegetation and landform patterns (Saco & Moreno-de las Heras, 2013).

2.3.2 Response of Geomorphic Systems to Varying Inputs

Fryirs et al. (2007) applied their *GIS- and field-based* 'effective catchment area' approach to explain catchment response to events of differing magnitude and frequency in the Weraamaia catchment in New Zealand, showing that connectivity changes as the magnitude of event changes. Baartman et al. (2013) used a landscape evolution *model* to link landscape complexity and sediment connectivity, simulating sediment connectivity and the effects of varying rainfall time series on geomorphic response of catchment systems with contrasting morphologies as being influenced by rainfall intensity. A direct relationship between rainfall magnitude and sediment connectivity has also been observed by Calsamiglia et al. (2020), who assessed changes in functional connectivity (FC) in a small *agricultural Mediterranean catchment* under contrasting rainfall events using unmanned aerial vehicles *(UAV)*. Moreover, some studies have shown a clear relationship between *climate change* inputs and sediment connectivity (in a *soil erosion* context, e.g., Nunes et al., 2009; Luetzenburg et al., 2020; in *(peri)glacial environments*, e.g., Lane et al., 2017; Mancini & Lane, 2020).

2.3.3 (Dis)connecting Landscape Features

The buffers, barriers and blankets concept Fryirs et al. (2007; see Section 2.1) has been applied in many (dis)connectivity studies in *fluvial geomorphology* (e.g., Souza et al., 2016; Schopper et al., 2019; Lisenby et al., 2020). Another approach to define (dis)connecting landscape features can be found in Messenzehl et al. (2014) who classified different types of topo-sequence according to their potential to affect sediment connectivity and sediment cascades in an *Alpine catchment* in Switzerland. The (potential) effects of different landscape features on sediment connectivity have also been assessed in studies using (variations of) the *index of connectivity* (Borselli et al., 2008; see also Chapter 10 'Indices'). Additionally, the effects of different types of *man-made landscape features* on sediment connectivity have been investigated in different contexts (river engineering, e.g., Cucchiaro et al., 2019; Marchi et al., 2019; Ondráčková & Máčka, 2019; Magilligan et al., 2021; topographic changes in catchments, e.g., Croke et al., 2005; Thomaz & Peretto, 2016; Calsamiglia et al., 2018; Llena et al., 2019).

2.3.4 Human Impact on the Behaviour of Geomorphic Systems

Coulthard and Van De Wiel (2017) used the CAESAR landscape evolution *model* to perform centennial scale simulations on the effects of deforestation and reforestation on basin scale sediment connectivity and geomorphic system behaviour, showing that basin geomorphology responds strongly to these *land use changes*. Similar effects of land use changes on connectivity and geomorphic system behaviour have been, for example, reported by Keesstra et al. (2005, 2009; cf. Pöppl et al., 2017) showing a direct effect of natural reforestation on lateral sediment connectivity, sediment delivery and channel response. The CAESAR model has further been used to model the effects of *river engineering* practices on connectivity and geomorphic channel behaviour. Pöppl et al. (2019) modelled the effects of dam construction and dam removal on channel evolution, sediment connectivity and sediment delivery in a multiple dam setting, revealing that geomorphic behaviour following dam construction and removal is complex in space and time, as being related to emerging feedback processes. Different types of *land management practices* such as tillage, or varying age and design have also shown to affect connectivity and the behaviour of geomorphic systems (e.g., by using plot measurements; Cerdà et al., 2017; Rodrigo-Comino et al., 2020).

As shown earlier, the concept of connectivity has been successfully applied in different contexts of geomorphology to study (complex) geomorphic systems. The concept has been used in different environments and process domains (i.e., hillslope, fluvial, aeolian, glacial, periglacial, coasts/deltas; see Part II 'Connectivity in

Process Domains'), as well in the contexts of climate change, human impact and environmental management (see Part IV 'Managing Connectivity'). The applied methods range from field appraisals (incl. measuring), GIS- and remote sensing approaches, indices, network analysis, to modelling approaches (see Part III 'Quantifying Connectivity in Geomorphology'). The concept of connectivity has therefore shown to have emerged as a key concept in geomorphology (cf. Gregory & Lewin, 2015).

2.4 Challenges

Notwithstanding the benefits of connectivity thinking for geomorphology, several challenges exist for its implementation. In Turnbull et al. (2018) different key challenges in using connectivity to understand complex systems have been identified. These are: (i) defining fundamental units (FU) of study, (ii) separating structural and FC and understanding emergent system behaviour and (iii) measuring connectivity. Based on considerations in Turnbull et al. (2018), the following paragraphs will discuss each of these key challenges in detail.

2.4.1 Defining Fundamental Units (FU) of Study

As pointed out by Pöppl and Parsons (2018; p. 1155), measurement of connectivity in geomorphic systems 'requires a set of entities to be defined that permit the connectivity among them to be quantified (termed fundamental units, FUs)', that need to be meaningful in the context of the spatial and temporal scale of investigation and the applied measurement technique. Otherwise, examination of the connectivity between system components will not yield useful insights into the characteristics and behaviour of the geomorphic system under study (Pöppl & Parsons, 2018). However, most connectivity studies in geomorphology have given no or only little attention to the issue of (meaningfulness of) FUs, which can, at least partly, be explained by a lack in underlying concepts.

Traditionally, geomorphologists have drawn structural boundaries between the units of study, defined by sharp changes in topography (e.g., between flats and slopes, Wooldridge, 1932; landforms as 'intuitive' FU, Heckmann et al., 2018). Later on, based on the 'site' (Bourne, 1931), 'land systems' (Christian & Steward, 1953) and 'land facets' (Brink et al., 1966) concepts, 'land elements' have been conceptualized as being units of study comprising uniform climate, parent material, topography, soil and vegetation characteristics (cf. Mabbutt, 1968). Size and demarcation of FU always depend on the respective spatial and temporal scales of study and the (key) processes involved which are often complex in space and time. Moreover, key processes typically tend to vary between spatial scales, leading to one of the key problems in geomorphology, that is, scaling up processes measured at small spatial and temporal scales to explain large-scale geomorphic evolution (cf. Bracken et al., 2015).

Based on the 'land element' concept, further integrating analogies from cell biology, Pöppl and Parsons (2018) defined so-called geomorphic cells as FU in geomorphological connectivity studies.

A geomorphic cell is conceptualized as being a three-dimensional body of the 'geomorphosphere' with distinct environmental characteristics that determine hydro-geomorphic boundary conditions (e.g., geology, soils, topography and/or vegetation), which is delimited from neighbouring cells and neighbouring spheres by different types of boundary and connection (called 'connecteins'). The geomorphic cell approach has, for example, been picked up by Pearson et al. (2020) to study coastal sediment connectivity. Other approaches to define FU in connectivity studies include the use of response units (e.g., Hydrological Response Unit (HRU); e.g., Busch et al., 1999); Geomorphological Response Units (GRU; e.g., Carr et al., 2015). Based on the HRU and the Volume to Breakthrough concept (Bracken & Croke, 2007; see also Table 2.1), Singh et al. (2017) defined so called Connectivity Response Units (CRU), i.e., spatial patterns of landscape units which have the potential of interconnection, to study hydrological connectivity. The CRU approach has, for example, been applied by Xie et al. (2020) to study hydrological connectivity in the Yellow River Delta. However, both the geomorphic cell and the CRU concepts have not been widely picked up by the connectivity community (so far), and their usefulness thus still remains to be seen.

2.4.2 Separating Structural and Functional Connectivity and Understanding Emergent System Behaviour

Structural connectivity (SC) and FC cannot be separated from each other in a meaningful way due to inherent feedbacks between them (cf. e.g., Wainwright et al., 2011; Bracken et al., 2015; Turnbull et al., 2018; see also Section 2.2), that is, process-form relationships. Process and form are inseparably intertwined, as landscapes exhibit a distinct type of memory, meaning that past geologic, anthropogenic and climatic controls influence contemporary process-form relationships (Brierley, 2010), that is, imprints of geomorphic processes (function) on form (structure) which govern future landscape processes (function; Turnbull et al., 2018). These process-form relationships are further acting on various spatial and temporal scales. A critical issue when separating SC and FC is determining the timescale at which a change in SC becomes dynamic (functional), which has profound implications for quantifying connectivity (see later and Part III 'Quantifying Connectivity in Geomorphology'; Turnbull et al., 2018). In many systems, the memory effects and the timescales at which structure-function feedbacks occur are too strong for separation of SC and FC (cf. Turnbull et al., 2018). A key challenge for any quantitative description of connectivity in geomorphic systems is thus how to incorporate such memory effects.

As landscapes are complex macroscopic features that emerge from a multitude of processes, structure–function interactions and (related) memory effects that shape them at different spatial and temporal scales, connectivity can be seen as an emergent property of geomorphic systems (Bracken & Croke, 2007; Wohl et al., 2019) which are organized in a hierarchical manner (Harrison, 2001; Turnbull et al., 2018). Emergence can be seen as the ability of complex systems to create dissipative structures (landforms) at certain scales due to self-organization of processes operating at different spatial and temporal scales which may not be understood by reductionist methodologies (Harrison, 2001). Emergent properties of geomorphic systems are for example, landforms such as dunes (e.g., Baas, 2002) or patterned ground which are shaped by myriad small-scale (microscopic) processes.

Due to methodical constraints prediction and reconstruction of landscape evolution is difficult (i.e., the problem of up and downscaling; Turnbull et al., 2018). To study emergent behaviour of geomorphic systems some researchers have applied complex systems theory and related approaches (e.g., via modelling and combined approaches, e.g., Coco & Murray, 2007; D'Alpaos et al., 2007; Castelle et al., 2010), comprising a promising strategy to study geomorphic system complexity. However, to date, most (if not all) connectivity concepts and related methods in geomorphology are grounded in general systems theory (cf. Bertalanffy, 1976) rather than in complex systems thinking and related approaches suitable to study emergent system behaviour (cf. Richards, 2002; Turnbull et al., 2018).

2.4.3 Measuring Connectivity

The key challenges when it comes to quantify connectivity are to define FU and the spatial and temporal scales over which connectivity should be assessed (i.e., scale problem in geomorphology; see also earlier). Another challenge is the lack of standardised protocols for the assessment of connectivity that would allow data comparison at different spatial and temporal scales (Larsen et al., 2012; Okin et al., 2015). Additionally, quantification of connectivity and thus data availability and comparability are constrained by the measurement design (incl. the technical equipment involved; Turnbull et al., 2018). Various methods are used to measure landscape structure and sediment fluxes to infer connectivity. Structural connectivity can be measured at high spatial and temporal resolutions, e.g., by using Structure-from-Motion photogrammetry (e.g., using UAV) and laser scanning technology. Sediment fluxes resulting from FC can be quantified at different spatial and temporal scales by using erosion plots, water sampling and sediment traps, tracer methods, laboratory experiments, or camera-based monitoring of sediment transport (see Chapter 9 'Measuring Connectivity: Methodologies for Assessing Connectivity as a Property of Soils and Landscapes'). However, the derived data only represent snapshots of fluxes lacking information on system dynamics over longer time scales that arise from structure–function relationships (FC; cf. Turnbull et al., 2018). Besides these measurement techniques different types of connectivity indices exist. Connectivity indices, such as the index of connectivity (Borselli et al., 2008), primarily use topography information to determine connectivity (see Chapter 10 'Indices'). Indices are generally static representations of connectivity, and thus do not provide information about fluxes or system dynamics (FC; Turnbull et al., 2018). Different kinds of models are used to study system dynamics, further providing information on fluxes used to infer FC (see Chapter 11 'Modelling').

References

Ali, G. A. & Roy, A. G. (2009). Revisiting hydrologic sampling strategies for an accurate assessment of hydrologic connectivity in humid temperate systems. *Geography Compass*, 3(1), 350–374.

Ambroise, B. (2004). Variable 'active' versus 'contributing' areas or periods: A necessary distinction. *Hydrological Processes*, 18, 1149–1155.

Aryal, S. K., Mein, R. G. & O'Loughlin, E. M. (2003). The concept of effective length in hillslopes: Assessing the influence of climate and topography on the contributing area of catchments. *Hydrological Processes*, 17, 131–151.

Baartman, J. E., Masselink, R., Keesstra, S. D. & Temme, A. J. (2013). Linking landscape morphological complexity and sediment connectivity. *Earth Surface Processes and Landforms*, 38(12), 1457–1471.

Baas, A. (2002). Chaos, fractals and self-organization in coastal geomorphology: Simulating dune landscapes in vegetated environments. *Geomorphology*, 48, 309–328.

Barling, R. D. (1992). *Saturation zones and ephemeral gullies on arable land in south-eastern Australia*. PhD thesis, University of Melbourne, VIC, Australia.

Baudry, J. & Merriam, G. (1988). Connectivity and connectedness: Functional versus structural patterns in landscapes. In K.-F. Schreiber, ed., *Connectivity in Landscape Ecology*. Proceedings of 2nd International Association for Landscape Ecology. Münstersche Geographische Arbeiten, 29, 23–38.

Beven, K. (1997). Topmodel: A critique. *Hydrological Processes*, 11, 1069–1085.

Borselli, L., Cassi, P., & Torri, D. (2008). Prolegomena to sediment and flow connectivity in the landscape: A GIS and field numerical assessment. *Catena*, 75(3), 268–277.

Bourne, R. (1931). Regional survey and its relation to stocktaking of the agricultural resources of the British Empire. *Oxford Forestry Memoirs*, 13, 16–18.

Bracken, L. J., & Croke, J. (2007). The concept of hydrological connectivity and its contribution to understanding runoff-dominated geomorphic systems. *Hydrological Processes*, 21(13), 1749–1763.

Bracken, L. J., Turnbull, L., Wainwright, J., & Bogaart, P. (2015). Sediment connectivity: A framework for understanding sediment transfer at multiple scales. *Earth Surface Processes and Landforms*, 40, 177–188.

Brierley, G., Fryirs, K., & Jain, V. (2006). Landscape connectivity: The geographic basis of geomorphic applications. *Area*, 38(2), 165–174.

Brierley, G. J. (2010). Landscape memory: The imprint of the past on contemporary landscape forms and processes. *Area*, 42, 76–85.

Brink, A. B., Mabbutt, J. A., Webster, R., & Beckett, P. H. T. (1966). *Military Engineering Experimental Establishment*, Christchurch, England. Report 940.

Brooks, C. P. (2003). A scalar analysis of landscape connectivity. *Oikos*, 102(2), 433–439.

Brunsden, D., & Thornes, J. B. (1979). Landscape sensitivity and change. *Transactions of the Institute of British Geographers*, 4(4), 463–484.

Brunsden, D. (1993). The persistence of landforms. *Zeitschrift für Geomorphologie*, 93, 13–28.

Brunsden, D. (2001). A critical assessment of the sensitivity concept in geomorphology. *Catena*, 42, 99–123.

Busch, G., Sutmoeller, J., Krüger, J., & Gerold, G. (1999). *Regionalization of Runoff Formation by Aggregation of Hydrological Response Units: A Regional Comparison*, IAHS Publication (International Association of Hydrological Sciences) 254., Wallingford: IAHS Press, 45–51.

Calsamiglia, A., Fortesa, J., García-Comendador, J., Lucas-Borja, M. E., Calvo-Cases, A., & Estrany, J. (2018). Spatial patterns of sediment connectivity in terraced lands: Anthropogenic controls of catchment sensitivity. *Land Degradation & Development*, 29(4), 1198–1210.

Calsamiglia, A., Gago, J., Garcia-Comendador, J., Bernat, J. F., Calvo-Cases, A., & Estrany, J. (2020). Evaluating functional connectivity in a small agricultural catchment under contrasting flood events by using UAV. *Earth Surface Processes and Landforms*, 45(4), 800–815.

Calvo-Cases, A., Arnau-Rosalén, E., Boix-Fayos, C., Estrany, J., Roxo, M. J., & Symeonakis, E. (2021). Eco-geomorphological connectivity and coupling interactions at hillslope scale in drylands: Concepts and critical examples. *Journal of Arid Environments*, 186, 104418.

Carr, M., Watkinson, D. A., Svendsen, J. C., Enders, E. C., Long, J. M., & Lindenschmidt, K. E. (2015). Geospatial modeling of the Birch River: Distribution of Carmine Shiner (*Notropis percobromus*) in Geomorphic Response Units (GRU). *International Review of Hydrobiology*, 100(5–6), 129–140.

Castelle, B., Ruessink, B. G., Bonneton, P., Marieu, V., Bruneau, N., & Price, T. D. (2010). Coupling mechanisms in double sandbar systems. Part 1: Patterns and physical explanation. *Earth Surface Processes and Landforms*, 35(4), 476–486.

Cerdà, A., Keesstra, S. D., Rodrigo-Comino, J., Novara, A., Pereira, P., Brevik, E., Giménez-Morera, A., et al.(2017). Runoff initiation, soil detachment and connectivity are enhanced as a consequence of vineyards plantations. *Journal of Environmental Management*, 202, 268–275.

Chin, A., Florsheim, J. L., Wohl, E., & Collins, B. D. (2014). Feedbacks in human-landscape systems. *Environmental Management*, 53(1), 28–41.

Chorley, R. J. (1962). *Geomorphology and General Systems Theory*. USGS Professional Paper 500-B. Washington, DC: United States Government Printing Office.

Chorley, R. J., & Kennedy, B. A. (1971). *Physical Geography: A Systems Approach*. London: Prentice-Hall.

Christian, C. S., & Stewart, G. A. (1953). *General report of the survey of the Katharine-Darwin region 1946*, Land Research Series No. 1. CSIRO Australia, Melbourne.

CoCo, G., & Murray, A. B. (2007). Patterns in the sand: From forcing templates to self-organization. *Geomorphology*, 91, 271–290.

Coulthard, T. J., & Van De Wiel, M. J. (2017). Modelling long term basin scale sediment connectivity, driven by spatial land use changes. *Geomorphology*, 277, 265–281.

Croke, J., Mockler, S., Fogarty, P., & Takken, I. (2005). Sediment concentration changes in runoff pathways from a forest road network and the resultant spatial pattern of catchment connectivity. *Geomorphology*, 68(3–4), 257–268.

Cucchiaro, S., Cazorzi, F., Marchi, L., Crema, S., Beinat, A., & Cavalli, M. (2019). Multi-temporal analysis of the role of check dams in a debris-flow channel: Linking structural and functional connectivity. *Geomorphology*, 345, 106844.

D'Alpaos, A., Lanzoni, S., Marani, M., & Rinaldo, A. (2007). Landscape evolution in tidal embayments: Modeling the interplay of erosion, sedimentation, and vegetation dynamics. *Journal of Geophysical Research: Earth Surface*, 112(F1). https://doi .org/10.1029/2006JF000537

Einstein, H. A. (1950). *The Bed-Load Function for Sediment Transportation in Open Channel Flows*. USDA, Soil Conservation Service Tech. Bull., 1026, Washington, DC: US Department of Agriculture.

Flügel, W. A. (1995). Delineating hydrological response units by geographical information system analyses for regional hydrological modelling using PRMS/MMS in the drainage basin of the River Bröl, Germany. *Hydrological Processes*, 9, 423–436.

Fryirs, K. A., Brierley, G. J., Preston, N. J., & Kasai, M. (2007). Buffers, barriers and blankets: The (dis)connectivity of catchment-scale sediment cascades. *Catena*, 70(1), 49–67.

Fryirs, K. A. (2017). River sensitivity: A lost foundation concept in fluvial geomorphology. *Earth Surface Processes and Landforms*, 42(1), 55–70.

Furbish, D. J., Haff, P. K., Roseberry, J. C., & Schmeeckle, M. W. (2012). A probabilistic description of the bed load sediment flux: 1. Theory. *Journal of Geophysical Research: Earth Surface*, 117(F3), F03031.

Gregory, K. J., & Lewin, J. (2015). Making concepts more explicit for geomorphology. *Progress in Physical Geography*, 39(6), 711–727.

Harrison, S. (2001). On reductionism and emergence in geomorphology. *Transactions of the Institute of British Geographers*, 26, 327–339.

Harvey, A. M. (2002). Effective timescales of coupling within fluvial systems. *Geomorphology*, 44, 175–201.

Heckmann, T., & Schwanghart, W. (2013). Geomorphic coupling and sediment connectivity in an alpine catchment – Exploring sediment cascades using graph theory. *Geomorphology*, 182, 89–103.

Heckmann, T., Cavalli, M., Cerdan, O., Foerster, S., Javaux, M., Lode, E., Smetanová, A., Vericat, D., & Brardinoni, F. (2018). Indices of sediment connectivity: Opportunities, challenges and limitations. *Earth-Science Reviews*, 187, 77–108.

Hewlett, J. D., & Hibbert, A. R. (1967). Factors affecting the response of small watersheds to precipitation in humid areas. In *Proceedings of 1st International Symposium on Forest Hydrology*, pp. 275–253.

Holling, C. S. (2001). Understanding the complexity of economic, ecological, and social systems. *Ecosystems*, 4, 390–405.

Hooke, J. M. (2003). Coarse sediment connectivity in river channel systems: A conceptual framework and methodology. *Geomorphology*, 56, 79–94.

Keesstra, S. D., van Huissteden, J., Vandenberghe, J., Van Dam, O., de Gier, J., & Pleizier, I. D. (2005). Evolution of the morphology of the river Dragonja (SW Slovenia) due to land-use changes. *Geomorphology*, 69, 191–207.

Keesstra, S. D., van Dam, O., Verstraeten, G., & van Huissteden, J. (2009). Changing sedimentgeneration due to natural reforestation in the Dragonja catchment, SW Slovenia. *Catena*, 78, 60–71.

Keesstra, S., Nunes, J. P., Saco, P., Parsons, T., Pöppl, R., Masselink, R., & Cerdà, A. (2018). The way forward: Can connectivity be useful to design better measuring and modelling schemes for water and sediment dynamics? *Science of the Total Environment*, 644, 1557–1572.

King, C. A. (1970). Feedback relationships in geomorphology. *Geografiska Annaler: Series A, Physical Geography*, 52(3–4), 147–159.

Kirkby, M. J., Bracken, L. J., & Reaney, S. (2002). The influence of landuse, soils and topography on the delivery of hillslope runoff to channels in SE Spain. *Earth Surface Landforms and Processes*, 27, 1459–1473.

Lane, S. N., Bakker, M., Gabbud, C., Micheletti, N., & Saugy, J. N. (2017). Sediment export, transient landscape response and catchment-scale connectivity following rapid climate warming and Alpine glacier recession. *Geomorphology*, 277, 210–227.

Larsen, L. G., Choi, J., Nungesser, M. K., & Harvey, J. W. (2012). Directional connectivity in hydrology and ecology. *Ecological Applications*, 22, 2204–2220.

Lexartza-Artza, I., & Wainwright, J. (2009). Hydrological connectivity: Linking concepts with practical implications. *Catena*, 79(2), 146–152.

Lisenby, P. E., Fryirs, K. A., & Thompson, C. J. (2020). River sensitivity and sediment connectivity as tools for assessing future geomorphic channel behavior. *International Journal of River Basin Management*, 18(3), 279–293.

Llena, M., Vericat, D., Cavalli, M., Crema, S., & Smith, M. W. (2019). The effects of land use and topographic changes on sediment connectivity in mountain catchments. *Science of the Total Environment*, 660, 899–912.

Lorenz, E. N. (1963). Deterministic nonperiodic flow. *Journal of Atmospheric Sciences*, 20(2), 130–141.

Luetzenburg, G., Bittner, M., Calsamiglia, A., Estrany, J., & Pöppl, R. (2020). Climate and land use change effects on soil erosion in two small 1 agricultural catchment systems: Fugnitz – Austria, Can Revull – Spain. *Science of the Total Environment*, 704, 135389.

Mabbutt, J. A. (1968). Review of concepts of land classification. In G. A. Stewart, ed., *Land Evaluation*. Melbourne: Macmillan of Australia, pp. 11–28.

Magilligan, F. J., Roberts, M. O., Marti, M., & Renshaw, C. E. (2021). The impact of run-of-river dams on sediment longitudinal connectivity and downstream channel equilibrium. *Geomorphology*, 376, 107568.

Mancini, D., & Lane, S. N. (2020). Changes in sediment connectivity following glacial debuttressing in an Alpine valley system. *Geomorphology*, 352, 106987.

Marchi, L., Comiti, F., Crema, S., & Cavalli, M. (2019). Channel control works and sediment connectivity in the European Alps. *Science of the Total Environment*, 668, 389–399.

Meade, R. H. (1982). Sources, sinks and storage of river sediment in the Atlantic drainage of the United States. *The Journal of Geology*, 90, 235–252.

Messenzehl, K., Hoffmann, T., & Dikau, R. (2014). Sediment connectivity in the high-alpine valley of Val Müschauns, Swiss National Park – Linking geomorphic field mapping with geomorphometric modelling. *Geomorphology*, 221, 215–229.

Metzger, J. P., & Décamps, H. (1997). The structural connectivity threshold: An hypothesis in conservation biology at the landscape scale. *Acta Oecologica*, 18(1), 1–12.

Nunes, J. P., Seixas, J., Keizer, J. J., & Ferreira, A. J. D. (2009). Sensitivity of runoff and soil erosion to climate change in two Mediterranean watersheds. Part II: Assessing impacts from changes in storm rainfall, soil moisture and vegetation cover. *Hydrological Processes*, 23(8), 1212–1220.

Okin, G. S., Parsons, A. J., Wainwright, J., Herrick, J. E., Bestelmeyer, B. T., Peters, D. C., & Fredrickson, E. L. (2009). Do changes in connectivity explain desertification? *BioScience*, 59(3), 237–244.

Okin, G. S., Moreno-de-las-Heras, M., Saco, P. M., Throop, H. L., Vivoni, E. R., Parsons, A. J., Wainwright, J., & Peters, D. P. C. (2015). Connectivity in dryland landscapes: Shifting concepts of spatial interactions. *Frontiers in Ecology and the Environment*, 13, 20–27

Ondráčková, L., & Máčka, Z. (2019). Geomorphic (dis) connectivity in a middle-mountain context: Human interventions in the landscape modify catchment-scale sediment cascades. *Area*, 51(1), 113–125.

Parsons, A. J., Wainwright, J., Powell, D. M., Kaduk, J., & Brazier, R. E. (2004). A conceptual model for determining soil erosion by water. *Earth Surface Processes and Landforms*, 29(10), 1293–1302.

Parsons, A. J., Brazier, R. E., Wainwright, J., & Powell, D. M. (2006). Scale relationships in hillslope runoff and erosion. *Earth Surface Processes and Landforms*, 31, 1384–1393.

Passalacqua, P. (2017). The Delta connectome: A network-based framework for studying connectivity in river deltas. *Geomorphology*, 277, 50–62.

Pearson, S. G., van Prooijen, B. C., Elias, E. P., Vitousek, S., & Wang, Z. B. (2020). Sediment connectivity: A framework for analyzing coastal sediment transport pathways. *Journal of Geophysical Research: Earth Surface*, 125(10), e2020JF005595.

Pöppl, R. E., Keesstra, S. D., & Maroulis, J. (2017). Conceptual connectivity framework for understanding geomorphic change in human-impacted fluvial systems. *Geomorphology*, 277, 237–250.

Pöppl, R. E., & Parsons, A. J. (2018). The geomorphic cell: A basis for studying connectivity. *Earth Surface Processes and Landforms*, 43(5), 1155–1159.

Pöppl, R. E., Coulthard, T., Keesstra, S. D., & Keiler, M. (2019). Modeling the impact of dam removal on channel evolution and sediment delivery in a multiple dam setting. *International Journal of Sediment Research*, 34(6), 537–549.

Pöppl, R.E., Fryirs, K.A., Tunnicliffe, J., Brierley, G.J. Managing sediment (dis)connectivity in fluvial systems. Science of the Total Environment 736, 139627

Prigogine, I., & Nicolis, G. (1977). *Self-Organization in Non-Equilibrium Systems*. Wiley

Pringle, C. M. (2001). Hydrologic connectivity and the management of biological reserves: A global perspective. *Ecological Applications*, 11, 981–998.

Richards, K. (1999). The magnitude-frequency concept in fluvial geomorphology: A component of a degenerating research programme? *Zeitschrift für Geomorphologie*, 115, 1–18.

Richards, A. (2002). Complexity in physical geography. *Geography*, 87(2), 99–107.

Rodrigo-Comino, J., Lucas Borja, M., Bertalan, L., & Cerdà, A. (2020). Integrating in situ measurements of an index of connectivity to assess soil erosion processes in vineyards. *Hydrological Sciences Journal*, 65(4), 671–679.

Saco, P. M., & Moreno-de las Heras, M. (2013). Ecogeomorphic coevolution of semiarid hillslopes: Emergence of banded and striped vegetation patterns through interaction of biotic and abiotic processes. *Water Resources Research*, 49(1), 115–126.

Schopper, N., Mergili, M., Frigerio, S., & Cavalli, M., Pöppl, R. (2019). Analysis of lateral sediment connectivity and its connection to debris flow intensity patterns at different return periods in the Fella River system in northeastern Italy. *Science of the Total Environment*, 658, 1586–1600.

Schumm, S. A., & Lichty, R. W. (1965). Time, space, and causality in geomorphology. *American Journal of Science*, 263, 110–119.

Schumm, S. A. (1973). Geomorphic thresholds and the complex response of drainage systems. In M. Morisawa, ed., *Fluvial Geomorphology*. Binghampton: State University of New York, pp. 299–310.

Singh, M., Tandon, S. K., & Sinha, R. (2017). Assessment of connectivity in a water-stressed wetland (Kaabar Tal) of Kosi-Gandak interfan, north Bihar Plains, India. *Earth Surface Processes and Landforms*, 42(13), 1982–1996.

Singh, M., Sinha, R., & Tandon, S. K. (2021). Geomorphic connectivity and its application for understanding landscape complexities: A focus on the hydro-geomorphic systems of India. *Earth Surface Processes and Landforms*, 46(1), 110–130.

Souza, J. O., Correa, A. C., & Brierley, G. J. (2016). An approach to assess the impact of landscape connectivity and effective catchment area upon bedload sediment flux in Saco Creek Watershed, Semiarid Brazil. *Catena*, 138, 13–29.

Taylor, P. D., Fahrig, L., Henein, K., & Merriam, G. (1993). Connectivity is a vital element of landscape structure. *Oikos*, 68(3), 571–573.

Thomaz, E. L., & Peretto, G. T. (2016). Hydrogeomorphic connectivity on roads crossing in rural headwaters and its effect on stream dynamics. *Science of the Total Environment*, 550, 547–555.

Trimble, S. W. (1983). A sediment budget for Coon Creek Basin in the Driftless Area, Wisconsin, 1853–1977. *American Journal of Science*, 283, 454–474.

Turnbull, L., Wainwright, J., & Brazier, R. E. (2008). A conceptual framework for understanding semi-arid land degradation: Ecohydrological interactions across multiple-space and time scales. *Ecohydrology: Ecosystems, Land and Water Process Interactions, Ecohydrogeomorphology*, 1(1), 23–34.

Turnbull-Lloyd, L., Parsons, A., Kininmonth, S., Pöppl, R. E., Huett, M., Keesstra, S. D., Tockner, K., Ioannides, A., & Masselink, R. (2018). Connectivity and complex systems: Learning from a multi-disciplinary perspective. *Applied Network Science*, 3(1), 1–49.

von Bertalanffy, L. (1976). *General System Theory: Foundations, Development, Applications* (rev. ed.). New York: George Braziller.

Wainwright, J., Parsons, A. J., Powell, D. M., & Brazier, R. (2001). A new conceptual framework for understanding and predicting erosion by water from hillslopes and catchments. In J.C. Ascough, D.C. Flanagan, eds., *Soil Erosion for the 21st Century*. St. Joseph, MI: American Society of Agricultural Engineers. pp. 607–610.

Wainwright, J., Turnbull, L., Ibrahim, T. G., Lexartza-Artza, I., Thomton, S. F., & Brazier, R. E. (2011). Linking environmental regimes, space and time: Interpretations of structural and functional connectivity. *Geomorphology*, 126, 387–404.

Walling, D. E. (1983). The sediment delivery problem. *Journal of Hydrology*, 65, 209–237.

Ward, J. V. (1989). The four-dimensional nature of lotic ecosystems. *Journal of the North American Benthological Society*, 8(1), 2–8.

Ward, J. V. (1997). An expansive perspective of riverine landscapes: Pattern and process across scales. *River Ecosystems*, 6, 52–60.

Wohl, E. (2017). Connectivity in rivers. *Progress in Physical Geography*, 41(3), 345–362.

Wohl, E., Brierley, G., Cadol, D., Coulthard, T., Covino, T., Fryirs, K., Grant, G., et al. (2019). Connectivity as an emergent property of geomorphic systems. *Earth Surface Processes and Landforms*, 44(1), 4–16.

Wolman, M. G., & Miller, J. P. (1960). Magnitude–frequency of forces in geomorphic processes. *Journal of Geology*, 68, 54–74.

Wooldridge, S. W. (1932). The cycle of erosion and the representation of relief. *Scottish Geographical Magazine*, 48, 30–36.

Xie, C., Cui, B., Xie, T., Yu, S., Liu, Z., Chen, C., Ning, Z., Wang, Q., Zou, Y., & Shao, X. (2020). Hydrological connectivity dynamics of tidal flat systems impacted by severe reclamation in the Yellow River Delta. *Science of the Total Environment*, 739, 139860.

Part II

Connectivity in Process Domains

3

Hillslope Processes

ANTHONY J. PARSONS

Most hillslopes are valley-side slopes. As Young (1972, p. 1) put it 'At a given point on the ground surface it is normally possible to follow the line of maximum slope downwards until a drainage channel is reached'. From the perspective of this chapter's review of connectivity, hillslopes may be regarded as conveyor belts transferring water (derived from the atmosphere) and sediment and nutrients (derived mainly from the weathering of their constituent materials) to other parts of the geomorphic system, of which rivers are probably the most important recipients. The movement of sediment from hillslopes by wind and ice is not covered in this chapter (but see Chapters 5 and 6). Ferguson (1981) used the expression 'jerky conveyor belt' to characterise sediment transport in rivers. The expression is equally apt as a metaphor for the connectivity in the movement of water, sediment and nutrients on hillslopes, which may be summarised in the question 'how far and how fast?'. In the long run, connectivity must equal 100%. All the water that arrives from the atmosphere must, in due course, leave the hillslope system. Likewise, all the eroded sediment must be transferred off the hillslope. The key issues here are 'what constitutes the "long run"?', and 'what happens up to that point?'

To some extent, the concept of connectivity is implicit in the earlier notions of sediment delivery and sediment budgets (Di Stefano & Ferro, 2017). A more explicit linking of sediment delivery and connectivity is given by Di Stefano and Ferro (2018) through the use of their sediment delivery distributed model. However, although this approach captures disparities between sediment movement on hillslopes and that in river channels, neither it nor the notion of sediment budgets, comes to terms with rates of transfer in the short term and the need to balance input and output in the long term (see Parsons et al., 2006; Parsons, 2012). It is in addressing this issue that the concept of connectivity has the potential to advance our understanding of hillslope geomorphology.

In this chapter, I will examine the mechanisms of, and the factors controlling, how far and how fast water, sediment and nutrients move along this jerky conveyor belt. For ease of presentation, I will discuss separately water movement in and on hillslopes, fluid-gravity and sediment-gravity movement of sediment and nutrient movement. However, it must be recognised that hillslope processes do not operate in isolation, and that, as well as considering the connectivity of transport by one processes, the interaction of connectivity among processes is also important. This interaction is particularly significant when assessing the importance of connectivity to understanding hillslopes within the context of landscape evolution, which forms the final section of this chapter.

3.1 Water Movement in and on Hillslopes

Water arriving at the surface of hillslopes may either infiltrate into the hillslope or travel downslope over the surface as overland flow. Conventionally, these two routes are treated separately, but there is an argument in favour of a more holistic view (Berkowitz & Zehe, 2020; Wolstenholme et al., 2020). Water arriving at the surface of a hillslope is subject to a single set of forces. The strengths of individual forces within that set determine the preferred pathway of the water.

3.1.1 Infiltration and Subsurface Flow

Although numerous equations exist to describe infiltration (e.g., Green & Ampt, 1911; Horton, 1933; Philip, 1957/58), in connectivity terms, what is needed is an understanding of the ease with which infiltrated water moves down the hillslope. Such an understanding is difficult to achieve because of the difficulty of observing subsurface flow, although a variety of techniques have been developed (for a review, see Blume & van Meersveld, 2015). Belter et al. (2020) proposed joint analysis of shallow near-stream groundwater and water level in streams as a way to infer subsurface hillslope-stream connectivity. Functional connectivity of subsurface flow has been investigated using graph theory (Phillips et al., 2011; Zuecco et al., 2019). Using a series of wells in the catchment as nodes, functional connectivity is deemed to exist between one node in the downslope flow direction of another if both wells contain water. The connectivity of the network, C, is then given as the ratio of active edges, E_A, and the total number of edges, E_T. As Phillips et al. point out, this connectivity index is analogous to a measure of antecedent moisture. Klaus and Jackson (2018) undertook an analysis of travel distance of interflow (i.e., infiltrated water that travels downslope above an impeding layer) before it percolated into the impeding layer, and showed that these travel distances ranged from around one metre to several hundred metres. Importantly, they showed that for the majority of the hillslopes they studied, this interflow travel distance was shorter than the hillslope length. Consequently, they argued, the existence of a perched saturated zone does not imply continuity of interflow from the hillslope to the stream channel. In an echo of previous research on partial area contribution (Hewlett & Hibbert, 1967), only interflow from the lower part of the hillslope contributes directly to streamflow. However, even if a perched saturated layer, as a whole, does not exhibit connectivity with streamflow, the existence of preferential flow paths may enhance such connectivity. Wilson et al. (2016) demonstrated in the catchments that they studied the existence of pipe networks that connected the uppermost parts of hillslopes with the catchment outlet that resulted in relatively high subsurface connectivity. In the modified hydrology of agricultural subsurface drains, connectivity of subsurface flow may be similarly enhanced, as Barkle et al. (2021) demonstrated for export of nutrients from pastoral fields.

Developing an understanding of the connectivity of subsurface flow is particularly challenging, not only because of the difficulty of observing and measuring it, but also because of the variety of routes that subsurface flow may take (cf. glaciers – see Chapter 6), and the temporal variation in connectivity that responds particularly to antecedent conditions. As Blume and van Meersveld (2015) point out, research on this topic has, so far, mainly concentrated on conditions at individual research sites with little progress towards a general understanding that might have widespread applicability.

3.1.2 Overland Flow

Water that does not infiltrate flows over the surface. The connectivity issues to be addressed are (i) the probability that water which fails to infiltrate at the point at which it arrives at the ground surface will continue its path over the surface to the bottom of the hillslope; (ii) the pathways that the water takes on its route to the bottom of the hillslope; (iii) the spatial and temporal variations in these pathways; and (iv) the effects of these pathway characteristics on the hillslope hydrograph.

Traditionally, overland flow discharge has been regarded as rainfall minus infiltration, leading to the characteristic run-off hydrograph (Figure 3.1). Inasmuch as infiltration decreases through time so overland flow discharge correspondingly increases through time. However, such a simple view of the relationship between infiltration and run-off fails to take account of spatial variability in infiltration, and interactions between infiltrated water and overland flow. Spatial variability in infiltration is well documented (e.g., Blackburn, 1975; Sharma et al., 1980), and some evidence indicates a log-normal distribution for infiltration losses (e.g., Murabayashi & Fox, 1979; Sharma et al., 1980), though others have found a normal distribution (e.g., Parsons et al., 1997). However, less attention has been given to the spatial arrangement of infiltration and its implications for connectivity. Parsons et al. (1997) compared run-off hydrographs obtained from a distributed

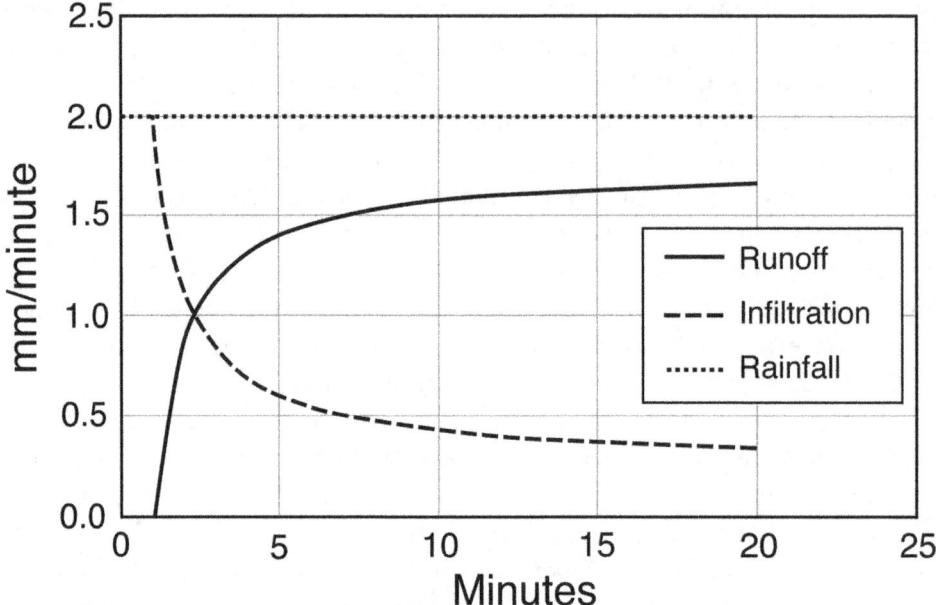

Figure 3.1   Hypothetical infiltration and run-off curves for a steady rainfall input
and modelling infiltration according to Green and Ampt (1911).

model using a deterministic model based upon measured ground-surface characteristics with stochastic distributions for both grassland and shrubland sites in
Arizona. They found, in general, stochastic spatial distributions of infiltration
parameters worsened model performance in terms of predicting the shape of the
run-off hydrograph, and that even where it led to better performance, the fit to
observed run-off hydraulics was worse. Working on small catchments in a similar
environment, Mueller et al. (2007) undertook a geostatistical analysis of vegetation
cover and ponded infiltration rate to provide input to a spatially distributed hydrological model. Their results demonstrated that 'the more connectivity, in the form
of connected pattern features, is included in the parameterization approach … the
better the modelling results'. These studies demonstrate the importance of structural connectivity (in these cases spatial patterns of infiltration and vegetation
cover) in determining runoff from semi-arid hillslopes. This contrast is illustrated
by the marked difference in run-off hydraulics (Figures 3.2 and 3.3) observed
between grassland and shrubland in Arizona, which can be attributed to the difference in spatial patterns and connectivity of bare patches in the two vegetation
communities. Whereas the grassland has minimal across-slope, but pronounced
downslope, microrelief that disperses run-off laterally, the shrubland has pronounced across-slope microrelief that forms connected downslope flow pathways (Figure 3.4). Comparable to the spatial patterns of microrelief, size and

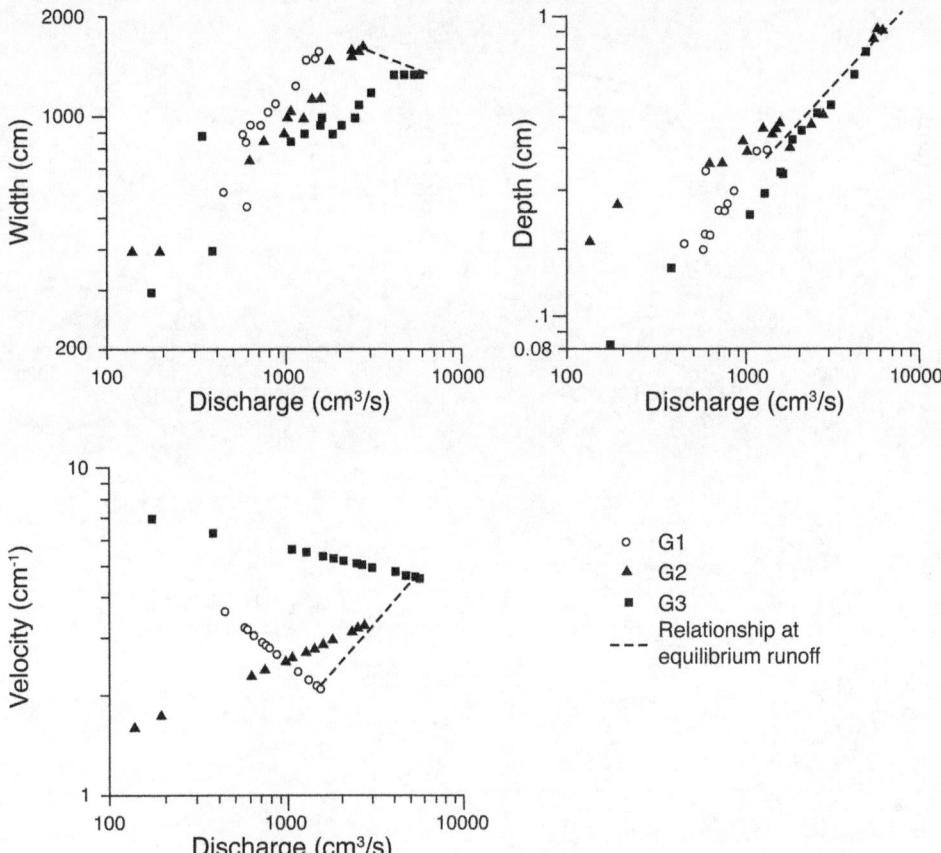

Figure 3.2   Run-off hydraulics on a grassland plot measured at three cross-sections and showing downslope trends in width, depth and velocity at equilibrium runoff (after Parsons et al., 1996).

spatial distribution of depressions also affect connectivity of runoff. Shook et al. (2021) undertook numerical experiments to study the effects of spatial and frequency distributions of depressions in the Canadian Prairie Pothole region and demonstrated that in areas of small numbers of depressions their size distribution and spatial arrangement are important in controlling the proportion of rainfall that becomes runoff. In contrast, where there are many depressions, their spatial arrangement is unimportant in controlling runoff. Preferential saturation of foot-slope areas (common in humid and upland landscapes) increases hydrological connectivity between hillslopes and channels (Antonelli et al., 2020; Iwasaki et al., 2020).

Hawkins (1982) drew attention to the positive relationship between infiltration rates and rainfall intensity, thereby demonstrating that spatial patterns of infiltration also affect the functional connectivity of hillslope runoff. Water that does not

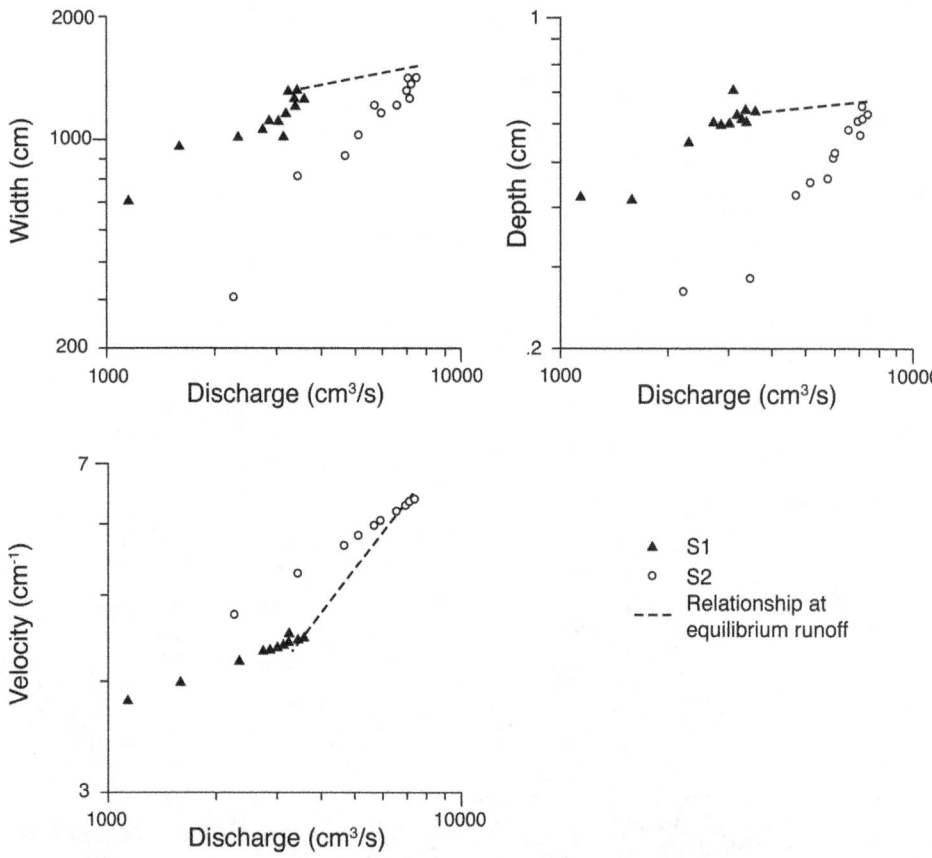

Figure 3.3   Run-off hydraulics on a shrubland plot measured at two cross-sections and showing downslope trends in width, depth and velocity at equilibrium runoff (after Parsons et al., 1996).

infiltrate at the point of arrival may do so if, as runoff-runon, it encounters an area of unsatisfied infiltration downslope. The probability that this will occur decreases with rainfall intensity. Nanda et al. (2019) showed that on infiltration-excess dominated hillslopes only high intensity events were able to generate runoff at the outlet and that run-off generated from the upper parts of hillslopes re-infiltrated in the middle part due to higher soil hydraulic conductivity. Reaney (2003; see also Reaney et al., 2007) developed a connectivity run-off model to investigate the effects of landscape morphology and rainfall input on hillslope hydrographs in semi-arid areas, and demonstrated that temporal rainfall patterns within a storm event had a marked effect on hillslope hydrographs. A similar idea is presented by Wolstenholme et al. (2020). Wainwright and Parsons (2002) also demonstrated that spatial patterns of infiltration (and consequent runoff-runon) coupled with temporal variations in rainfall intensity would lead to the commonly observed

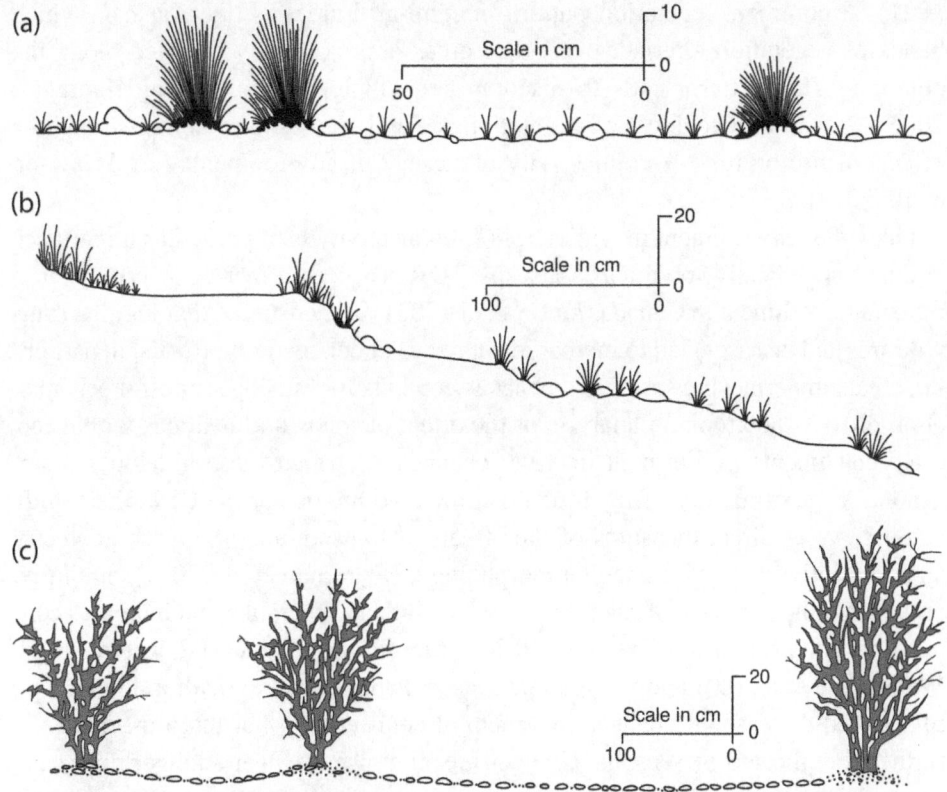

Figure 3.4 (a) Across-slope microrelief on the grassland, (b) downslope microrelief on the grassland and (c) across-slope microrelief on the shrubland (after Parsons et al., 1996).

scale dependency of runoff whereby runoff coefficient decreases as size of area increases. The importance of rainfall frequency to run-off connectivity was demonstrated by Zimmermann et al. (2014), who showed that more frequent rainstorms increase antecedent soil moisture which, in turn, increases connectivity of flow lines in their forested environment.

Many of the studies that have shown the importance of both structural and functional connectivity in controlling hillslope runoff have focused on semi-arid environments. There are two reasons why this is so. First, these environments are particularly prone to flash flooding, so understanding the mechanisms of such flooding is especially important. Second, these environments have very marked spatial discontinuity of vegetation, typically leading to areas of high infiltration under vegetation and areas of low infiltration in bare patches between vegetation. Consequently, connectivity of areas of low infiltration is particularly important in controlling hillslope hydrographs and system resources (Calvo-Cases et al.,

2021). A common vegetation pattern in semi-arid areas is tiger bush in which bands of vegetation alternate with bare areas approximately oriented along the contours. This pattern leads to disconnected hillslope runoff (Puigdefabregas, 2005). Linked to patchiness of vegetation, several studies have demonstrated the effects of fire on run-off connectivity in a range of environments (e.g., Cawson et al., 2013).

One other environment in which run-off connectivity is of particular importance is anthropogenically modified landscapes. Disturbances affect connectivity. In a mountain catchment in Canada, Jautzy et al. (2021) demonstrated that logging (plus wildfires and diseases) led to increased pathways of connectivity. Roads, in particular, create impermeable surfaces that act as conduits for hillslope runoff. Pechenick et al. (2014) undertook an analysis of the effect of roads on hillslopes within forested catchments in Vermont on river channels to which these hillslopes drain, using a range of metrics. They found that the road metrics proposed in their study could serve as direct measures of the effects of transportation network geometry and connectivity on river channel morphology. Boardman et al. (2019), similarly, argued that the presence of roads, culverts and ditches on arable land increases connectivity and promotes delivery of hillslope runoff (and sediment) to watercourses. Wainwright et al. (2011) observed that seasonal changes on agricultural land (e.g., tillage) led to seasonal changes in structural connectivity, but that this change is further complicated by seasonal changes in soil moisture. Over a longer timescale, Lopez-Vicente et al. (2017) documented changes in structural connectivity due to changing land use. These changes, similarly, affect flow pathways.

Because hillslopes comprise material that is erodible by runoff, runoff connectivity evolves in response to run-off hydraulics, since hydraulics determine shear stress and, hence, flow detachment. The formation of rills and gullies by flow detachment leads to greater connectivity of flow pathways, greater concentration of flow and, hence, more rapid hillslope runoff. In line with the holistic viewpoint (Berkowitz & Zehe. 2020), concentrated flow paths are preferred because resistance to flow in such pathways is lower. Hence, there is a positive feedback. Concentrated flowpaths are preferred, hence leading to greater shear stress and flow detachment. However, other factors are important. Parsons and Wainwright (2006) argued that whether or not rills do develop from unconcentrated overland flow depends on the combined influence of (i) the probability that the shear stress caused by a burst event in turbulent overland flow will exceed the local shear strength of the soil, (ii) the spatial variability of soil shear strength and (iii) the strength of the diffusive effect of soil detachment by raindrops which inhibits the formation of such pathways. Luk ct al. (1993) cstimatcd for a shrubland site in Arizona that a storm with a recurrence interval of 3–4 years was necessary to shift the balance in favour of flow detachment over diffusive infilling of concentrated flow pathways.

3.1.3 Interactions between Infiltration and Overland Flow

Although it is convenient to separate water into that which infiltrates and that which does not, in reality the two are not so readily separated. The case of overland flow that subsequently infiltrates as runoff-runon has been dealt with above. Here the other half of the interaction is addressed. Soils are often saturated on the lower parts of hillslopes or where soils are thin such that water that has infiltrated in the upslope areas may exfiltrate as return flow (Betson & Marius, 1969). Because velocity of surface flow is typically greater than that of subsurface flow, the existence of return flow, particularly where it occurs close to the base of the hillslope, increases connectivity of hillslopes to channels in the sense of increasing its 'how fast' element. Mapping of soil thickness and monitoring of antecedent moisture conditions can yield successful modelling of return flow connectivity (e.g., Deb et al., 2006).

3.2 Sediment Movement on Hillslopes

3.2.1 Fluid-gravity Processes

Fluid-gravity movement of sediment is that movement of sediment that is induced by the exertion of stress on hillslope materials by the movement of water. For the sake of completeness, that must include stress exerted by falling raindrops as well as that exerted by flowing water. Acting alone, raindrops diffuse sediment around the point of impact, but connectivity of sediment transport is effectively zero, other than there being a net downslope transport as a function of gradient. In conjunction with shallow overland flow that, itself, exerts too little shear stress to detach sediment, transport distance, and hence connectivity, increases markedly. Conventionally, it has been assumed that sediment travels in suspension in, and with the same velocity as, such flow, though with little justification (Bennett, 1974, Wainwright et al., 2008). Parsons et al. (1998) demonstrated that such raindrop-detached sediment typically travels as bedload, and developed a transport-distance equation for sediment under rain-impacted overland flow

$$ML = (RE.FE)^{1.6363},$$

in which M is particle mass (g), L is distance moved in unit time (cm min^{-1}), RE is rainfall energy (Jm^{-2}s^{-1}) and FE is flow energy (Jm^{-2}s^{-1}). Subsequently, Wainwright et al. (2008) argued that particle travel distance in unit time could be expressed more usefully as virtual velocity. By analysing the virtual velocities of particles over a range of different time periods, Parsons et al. (2015) assessed the connectivity of sediment movement at event, seasonal and decadal scales. Their data do not come from the same, or even comparable, sites but, nonetheless, show an expected reduction in virtual velocity with increased timescale, ranging from 100 mm h^{-1} at the event scale to 11.25 mm a^{-1} at the decadal scale.

Where overland flow exerts sufficient shear stress to detach sediment, connectivity of sediment movement may be expected to increase. However, little effort has so far been made to understand the connectivity of sediment movement, other than to apply approaches from fluvial geomorphology in which sediment is divided into bedload and suspended load, and the two treated separately. Again, the assumption has been made that suspended load travels with the same velocity as the water flow, whereas bedload travels intermittently and with a virtual velocity that takes account of periods at rest on the bed. Based on a laboratory flume experiment, Parsons et al. (2018) demonstrated that so-called suspended sediment also possesses a virtual velocity that is less than water velocity, and, combining their data with data from bedload transport in rivers, proposed a general equation for the virtual velocity of all sediment transported in water

$$V = 58209\tau^{3.4434},$$

in which V is sediment velocity (m h^{-1}) and τ is dimensionless shear stress. The equation is based on very limited data, but if such an equation can be shown to be supported by more data, then it offers the possibility to understand the connectivity of flow-detached sediment on hillslopes, in a way similar to that discussed earlier for raindrop-detached sediment.

The laboratory studies quoted earlier all deal with fully connected flow paths, whereas at the hillslope scale, such connectivity of flow paths will not always exist (as discussed in Section 3.1.2). A full understanding of the connectivity of fluid-gravity sediment movement on hillslopes requires a combination of the data discussed in this section with the data on connectivity of water flow discussed in Section 3.1.2. So far, that integration has been limited (see, for example, Medeiros et al., 2014; Sun et al., 2019; Moreno-de-las-Heras et al., 2020), and remains a significant gap in our understanding of the connectivity of sediment movement by fluid-gravity processes. As with runoff, in anthropogenically modified environments, the presence of structures can significantly affect fluid-gravity controlled sediment connectivity (see, for example, Batista et al., 2021).

3.2.2 Sediment-gravity Processes

Sediment-gravity processes are those processes whereby

$$\rho ghs > c + \sigma' \tan \phi,$$

in which ρ is sediment density (kg m^{-3}), g is acceleration due to gravity (m s^{-2}), h is sediment depth (m) and s is slope (m m^{-1}), c is sediment cohesion (kPa), σ' is effective shear strength (kPa) and ϕ is the angle of internal friction.

As this inequality is established sediment begins to move downslope. Terms in this inequality may change to reverse the inequality. For example, sediment may run

Figure 3.5   An almost fully connected pathway delivering sediment from near the top of the hillslope to a point adjacent to a river channel (photo by the author). A black and white version of this figure will appear in some formats. For the colour version, refer to the plate section.

out onto a lower-angled footslope, or de-water during movement. Consequently, it is common for sediment-gravity processes to lead to limited travel distance downslope. Occasionally, a mass of deposited material reaches the foot of the hillslope (Figure 3.5), where it may subsequently be entrained by a river (thus giving rise to a connected sediment pathway to an adjacent part of the geomorphic system, but also destabilise that part of the system). Germain et al. (2020) used tree-ring chronology to assess the frequency with which debris flows connected with the stream channel in the Mazeri Valley, Georgia and found that such events had a return period of 5.4 years. More commonly, however, the material is merely deposited further downslope (Figure 3.6), where it may be re-activated at a later time, or provide input to fluid-gravity processes. It has been estimated that between 20% and 70% of sediment travels from the source area to a river channel (Cislaghi & Bischetti, 2019). Various formulae have been proposed to estimate the travel distance, L, of landslides such as

$$L = aV^b H^c,$$

Figure 3.6 A landscape of disconnected mass movements in which sediment-gravity processes move material only part way down the hillslopes (photo by the author). A black and white version of this figure will appear in some formats. For the colour version, refer to the plate section.

in which *V* is landslide volume, *H* is the vertical drop, and a, b, c are empirical constants. However, the use of such empirical constants means that no generally applicable prediction for *L* exists (Cislaghi & Bischetti, 2019). By comparing this distance to the average distance to the channel throughout the catchment, a probability for a connected pathway may be developed.

However, a full understanding of the connectivity of sediment-gravity processes depends not only on their travel distances, but also on the timescales of their activity, the probability of their remobilisation, or the timescale necessary for fluid-gravity processes to complete the connectivity of sediment movement through the hillslope system. A few studies point the way to such an understanding. In a detailed study of the Schimbrig catchment in the Central Swiss Alps, Clapuyt et al. (2019) undertook analysis at annual, decadal and millennial timescales of the connectivity of one landslide to the stream network. The discrepancy between decadal and millennial fluxes indicated short periods of connectivity between the landslide and the river separated by long periods during which sediment mobilised by the landslide was retained on the hillslope as colluvial deposits. Bennett et al. (2014) focused on the stochastic nature of sediment-gravity processes to develop a model for hillslope sediment output in a mountain catchment. They applied this model to the Illgraben catchment in southwestern Switzerland and successfully predicted the observed temporal distribution of sediment yield from the catchment. The model produced a complex sediment discharge based upon knowledge of availability of sediment and water and the role of history and climatic triggering events. There is no doubt that the role of connectivity of sediment-gravity processes to the sediment output of hillslopes is complex in both space and time. Inasmuch as these processes may also deliver large quantities of sediment to the adjacent fluvial system over very short time periods, the ability of the fluvial system to deal with this sudden, large influx of sediment must also be included in a full understanding of the connectivity of sediment-gravity processes.

3.2.3 Nutrient Movement

Nutrients are mainly derived from weathering of hillslope materials but also include those that are deposited from the atmosphere. They may move down hillslopes either in dissolved form, travelling with surface or sub-surface flow, or adsorbed onto moving sediment particles, typically clays. Consequently, connectivity of nutrient movement is governed by connectivity of surface and sub-surface flow, and size selectivity in sediment transport distance under fluid-gravity flow. For surface runoff, the relative importance of sediment-bound and dissolved nutrient fluxes has been shown to be a function of the availability and concentration of different nutrient species in the soil or rainfall (Michaelides et al., 2012) and the magnitude of sediment-bound nutrient flux to be determined by the grain-size distribution of the eroded sediment and the total sediment yield. In sub-surface flow, Laine-Kaulio and Koivusalo (2018) used a dye-tracing experiment to show that the transport of solutes was strongly dependent on connected, preferential flowpaths and the linkages between these preferential flowpaths and the soil matrix. The presence of such connected, preferential flowpaths can significantly increase nutrient movement. Weihrauch et al. (2021) combined soil sampling with network analysis to identify sources and sinks of subsurface phosphorus movement.

3.3 Hillslope Evolution

Up until the middle of the twentieth century, there was an emphasis on understanding hillslopes from the point of view of landscape evolution. From the middle of that century hillslope research in focused on the mechanics of hillslope processes. A renewed interest in landscape evolution (e.g., Willgoose et al., 1991; Howard, 1994; Coulthard et al., 2002) has benefited from the resultant increased process understanding coupled with advances in computing power. However, the more recent recognition of the importance of connectivity in the effectiveness of hillslope processes has yet to make a significant impact on landscape evolution models. Yet it is clear from the analysis presented in this chapter that without the incorporation of connectivity, models aiming to predict not only run-off hydrographs and soil erosion, but also landscape evolution, may be unsuccessful. Nuñes et al. (2018) argue that models that attempt to predict at spatial and temporal scales different from those of the empirical data that drive the models need to take account of emergent properties of structural and functional connectivity that control how processes operate and how their operation evolves through time. Typically, with few exceptions, models of soil erosion over relatively short timescales fail to do this. Hewett et al. (2014) developed a version of Mahleran (see Wainwright et al., 2008) that incorporated topographic updating, and hence implicitly emergent properties of connectivity, but so far little evaluation of the importance of such updating has been presented. More explicit treatment of changes in connectivity is provided by Rillgrow (Favis-Mortlock et al., 2000) which is based upon conceptualising the hillslope erosional system as a self-organizing dynamic system.

In contrast to short-term modelling of soil loss and run-off hydrographs, long-term models of hillslope evolution do incorporate topographic change since that is their very purpose. However, as Hewett et al. (2014) point out, they typically do so at the expense of very simplified process models. Whether detachment-limited (e.g., Howard, 1994), or transport-limited (e.g., Willgoose et al., 1991) such models assume a continuity of sediment transport across hillslopes that is at odds with notions of connectivity and dynamic feedback of process interactions. Nonetheless, notwithstanding the argument put forward by Nuñes et al. (2018), it remains unclear how far long-term hillslope evolution would differ if the evolution of processes through time were to be incorporated in models. In the short-term, however, the evolution of rills and other channel networks, as well as stochastic patterns of sediment-gravity processes, has a significant impact on the transfer of water and sediment through the hillslope system.

3.4 Concluding Remarks

This chapter has focused on the spatial connectivity of hillslope processes in addressing the question of 'how far, and how fast', and it has taken a reductionist approach by considering individual processes largely in isolation. But, as already noted, connectivity in hillslope processes cannot be fully addressed without consideration of the connectivity among these processes, as illustrated by Hancock and Lowry (2021). The temporal dimension of connectivity needs also to be considered. This dimension is bound up with the concept of magnitude and frequency (Wolman & Miller, 1960) in addressing both which events exhibit high and low connectivity and the extent to which connectivity during a particular event is conditioned on the past sequence of events. In broad terms, high magnitude events may be expected to exhibit greater connectivity, as illustrated by the studies of Luk et al. (1993), Masselink et al. (2017) and Moreno-de-las-Heras et al. (2020). A full description of the connectivity of hillslope processes will require combined knowledge of both the magnitude–connectivity relationship, the probability distribution of event magnitudes and, to explain specific cases of functional connectivity, the actual sequence of events (as illustrated by Blampied et al., 2018; Reulier et al., 2017).

There can be no doubt that in recent years there has been a growing recognition of the importance of connectivity in understanding the effects of hillslope processes. At best, however, that understanding remains patchy and incomplete. To conclude this chapter as it began, with a metaphor: we currently have some bits of the jigsaw, but do not necessarily know how these bits fit together and nor do we know what the full picture looks like. Much remains to be done.

References

Antonelli, M., Glaser, B., Teuling, A. J., Klaus, J. & Pfister, L. (2020). Saturated areas through the lens: 2. Spatio-temporal variability of streamflow generation and its relationship with surface saturation. *Hydrological Processes*, 34, 1333–1349.

Barkle, G., Stenger, R., Moorhead, B. & Clague, J. (2021). The importance of the hydrological pathways in exporting nitrogen from grazed artificially drained land. *Journal of Hydrology*, 597. doi:10.1016/j.hydrol.2021.126218.

Batista, P. V. G., Fiener, P., Scheper, S. & Alewell, C. (2021). A conceptual model-based sediment connectivity assessment for patchy agricultural catchments. *Hydrology and Earth System Sciences, Discussions*. doi:10.5194/hess-2021-231.

Belter, D., Wwller, M. & Blume, T. (2020). Characterising hillslope-stream connectivity with a joint analysis of stream and groundwater levels. *Hydrology and Earth System Sciences*, 24, 5713–5744.

Bennett, G. L., Molnar, P., McArdell, B. W. & Burlando, P. (2014). A probabilistic sediment cascade model of sediment transfer in the Illgraben. *Water Resources Research*, 50, 1225–1244.

Bennett, J. P. (1974). Concepts of mathematical modelling of sediment yield. *Water Resources Research* 10, 485–492.

Berkowitz, B. & Zehe, E. (2020). Surface water and groundwater: Unifying conceptualization and quantification of two 'water worlds'. *Hydrology and Earth System Sciences*, 24, 1831–1858.

Betson, R. P. & Marius, J. B. (1969). Source areas of storm runoff. *Water Resources Research*, 5, 574–582.

Blackburn, W. H. (1975). Factors influencing infiltration and sediment production of semi-arid rangelands in Nevada. *Water Resources Research*, 11, 929–937.

Blampied, J., Carozza, J. M. & Antoine, J. M. (2018). Sediment connectivity in the high Pyrenees mountain range by 2013 flood analysis: role of surficial sediment storages. *Geomorphologies-Relief Proccessus Environnement*. 24, 389–402.

Blume, T & van Meerveld, H. J. (2015). From hillslopes to stream: methods to investigate subsurface connectivity. *Wiley Interdisciplinary Reviews – Water* 2015, 2: 177–198.

Boardman, J., Vandaele, K., Evans, R. & Foster, I. D. L. (2019). Off-site impacts of soil erosion and runoff: Why connectivity is more important than erosion rates. *Soil Use and Management*, 35, 245–256.

Brouwer, J., Barker, A. P., Gaze, S. R., Valentin, C., Bromley, J. & Valentine, C. (1997). The role of surface water redistribution in an area of patterned vegetation in a semi-arid environment, south-west Niger. *Journal of Hydrology*, 198, 1–29.

Calvo-Cases, A., Arnau-Rosalen, E., Boix-Fayos, C., Estrany, J., Roxo, M. J. & Symeonakis. (2021). Eco-geomorphological connectivity and coupling interactions at hillslope scale in drylands: concepts and critical examples. *Journal of Arid Environments*, 186, doi:10.1016/jaridenv.2020.104418.

Cawson, J. G., Sheridan, G. J., Smith, H. G. & Lane, P. N. J. (2013). Effects of fire severity and burn patchiness on hillslope-scale surface runoff, erosion and hydrologic connectivity in a prescribed burn. *Forest Ecology and Management*, 310, 219–233.

Cislaghi, A. & Bischetti, G. B. (2019). Source areas, connectivity, and delivery rate of sediments in mountainous forested hillslopes: A probabilistic approach. *Science of the Total Environment*, 652, 1168–1186.

Clapuyt, F., Vanacker, V., Christi, M., Van Oost, K. & Schlunegger, F. (2019). Spatio-temporal dynamics of sediment transfer systems in landslide prone Alpine catchments. *Solid Earth*, 10, 1480–1503.

Coulthard, T. J., Macklin, M. G. & Kirkby, M. J. (2002). A cellular model of Holocene upland river basin and alluvial fan evolution. *Earth Surface Processes and Landforms* 27, 269–288.

Deb, K., Miyazaki, T., Mizoguchi, M. & Seki, K. (2006). Return flow generating point on a hillslope layered with traffic pan. *Transactions of Japanese Society of Irrigation, Drainage and Reclamation Engineering*, 241, 1–11, doi:10.11408/jsidre1965.2006.1.

Di Stefano, C. & Ferro, V. (2017). Testing sediment connectivity at the experimental SPA2 basin, Sicily (Italy). *Land Degradation & Development*, 28, 1557–1567.

Di Stefano, C. & Ferro, V. (2018). Modelling sediment delivery using connectivity components at the experimental SPA2 basin, Sicily (Italy). *Journal of Mountain Science*, 15, 1868–1880.

Favis-Mortlock. D. T., Boardman, J., Parsons, A. J. & Lascelles, B. (2000). Emergence and erosion: a model for rill initiation and development. *Hydrological Processes*, 14, 2173–2205.

Ferguson, R. I. (1981). Channel form and channel changes. In J. Lewin (ed.) *British Rivers*. London: Allen and Unwin, pp. 90–125.

Germain, D., Gavril, I-G., Elizbarashvili, M. & Pop, O. M. (2020). Multidisciplinary approach to sediment connectivity between debris-flow channel network and the Doira River, Mazeri Valley, Southern Caucasus, Georgia. *Geomorphology*, 371, doi:10.1016/j/geomorph.2020.107455.

Green, W. H. & Ampt, G. A. (1911). Studies on soil physics. 1. The flow of air and water through soils. *Journal of Agricultural Soils*, 4, 1–24.

Hancock, G. & Lowry, J. (2021). Quantifying the influence of rainfall, vegetation and animals on soil erosion and hillslope connectivity in the monsoonal tropics of northern Australia. *Earth Surface Processes and Landforms*, 46, 2110–2123.

Hawkins, R. H. (1982). *Interpretations of Source-area Variability in Rainfall-runoff Relations*. Littleton, CO: Water Resources Publications, pp. 303–324.

Hewett, C. J. M., Wainwright, J., Parsons, A. J. Cooper, J. R., Kitchener, B., Hargrave, G. K., Long, E. J., Onda, Y. & Patin, J. (2014). The importance of simulating changes in topography in process-based soil erosion modelling: implications for landscape evolution modelling. *Geophysical Research Abstracts*, 16, EGU2014-5422.

Hewlett, J. D. & Hibbert, A. R. (1967). Factors affecting the response of small water-sheds to precipitationin humid areas. In *1965 International Symposium on Forest Hydrology, Pennsylvania State University*, New York: Pergamon, pp. 275–90.

Horton, R. E. (1933). The role of infiltration in the hydrological cycle. *Transactions of the American Geophysical Union*, 14, 446–460.

Howard, A. D. (1994). A detachment limited model for drainage basin evolution. *Water Resources Research*, 30, 2261–2285.

Iwasaki, K., Katsuyama, M. & Tani, M. (2020). Factors affecting dominant peak-flow run-off generation mechanisms among five neighbouring granitic headwater catchments. *Hydrological Processes*, 34, 1154–1166.

Jautzy, T., Maltaid, M. & Buffin-Belanger, T. (2021). Interannual evolution and hydrosed-imenary connectivity induced by forest cover change in a snow-dominated moun-tainous catchment. *Land Degradation and Development*, 32, doi: 10.1002/ldr.3902.

Klaus, J. & Jackson, C. R. (2018). Interflow is not binary: a continuous perched layer does not imply continuous connectivity. *Water Resources Research*, 54, 5921–5932.

Laine-Kaulio, H. & Koivusalo, H. (2018). Model-based exploration of hydrological connectivity and solute transport in a forested hillslope. *Land Degradation and Development*, 29, 1176–1189.

Lopez-Vicente, M., Nadal-Romero, E. & Cammeraat, E. L. H. (2017). Hydrological con-nectivity does change over 70 years of abandonment and afforestation in the Spanish Pyrenees. *Land Degradation and Development*, 28, 1298–1310.

Luk, S.-H., Abrahams, A. D. & Parsons, A. J. (1993). Sediment sources and sediment transport by rill flow and interrill flow on a semi-arid piedmont slope, Southern Arizona. *Catena*, 20, 93–111.

Madeiros, P. H. A., de Araujo, J. C., Marmede, G. L., Creutzfeldt, B., Guntner, A. & Bronstert, A. (2014). Connectivity of sediment transport in a semiarid environment: a synthesis for the Uppeer Jaguaribe Basin, Brazil. *Journal of Soils and Sediments*, 14, 1938–1948.

Masselink, R. J. H., Temme, A. J. A. M., Gimenez, R., Casali, J. & Keesstra, S. D. (2017). Assessing hillslope-channel connectivity in an agricultural catchment using rare-earth oxide tracers and random forests models. *Cuadernos de Investigacion Geografica*, 43, 19–39.

Michaelides, K., D. Lister, J. Wainwright, and A. J. Parsons (2012), Linking runoff and erosion dynamics to nutrient fluxes in a degrading dryland landscape, *J. Geophys. Res.*, 117, G00N15, doi:10.1029/2012JG002071.

Moreno-de-las-Heras, M., Merino-Martin, L., Saco, P. M., Espigares, T., Gallart, F. & Nicolau, J. M. (2020). Structural and functional control of surface-patch to hillslope runoff and sediment connectivity in Mediterranean dry reclaimed slope systems. *Hydrology and Erath System Sciences*, 24, doi: 10.5194/hess-24-2855-2020. Southern Arizona.

Mueller, E. N., Wainwright, J. & Parsons, A. J. (2007). Impact of connectivity on the mod-elling of overland flow within semi-arid shrubland environments. *Water Resources Research*, 43, W09412.

Murabayashi, E. T. & Fox, Y.-S. (1979). *Urbanization-induced impacts on infiltration capacity and rainfall-runoff relation in Hawaiian urban area*. Technical Report 27, Water Resources Research Center, Honolulu, Hawaii.

Nanda, A., Sen, S. & McNamara J. P. (2019). How spatiotemporal variation in soil mois-ture can explain hydrological connectivity of infiltration-excess dominated hillslope: observations from lesser Himalyan landscapes. *Journal of Hydrology*, 579, doi: 10.1016/j.hydrol,2019.124146.

Nunes, J. P., Wainwright, J., Bielders, C. L., Darboux, F., Fiener, P., Finger, D. & Turnbull, L. (2018). Better models are more effectively connected models. *Earth Surface Processes and Landforms*, 43, 1355–1360. doi:10.1002/esp.4323.

Parsons, A. J., Wainwright, J. & Abrahams, A. D. (1996). Runoff and erosion on semi-arid hillslopes. In M. G. Anderson & S. M. Brooks (eds.) *Advances in Hillslope Processes*. Chichester: John Wiley & Sons, pp. 1061–1078.

Parsons, A. J., Stromberg, S. G. L. & Greener, M. (1998). Sediment-transport competence of rain-impacted interrill overland flow. *Earth Surface Processes and Landforms*, 23, 365–375.

Parsons, A. J. & Wainwright, J. (2006). Depth distribution of interrill overland flow and the formation of rills. *Hydrological Processes*, 20, 1511–1523.

Parsons, A. J., Wainwright, J., Powell, D. M., Brazier, R. E. (2006). Is sediment delivery a fallacy? *Earth Surface Processes and Landforms* 31, 1325–1328.

Parsons, A. J. (2012). How useful are catchment sediment budgets? *Progress in Physical Geography*, 36, 60–71.

Parsons, A., Cooper, J., Onda, Y., Patin, J. & Wainwright, J. (2015). Conceptualizing hillslope sediment connectivity as virtual velocity. *Geophysical Research Abstracts*, 17, EGU2015-1444.

Parsons, A. J., Wainwright, J., Abrahams, A. D. & Simanton, J. R. (2016) Distributed dynamic modelling of interrill overland flow. *Hydrological Processes*, 11, 1833–1859.

Parsons, A. J., Cooper, J. R., Wainwright, J. & Sekiguchi, T. (2018). Virtual velocity of sand transport in water. *Earth Surface Processes and Landforms*, 43, 755–761.

Phillips, R. W., Spence, C. & Pomeroy, J. W. (2011). Connectivity and runoff dynamics in heterogeneous basins. *Hydrological Processes*, 25, 3061–3075.

Pechenick, A. M., Rizzo, D. M., Morrissey, L. A., Garvey, K. M., Underwood, K. L. & Wemple, B. C. (2014). A multi-scale statistical approach to assess the effects of connectivity of road and stream networks on geomorphic channel condition. *Earth Surface Processes and Landforms*, 39, 1538–1549.

Philip, J. R. (1957/8). The theory of infiltration. *Soil Science*, 83, 345–357 and 435–458.

Puigdefabregas, J. (2005). The role of vegetation patterns in structuring runoff and sediment fluxes in drylands. *Earth Surface Processes and Landforms*, 30, 133–147.

Reaney S. M. 2003. *Modelling runoff generation and connectivity for semiarid hillslopes and small catchments*. PhD thesis, University of Leeds.

Reaney, S. M., Bracken, L. J. & Kirkby, M. J. (2007). Use of the Connectivity of Runoff Model (CRUM) to investigate the influence of storm characteristics on runoff generation and connectivity in semi-arid areas. *Hydrological Processes*, 21, 894–906.

Reulier, R., Delahaye, D., Viel, V. & Davidson, R. (2017). Hydro-sedimentary connectivity in a small agricultural watershed in French northwest: from field expertise to multi-agent system modelling. *Geomorphologie-Relief Processus Environnement*, 23, 327–340.

Sharma, M. L., Gander, G. A. & Hunt, C. G. (1980). Spatial variability of infiltration in a watershed. *Journal of Hydrology*, 45, 101–122.

Shook, K., Papalexiou, S. & Pomeroy, J. W. (2021). Quantifying the effects of Prairie depressional storage complexes on drainage basin connectivity. *Journal of Hydrology* 593, doi: 10.1016/j.jhydrol.2020.125846.

Sun, W. Y., Mu, X. M., Gao, P., Zhao, G. J., Li, J. Y., Zhang, Y. Q. & Francis, C. (2019). Landscape patches influencing hillslope erosion processes and flow dynamics. *Geoderma*, 353, 391–400.

Wainwright J & Parsons A. J. (2002). The effect of temporal variations in rainfall on scale dependency in runoff coefficients. *Water Resources Research*, 38(12), 7-1–7-10.

Wainwright, J., Parsons, A. J., Mueller, E., Brazier, R., Powell, D. M. & Fenti, B. (2008). A transport-distance approach to scaling erosion rates: 1. Background and model development. *Earth Surface Processes and Landforms*, 33, 813–826.

Wainwright, J., Turnbull, L., Ibrahim, T. G., Lexartza-Artza, I., Thornton, S. F. & Brazier, R. E. (2011). Linking environmental régimes, space and time: interpretations of structural and functional connectivity. *Geomorphology*, 126, 387–404.

Weihrauch, C., Weber, C. J. & von Sperber, C. (2021). A Soilscaple network approach (SNAp) to investigate subsurface phosphorus translocation along slopes. *Science of the Total Environment*, 784, doi:10.1016/j.scitotenv.2021.147131.

Willgoose, G., Bras, R. L. & Rodriguez-Iturbe, I. (1991). A coupled channel network growth and hillslope evolution model 1. Theory. *Water Resources. Research*, 27, 1671–1684.

Wilson, G., Rigby, J. R., Ursic, M. & Dabney, S. M. (2016). Soil pipe flow tracer experiments: 1. Connectivity and transport characteristics. *Hydrological Processes*, 30, 1265–1279.

Wolman, M. G. & Miller, J. P. (1960). Magnitude and frequency of forces in geomorphic processes. *Journal of Geology*, 68, 54–74.

Wolstenholme, J. M., Smith, M. W., Baird, A. J. & Sim, T. G. (2020). A new approach for measuring surface hydrological connectivity. *Hydrological Processes*, 34, 538–552.

Young, A. (1972). *Slopes*. Edinburgh: Oliver & Boyd.

Zimmermann, B. Zimmermann, A., Turner, B. L., Francke, T. & Elsenbeer, H. (2014). Connectivity of overland flow by drainage network expansion in a rain forest catchment. *Water Resources Research*, 50, 1457–1473.

Zuecco, G., Rinderer, M., Penna, D., Borga, M. & van Meerveld, H. J. (2019). Quantification of subsurface hydrologic connectivity in four headwater catchments using graph theory. *Science of the Total Environment*, 646, 1265–1280.

4

Fluvial Processes

JANET HOOKE

4.1 Introduction

Analysis of fluvial processes is fundamental to understanding fluvial systems, the spatial variation in their characteristics, how they change and evolve over various timescales, the effects of changes in controls and conditions, and the applications for management of catchments and river systems. Fluvial processes of water flow and sediment transmission have long been studied through direct measurement and indirect monitoring, through tracking and tracing of materials and fluxes, through statistical analysis of characteristics, and through modelling of process components. Overall approaches and conceptual frameworks have included flood routing, sediment budgeting and the recognition of sources and sinks and their identification by fluvial audits. It was long assumed that catchments and channels acted as systems with a continuity of flows of water and sediment (Toone et al., 2014) and that, in most systems, the major sources are in the upland or upper parts of catchments (Schumm, 1977). Processes of water flow involve understanding fundamental hydraulics; conveyance of sediment has been considered through formulation of sediment entrainment, transport and deposition mechanisms. Much research has focused on analysing statistical relations between fluxes and characteristics of catchments or areas to identify controls and influence of factors. The idea that sediment transport may not be spatially continuous was recognised in the concept of the Sediment Delivery Ratio (Roehl, 1962; Walling, 1983, Fryirs, 2013). Ferguson (1981) stated that river systems may work as jerky conveyor belts and comprise a series of sources, transfers and stores. Evidence of the key role and high magnitude of intervening storages was provided by Trimble's (1983) research on the Upper Mississippi system, within the framework of the sediment budgeting approach.

Connectivity as an explicit concept and framework for analysis of fluvial geomorphic systems developed in the late 1990s and early 2000s, partly from Brierley and Fryirs' research in eastern Australia (Brierley & Fryirs, 1999; Fryirs & Brierley, 1999, 2001) showing that the legacy of increased erosion and sediment loads due to European settlement had not been transmitted right through the system, which was seminal in exposing the fallacy of many previous assumptions and in documenting the patterns of connectivity and lack of it. Harvey (1996) and Poesen and Hooke (1997) had previously recognised connectivity as a key characteristic and Cammeraat (2002) examined the processes of connectivity at different scales in two contrasting environments of SE Spain and Luxembourg. Since then both the conceptual development and techniques of quantification, modelling and analysis have burgeoned, with accelerating adoption and application of the framework to understand the successive influence of one part on another in fluvial systems. This chapter discusses what is different about the connectivity approach from preceding research, what difference it makes to our understanding of fluvial processes, how it operates and its controls, and how our knowledge of mechanisms and dynamics of processes fits into this framework.

Very many definitions of connectivity and several reviews at successive stages of development have been produced, including more recently those of Najafi et al. (2021) and Singh et al. (2021) who tabulate selected origins and meanings of alternative definitions and associated concepts. Singh et al. (2021) also provide a useful summary of the components and relations of fluvial connectivity and the role of fluvial processes within that (Figure 4.1).

The focus here is on the longitudinal connectivity through river systems, mainly in large catchments and river channels. However, a high proportion of the published research on connectivity in fluvial systems is at the hillslope and small catchment scale (see Chapters 9 and 10). Much of the attention is on sediment connectivity, since the dynamics and processes of run-off generation and water flow are part of the much wider considerations of hydrological and flood analysis and modelling. The definition used here is that of Hooke (2003, p. 79), the first paper entitled sediment connectivity and one of the first on connectivity as a conceptual framework: 'the physical linkage of sediment through the channel system, which is the transfer of sediment from one zone or location to another and the potential for a specific particle to move through the system'. The distinctive component of the connectivity approach is that it examines the linkage along pathways between every part of the system. This approach presents a major challenge in that it requires documentation and analysis of the pathways and links throughout the whole network. Longitudinal connectivity down main river systems cannot be understood without also considering lateral connectivity since the connectivity with floodplains influences the longitudinal transmission of water and sediment.

The concept of connectivity implies that some parts are less connected and thus the concept of disconnectivity. Crucial questions therefore relate to what connects

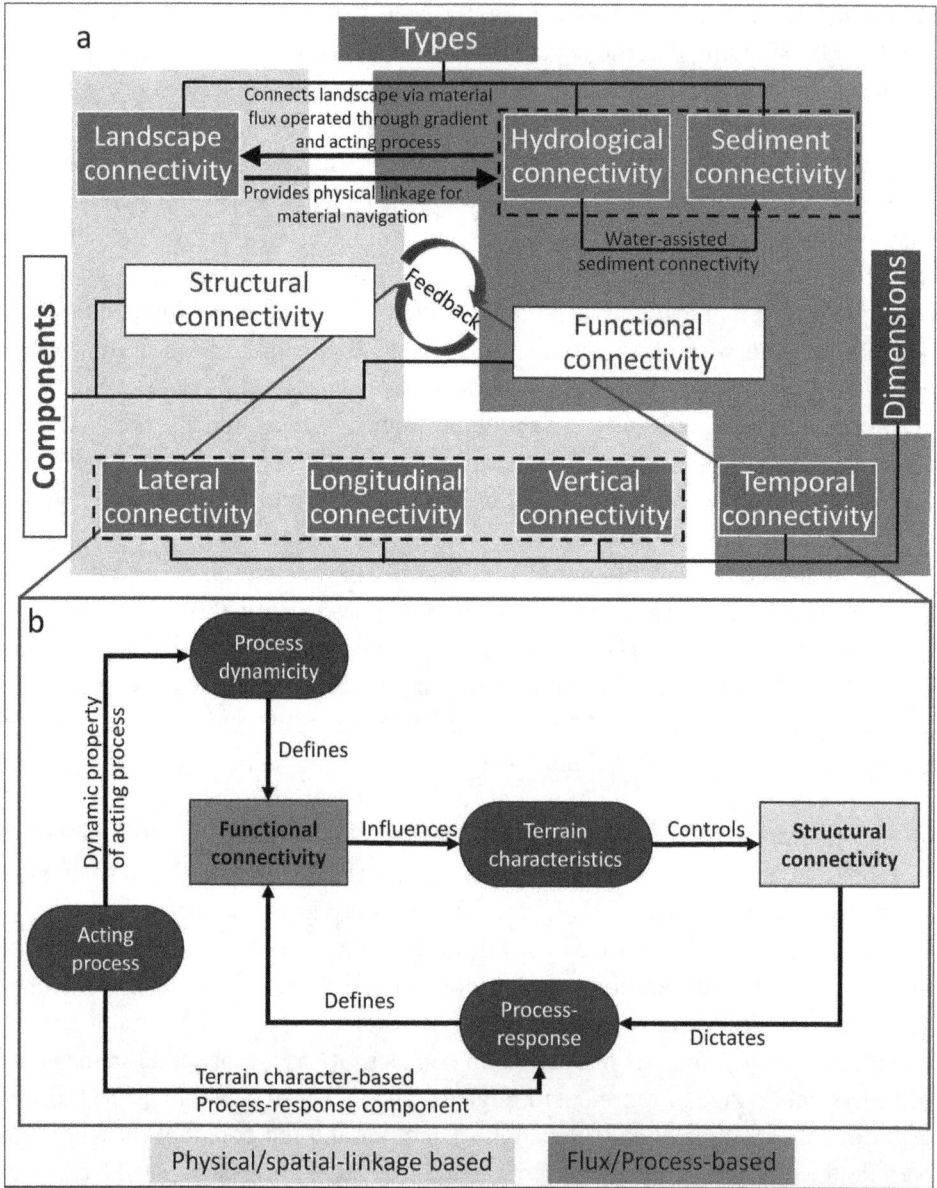

Figure 4.1   (a) The interdependency and interactions between different factors of geomorphic connectivity. (b) the feedback system among components of connectivity and their driving factors. A black and white version of this figure will appear in some formats. For the colour version, refer to the plate section.

or disconnects parts of fluvial systems and the identification and location of those disconnectors. Fryirs et al. (2007a) identified three types: buffers, barriers and blankets. However, disconnectors may not be simple on/off switches and various classifications and degrees of connectivity have been suggested (Hooke, 2003;

Jain & Tandon, 2010; Lewin & Ashworth, 2014). The initial stages in connectivity analysis involve identification of the state of links between each part and, preferably, their quantification (Hooke & Souza, 2021). Distinction is commonly made between structural and functional connectivity and most studies have been structural, though the goal for many purposes would be functional connectivity. Heckmann et al. (2018) considered that 'limited attention' had been paid to process-based, functional connectivity. This is rapidly changing.

4.2 Approaches and Associated Concepts

A key part of understanding the connectivity in a fluvial system is the identification of the degree and locations of connection between the hillslope part of the system and the main channel. This is the concept known as coupling, developed by Brunsden and Thornes (1979), Brunsden (1993) and Harvey (1997). The status of this link has a profound influence on the longitudinal transmission of fluxes and has mostly been approached by direct mapping of evidence of connections of water and/or sediment. These connections, of course, vary with events and can vary over time (Harvey, 2002). Differences in coupling were found to influence recovery from a major flood event (Harvey, 2007) and Al Farraj and Harvey (2010) show the differences in the morphological and sedimentological characteristics of two reaches with differing degrees of coupling and sediment sources.

A lack of connection or disconnectivity implies a lack of influence on downstream leading to the development of the concept and approach of effective catchment area (ECA) (Fryirs et al., 2007b; Fryirs, 2013). It means that if a complete disconnector between a tributary and a main channel is found and the focus is on sediment sources and flux to a downstream point, then the area upstream of the disconnector is not contributing. Fryirs et al. (2007b), under simulated conditions for differing sub-catchments in the upper Hunter, found the ECA, as the proportion of a catchment with potential to contribute sediment to the channel network, varies from 73% to just 3% of the total catchment area; in the Richmond catchment, the effect of buffers on ECA is illustrated (Figure 4.2) (Khan et al., 2021).

The key role of storage zones in influencing connectivity means that such zones need to be available in river systems and this is recognised in the concept of accommodation space (Brierley et al., 2006). Such occurrence has been found to be closely related to valley width and degree of confinement. However, apparent space at a height of present river flows implies a connectivity of that floodplain with the river channel and often this is constrained, mainly due to human activities such as construction of embankments and channelisation. Therefore, a major issue in relation to the degree and locations of longitudinal connectivity is the lateral connection to floodplains and wetlands. A major motivation to connectivity analysis is the aim

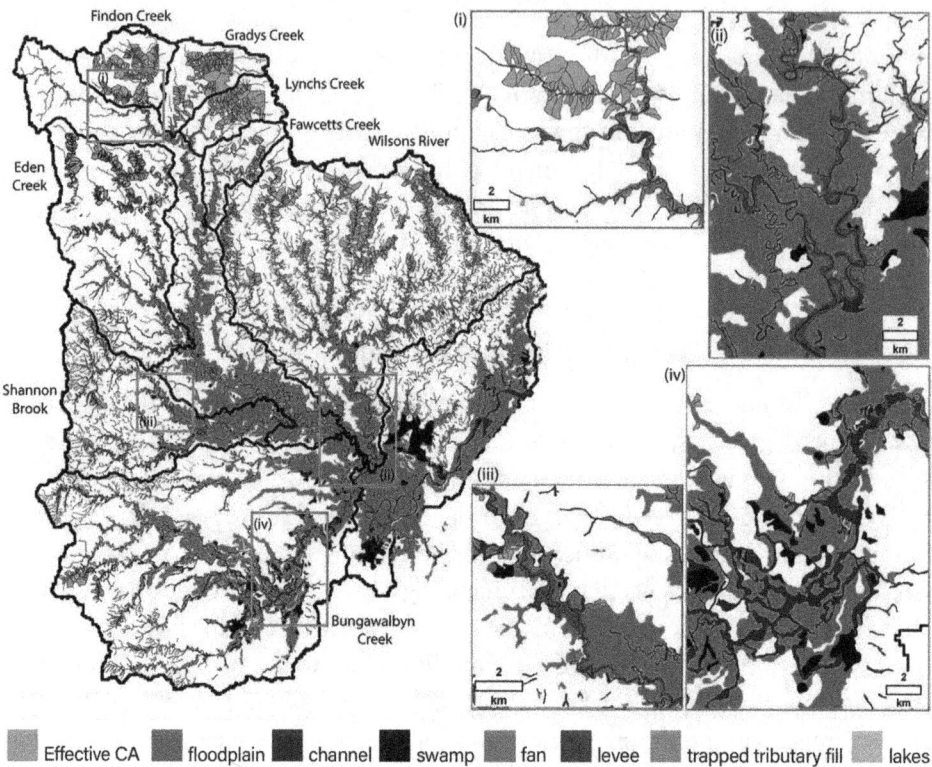

Effective CA ■ floodplain ■ channel ■ swamp ■ fan ■ levee ■ trapped tributary fill ■ lakes

Figure 4.2   Spatial distribution of the ECA and buffers in the Richmond catchment
(Khan et al., 2021). A black and white version of this figure will appear in some
formats. For the colour version, refer to the plate section.

of restoring lateral connectivity of floodplains and wetlands (Gumiero et al., 2013;
Hohensinner et al., 2014; Diaz-Redondo et al., 2018; Fuller & Death, 2018).

4.3 Methods and Techniques

Various approaches to documenting and quantifying connectivity have been developed (Heckmann et al., 2018; Hooke & Souza, 2021), which are also discussed in
Chapters 9 and 10. Najafi et al. (2021) consider that 'disagreement and confu-
sion seem to remain about conceptual and especially methodological approaches,
related indices, methods of quantifying sediment connectivity at a range of spatial and temporal scales'. Those applied to large systems and main river channels
are briefly identified and exemplified here. Connectivity is essentially a spatial
concept and involves analysis of the spatial relations of every part of the system
being studied. Many of the approaches entail division of the channel network into
reaches or links. This may be on the basis of homogenous characteristics or may
be between tributary contributions.

Identification of the connectivity in a system requires some kind of mapping of the status of each link. This identification may be done directly by field observation and measurement, by use of remote sensing sources, or by modelling. Hooke (2003, 2004) and Sandercock and Hooke (2006) developed a system for mapping the characteristics of reaches and surmising their connectivity status (Figure 4.3). This method is based on assumptions of the process balance and the features this will produce. A source zone will contain erosional features of bed and/or banks and much lesser signs of deposition. A transfer zone will tend to have little sign of sediment accumulation but must have supply in order for connectivity to occur (Sandercock & Hooke, 2006, 2011). If there is lack of supply, then the zone may have potential for transport from the stream power but will not connect unless there is a source. Net deposition of sink zones will have evidence of sediment accumulation, which may be within the channel or on the floodplain or other features such as a fan, delta or water body. Several authors, using other approaches, such as remote sensing or modelling, and then validating the results have concluded that field checking/mapping is essential (e.g., Lexartza-Artza & Wainwright, 2011; Messenzehl et al., 2014; Nicoll & Brierley, 2017; Mahoney et al., 2018; Fressard & Cossart, 2019; Hooke & Souza, 2021). Such approaches are very demanding of time and very challenging for large systems. They also require some expert interpretation. Hierarchical and nested approaches to mapping have been recommended to overcome problems of scale and identify emergent properties (e.g., Brierley et al., 2006; Sandercock & Hooke, 2006, 2011; Hooke & Sandercock, 2012, 2017; Sidle et al., 2017; Hooke & Souza, 2021). Ideally, the inputs, fluxes and stores should be quantified but such quantification would be needed for every reach and thus modelling or reconnaissance-level techniques have been developed. Remote sensing techniques and sources of evidence have helped to overcome the mapping challenges, facilitated by technological advances. For example, Cabre et al. (2020) developed a change detection technique using Copernicus Sentinel data that shows the relation between sediment sources and sink pathways in arid areas. Both field and remote sensing data are commonly fed into GIS for data manipulation and calculations of spatial relations (e.g., Ondrackova & Macka, 2019), including assessment of indices (e.g., Grauso et al., 2018; Nicoll & Brierley, 2017).

Major progress in advancing use of the connectivity framework was made with the development of calculation techniques that produce a connectivity index (IC) along the channel network. Details of the main indices are provided elsewhere (Chapter 10) but the major indices produced originally (Borselli et al., 2008; Cavalli et al., 2013; Crema & Cavalli, 2018) model the ability for flow and sediment to be transmitted based on the pathway characteristics, notably of topography (slope), and roughness (vegetation or surface roughness). In relation to application of indices to main river channel systems then the development of two different

a)

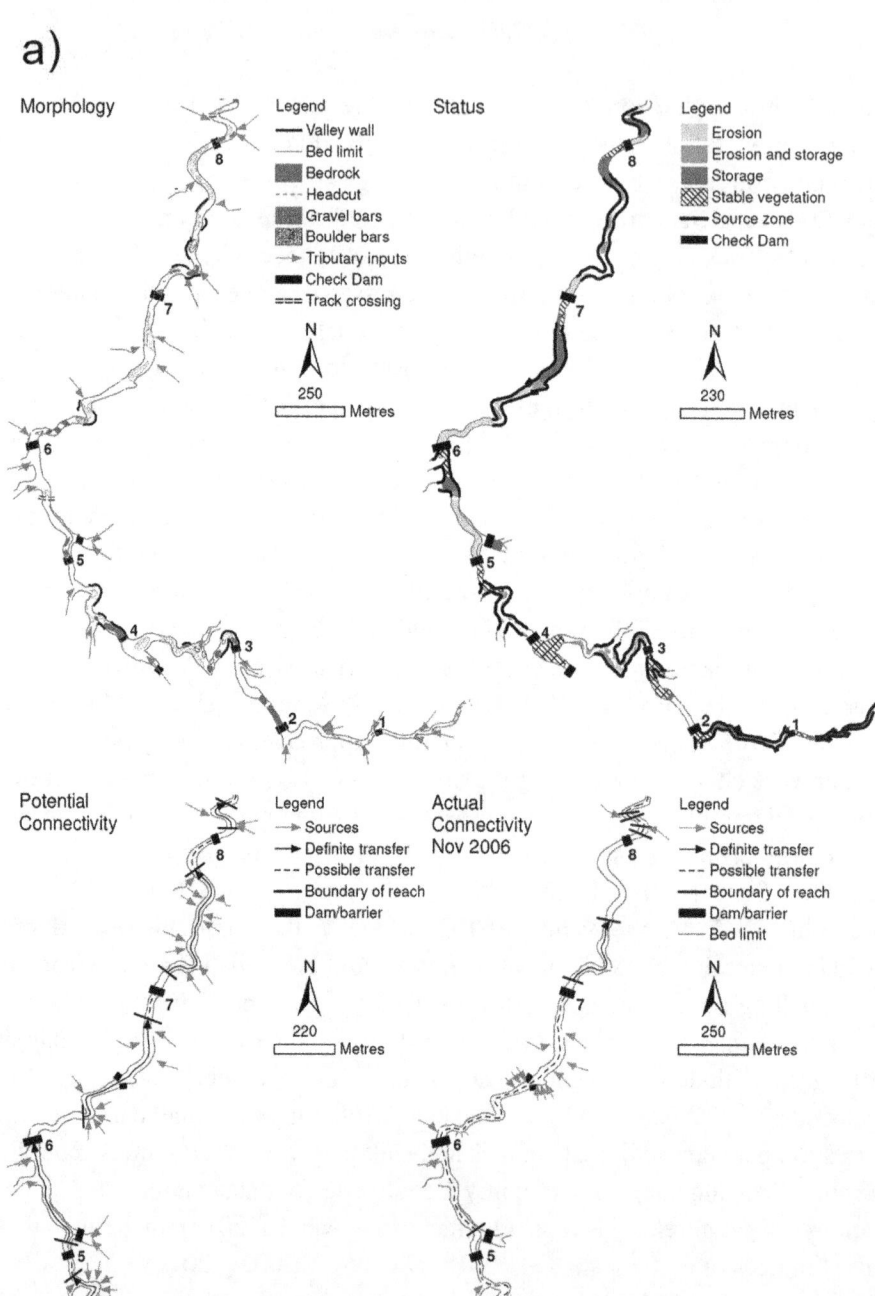

Figure 4.3   (a) Connectivity mapping of the Carcavo channel, SE Spain (after Sandercock & Hooke, 2006), and comparison of potential (structural) and actual (functional) connectivity in an event (after Sandercock & Hooke, 2011), (b) connectivity mapping of the River Dane channel, NW England (after Hooke, 2003). A black and white version of this figure will appear in some formats. For the colour version of panel 4.3 (b), refer to the plate section.

b)

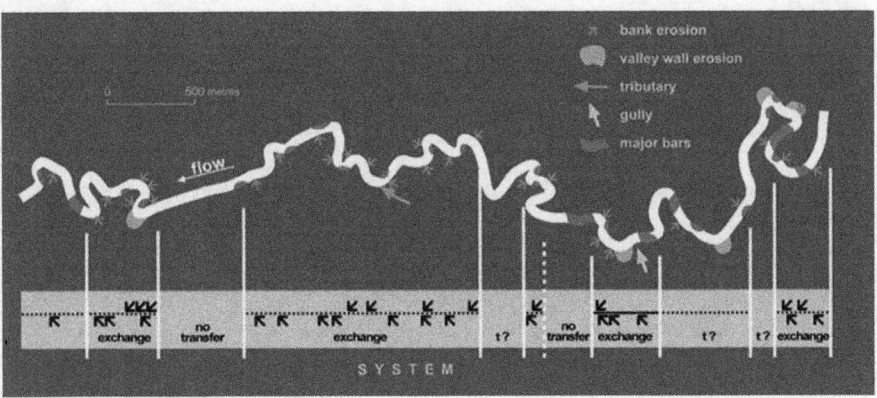

Figure 4.3   (cont.)

indices, IC_Channel and IC_Outlet, depending on the targeted outlet selected, has facilitated analysis of the distinction between hillslope connectivity and main channel connectivity (e.g., Mishra et al., 2019) (Figure 4.4). Many refinements and developments of more sophisticated indices are now taking place to incorporate other factors such as soil characteristics and to model functional connectivity by factors that vary in state over time (e.g., Chartin et al., 2017; Persichillo et al., 2018; Cislaghi & Bischetti, 2019; Kalantari et al., 2019; López-Vicente & Ben-Salem, 2019; Zingaro et al., 2019; Bombino et al., 2020).

An alternative approach to connectivity analysis is that based on networks and graph theory. Network-scale approaches and the importance of analysis at catchment or whole-network scale and of using a range of scales, especially to identify emergent properties, is highlighted in much research (e.g., Czuba and Foufoula-Georgiou, 2015; Covino, 2017; Wohl et al., 2019). Gootman et al. (2020) imply a connectivity approach by advocating a move away from simply studying process at multiple sites in a river system. Conventional techniques and measures of network analysis involving nodes and links or edges can be applied (e.g., Mayor et al., 2008; Puttock et al., 2013; Marchamalo et al., 2016). Development of graph theory to produce a IC was pioneered by Heckmann and Schwanghart (2013), and further developed and applied by Cossart and Fressard (2017) and Fressard and Cossart (2019) to estimate the sediments source contributions of different parts of a basin and to assess the pathway of sediment flux.

Explicit incorporation of processes in connectivity assessment and modelling in large river systems at network scale has come from work by Czuba and Foufoula-Georgiou (2015) and by Schmitt et al. (2016) and subsequent research. The former developed a dynamic connectivity model to calculate zones of imbalance using the

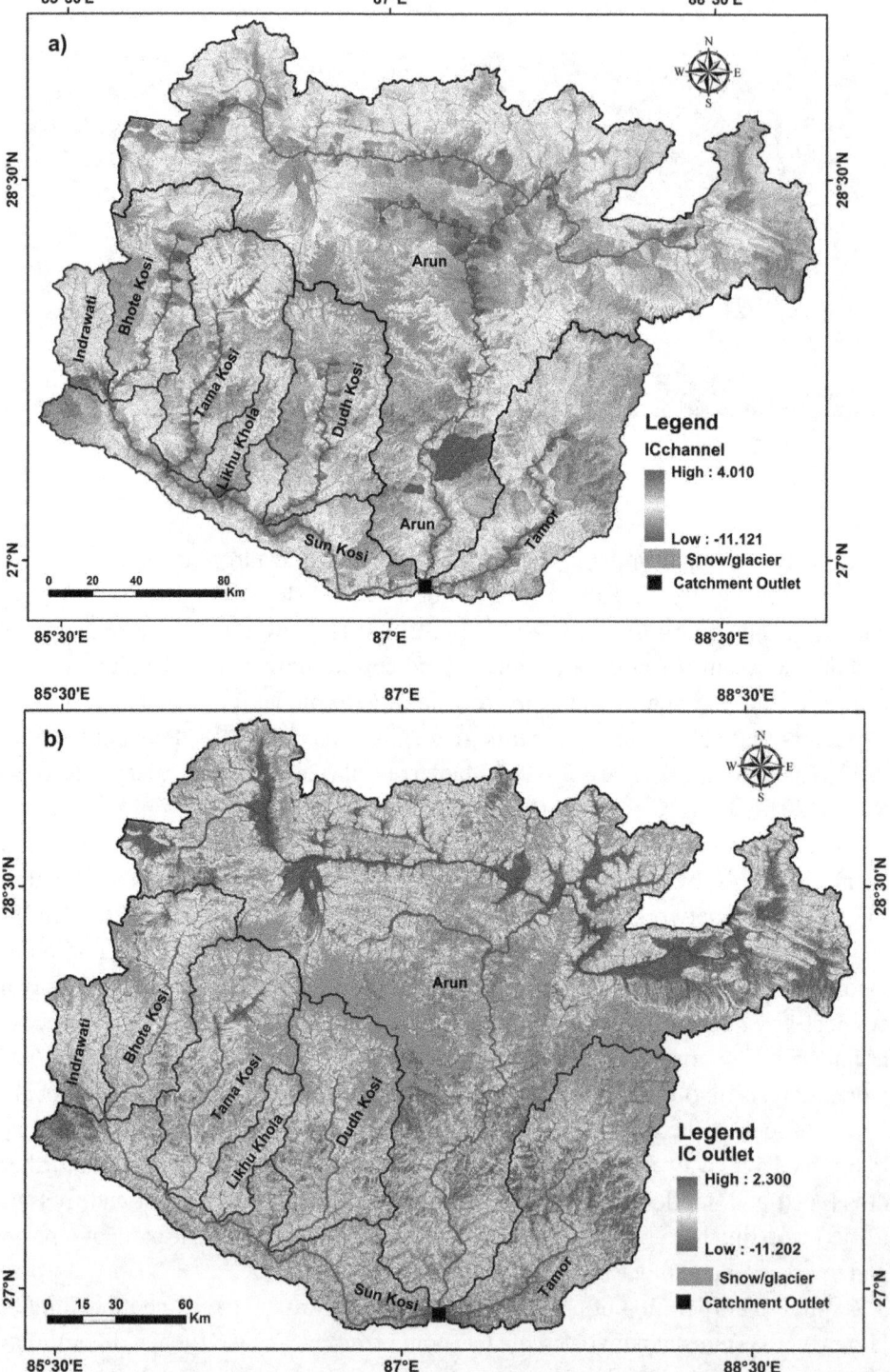

Figure 4.4   (a) IC_channel map of upper Kosi basin, (b) IC_outlet map of upper
Kosi basin (after Mishra et al., 2019). A black and white version of this figure will
appear in some formats. For the colour version, refer to the plate section.

network structure and a sediment transport model to track fluxes so that clusters of accumulation in sand transport emerge, and applied it to the Greater Blue River in Minnesota at various timescales. Likewise, Schmitt et al. (2016) model transfers between sources and sinks, treating each source as a suite of individual cascading processes in their CASCADE model, later developed as a toolbox, available as open-source software (Tangi et al., 2019). Calculations of sediment connectivity are based on empirical sediment transport formulae and disconnection can occur if local transport capacity becomes insufficient. The model has been applied to large river systems in SE Asia (Schmitt et al., 2016; Schmitt et al., 2018), revealing patterns of connectivity and disconnectivity. In other developments, Liu et al. (2020) also model fluxes of materials to assess functional longitudinal connectivity and Mahoney et al.'s (2020) model combines the probability of connectivity and sediment transport formulae. Modelling that addresses disconnectivity has recently been reviewed by Brierley et al. (2022). Some analyses and modelling of process-based connectivity have been complemented or validated by combination with techniques of measurements of sediment characteristics to assess provenance, for example, fingerprinting (D'Haen et al., 2013; Miller et al., 2013; Upadhayay et al., 2020); XRF and XRD (Fryirs & Gore, 2013); mineral magnetics (Rowntree & Foster, 2012); OSL (Oldknow & Hooke, 2017); radiocarbon dating (Chiverrell et al., 2009).

4.4 Patterns of Connectivity in River Systems

Connectivity analysis has been applied to large river systems, using direct evidence of fluxes and deposits, for example, in the Bega catchment, Eastern Australia (Fryirs & Brierley, 2001). Modelling studies at catchment and network scale are helping to indicate locations of hotspots of erosion, sedimentation and flooding in large river systems, for example, in Schmitt et al.'s (2016) Cascade modelling of the Mekong system. Liu et al. (2020) modelled the connectivity patterns in the river-lake-marsh system of the Baiyangdian Basin, China, comparing structural with functional patterns related to different events (Figure 4.5).

Analysis at large catchment scale demonstrates the influence of reaches of different morphological characteristics, represented for example, by types of zones in River Styles (Brierley & Fryirs, 2005). Sequences of connectivity emerge, relating to variation in valley width, with more efficient sediment conveyance in narrow valleys relative to wider valleys in which piedmonts, terraces, fans and extensive floodplains impede connectivity by greater lateral connectivity and space for stores and sinks (e.g., Fryirs et al., 2007b; Croke et al., 2013) (Figure 4.6a). Much research has shown the fundamental effects of coupled and non-coupled tributaries on sediment supply and onward connection and transmission downstream. Harvey (2012) has reviewed the key role of alluvial fans in tributary-main channel connection

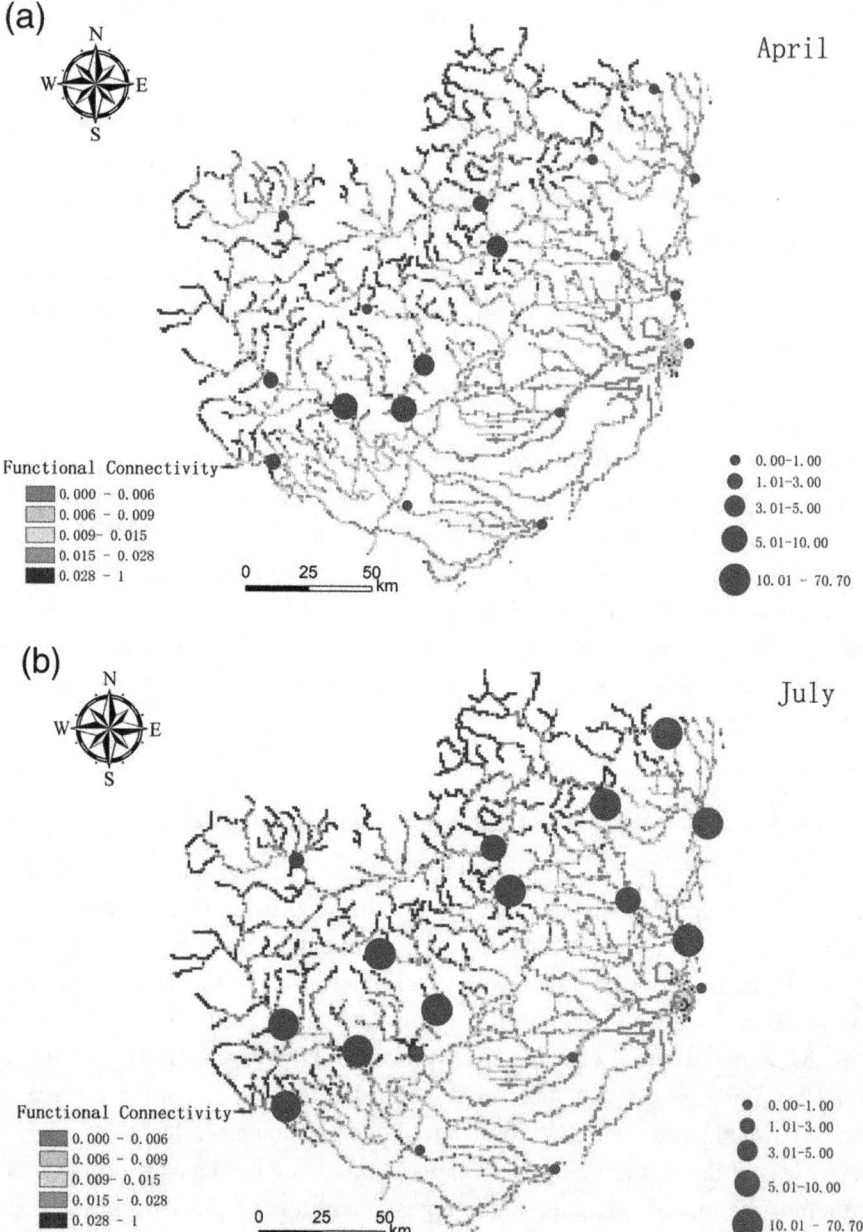

Figure 4.5   Spatial pattern of water and sediment functional connectivity and discharge measured at hydrological stations in the Baiyangdian Basin in (a) April, (b) July and (c) December 2016 (after Liu et al., 2020). A black and white version of this figure will appear in some formats. For the colour version, refer to the plate section.

and discusses the mechanisms of connectivity and their variation spatially and temporally. Khan et al. (2021) apply the CASCADE model to the Richmond River Catchment, New South Wales, Australia (Figure 4.2) and demonstrate the contrast

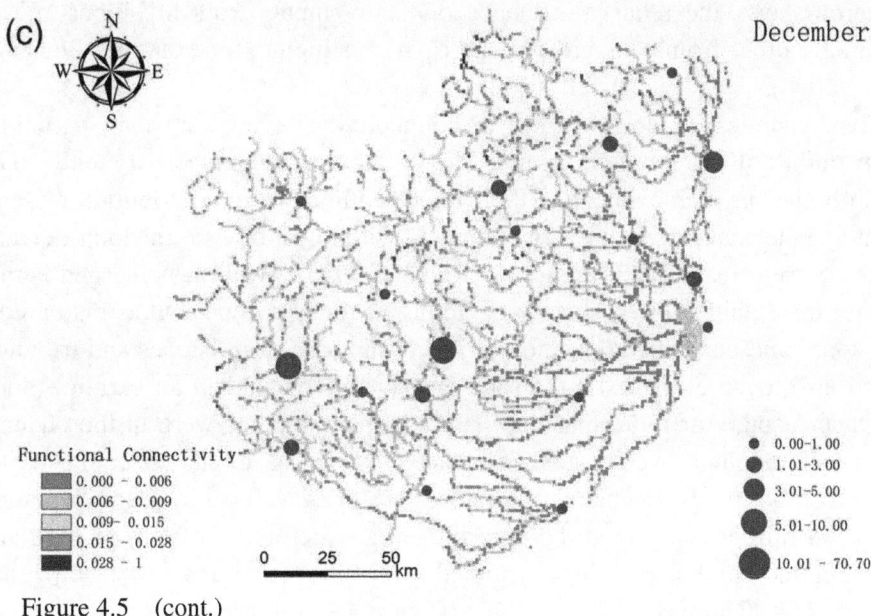

(c) N
W E
S

December

Functional Connectivity
◼ 0.000 ~ 0.006
◻ 0.006 ~ 0.009
◻ 0.009~ 0.015
◻ 0.015 ~ 0.028
◼ 0.028 ~ 1

0   25   50
km

• 0.00–1.00
• 1.01–3.00
● 3.01–5.00
● 5.01–10.00
● 10.01 ~ 70.70

Figure 4.5   (cont.)

in coupled and decoupled or disconnected sub-catchments, influenced by the presence of buffers, differentiating the effects on transport-limited and supply-limited sediment transport.

Disconnectors can be both natural and anthropogenic and profoundly influence the pattern, including in main channels and large catchments. Major rock steps and natural structures can create depocenters in the longer-term, which fill before connectivity of sediment is extended downstream (Chiverrell et al., 2010; Oldknow & Hooke, 2017). Alternating bedrock and alluvial reach rivers show particular patterns of disconnectivity (Toone et al., 2014). Miller et al. (2012) found that in central Nevada unincised valley segments disconnect large sections of the drainage basin from channelised flows and Marcal et al. (2017), in the Macae Basin, Brazil, document the increased connectivity in downstream reaches produced by incision and channelisation. Major human structures include dams and reservoirs, that reduce connectivity of both water and sediment, though only analysed more recently through a connectivity framework (e.g., Pöppl et al., 2017; Chen & Wang, 2019; Pöppl et al., 2019; Zhao et al., 2020).

Increasing emphasis on analysing the connectivity in systems has helped to demonstrate that main channel erosion sources can be much more important than tended to be assumed in earlier literature, especially for coarse sediment. Ondrackova and Macka (2019) indicate the importance of in-channel sources, both cut banks and incising reaches, and Gao and Zhang (2016) state that in large catchments channel process contributions should not be underestimated. Other research shows the relative influence of lateral inputs from hillslopes and of channel sources from bank erosion and riparian slipping, for example, Cienciala et al. (2020).

The dynamics of the processes of connectivity have emerged in particular from studies of extreme flood events, though obviously connectivity tends to be much higher in such events and their relative influence on distribution of sediment and the connectivity pattern compared with lower flows is the long-debated issue of event magnitude-frequency (Rhoads, 2020). What has emerged is that upland floodplain valleys may not be acting as transfer zones but as major storage zones and disconnectors, and not just at the base of hillslopes and tributary junctions (Joyce et al., 2018). Croke et al. (2013) found in an extreme flood in Queensland, Australia, that the main sediment sources were in the channel and the floodplains were sinks, influenced by degree of lateral connectivity. There can be trends downstream as stores are filled, related to accommodation space; in further analysis at the reach scale, an alternating pattern of high and low longitudinal connectivity associated with contraction and expansion zones was evident (Thompson et al., 2016) (Figure 4.6b). Overall, the efficiency of sediment transfer downstream decreased exponentially, with the strength of lateral connectivity increasing for each expansion reach. It has, however, also been found that previous, longer-term stores may be reworked and become a major source, particularly in extreme events. For example, Cabre et al. (2020) found that the devastating consequences of a major rainfall event in an arid area were due to valley-fill erosion in the trunk valleys by channel widening and incision, whereas the hillslopes remained intact. Postglacial sediment stores were important sources and highly connected in a flood in the Pyrenees (Blanpied et al., 2018).

Many studies seek to understand the effects of land-use changes on connectivity (Coulthard & Van De Wiel, 2017). For example, land use change was found to increase connectivity on the Mzimvubu River, South Africa, through channel incision reducing the channel-valley fill connectivity due to more efficient drainage arising from increased runoff (van der Waal & Rowntree, 2018). The timing and responses to changes in runoff and sediment load may also be modified by the connectivity patterns and states so that the relation between land-use change and sediment-flux response is lagged (Rowntree & Foster, 2012). Connectivity may change over time, and changes in inputs and fluxes may produce feedbacks on the connectivity through the buffers and stores (e.g., Lane et al., 2017). In a study extending over historical timescales, Godfrey et al. (2008) concluded that on the Fremont River, Utah, connection of sediment transport through the floodplain was low on a decadal time scale, but increases in the century and millennial time scales, dependent on supply from arroyo cutting and filling.

(a)

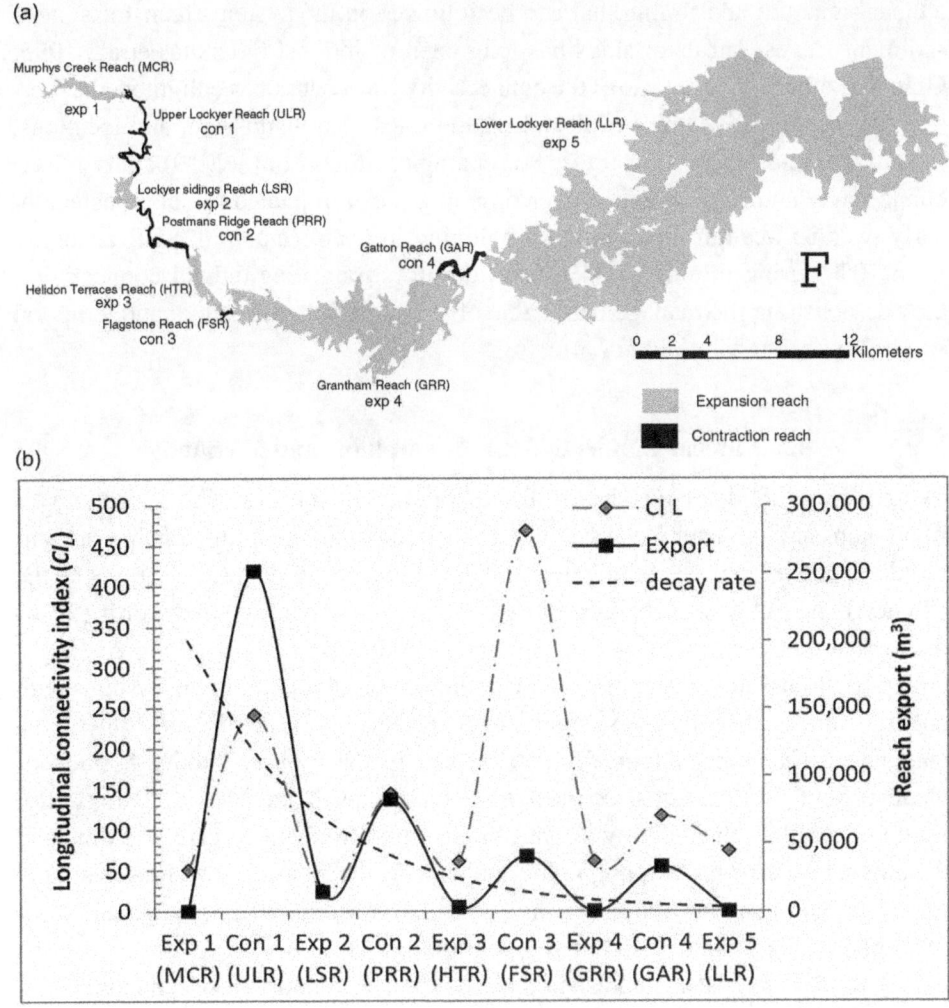

Figure 4.6 (a) Nine reaches in the Lockyer Valley, Queensland that display variations in channel–floodplain hydrological connectivity, with limited floodplain inundation in the Contraction (con) reaches and extensive channel–floodplain connectivity in the Expansion (exp) reaches; (b) Sediment connectivity between zones in Figure 4.6a represented by the longitudinal connectivity index, CIL and reach net sediment export. CIL values greater than 100 imply reach erosion exceeds reach deposition and vice versa for values less than 100. The dashed line is an exponential decay function fit to the reach transfer volume. Decay rate $(k) = 0.5$ (Thompson et al., 2016).

Legacy sediments have been the subject of several studies, partly because of the geomorphological interest but also the management need to understand what has happened to sediment from past major inputs, such as from European settlement or from mining or major land-use changes. These sediments are also useful in having a clear signature and timing that can be followed in the system. Transmission of sediment waves, pulses or slugs has long been of interest (Nicholas et al., 1995; Gran & Czuba, 2017) but use of the connectivity framework has enlightened understanding of these dynamics of distances and rates of transmission, and locations, magnitudes and timing of storage. For example, James et al. (2019) analyse how connectivity and SDR changed over time in a basin impacted by late nineteenth/early twentieth-century hydraulic gold mining in California, with SDR changing from 70% during mining to 84% now, due to strong longitudinal connectivity; they demonstrate the management value of such insights into spatial and temporal variability and transmission of effects.

4.5 Lateral Connectivity in Floodplains and Wetlands

The earlier examples have shown the importance of the extent and locations of floodplains and the lateral connectivity to the main channel in influencing the longitudinal connectivity of both water and sediment. More detailed analyses of the characteristics of the exchanges have been made. Lewin and Ashworth (2014) distinguish between hydrological and geomorphic connectivity in big river systems and discuss how their interactions influence biodiversity. In big systems with anabranching and braiding as well as former channels in floodplains, only some channels are geomorphologically active even in the highest floods. A study of channel-floodplain seasonal connectivity along the Amazon River (Park, 2020) found that within-channel flows control inundation frequency, residence time and development of positive topographic features in the floodplain, while overbank flows are the main contributors to the floodplain seasonal water storage and sediment budget, tending to smooth the topography.

At a more localised scale, research has examined the links between main channels and floodplain features and water bodies such as former channels (cut-offs), mainly because of their importance ecologically and the need for restoration or management (e.g., Citerrio & Piegay, 2009). For example, on the Drava River, Dezso et al. (2019) demonstrated how oxbow lakes in former channels had become disconnected and that groundwater was now the main connecting process. The processes and effects of lateral connectivity, or lack of it, on 'terrestrialisation' of floodplains from different kinds of modification were examined by Tena et al. (2020) on reaches of the Rhone river where channelisation and flow regulation had taken place. They found the channels had narrowed, number of flowing channels decreased, and pattern evolved from anabranching to single-thread. Sedimentation and frequency of overbank flooding declined but timing of these changes differed with their longitudinal position. Pöppl et al. (2012) show how riparian vegetation influences lateral sediment connectivity, and Thompson et al. (2016) found that the lateral connectivity in the extreme Queensland floods was modulated by the vegetation, with implications for its use in erosion and flood management. Embankment height relative to floodplain elevation influences the extent of hydrological connectivity in the lower Mississippi (Hudson et al., 2013), and on the Missouri River, USA, Jacobson et al. (2011) emphasise the need to analyse water-surface elevations and floodplain topography for restoration planning. Fritz et al. (2018), likewise, point out the complexity and variability of the influence of embankments and levels and channelisation on lateral connectivity.

Wetlands are of particular importance because of their ecological value and high biodiversity naturally. However, many have been degraded by the anthropogenic modifications of floodplains and of connectivity downstream to lakes and water-bodies so management of wetlands is of high concern. However, numbers of studies of wetland-scale hydro-geomorphic dynamics within an explicitly geomorphic connectivity framework are still somewhat limited (Singh et al., 2021). Singh et al. (2017) and Singh and Sinha (2019) analyse the impacts of changing land use on the connectivity structure of a wetland in Bihar, northern India (Figure 4.7), and Singh and Sinha (2020) distinguish the role of fluvial scouring and channel abandonment at the regional scale from that of cut and fill on the local scale in the evolution of wetlands on the Ganga floodplains. In central Nevada, persistence of wet meadows is related to low connectivity zones and unincised reaches of the main channel and, because the meadows provide recharge sites, keeping groundwater levels high is essential (Miller et al., 2012). Most restoration of wetlands aims to increase or restore hydrological connectivity but Singh and Sinha (2020) warn that a high sediment connectivity may be detrimental. Miller et al. (2013) showed how the construction of a drainage ditch through a previously unchanneled wetland altered the hydrologic connectivity of the catchment, allowing sediment to be transported from the headwaters to the lower basin where much of it was deposited within riparian wetlands. Restoration is also often constrained by limitations, particularly relating to flooding and navigation as Diaz-Redondo et al. (2018) recognised in modelling strategies of restoration of lateral connectivity of the Upper River Rhine, involving embankment removal, bank lowerings and side channel widenings. Biogeochemical and nutrient fluxes have been a major focus of study in wetlands and floodplain-river connectivity. Covino (2017) reviewed how hydrologic connectivity controls fluxes of organic material and nutrients. In a detailed analysis of biogeochemical fluxes in the Congo River, Borges et al. (2019) conclude it is unlikely that the CO_2 emissions from the river network were sustained by the hydrological carbon export from terrestrial soils but were sustained by the much higher hydrological connectivity from the wetlands to the rivers.

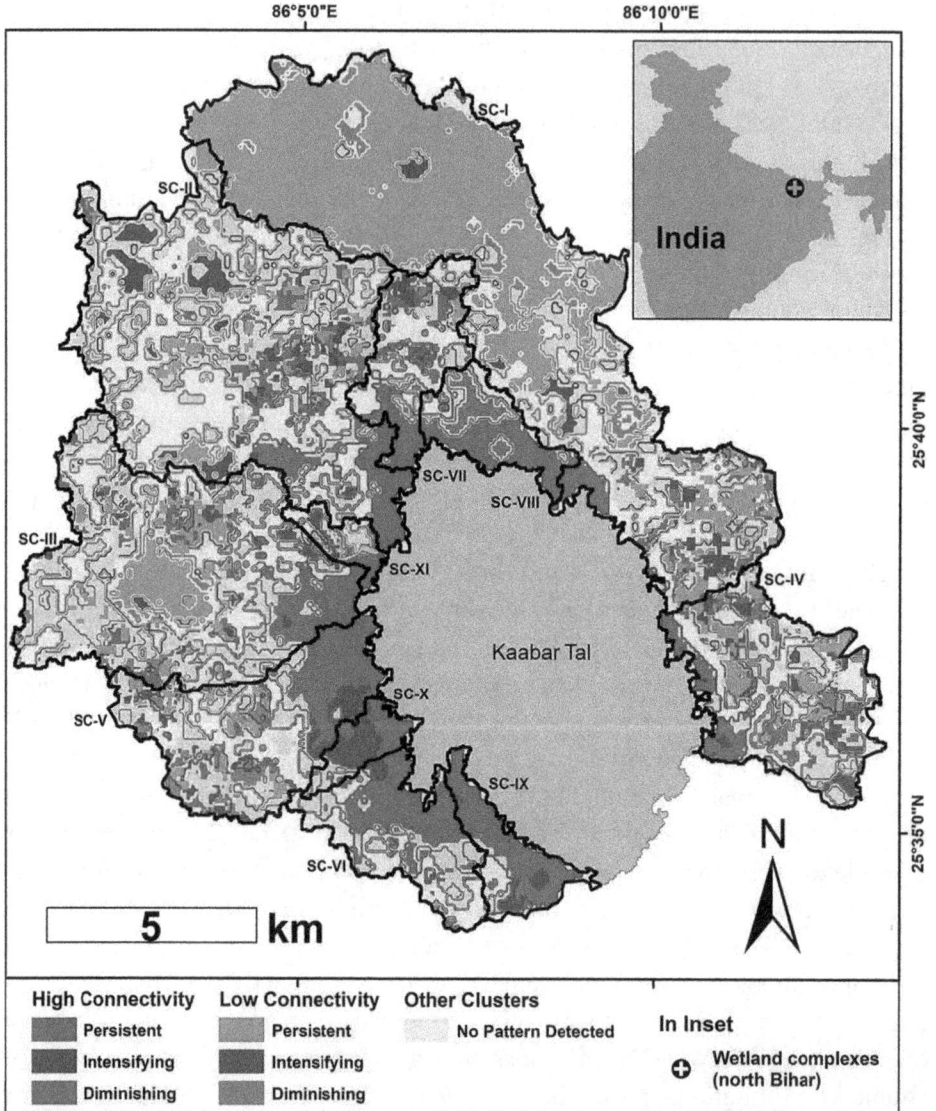

Figure 4.7   The dynamicity of the wetland-catchment connectivity as a function of land-use/land-cover (LULC) over approximately three decades for a wetland complex in north Bihar, Kaabar Tal, for the post-monsoon season (modified after Singh & Sinha, 2019) (Singh et al., 2021). A black and white version of this figure will appear in some formats. For the colour version, refer to the plate section.

4.6 Processes and Controls of Fluvial Connectivity

Some debate on the influence of structural connectivity on functional connectivity emerges from recent research. In examining a river-lake-marsh system, Liu et al. (2020) found that structural connectivity did not completely determine functional connectivity; poor functional connectivity could occur in systems with good structural connectivity and vice versa. Mahoney et al. (2021) couple a connectivity approach with RUSLE soil erosion modelling and demonstrate the importance of structural control, arguing this 'is contrary to recent suggestions that functional, dynamic processes control sediment connectivity in all landscapes'.

Various analyses have demonstrated that the spatial patterns and spatial arrangements of characteristics of terrain, such as relief, lithology, land use, channel morphology and the location of impediments influence the pattern and sequence of connectivity. For example, Vanacker et al. (2005) argue that responses are not necessarily commensurate with land-use change in a catchment due to the large effect of position of the land-use changes. Liu et al. (2020) consider that assessment of functional connectivity in a river-lake-marsh system should take into account landscape patterns and de Souza and Correa (2020), in a semi-arid watershed, showed that the position of the area in which land cover was modified has different effects on the connectivity, through incision or valley/channel filling.

Network pattern or configuration can influence connectivity and sediment delivery. For example, Walley et al. (2018) found a difference in longitudinal and lateral connectivity between a dendritic and elongate, herringbone structure in neighbouring catchments in New Zealand, with connectivity low and lateral tributary storage dominant in the former, whereas connectivity is high in the latter, with deposition in the mainstem channel. Network structure is also a crucial component of Gran and Czuba's (2017) simulation of sediment wave translation, but the effects vary with locations and magnitude of sediment supply.

Within main channels, a major control that has emerged is that of degree of valley confinement (Figure 4.6), which is largely influenced by long-term history and landscape evolution of the channel system, with conveyance of water and sediment high in confined reaches. In Taiwan, where valleys are mostly very steep and narrow this limits storage to lateral bars; the 18% of the network that is less confined stores 95% of the sediment (Kuo & Brierley, 2013). Major controls are the presence of expansion zones but, at a more local scale, vegetation can alter the lateral and longitudinal connectivity (Sandercock & Hooke, 2011; Thompson et al., 2016). Lithology and bedrock structure profoundly affect the valley evolution and morphology, of course, but can also have an influence directly on connectivity (Harries et al., 2021; Mather & Stokes, 2018). Process modelling has helped to identify locations of hydromorphological controls of accumulation zones, as shown in application of the Cascade model to the Mekong system (Schmitt et al., 2016) (Figure 4.8).

Disconnectors vary from natural to human constructions and alterations, operating at various temporal and spatial scales, and include valley dimensions, natural dams, such as rock steps, landslides, and lava flows, ecological factors such as

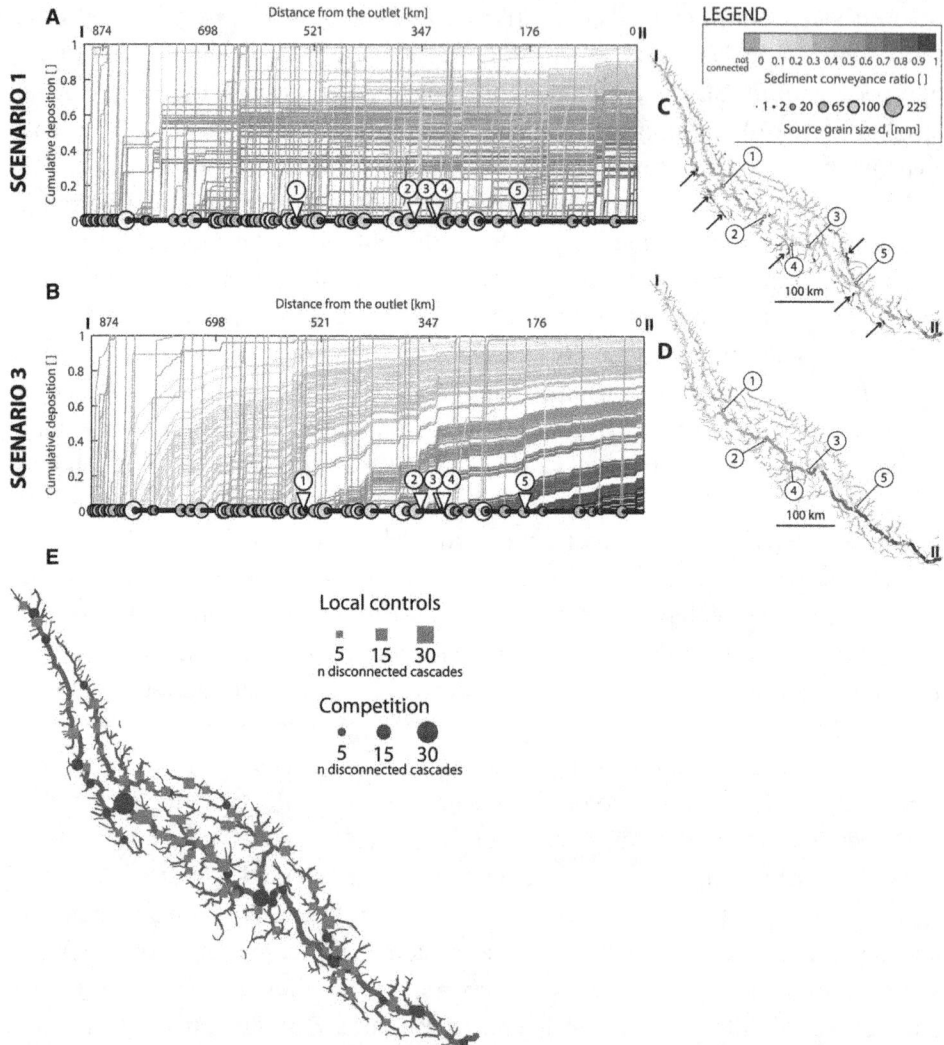

Figure 4.8   Network connectivity for two scenarios in the River Mekong system. (A) and (B) show deposition trajectories (y-axis), respectively of the sediment conveyance ratio along the main stem of the Da River. Dots indicate the source grain size d1 of each cascade (identical between scenarios). Numbers and triangles indicate the location of major tributaries. (C) and (D) are the sediment conveyance ratio mapped on the network throughout the river basin. Arrows in Figure 4.8C indicate some hotspots of sediment recruitment. (E) Hotspots of disconnectivity for scenario 3. Red squares and blue dots indicate edges where multiple cascades are interrupted either due to local hydromorphologic controls or competition. The marker size indicates the number of interrupted cascades. (Schmitt et al., 2016). A black and white version of this figure will appear in some formats. For the colour version, refer to the plate section.

wood and beavers, and artificial structures, notably embankments, weirs and dams. Much research has shown the fundamental effects of coupled and non-coupled tributaries on sediment supply, and major sediment buffering can influence fining of coarse sediment downstream (Lisenby & Fryirs, 2017). Tributary position and shape influence the extent of buffers and in turn the ECA and the sedimentological significance of each tributary. Marteau et al. (2020), conversely, examined the effects on sediment flux of the reconnection of a diverted tributary.

Studies of the translation and dynamics of the movement of sediment slugs/ waves/pulses have been reviewed by Gran and Czuba (2017) who point out that most have been at reach scale, though some studies of mining did examine major parts of systems (Knighton, 1998). Gran and Czuba (2017) in their simulation model applied to the Greater Blue Earth River, Minnesota, show that it is the spatial pattern of transport capacity and their relations between each reach which is a primary control. Analysis and validation of distinction between coarse and fine sediment is still limited but this style of modelling of sediment processes, as with the CASCADE model (Schmitt et al., 2016, 2018; Tangi et al., 2019) allows input of a range of sediment sizes, with connectivity controlled by the transport competence and capacity.

Sediment supply is somewhat neglected or proving challenging to track so far in connectivity analyses. Zanandrea et al. (2021) developed functional components of the IC by adding precipitation and surface run-off characteristics but found that it was incapable of representing the sediment exhaustion in a subtropical catchment in Brazil. Khan et al. (2021) apply graph theory and the Cascade model along with the ECA and buffer concepts to segregate supply-limited and transport-limited reaches, facilitating network-scale identification of potential hotspots of geomorphic adjustment. In the case of glacier recession due to climate change, Lane et al. (2017) have shown that the effects may be complex due to feedback effects between sediment supply and connectivity.

The mechanisms of connectivity have also been studied at reach scale in analyses of the patterns and dynamics of movement of sediment from bar to bar. Hooke (2003) took the presence of bars to be an indicator of sediment availability and transfer and questioned whether absence of bars indicated a lack of connectivity. Skarpich et al. (2019), and Lehotsky et al. (2018) both focus on bar-to-bar connectivity in channel systems and Parida et al. (2019) examined structural connectivity effects on grain size distributions in Himalayan valleys. Analysis of sediment transport distance may be fundamental to understanding sediment connectivity (Bracken et al., 2015).

Channelisation is a deliberate creation of high connectivity reaches (Heritage & Entwistle, 2020) and much activity now aims to reduce connectivity by reversing that and remeandering channels (Fuller & Death, 2018). Ditches and drainage structures may have an influence in lowland areas and floodplains, for example, on the Durance, southern France (Warner, 2006). Impacts of dams up and downstream have long been studied but are now situated more explicitly in connectivity frameworks. Magilligan et al. (2021) examined actual movement of particles in channels with run-of-river dams and find that this type of dam has little greater effect than no structure being present. However, Wang et al. (2021) indicate that even very small hydro-power dams can have an effect on connectivity. In analysing connectivity functioning under different rainfall events, Souza et al. (2016) found effects modulated by presence of dams. Other structural influences include that of roads and tracks, mostly studied at small catchment scale, but Pechenick et al. (2014) try to assess their overall downstream effects.

4.7 Application and Evaluation of Fluvial Connectivity Research

The need for management and restoration of river systems has been a primary motivation of much research and applied studies of connectivity (see Chapter 13). Li et al. (2021) found that ecosystem, climate change, and management feature highly as keywords in hydrological connectivity. Analysis of disconnectivity can provide a framework to forecast impact of environmental changes (Fryirs, 2013; Pöppl et al., 2020). Many researchers emphasise how a catchment scale approach to connectivity analysis is needed for management of sediment delivery in a system (e.g., Lane et al., 2008; Brierley & Fryirs, 2009). Connectivity approaches are being combined with the concept of river sensitivity to identify hotspots and key zones for targeting management or predicting future river scenarios (Lisenby et al., 2020). Zingaro et al. (2020) find that applying a sediment flow IC produced a good correspondence with flood occurrence to help identify hotspots. Gilbert and Wilcox (2020) proposed that modelling sediment balance and connectivity by a sediment floodplain exchange routing program could be used to track effects of catchment disturbances. In dryland systems Saco et al. (2020) use the connectivity framework at a range of spatial and temporal scales, emphasising the coevolution of system structures and function, and demonstrate how they may control threshold behaviour in both semi-arid rangelands and semi-arid wetlands, leading to desertification and degradation. Restoration strategies for removal of embankments and remeandering of channelised sections benefit from analyses of connectivity at a range of scales but increased connectivity can have detrimental effects (Fuller & Death, 2018). In relation to dams and weirs, the movement is now towards removal of these features in many places and that means prediction of the effects of increased connectivity downstream, though effects may be complex (Pöppl et al., 2019, 2020).

A large EU project, RECONDES, was specifically based on connectivity and designed to produce spatial strategies for reduction of longitudinal connectivity in areas prone to high soil erosion, using vegetation as a more sustainable method of management than engineering (Sandercock & Hooke, 2006; Hooke, 2006; Hooke & Sandercock, 2012, 2017) and was applied at a range of scales. Within that,

Borselli et al.'s (2008) research on connectivity modelling and indices was specifically designed to include the effects of vegetation cover. Pöppl et al. (2019) has tested various soil erosion models and connectivity approaches in quantifying lateral sediment input from agricultural areas and find they vary in their effectiveness in assessing aspects of the lateral fluxes.

Much of the aim of recent work has been to develop specific tools for management. Czuba et al.'s work (Czuba & Foufoula-Georgiou, 2015; Gran & Czuba, 2017) and research by Schmitt et al. (Schmitt et al., 2016; Schmitt et al., 2018; Tangi et al., 2019) that resulted in the CASCADE software was specifically designed to identify hotspots of erosion, sedimentation and channel change in large river systems and has been applied to the Mekong and other rivers of SE Asia. Maxwell et al. (2021) developed a decision-support tool, that quantitatively provides guidance to increase buffering functions of increasing infiltration and reducing flow energy and thus flood flow connectivity. Tan et al. (2021) also developed a tool that quantifies hydrological connectivity for ecological purposes. The U.S. Department of Agriculture and the U.S. Environmental Protection Agency (Arnold et al., 2021) have proposed a new connectivity framework that 'has the potential to dramatically improve national conservation and environmental assessments'. It hydrologically connects across a hierarchy of subdivisions from fields though to main channels.

Many of the issues of fluvial processes are of longstanding interest in fluvial geomorphology but reanalysis or adoption of a connectivity framework has provided new insights and different interpretations. For example, Upadhayay et al. (2020) found that incorporation of a sediment IC in a catchment assessment in the Himalayas altered apportionment of sediment sources from 66% from forest soils, 19% lowland and 15% upland terraces, to 90%, 9% and 1% respectively, albeit in a small catchment. Gaps in research still remain and these include little research on mechanisms of filling and effects of barriers in the longer-term, differentiation of fine and coarse sediment connectivity, and the effects of sediment supply. Much more needs to be done on modelling dynamics and functional connectivity, especially incorporating feedbacks, and in providing the validation.

4.8 Conclusions

Analysis of the connectivity patterns and sequences in river systems is now complementing longstanding research on fluvial processes and dynamics and new insights are emerging that are replacing earlier assumptions of continuity within fluvial systems. The connectivity concept and approach focus attention on the pathways and emphasise the spatial patterns and actual links between every part of the network in a catchment. Types and locations of disconnectors have been identified in various environments and are spatially varied in occurrence. Major controls on large catchment scale connectivity include valley confinement and degree of lateral connection with floodplains that provide space for storage. Connectivity analysis has provided insights into the effects and transmission of changes in inputs, particularly due to land use change and other legacy sediments, but it is shown that feedback can alter the connectivity and responses may be complex and that needs more research. Much research has been motivated by management and conservation needs and studies are increasingly demonstrating how the connectivity framework can provide an effective basis for restoration. It is essential for understanding the transmission of effects and future responses and for prediction of likely future impacts of climate change, land use alterations and human actions. Challenges remain in validating modelling outcomes and assessing connectivity in specific, especially large, systems. Future research needs to link our understanding of process mechanisms, derived over many decades, to the functioning within a connectivity framework.

References

Al Farraj, A., & Harvey, A. (2010). Influence of hillslope-to-channel and tributary-junction coupling on channel morphology and sediments: Bowderdale Beck, Howgill Fells, NW England. *Zeitschrift Fur Geomorphologie*, 54(2), 203–224. doi:10.1127/0372-8854/2010/0054-0018

Arnold, J. G., White, M. J., Allen, P. M., Gassman, P. W., & Bieger, K. (2021). Conceptual framework of connectivity for a national agroecosystem model based on transport processes and management practices. *Journal of the American Water Resources Association*, 57(1), 154–169.

Blanpied, J., Carozza, J. M., & Antoine, J. M. (2018). Sediment connectivity in the high Pyrenees mountain range by 2013 flood analysis: role of surficial sediment storages. *Geomorphologie-Relief Processus Environnement*, 24(4), 389–402. doi:10.4000/geomorphologie.12718

Bombino, G., Boix-Fayos, C., Cataldo, M. F., D'Agostino, D., Denisi, P., de Vente, J., Labate, A., & Zema, D. A. (2020). A modified catchment connectivity index for applications in semi-arid torrents of the Mediterranean environment. *River Research and Applications*, 36 (5). https://doi.org/10.1002/rra.3606

Borges, A. V., Darchambeau, F., Lambert, T., Morana, C., Allen, G. H., Tambwe, E., ... Bouillon, S. (2019). Variations in dissolved greenhouse gases (CO_2, CH_4, N_2O) in the Congo River network overwhelmingly driven by fluvial-wetland connectivity. *Biogeosciences*, 16(19), 3801–3834.

Borselli, L., Cassi, P., & Torri, D. (2008). Prolegomena to sediment and flow connectivity in the landscape: a GIS and field numerical assessment. *Catena*, 75(3), 268–277.

Bracken, L. J., Turnbull, L., Wainwright, J., & Bogaart, P. (2015). Sediment connectivity: a framework for understanding sediment transfer at multiple scales. *Earth Surface Processes and Landforms*, 40(2), 177–188. doi:10.1002/esp.3635

Brierley, G. J., & Fryirs, K. A. (2005). *Geomorphology and River Management*. Blackwell.

Brierley, G., & Fryirs, K. (2009). Don't fight the site: three geomorphic considerations in catchment-scale river rehabilitation planning. *Environmental Management*, 43(6), 1201–1218. doi:10.1007/s00267-008-9266-4

Brierley, G., Fryirs, K., & Jain, V. (2006). Landscape connectivity: the geographic basis of geomorphic applications. *Area*, 38(2), 165–174. doi:10.1111/j.1475-4762.2006 .00671.x

Brierley, G. J., & Fryirs, K. (1999). Tributary-trunk stream relations in a cut-and-fill landscape: a case study from Wolumla catchment, New South Wales, Australia. *Geomorphology*, 28(1–2), 61–73. doi:10.1016/s0169-555x(98)00103-2

Brierley, G., Tunnicliffe, J., Bizzi, S., Lee, F. Perry, G., Pöppl, R., & Fryirs, K. (2022). Quantifying sediment (Dis)connectivity in the modeling of river systems. *Treatise on Geomorphology* (2nd ed.) 10, 206–224.

Brunsden D, & Thornes J. B. (1979). Landscape sensitivity and change. *Transactions of the Institute of British Geographers*, NS4, 463–484.

Brunsden D. (1993). Barriers to geomorphological change. In Thomas D. S. G, & Allison R. J. (eds), *Landscape Sensitivity*. Chichester: John Wiley & Sons, pp. 7–12.

Cabre, A., Remy, D., Aguilar, G., Carretier, S., & Riquelme, R. (2020). Mapping rainstorm erosion associated with an individual storm from InSAR coherence loss validated by field evidence for the Atacama Desert. *Earth Surface Processes and Landforms*, 45(9), 2091–2106. doi:10.1002/esp.4868

Cammeraat, L. H. (2002). A review of two strongly contrasting geomorphological systems within the context of scale. *Earth Surface Processes and Landforms*, 27, 1201–1222.

Cavalli, M., Trevisani, S., Comiti, F., & Marchi, L. (2013). Geomorphometric assessment of spatial sediment connectivity in small Alpine catchments. *Geomorphology*, 188, 31–41.

Chartin, C., Evrard, O., Laceby, J. P., Onda, Y., Ottl'e, C., Lef'evre, I., & Cerdan, O. (2017). The impact of typhoons on sediment connectivity: lessons learnt from contaminated coastal catchments of the Fukushima Prefecture (Japan). *Earth Surface Processes and Landforms*, 42 (2). https://doi.org/10.1002/esp.4056

Chen, Y., & Wang, Y. G. (2019). Changes in river connectivity indexes in the lower Yellow River between 1960 and 2015. *River Research and Applications*, 35(9), 1377–1386.

Chiverrell, R. C., Foster, G. C., Marshall, P., Harvey, A. M., & Thomas, G. S. P. (2009). Coupling relationships: Hillslope-fluvial linkages in the Hodder catchment, NW England. *Geomorphology*, 109 (3–4), 222–235. doi:10.1016/j.geomorph.2009.03.004

Chiverrell, R. C., Foster, G. C., Thomas, G. S. P., & Marshall, P. (2010). Sediment transmission and storage: the implications for reconstructing landform development. *Earth Surface Processes and Landforms*, 35(1), 4–15. doi:10.1002/esp.1806

Cienciala, P., Nelson, A. D., Haas, A. D., & Xu, Z. W. (2020). Lateral geomorphic connectivity in a fluvial landscape system: unraveling the role of confinement, biogeomorphic interactions, and glacial legacies. *Geomorphology*, 354. doi:10.1016/j .geomorph.2020.107036

Cislaghi, A., Bischetti, G. B., 2019. Source areas, connectivity, and delivery rate of sediments in mountainous-forested hillslopes: a probabilistic approach. *Science of the Total Environment*, 652. https://doi.org/10.1016/j.scitotenv.2018.10.318

Citterio, A., & Piegay, H. (2009). Overbank sedimentation rates in former channel lakes: characterization and control factors. *Sedimentology*, 56(2): 461–482.

Cossart, É., & Fressard, M. (2017). Assessment of structural sediment connectivity within catchments: Insights from graph theory. *Earth Surface Dynamics*, 5(2). https://doi .org/10.5194/esurf-5-253-2017

Coulthard, T. J., & Van De Wiel, M. J. (2017). Modelling long term basin scale sediment connectivity, driven by spatial land use changes. *Geomorphology*, 277, 265–281. doi:10.1016/j.geomorph.2016.05.027

Covino, T. (2017). Hydrologic connectivity as a framework for understanding biogeochemical flux through watersheds and along fluvial networks. *Geomorphology*, 277, 133–144. doi:10.1016/j.geomorph.2016.09.030

Crema, S., & Cavalli, M. (2018). SedInConnect: a stand-alone, free and open source tool for the assessment of sediment connectivity. *Computers and Geosciences*, 111, 39–45.

Croke, J., Fryirs, K., & Thompson, C. (2013). Channel-floodplain connectivity during an extreme flood event: implications for sediment erosion, deposition, and delivery. *Earth Surface Processes and Landforms*, 38(12), 1444–1456. doi:10.1002/esp.3430

Czuba, J. A., & Foufoula-Georgiou, E. (2015). Dynamic connectivity in a fluvial network for identifying hotspots of geomorphic change. *Water Resources Research*, 51(3): 1401–1421.

de Souza, J. O. P., & Correa, A. C. D. (2020). Evolution scenarios of landscape connectivity in semiarid environment – Saco Creek watershed, Serra Talhada/pe – Brazil. *Revista Brasileira De Geomorfologia*, 21(1), 63–77. doi:10.20502/rbg.v21i1.1529

Dezso, J., Loczy, D., Salem, A. M., & Nagy, G. (2019). Floodplain connectivity. In Loczy, D. (ed.), *Drava River: Environmental Problems and Solutions*, pp. 215–230.

D'Haen, K., Dusar, B., Verstraeten, G., Degryse, P., & De Brue, H. (2013). A sediment fingerprinting approach to understand the geomorphic coupling in an eastern Mediterranean mountainous river catchment. *Geomorphology*, 197, 64–75.

Diaz-Redondo, M., Egger, G., Marchamalo, M., Damm, C., de Oliveira, R. P., & Schmitt, L. (2018). Targeting lateral connectivity and morphodynamics in a large river-floodplain system: the upper Rhine River. *River Research and Applications*, 34(7), 734–744. doi:10.1002/rra.3287

Ferguson R. I. (1981). Channel forms and channel changes. In Lewin, J. (ed.), *British Rivers*. London: Allen and Unwin, pp. 90–125.

Fressard, M., Cossart, E., 2019. A graph theory tool for assessing structural sediment connectivity: Development and application in the Mercurey vineyards (France). *The Science of the Total Environment*, 651, 2566–2584.

Fritz, K. M., Schofield, K. A., Alexander, L. C., McManus, M. G., Golden, H. E., Lane, C. R., ... Pollard, A. I. (2018). Physical and chemical connectivity of streams and riparian wetlands to downstream waters: a synthesis. *Journal of the American Water Resources Association*, 54(2), 323–345. doi:10.1111/1752-1688.12632

Fryirs, K. (2013). (Dis)Connectivity in catchment sediment cascades: a fresh look at the sediment delivery problem. *Earth Surface Processes and Landforms*, 38(1), 30–46. doi:10.1002/esp.3242

Fryirs, K., & Brierley, G. J. (1999). Slope-channel decoupling in Wolumla catchment, New South Wales, Australia: the changing nature of sediment sources following European settlement. *Catena*, 35(1), 41–63. doi:10.1016/s0341-8162(98)00119-2

Fryirs, K., & Brierley, G. J. (2001). Variability in sediment delivery and storage along river courses in Bega catchment, NSW, Australia: implications for geomorphic river recovery. *Geomorphology*, 38(3–4), 237–265.

Fryirs, K and Gore, D., 2013. Sediment tracing in the upper Hunter catchment using elemental and mineralogical compositions: implications for catchment-scale suspended sediment (dis)connectivity and management. *Geomorphology*, 193, 112–121.

Fryirs, K. A., Brierley, G. J., Preston, N. J., & Kasai, M. (2007a). Buffers, barriers and blankets: the (dis)connectivity of catchment-scale sediment cascades. *Catena*, 70(1), 49–67. doi:10.1016/j.catena.2006.07.007

Fryirs, K. A., Brierley, G. J., Preston, N. J., & Spencer, J. (2007b). Catchment-scale (dis) connectivity in sediment flux in the upper Hunter catchment, New South Wales, Australia. *Geomorphology*, 84(3–4), 297–316. doi:10.1016/j.geomorph.2006.01.044

Fuller, I. C., & Death, R. G. (2018). The science of connected ecosystems: What is the role of catchment-scale connectivity for healthy river ecology? *Land Degradation & Development*, 29(5), 1413–1426. doi:10.1002/ldr.2903

Gao, P., & Zhang, Z. R. (2016). Spatial patterns of sediment dynamics within a medium-sized watershed over an extreme storm event. *Geomorphology*, 267, 25–36. doi:10.1016/j.geomorph.2016.05.025

Gilbert, J. T., & Wilcox, A. C. (2020). Sediment routing and floodplain exchange (SeRFE): a spatially explicit model of sediment balance and connectivity through river networks. *Journal of Advances in Modeling Earth Systems*, 12(9). doi:10.1029/2020ms002048

Godfrey, A. E., Everitt, B. L., & Duque, J. F. M. (2008). Episodic sediment delivery and landscape connectivity in the Mancos Shale badlands and Fremont River system, Utah, USA. *Geomorphology*, 102(2), 242–251. doi:10.1016/j.geomorph.2008.05.002

Gootman, K. S., Gonzalez-Pinzon, R., Knapp, J. L. A., Garayburu-Caruso, V., & Cable, J. E. (2020). Spatiotemporal variability in transport and reactive processes across a first- to fifth-order fluvial network. *Water Resources Research*, 56(5). doi:10.1029/2019wr026303

Gran, K. B., & Czuba, J. A. (2017). Sediment pulse evolution and the role of network structure. *Geomorphology*, 277, 17–30. doi:10.1016/j.geomorph.2015.12.015

Grauso, S., Pasanisi, F., & Tebano, C. (2018). Assessment of a simplified connectivity index and specific sediment potential in river basins by means of geomorphometric tools. *Geosciences*, 8(2). doi:10.3390/geosciences8020048

Gumiero, B., Mant, J., Hein, T., Elso, J., & Boz, B. (2013). Linking the restoration of rivers and riparian zones/wetlands in Europe: sharing knowledge through case studies. *Ecological Engineering*, 56, 36–50. doi:10.1016/j.ecoleng.2012.12.103

Harries, R. M., Gailleton, B., Kirstein, L. A., Attal, M., Whittaker, A. C., & Mudd, S. M. (2021). Impact of climate on landscape form, sediment transfer and the sedimentary record. *Earth Surface Processes and Landforms*, 46(5), 990–1006. doi:10.1002/esp.5075

Harvey, A. M. (1996). The role of alluvial fans in the mountain fluvial systems of southeast Spain: implications of climatic change. *Earth Surface Processes and Landforms*, 21(6), 543–553.

Harvey, A. M. (1997). Coupling between hillslope gully systems and stream channels in the Howgill Fells, northwest England: temporal implications. *Geomorphologie: Relief, Processus, Environnement*, 1, 3–20.

Harvey, A. M. (2002). Effective timescales of coupling within fluvial systems. *Geomorphology*, 44(3–4), 175–201. doi:10.1016/s0169-555x(01)00174-x

Harvey, A. M. (2007). Differential recovery from the effects of a 100-year storm: Significance of long-term hillslope-channel coupling; Howgill Fells, northwest England. *Geomorphology*, 84(3–4), 192–208. doi:10.1016/j.geomorph.2006.03.009

Harvey AM. (2012). The coupling status of alluvial fans and debris cones: a review and synthesis. *Earth Surface Processes and Landforms*, 37, 64–76.

Heckmann, T., Cavalli, M., Cerdan, O., Foerster, S., Javaux, M., Lode, E., … Brardinoni, F. (2018). Indices of sediment connectivity: opportunities, challenges and limitations. *Earth-Science Reviews*, 187, 77–108. doi:10.1016/j.earscirev.2018.08.004

Heckmann, T., & Schwanghart, W. (2013). Geomorphic coupling and sediment connectivity in an alpine catchment – exploring sediment cascades using graph theory. *Geomorphology*, 182, 89–103. doi:10.1016/j.geomorph.2012.10.033

Heritage, G., & Entwistle, N. (2020). Impacts of river engineering on river channel behaviour: implications for managing downstream flood risk. *Water*, 12(5). doi:10.3390/w12051355

Hohensinner, S., Jungwirth, M., Muhar, S., & Schmutz, S. (2014). Importance of multi-dimensional morphodynamics for habitat evolution: Danube River 1715–2006. *Geomorphology*, 215, 3–19. doi:10.1016/j.geomorph.2013.08.001

Hooke, J. (2003). Coarse sediment connectivity in river channel systems: a conceptual framework and methodology. *Geomorphology*, 56(1–2), 79–94. doi:10.1016/s0169-555x(03)00047-3

Hooke, J. M., (2004). Analysis of coarse sediment connectivity in semi-arid river channels. In *IAHS Publn 288, Sediment Transfer through the Fluvial System*, Moscow Conference August 2004, pp. 269–275.

Hooke, J. M. (2006). Human impacts on fluvial systems in the Mediterranean region. *Geomorphology*, 79(3–4), 311–335. doi:10.1016/j.geomorph.2006.06.036

Hooke, J., & Sandercock, P. (2012). Use of vegetation to combat desertification and land degradation: recommendations and guidelines for spatial strategies in Mediterranean lands. *Landscape and Urban Planning*, 107, 389–400.

Hooke, J. M., & Sandercock, P. J. (eds.) (2017). *Combating Desertification and Land Degradation: Spatial Strategies Using Vegetation*. Springer Briefing, 110 pp. Cham: Springer.

Hooke, J., & Souza, J. (2021). Challenges of mapping, modelling and quantifying sediment connectivity. *Earth-Science Reviews*, 223. doi:10.1016/j.earscirev.2021.103847

Hudson, P. F., Sounny-Slittine, M. A., & LaFevor, M. (2013). A new longitudinal approach to assess hydrologic connectivity: Embanked floodplain inundation along the lower Mississippi River. *Hydrological Processes*, 27(15), 2187–2196. doi:10.1002/hyp.9838

Jacobson, R. B., Janke, T. P., & Skold, J. J. (2011). Hydrologic and geomorphic considerations in restoration of river-floodplain connectivity in a highly altered river system, Lower Missouri River, USA. *Wetlands Ecology and Management*, 19(4), 295–316.

Jain, V., & Tandon, S. K. (2010). Conceptual assessment of (dis)connectivity and its application to the Ganga River dispersal system. *Geomorphology*, 118(3–4), 349–358. doi:10.1016/j.geomorph.2010.02.002

James, L. A., Monohan, C., & Ertis, B. (2019). Long-term hydraulic mining sediment budgets: Connectivity as a management tool. *Science of the Total Environment*, 651, 2024–2035. doi:10.1016/j.scitotenv.2018.09.358

Joyce, H. M., Hardy, R. J., Warburton, J., & Large, A. R. G. (2018). Sediment continuity through the upland sediment cascade: geomorphic response of an upland river to an extreme flood event. *Geomorphology*, 317, 45–61. doi:10.1016/j.geomorph.2018.05.002

Kalantari, Z., Ferreira, C. S. S., Koutsouris, A. J., Ahmer, A. K., Cerd'a, A., Destouni, G. (2019). Assessing flood probability for transportation infrastructure based on catchment characteristics, sediment connectivity and remotely sensed soil moisture. *Science of the Total Environment*, 661. https://doi.org/10.1016/j.scitotenv.2019.01.009.

Khan, S., Fryirs, K., & Bizzi, S. (2021). Modelling sediment (dis)connectivity across a river network to understand locational-transmission-filter sensitivity for identifying hotspots of potential geomorphic adjustment. *Earth Surface Processes and Landforms*. doi:10.1002/esp.5213

Knighton, D. (1998). *Fluvial Forms and Processes: A New Perspective*. Arnold.

Kuo, C. W., & Brierley, G. J. (2013). The influence of landscape configuration upon patterns of sediment storage in a highly connected river system. *Geomorphology*, 180, 255–266. doi:10.1016/j.geomorph.2012.10.015

Lane, S. N., Bakker, M., Gabbud, C., Micheletti, N., & Saugy, J. N. (2017). Sediment export, transient landscape response and catchment-scale connectivity following rapid climate warming and Alpine glacier recession. *Geomorphology*, 277, 210–227. doi:10.1016/j.geomorph.2016.02.015

Lane, S. N., Reid, S. C., Tayefi, V., Yu, D., & Hardy, R. J. (2008). Reconceptualising coarse sediment delivery problems in rivers as catchment-scale and diffuse. *Geomorphology*, 98(3–4), 227–249. doi:10.1016/j.geomorph.2006.12.028

Lehotsky, M., Rusnak, M., Kidova, A., & Dudzak, J. (2018). Multitemporal assessment of coarse sediment connectivity along a braided-wandering river. *Land Degradation & Development*, 29(4), 1249–1261. doi:10.1002/ldr.2870

Lewin, J., & Ashworth, P. J. (2014). Defining large river channel patterns: Alluvial exchange and plurality. *Geomorphology*, 215, 83–98. doi:10.1016/j.geomorph.2013.02.024

Lexartza-Artza, I., & Wainwright, J. (2011). Making connections: changing sediment sources and sinks in an upland catchment. *Earth Surface Processes and Landforms*, 36(8). https://doi.org/10.1002/esp.2134.

Li, B. W., Yang, Z. F., Cai, Y. P., & Li, B. (2021). The frontier evolution and emerging trends of hydrological connectivity in river systems: a scientometric review. *Frontiers of Earth Science*, 15(1), 81–93. doi:10.1007/s11707-020-0852-y

Lisenby, P. E., & Fryirs, K. A. (2017). Sedimentologically significant tributaries: catchment-scale controls on sediment (dis) connectivity in the Lockyer Valley, SEQ, Australia. *Earth Surface Processes and Landforms*, 42(10), 1493–1504. doi:10.1002/esp.4130

Lisenby, P. E., Fryirs, K. A., & Thompson, C. J. (2020). River sensitivity and sediment connectivity as tools for assessing future geomorphic channel behavior. *International Journal of River Basin Management*, 18(3), 279–293. doi:10.1080/15715124.2019.1672705

Liu, Y. L., Cui, B. S., Du, J. Z., Wang, Q., Yu, S. L., & Yang, W. (2020). A method for evaluating the longitudinal functional connectivity of a river-lake-marsh system and its application in China. *Hydrological Processes*, 34(26), 5278–5297. doi:10.1002/hyp.13946

López-Vicente, M., Ben-Salem, N., 2019. Computing structural and functional flow and sediment connectivity with a new aggregated index: a case study in a large Mediterranean catchment. *Science of the Total Environment*, 651. https://doi.org/10.1016/j.scitotenv.2018.09.170

Magilligan, F. J., Roberts, M. O., Marti, M., & Renshaw, C. E. (2021). The impact of run-of-river dams on sediment longitudinal connectivity and downstream channel equilibrium. *Geomorphology*, 376. doi:10.1016/j.geomorph.2020.107568

Mahoney, D. T., Fox, J. F., al Aamery, N. (2018). Watershed erosion modeling using the probability of sediment connectivity in a gently rolling system. *Journal of Hydrology*, 561. https://doi.org/10.1016/j.jhydrol.2018.04.034

Mahoney, D., Blandford, B., & Fox, J. (2021). Coupling the probability of connectivity and RUSLE reveals pathways of sediment transport and soil loss rates for forest and reclaimed mine landscapes. *Journal of Hydrology*, 594. doi:10.1016/j.jhydrol.2021.125963

Mahoney, D. T., Fox, J., Al-Aamery, N., & Clare, E. (2020). Integrating connectivity theory within watershed modelling part I: Model formulation and investigating the timing of sediment connectivity. *Science of the Total Environment*, 740. doi:10.1016/j.scitotenv.2020.140385

Marcal, M., Brierley, G., & Lima, R. (2017). Using geomorphic understanding of catchment-scale process relationships to support the management of river futures: Macae Basin, Brazil. *Applied Geography*, 84, 23–41. doi:10.1016/j.apgeog.2017.04.008

Marchamalo, M., Hooke, J. M., & Sandercock, P. J. (2016). Flow and sediment connectivity in semi-arid landscapes in SE Spain: patterns and controls. *Land Degradation and Development*, 27(4), 1032–1044.

Marteau, B., Gibbins, C., Vericat, D., & Batalla, R. J. (2020). Geomorphological response to system-scale river rehabilitation I: sediment supply from a reconnected tributary. *River Research and Applications*, 36(8), 1488–1503. doi:10.1002/rra.3683

Mather, A. E., & Stokes, M. (2018). Bedrock structural control on catchment-scale connectivity and alluvial fan processes, High Atlas Mountains, Morocco. In D. Ventra, & L. E. Clarke (eds.), *Geology and Geomorphology of Alluvial and Fluvial Fans: Terrestrial and Planetary Perspectives*, vol. 440, pp. 103–128.

Maxwell, C. M., Fernald, A. G., Cadol, D., Faist, A. M., & King, J. P. (2021). Managing flood flow connectivity to landscapes to build buffering capacity to disturbances: An ecohydrologic modeling framework for drylands. *Journal of Environmental Management*, 278. doi:10.1016/j.jenvman.2020.111486

Mayor, Á. G., Bautista, S., Small, E. E., Dixon, M., & Bellot, J. (2008). Measurement of the connectivity of runoff source areas as determined by vegetation pattern and topography: a tool for assessing potential water and soil losses in drylands. *Water Resources Research*, 44, W10423.

Messenzehl, K., Hoffmann, T., & Dikau, R. (2014). Sediment connectivity in the high-alpine valley of Val Müschauns, Swiss National Park – linking geomorphic field mapping with geomorphometric modelling. *Geomorphology*, 221. https://doi.org/10.1016/j.geomorph.2014.05.033

Miller, J. R., Lord, M. L., Villarroel, L. F., Germanoski, D., & Chambers, J. C. (2012). Structural organization of process zones in upland watersheds of central Nevada and its influence on basin connectivity, dynamics, and wet meadow complexes. *Geomorphology*, 139, 384–402. doi:10.1016/j.geomorph.2011.11.004

Miller, J. R., Mackin, G., Lechler, P., Lord, M., & Lorentz, S. (2013). Influence of basin connectivity on sediment source, transport, and storage within the Mkabela Basin, South Africa. *Hydrology and Earth System Sciences*, 17(2), 761–781. doi:10.5194/hess-17-761-2013

Mishra, K., Sinha, R., Jain, V., Nepal, S., & Uddin, K. (2019). Towards the assessment of sediment connectivity in a large Himalayan river basin. *Science of the Total Environment*, 661, 251–265. doi:10.1016/j.scitotenv.2019.01.118

Najafi, S., Dragovich, D., Heckmann, T., & Sadeghi, S. H. (2021). Sediment connectivity concepts and approaches. *Catena*, 196. doi:10.1016/j.catena.2020.104880

Nicholas, A.P., Ashworth, P.G., Kirkby, M., Macklin, M.G. & Murray, T. (1995). Sediment slugs: large-scale fluctuations in fluvial sediment transport rates and storage volumes. *Progress in Physical Geography Earth and Environment*, 19, 500–519. doi:10.1177/030913339501900404

Nicoll, T., & Brierley, G. (2017). Within-catchment variability in landscape connectivity measures in the Garang catchment, upper Yellow River. *Geomorphology*, 277, 197–209. doi:10.1016/j.geomorph.2016.03.014

Oldknow, C. J., & Hooke, J. M. (2017). Alluvial terrace development and changing landscape connectivity in the Great Karoo, South Africa. Insights from the Wilgerbosch River catchment, Sneeuberg. *Geomorphology*, 288, 12–38. doi:10.1016/j.geomorph.2017.03.009

Ondrackova, L., & Macka, Z. (2019). Geomorphic (dis)connectivity in a middle-mountain context: human interventions in the landscape modify catchment-scale sediment cascades. *Area*, 51(1), 113–125. doi:10.1111/area.12424

Parida, S., Singh, V., & Tandon, S. K. (2019). Sediment connectivity and evolution of gravel size composition in Dehra Dun – an Intermontane Valley in the frontal zone of NW Himalaya. *Zeitschrift Fur Geomorphologie*, 62(2), 83–105. doi:10.1127/zfg/2019/0568

Park, E. (2020). Characterizing channel-floodplain connectivity using satellite altimetry: mechanism, hydrogeomorphic control, and sediment budget. *Remote Sensing of Environment*, 243. doi:10.1016/j.rse.2020.111783

Pechenick, A. M., Rizzo, D. M., Morrissey, L. A., Garvey, K. M., Underwood, K. L., & Wemple, B. C. (2014). A multi-scale statistical approach to assess the effects of connectivity of road and stream networks on geomorphic channel condition. *Earth Surface Processes and Landforms*, 39(11), 1538–1549. doi:10.1002/esp.3611

Persichillo, M. G., Bordoni, M., Cavalli, M., Crema, S., & Meisina, C. (2018). The role of human activities on sediment connectivity of shallow landslides. *Catena*, 160. https://doi.org/10.1016/j.catena.2017.09.025

Pöppl, R. E., Dilly, L. A., Haselberger, S., Renschler, C. S., & Baartman, J. E. M. (2019). Combining soil erosion modeling with connectivity analyses to assess lateral fine sediment input into agricultural streams. *Water*, 11(9). doi:10.3390/w11091793

Pöppl, R. E., Fryirs, K. A., Tunnicliffe, J., & Brierley, G. J. (2020). Managing sediment (dis) connectivity in fluvial systems. *Science of the Total Environment*, 736. doi:10.1016/j .scitotenv.2020.139627

Pöppl, R. E., Keiler, M., von Elverfeldt, K., Zweimueller, I., & Glade, T. (2012). The influence of riparian vegetation cover on diffuse lateral sediment connectivity and biogeomorphic processes in a medium-sized agricultural catchment, Austria. *Geografiska Annaler Series a-Physical Geography*, 94A(4), 511–529. doi:10.1111/j.1468-0459.2012.00476.x

Pöppl, R. E., Keesstra, S. D., & Maroulis, J. (2017). A conceptual connectivity framework for understanding geomorphic change in human-impacted fluvial systems. *Geomorphology*, 277, 237–250. doi:10.1016/j.geomorph.2016.07.033

Pöppl, R. E., Coulthard, T., Keesstra, S. D., & Keiler, M. (2019). Modeling the impact of dam removal on channel evolution and sediment delivery in a multiple dam setting. *International Journal of Sediment Research*, 34(6), 537–549. doi:10.1016/j .ijsrc.2019.06.001

Poesen, J. W. A., & Hooke, J. M. (1997). Erosion, flooding and channel management in Mediterranean environments of southern Europe. *Progress in Physical Geography-Earth and Environment*, 21(2), 157–199. doi:10.1177/030913339702100201

Puttock, A., Macleod, C. J. A., Bol, R., Sessford, P., Dungait, J., & Brazier, R. E. (2013). Changes in ecosystem structure, function and hydrological connectivity control water, soil and carbon losses in semi-arid grass to woody vegetation transitions. *Earth Surface Processes and Landforms*, 38(13). https://doi.org/10.1002/esp.3455

Rhoads, B. L. (2020). *River dynamics: Geomorphology to Support Management*. Cambridge: Cambridge University Press.

Roehl, J. W. (1962). Sediment source areas, delivery ratios and influencing morphological factors. *In Symposium of Bari*, 59 (59).

Rowntree, K., & Foster, I. (2012). A reconstruction of historical changes in sediment sources, sediment transfer and sediment yield in a small, semi-arid Karoo catchment, South Africa. *Zeitschrift Fur Geomorphologie*, 56, 87–100. doi:10.1127/0372-8854/2012/s-00074

Saco, P. M., Rodriguez, J. F., Moreno-de las Heras, M., Keesstra, S., Azadi, S., Sandi, S., … Rossi, M. J. (2020). Using hydrological connectivity to detect transitions and degradation thresholds: Applications to dryland systems. *Catena*, 186. doi:10.1016/j .catena.2019.104354

Sandercock, P., & Hooke, J. (2006). Strategies for reducing sediment connectivity and land degradation in desertified areas using vegetation: The RECONDES project. *IAHS-AISH Publication*, 306, 127–135.

Sandercock, P. J., & Hooke, J. M. (2011). Vegetation effects on sediment connectivity and processes in an ephemeral channel in SE Spain. *Journal of Arid Environments*, 75 (3), 239–254.

Schmitt, R. J. P., Bizzi, S., & Castelletti, A. (2016). Tracking multiple sediment cascades at the river network scale identifies controls and emerging patterns of sediment connectivity. *Water Resources Research*, 52(5), 3941–3965. doi:10.1002/2015wr018097

Schmitt, R. J. P., Bizzi, S., Castelletti, A. F., & Kondolf, G. M. (2018). Stochastic modeling of sediment connectivity for reconstructing sand fluxes and origins in the unmonitored Se Kong, Se San, and Sre Pok tributaries of the Mekong river. *Journal of Geophysical Research-Earth Surface*, 123(1), 2–25. doi:10.1002/2016jf004105

Schumm, S. A. (1977). *The Fluvial System*. New York: Wiley, 338 pp.

Sidle, R. C., Gomi, T., Usuga, J. C. L., & Jarihani, B. (2017). Hydrogeomorphic processes and scaling issues in the continuum from soil pedons to catchments. *Earth-Science Reviews*, 175, 75–96.

Singh M, & Sinha R. (2019). Evaluating dynamic hydrological connectivity of a floodplain wetland in North Bihar, India using geostatistical methods. *Science of the Total Environment*, 651, 2473–2488.

Singh, M., & Sinha, R. (2020). Distribution, diversity, and geomorphic evolution of floodplain wetlands and wetland complexes in the Ganga plains of north Bihar, India. *Geomorphology*, 351. doi:10.1016/j.geomorph.2019.106960

Singh, M., Sinha, R., & Tandon, S. K. (2021). Geomorphic connectivity and its application for understanding landscape complexities: a focus on the hydro-geomorphic systems of India. *Earth Surface Processes and Landforms*, 46(1), 110–130. doi:10.1002/esp.4945

Singh, M., Tandon, S. K., & Sinha, R. (2017). Assessment of connectivity in a water-stressed wetland (Kaabar Tal) of Kosi-Gandak interfan, north Bihar Plains, India. *Earth Surface Processes and Landforms*, 42(13), 1982–1996. doi:10.1002/esp.4156

Skarpich, V., Galia, T., Ruman, S., & Macka, Z. (2019). Variations in bar material grain-size and hydraulic conditions of managed and re-naturalized reaches of the gravel-bed Becva River (Czech Republic). *Science of the Total Environment*, 649, 672–685. doi:10.1016/j.scitotenv.2018.08.329

Souza, J. O. P., Correa, A. C. B., & Brierley, G. J. (2016). An approach to assess the impact of landscape connectivity and effective catchment area upon bedload sediment flux in Saco Creek Watershed, Semiarid Brazil. *Catena*, 138, 13–29. doi:10.1016/j.catena.2015.11.006

Tan, Z. Q., Li, Y. L., Zhang, Q., Liu, X. G., Song, Y. Y., Xue, C. Y., & Lu, J. Z. (2021). Assessing effective hydrological connectivity for floodplains with a framework integrating habitat suitability and sediment suspension behavior. *Water Research*, 201. doi:10.1016/j.watres.2021.117253

Tangi, M., Schmitt, R., Bizzi, S., & Castelletti, A. (2019). The CASCADE toolbox for analyzing river sediment connectivity and management. *Environmental Modelling & Software*, 119, 400–406. doi:10.1016/j.envsoft.2019.07.008

Tena, A., Piegay, H., Seignemartin, G., Barra, A., Berger, J. F., Mourier, B., & Winiarski, T. (2020). Cumulative effects of channel correction and regulation on floodplain terrestrialisation patterns and connectivity. *Geomorphology*, 354. doi:10.1016/j.geomorph.2020.107034

Thompson, C. J., Fryirs, K., & Croke, J. (2016). The disconnected sediment conveyor belt: patterns of longitudinal and lateral erosion and deposition during a catastrophic flood in the Lockyer valley, south east Queensland, Australia. *River Research and Applications*, 32(4), 540–551. doi:10.1002/rra.2897

Toone, J., Rice, S. P., & Piegay, H. (2014). Spatial discontinuity and temporal evolution of channel morphology along a mixed bedrock-alluvial river, upper Drome River, southeast France: Contingent responses to external and internal controls. *Geomorphology*, 205, 5–16. doi:10.1016/j.geomorph.2012.05.033

Trimble S. W. (1983). A sediment budget for Coon Creek Basin in the Driftless Area, Wisconsin, 1853–1977. *American Journal of Science*, 283, 454–474.

Upadhayay, H. R., Lamichhane, S., Bajracharya, R. M., Cornelis, W., Collins, A. L., & Boeckx, P. (2020). Sensitivity of source apportionment predicted by a Bayesian tracer mixing model to the inclusion of a sediment connectivity index as an informative prior: Illustration using the Kharka catchment (Nepal). *Science of the Total Environment*, 713. doi:10.1016/j.scitotenv.2020.136703

van der Waal, B., & Rowntree, K. (2018). Landscape connectivity in the upper Mzimvubu River catchment: an assessment of anthropogenic influences on sediment connectivity. *Land Degradation & Development*, 29(3), 713–723. doi:10.1002/ldr.2766

Vanacker, V., Molina, A., Govers, G., Poesen, J., Dercon, G., & Deckers, S. (2005). River channel response to short-term human-induced change in landscape connectivity in Andean ecosystems. *Geomorphology*, 72(1–4), 340–353. doi:10.1016/j.geomorph.2005.05.013

Wang, D. C., Wang, X., Huang, Y., Zhang, X., Zhang, W., Xin, Y., … Cao, Z. J. (2021). Impact analysis of small hydropower construction on river connectivity on the upper reaches of the great rivers in the Tibetan Plateau. *Global Ecology and Conservation*, 26. doi:10.1016/j.gecco.2021.e01496

Walley, Y., Tunnicliffe, J., & Brierley, G. (2018). The influence of network structure upon sediment routing in two disturbed catchments, East Cape, New Zealand. *Geomorphology*, 307, 38–49. doi:10.1016/j.geomorph.2017.10.029

Walling D. E. (1983). The sediment delivery problem. *Journal of Hydrology*, 65, 209–237.

Warner, R. F. (2006). Natural and artificial linkages and discontinuities in a Mediterranean landscape: Some case studies from the Durance Valley, France. *Catena*, 66(3), 236–250. doi:10.1016/j.catena.2006.02.004

Wohl, E., Brierley, G., Cadol, D., Coulthard, T. J., Covino, T., Fryirs, K. A., … Sklar, L. S. (2019). Connectivity as an emergent property of geomorphic systems. *Earth Surface Processes and Landforms*, 44(1), 4–26. doi:10.1002/esp.4434

Zanandrea, F., Michel, G. P., Kobiyama, M., Censi, G., & Abatti, B. H. (2021). Spatial-temporal assessment of water and sediment connectivity through a modified connectivity index in a subtropical mountainous catchment. *Catena*, 204. doi:10.1016/j.catena.2021.105380

Zhao, L., Liu, Y., & Luo, Y. (2020). Assessing hydrological connectivity mitigated by reservoirs, vegetation cover, and climate in Yan River Watershed on the Loess Plateau, China: the network approach. *Water*, 12(6). doi:10.3390/w12061742

Zingaro, M., Refice, A., Giachetta, E., D'Addabbo, A., Lovergine, F., de Pasquale, V., Pepe, G., Brandolini, P., Cevasco, A., & Capolongo, D. (2019). Sediment mobility and connectivity in a catchment: a new mapping approach. *Science of the Total Environment*, 672. doi:10.1016/j.scitotenv.2019.03.461

Zingaro, M., Refice, A., D'Addabbo, A., Hostache, R., Chini, M., & Capolongo, D. (2020). Experimental application of sediment flow connectivity index (SCI) in flood monitoring. *Water*, 12(7). doi:10.3390/w12071857

5

Aeolian Processes

GREGORY S. OKIN

5.1 Introduction

Aeolian processes refer to movement of air – wind – and the movement of material by wind – aeolian transport. This chapter will discuss the impact and feedbacks of connectivity on aeolian processes. Due to low and discontinuous cover, which allows the force of the wind to reach the soil surface, aeolian transport is especially prevalent in deserts. Although the transport of material by both wind and water occurs during specific events, rain events large enough to transport material can be less frequent than aeolian transport events. The wind blows every month of every year, whereas arid regions often go months or even years without precipitation events large enough to produce significant runoff or hillslope erosion (Okin et al., 2018).

Aeolian transport occurs when the erosivity of the wind exceeds the erodibility of the soil (Bagnold, 1941). There are three ways that particles move in response to wind. The term "creep" generally refers to the slow motion of grains that are too large to be lifted by the wind (Greeley & Iversen, 1985, p. 293). Reptation includes creep but may also include grains that are small enough to be lifted by the wind. Both creep and reptation occur as a result of impacts by airborne particles, which impart some of their forward kinetic energy to particles on the surface (Anderson, 1987). Saltation is the jumping, arcuate motion of particles small enough to be entrained directly by the wind but too large to stay airborne (Bagnold, 1941). Suspension is transport of fine particles, with settling velocities less than the vertical motion caused by turbulent eddies, that thus remain entrained in the air and can be transported, in some cases, thousands of kilometers (e.g., Prospero, 1999). Horizontal aeolian flux typically refers to the sum of these modes of transport in the direction of the wind. Vertical aeolian flux typically refers to the rate of emission of suspended particles.

In terms of mass, horizontal aeolian flux is dominated by saltation; although they are large, the rate of movement of reptating particles is small relative to the speed of saltating particles, which travel just slower than the wind. And, although suspended particles travel faster than saltating particles on average, they are small enough that they make a minimal contribution to total mass flux. On purely geomorphic terms, saltation is therefore generally responsible for aeolian landforms (landforms caused by horizontal aeolian transport), from mega-dunes to ripples. Both saltation and suspension are important in terms of ecological impacts of aeolian transport. Wind erosion and deposition of saltation-sized particles lead to changes in surface height which lead to exposure of roots and removal of soil crusts in the former case and burial of plants and soil crusts in the latter case (Okin et al., 2006). Saltating particles can also abrade plant tissue, impacting plant physiology (Armbrust & Retta, 2000). Suspension, in contrast, removes fines from the soil surface. These fines carry the bulk of soil nutrients and cation exchange capacity and also contribute to increasing soil water holding capacity in sandy or loamy soils (Li et al., 2007). Fines can be redeposited a considerable distance downwind of where they were emitted or, scrubbed from the air by striking vegetation, can be transferred from plant interspaces to the soil beneath plant canopies (e.g., Raupach et al., 2001), in both cases contributing nutrients and water holding capacity (Li et al., 2007, 2009a). This redistribution of fines beneath plant canopies contributes to the so-called "island of fertility" effect that has been cited as a cause for the encroachment of shrubs into grasslands worldwide (Schlesinger et al., 1990).

5.2 Basics of Aeolian Transport

Aeolian transport begins when the shear stress exerted by the wind exceeds a threshold. Sand particles with diameters of $70 - 100 \, \mu m$ have the lowest threshold shear stress and are sometimes called "efficient saltators." They can be made of individual clasts or aggregates (Iversen & White, 1982). When these saltating particles strike the surface, some of their kinetic energy goes into breaking soil aggregates (in the soil or the saltator itself) containing fine particles held together by interparticle forces (Iversen et al., 1976), thus emitting suspension-sized particles (smaller than around $50 \, \mu m$) through a process of sandblasting (Gillette, 1974). As a result, vertical flux (i.e., emission of fines or sometimes "dust flux" depending on the context) is often considered to be linearly related to horizontal flux by a constant that is determined by the proportion of clay in the soil (Gillette, 1974, 1977; Marticorena & Bergametti, 1995).

For the purpose of characterizing aeolian transport, the strength of the wind on the surface, is usually represented as shear stress, τ, or shear speed, u_*, which are related by $\tau = \rho u_*^2$ where ρ is the density of air. u_* is sometimes called the "shear velocity," but since only the magnitude of the velocity vector is usually considered "shear speed" will be used here. u_* is typically related to wind speed at height z, $U(z)$ by the Law of the Wall:

$$U(z) = \frac{u_*}{\kappa} \ln\left(\frac{z}{z_o}\right),$$

where κ is the von Kármán constant (~ 0.4, Foken, 2006) and z_o is the roughness length (Von Kármán, 1931). For a bare surface, transport is initiated when u_* exceeds the threshold shear speed, u_{*_t}, which has a minimum value of ~ 0.25 m s^{-1} for the most easily moved particles on Earth (Iversen & White, 1982). A number of factors control u_{*_t} including soil texture, crusting, and wetness (e.g., Gillette et al., 1980; Gillette, 1988; Fecan et al., 1999). There are many formulations for the rate of horizontal transport, Q, when the wind is above threshold (Li et al., 2013), but nearly all have Q increasing with roughly the cube of excess shear speed (i.e., $u_* - u_{*_t}$, Bagnold, 1941, p. 106).

For surfaces with roughness, whether the roughness is created by stones, vegetation, or man-made objects, total wind shear is partitioned into that acting on components of the surface:

$$\tau = \tau_r + \tau_s,$$

where τ_r and τ_s, respectively are the shear stress acting on the roughness elements and the soil, respectively (Raupach, 1992). For these complex surfaces, transport is initiated when the shear speed acting on the soil surface, u_{*_s}, exceeds u_{*_t}. Although expressions have been derived to express the average shear speed acting on the soil surface (Raupach et al., 1993), in reality, the shear speed of the wind at the soil surface is spatially heterogeneous (Gillies et al., 2014; Webb et al., 2014).

5.3 Structural Connectivity and Aeolian Processes

An important difference between aeolian and fluvial transport is that roughness elements remove momentum directly from the wind whereas such elements are either obstacles to flow or shape the microtopography for water movement. Unlike water, flowing air is also compressible. Therefore, movement of the air around roughness elements manifests as changes in the air pressure at the edges of the objects (e.g., Walter et al., 2012; Mayaud et al., 2016) rather than a wholescale change in the direction of flow, as happens with water. This is why, in terms of aeolian transport, we are rarely concerned about metric such as sinuosity or tortuosity, which convey something of the convoluted path that water follows when not on or in heterogeneous media.

There are a range of scales relevant to aeolian transport. Each of these spatial scales is also associated with a temporal scale over which transport through a connected system can occur. In the three Boxes in this chapter, calculations that link the spatial and temporal scales of transport are introduced. In each of the Boxes, one spatial scale of connectivity is considered, starting from the finest, where individual events provide the appropriate temporal scale, to the coarsest, where the scale of millennia is more applicable. Sand dune systems are not discussed in this chapter because the concept of connectivity is not well established within the study of dunes (outside of the case of *nebkhas*, sometimes also called coppice dunes), perhaps because sand dunes are often considered to be entirely connected, especially if free of vegetation. That is to say, there are no impediments in vegetation-free sand dunes that ultimately block the transport by wind and interrupt connected pathways for material transport. Dunes themselves do not break connectivity; almost by definition, sand can be transported up and over sand dunes.

5.3.1 Connectivity at the "Gap" Scale

A "gap" is considered as a patch of bare soil with roughness elements (typically vegetation, but any object that is higher than roughly the height of saltation, $0.5-1$ m). Connectivity at the gap scale is the basic building block from which connectivity at other scales is observed. This is the scale at which we can envision the action of the wind's shear on the soil in a gap to produce saltation, either by direct aerodynamic entrainment of particles or through ejection of particles by other saltators.

There are two ways that gaps define structural connectivity. First, they provide the actual pathways in which aeolian transport occurs. Although wind streamlines bend around objects (Walter et al., 2012; Mayaud et al., 2016) and spiral vortices can be spawned at the edge of objects (Wolfe & Nickling, 1993) the average motion and bulk of the horizontal transport is parallel with the *instantaneous* direction of the wind. Second, gaps allow the momentum of the wind to reach soil surface. Roughness elements protect the soil directly beneath them (e.g., under a plant canopy), but they also remove momentum from the wind (in the case of porous roughness elements, such as many types of vegetation, Wolfe & Nickling, 1996) or block the wind entirely (in the case of a solid roughness element, such as a rock or hay bale, Gillies et al., 2015). The combined effects of blocking the wind and removing momentum from it produce a wake zone in the lee of the roughness element where shear speed is depressed or even reversed relative to the overall wind direction (Wolfe & Nickling, 1993). Except for objects that are much higher than they are wide (e.g., telephone poles, fence posts, saguaro cactuses), the size of this wake zone is defined in relation to the height of the roughness element immediately upwind of the gap. Specifically, the shear speed in the wake of the object returns to the value it would have had in the absence of that object over a distance of about 8–10 heights (Bradley & Mulhearn, 1983). For objects that are considerably taller than they are wide, the width of the object defines the size of the wake zone.

The first definition of connectivity with regard to aeolian transport defined it in terms of the gap size measured along straight lines (Okin & Gillette, 2001). This study also highlighted an important component of structural connectivity in some environments: It can be anisotropic whereby the gap size and therefore connectivity at the gap scale depends on direction (see also, Zhang et al., 2021a). Thus, structural connectivity for aeolian transport tends to be defined as the straight-line lengths of exposed soil between roughness elements, scaled by their height, and should be measured along multiple directions. In the field, gaps are measured along a straight transect tape along with measurements of plant height (Herrick et al., 2005). In this method, small plants or other objects that serve to interrupt the length of the gap along the tape, or "gap breakers," are given a minimum size (e.g., 25 cm). That is, if two gaps are separated by a gap breaker than less than this minimum, the two gaps and the gap breaker are considered a single gap. More detailed approaches, which measure the height of each roughness element and scale the size of the adjacent gap accordingly have also been developed (Li et al., 2013).

Connectivity at the gap scale is the basic building block of connectivity observed at other scales because of basic characteristics of saltation. Even during large events, only a fraction of the particles at the surface move (Box 5.1), and when they do, they typically only move once. The length of a single saltating hop depends on windspeed but is typically a fraction of a metre (~ 40 cm, Belly, 1964). In other words, within a bare soil gap, a relatively small number of particles travel a fraction of a gap size during an event. This relationship fundamentally links connectivity at the gap scale in a vegetated landscape with the timescale of a single wind event.

Although the literature on aeolian transport on beaches is somewhat separate from literature examining transport through landscapes with roughness elements, the former can still be seen as having this smallest scale of connectivity. Specifically, the distance between the point on the beach where the sand is too wet to be transportable (i.e., the highest point of the swash zone) and dunes, roughness elements, or other barriers to transport can be considered a single gap for our present purposes (Houser, 2009).

5.3.2 Connectivity at the Patch Scale

The "patch" scale corresponds roughly to the hillslope scale in fluvial systems. It is the scale at which there are many gaps and roughness elements of roughly the same size (so ~ 10 – 100 m). In fluvial systems, there are well-defined hillslopes with a more or less clear top (e.g., crest of an interfluve) and bottom (e.g., channel). In aeolian systems, there are rarely such clearly defined boundaries but, just

---

Box 5.1
Timescales of Movement #1

The time it takes aeolian transport to move sediments depends on the rate of horizontal transport, Q, the distance it travels, and the amount of material considered. With just a few variables we can make order of magnitude estimates of many components of transport that relate space scale with time scale.

Let V be the volume of a $1\,m^2$ area on the surface with a depth of z. Let ρ_B be the bulk density of the soil ($\sim 1.25\ g\ cm^3$ for a sandy soil) whereas ρ_P is the density of the saltators ($\sim 2.65\ g\ cm^3$ for quartz). Let L be the average distance that a saltator moves in a single hop ($\sim 0.4\ m$, Belly, 1964 pp. 28–29).

What fraction of saltators at the soil surface move during a single event?

Assuming particles with diameter, d ($100\ \mu m$), the mass of each particle is given by $N_p = \frac{d\rho_B}{M_p}\ M_p = \frac{\pi \rho_p d^3}{6}$. Let Q_E be the horizontal mass flux occurring during a single event $(g\ m^{-1}\ event^{-1})$. The total number of particles that are detached during an event is then $N_d = \frac{Q_E}{LM_p}$. If we consider just the top layer of particles, then $z = d$, and the total number of particles in this layer is $N_p = \frac{d\rho_B}{M_p}$. Thus the fraction of surface saltators that move during a single event is given by $f_d = \frac{N_d}{N_p} = \frac{Q_E}{Ld\rho_B}$.

In a vegetated landscape with erodible sandy loam soil, a reasonable rate of transport is on the order of $1\ g\ cm^{-1}\ d^{-1}$ over the course of a year (Gillette & Pitchford, 2004). If this rate of flux occurs during 10 events over the course of a year, then $Q_E = 3650\ g\ m^{-1}\ event^{-1}$ and $f_d = 0.73$. Meaning that during one of these transport events, only $\sim 3/4$ of the particles on the surface move.

---

as there are several hillslopes in a watershed, there are many patches within a landscape.

Although horizontal aeolian transport at the gap scale occurs mainly in single straight-line saltation hops, horizontal transport at the patch scale occurs as a series of such hops along curved paths around and through roughness elements. The sinuous paths at the patch scale are a result of wind events from different directions that can shepherd particles in complex landscape. Over the timescale of one to several years, this transport might have an overall net direction or, in the case of a system with winds from opposing direction, may move particles to-and-fro without much net transfer through the patch. Fryberger and Dean (1979) introduced the concept of "drift potential (DP)" to characterize the directional variability in the wind's capacity to transport material. Recalling that the rate of horizontal transport scales with roughly the cube of the excess wind speed (Bagnold, 1941), DP is defined as:

$$DP_i = \sum_U U_i^2 (U_i - U_t) T_i^U,$$

where U_i is the windspeed at direction i, U_t is the windspeed at which aeolian transport starts, and T_i^U is the amount of time that the wind blows toward direction i at windspeed U. DP is therefore a function of wind direction, and DP "roses" can be constructed much like wind roses. Resultant drift potential (RDP) is the vector sum of DP_i and represents the net potential aeolian transport over a period of time. In later work, Marsham et al. (2011) expanded on the ideas of Fryberger and Dean (1979) in introducing the dust uplift potential, which is similar in form to DP (using a different equation for horizontal transport after Kawamura, 1951), but is independent of wind direction and which should not be confused with RDP.

This spatial heterogeneity in surface shear speed that results from the distribution of roughness elements leads to patterns of sediment mobilization that are controlled by the patterning of the roughness elements. Okin (2008) proposed a model (further evaluated by Li et al., 2013) for this spatial patterning based on the size distribution of unvegetated gaps within a patch. In this model, the wind shear speed acting on the surface is depressed in the immediate lee of vegetation and recovers to the value it would be without any vegetation over a characteristic exponential length scale represented in units of roughness element height (Figure 5.1).

The asymptotic recovery of wind shear speed in the lee of vegetation can be written as

$$u_{*_s} = u_* \left(\left(\frac{u_{*_s}}{u_*} \right)_{x=0} + \left(1 - \left(\frac{u_{*_s}}{u_*} \right)_{x=0} \right) \left(1 - e^{-C/x} \right) \right),$$

where u_{*_s} is the effective wind shear speed at the soil surface, $\left(\frac{u_{*_s}}{u_*} \right)_{x=0}$ is the ratio of u_{*_s} to u_* in the immediate lee of the roughness element $(x = 0)$, x is the distance to the nearest upwid roughness element (e.g., plant or rock) divided by its height, and C is the e-folding distance (also in units of plant height) for the recovery of u_{*_s}. Where u_{*_s} exceeds the soil's threshold u_{*_t}, horizontal flux, q, occurs and can be calculated using one of several equations. Li et al. (2013) found the approach of Kawamura (1951) to provide reasonable estimates:

$$q = f_s A \frac{\rho}{g} u_{*_s}^3 \left(1 - \frac{u_{*_t}^2}{u_{*_s}^2} \right) \left(1 - \frac{u_{*_t}}{u_{*_s}} \right),$$

where f_s is the fraction of bare, erodible soil at the surface, A is a scaling constant, ρ is the density of air, and g is the acceleration due to gravity. With the Kawamura (1951) flux equation, the best parameter values are: $\left(\frac{u_{*_s}}{u_*} \right)_{x=0} = 3.4$, $C = 5.1$, and $A = 4.3 \times 10^{-3}$. In practice q is evaluated for all possible values of x with their

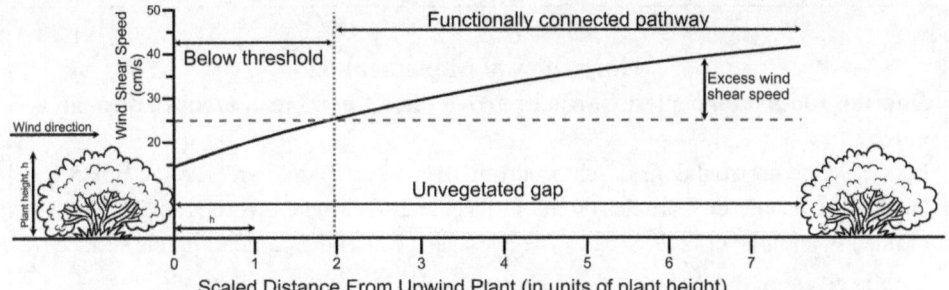

Figure 5.1 Schematic of an unvegetated gap showing how the wind shear speed is depressed in the lee of an upwind plant, with the distance measured in units of plant height assuming. In this example, wind shear speed would be 50 cm/s in the absence of plants and the threshold wind shear speed above which transport can occur is 25 cm/s (dashed line). In the area closest to the plant, the wind is below threshold but the excess wind shear speed (the wind shear speed minus the threshold wind shear speed) increases downwind of this below-threshold region.

respective values of u_{*_s} to give $q_x^{u_*}$, and these are used to estimate the flux from all x at for a particular u_*:

$$Q^{u_*} = \sum_x P_x q_x^{u_*},$$

where P_x is the probability of any point being x from the nearest upwind plant. P_x can be estimated from in situ gap size distribution data (Li et al., 2013) or analytical expressions for gap size distributions (Okin, 2008). In this formulation total horizontal flux, Q, is the sum of all of the fluxes (zero or nonzero) in the system, weighted by their likelihood:

$$Q = \sum_{u_*} P_{u_*} q_x^{u_*},$$

where P_{u_*} is the likelihood of a value of u_* derived from the histogram of wind speeds and the Law of the Wall (when using the paramters above, zo = 0.77 m should be used).

There are important feedbacks between vegetation and aeolian processes at the patch scale (Okin et al., 2006; Ravi et al., 2011). In addition to the interactions between plant interspaces (gaps) and adjacent plants that produces fertile islands and nebkhas, saltation through sparsely vegetated patches can damage both woody and herbaceous vegetation (Armbrust & Retta, 2000; Okin et al., 2001a, 2001b). Emission and loss of dust causes winnowing of fines in the soil that support important soil functions such as cation exchange capacity and plant-available water holding capacity (e.g., Li et al., 2009b). In addition, erosion can damage plants by exposing their roots (pedestaling) or through burial (Webb et al., 2020). The combined effect of these processes constitute strong positive feedbacks that enhance aeolian connectivity and support growth and expansion of desertified shrublands

---

Box 5.2
Timescales of Movement #2
How long does it take all the surface particles in a 1 m × 1 m area move through a patch?
Let Q_m be the horizontal mass flux occurring over a year (g m^{-1} yr^{-1}). As in Box 5.1, let us just consider the particles on the surface, so $z = d$ where d is 100 µm. The mass of this 1 m × 1 m layer is $M_z = d\rho_B$. The average speed of the particles in this surface layer is given by $V_z = \frac{Q_M L}{M_z}$.

In a vegetated landscape with erodible sandy loam soil, a reasonable average rate of transport is on the order of 1 g m^{-1} d^{-1} (Gillette & Pitchford, 2004), making $Q_m = 36500$ g m^{-1} yr^{-1}. Thus $V_z = 117$ m yr^{-1}. Therefore, it takes approximately one year for the particles at the surface to move through a patch roughly 100 m across.

(e.g., Okin et al., 2009). Ludwig et al. (2002, 2006, 2007); McGlynn and Okin (2006) and Karl et al. (2011) have suggested various indices of connectivity at the patch scale that are relevant for characterizing patch-scale connectivity, particularly in the context of land degradation or desertification.

Although there is no exact size that defines a patch, they are envisioned here as being on the scale of ~ 100 m in diameter. Given the spatial variability that characterize deserts, it is rare for homogenous patches of vegetation to be much larger than this, though fine-scale variations in the underlying geomorphic template can create patches that are considerably smaller. Nonetheless, the timescale of transport of the surface grains through a patch is on the order of ~ 1 y (Box 5.2).

5.3.3 Connectivity at the Landscape Scale

A "landscape" is considered here to be a mosaic of patches typically occurring on a single geomorphic unit. Geomorphic unit is useful here as a way to demarcate landscapes because the differences between origin, parent material, and age of a geomorphic units can result in surface soils with different textures, crusting, and sediment availability. Soil texture and crusting are important determinants of soil threshold (e.g., Gillette et al., 1980; Gillette, 1988; Belnap et al., 2007). Sediment availability depends on both crusting (physical as well as chemical, e.g., Webb et al., 2016) and characteristics of the surface that derive from its geomorphology to determine the amount of erodible particles that are present (e.g., de Vries et al., 2014). As important as vegetation is within patches for determining the distribution of wind shear stress acting on the surface, threshold shear speed is the primary control of the instantaneous magnitude transport in supply unlimited systems. The distribution of shear speed due to vegetation is irrelevant, for example, if the threshold of the soil is higher than that experienced anywhere on the surface.

Sediment supply, though less frequently characterized on surfaces and harder to measure, is also a first-order control on the long-term rate of transport. The presence of a limited amount of sediment at the surface that is capable of transport fundamentally caps the amount of transport that can happen in a system (e.g., Gillette et al., 2001; de Vries et al., 2014), regardless of how easily detachable it is (threshold) or what the distribution of vegetation is.

The distribution of patches within a landscape does not have to be only related to natural environmental factors that affect the distribution of vegetation. Human activities frequently impose patchiness within a landscape. In an agricultural setting, individual fields can be considered patches within an agricultural landscape since the transport across a field is relatively homogenous and decoupled from the transport in neighboring patches with different crop or fallow characteristics (Hagen, 2004; Feng & Sharratt, 2007). Likewise, the imprint of field agriculture within native vegetation generates patchiness at the landscape scale with areas of agriculture (whether active, fallow, or abandoned) behaving in a different manner than native vegetation (Okin et al., 2001a; Li et al., 2014).

There can be consequential interactions between adjacent patches in a landscape due to the transport of one patch onto another. Okin and Painter (2004), for example, showed the effect of deposition of material from abandoned agricultural fields on the surface soil texture in the area of native vegetation immediately downwind. Similarly, Okin et al. (2001a, 2001b) showed that blowing sand from abandoned agriculture or bare areas can damage native vegetation in adjacent patches due to sandblasting. Inter-patch dynamics are not limited to relationships between within and between agriculture and native vegetation patches. Hernández-Calvento et al. (2014), for example, have demonstrated that patches of developed land within dune systems impact dune dynamics in adjacent undeveloped areas.

Windbreaks and shelter belts are designed to disrupt inter-patch connectivity. They do this through two mechanisms: reducing the wind shear speed to reduce flux and directly capturing windblown materials. Windbreaks and shelterbelts have been used in a wide range of geographic and historical contexts (Bradley & Mulhearn, 1983; Brandle et al., 2004; Nordstrom & Hotta, 2004; Cornelis & Gabriels, 2005; Sterk et al., 2012; Ma et al., 2020; Torshizi et al., 2020).

There are additional forms of aeolian connectivity that are also perhaps best considered to exist at the landscape scale. For example, there is clear connectivity between beaches and foredunes in coastal systems (e.g., Nordstrom & Jackson, 1992; Sherman & Lyons, 1994; Psuty, 1996; Marqués et al., 2001, Hesp, 2002, Bauer et al., 2009; Houser, 2009; Houser & Mathew, 2011; Anthony & Aagaard, 2020) with beaches and foredunes plausibly considered as different patches. Other source-bordering dunes, including those where sediment is supplied by fluvial systems (e.g., Page, 1971; Page et al., 2001; Ivester & Leigh, 2003; Rendell et al.,

2003; Han et al., 2007; Draut, 2012; Muhs et al., 2013; Al-Masrahy & Mountney, 2015; Bogle et al., 2015, Sankey et al., 2018a, 2018b) and sandy lake shorelines (e.g., Wopfner & Twidale, 1988; Davidson-Arnott & Law, 1990; Kocurek & Lancaster, 1999; Yu et al., 2013) display landscape-scale connectivity between the sediment source (patches of loose sediment delivered by the aqueous source) and the derivative aeolian deposits.

The timescale for transport at the landscape scale depends on the system in question. However, since transport at this scale, as defined here, essentially means transport through adjacent patches, the timescale is considerably longer than that for the patch scale. Indeed, the timescale should increase roughly linearly with the total length (adjacent or not) of transport through patches with relatively homogenous patch characteristics. However, transport may occur through patches with more or less resistance to transport due to the soil and vegetation characteristics within each type.

5.3.4 Connectivity at the Basin Scale

A "basin" is considered here in its primary geomorphic context, an area of lower topography separated by higher topography that may or may not have an outlet for water. Large "aeolian systems" (Kocurek & Havholm, 1993; Lancaster, 1997; Thomas & Wiggs, 2008) must include a sand sink (usually dunes or sandsheets) and a sand source. The sand source may be a depositional basin (Kocurek et al., 2007) or a local site of deposition/sediment storage that do not represent terminal fluvial sinks (Kocurek & Lancaster, 1999). Because water flows downhill but wind does not, the spatial separation at the basin scale of fluvial sinks (aeolian sources) and aeolian sinks can be substantial. Therefore, in addition to the aeolian sink and source, a corridor through which aeolian transport occurs is also a requirement for basin-scale aeolian connectivity in all cases where the source and sink do not abut one another.

Any of this triumvirate – aeolian source, transport corridor, and sink – may be vegetated and therefore, the connectivity at gap, patch, and landscape scales (as previously defined) are important determinants of the functioning of a basin-scale aeolian system. Vegetation dynamics, indeed, can have an important control on the long-term dynamics of aeolian systems, particularly with regard to the control that climate has on their stability and evolution (e.g., Lancaster, 1997; Kocurek & Lancaster, 1999).

Due to the large distances and quantities of sediment involved, the timescales for transport at the basin scale are much longer than those at other scales considered here. Whereas patch- and landscape-scale transport may occur on the scale of tens to hundreds of years, appreciable basin-scale transport capable of producing mature aeolian systems likely takes on the order of tens of thousands of years or longer, depending on the system (Box 5.3).

Box 5.3
Timescales of Movement #3
How long does it take a 1 m × 1 m column of sand move across a basin?
Let Q_m be the horizontal mass flux occurring over a year (g m^{-1} yr^{-1}). Let us consider a 10 m deep column of sand in a deep sand sheet, so $z = 10$ m. The mass and average speed of the particles in this volume are given by the same equations as in Box 5.2.

In bare, sandy soil, a reasonable rate of transport is on the order of $1000 - 2000$ g cm^{-1} d^{-1} (Dong et al., 2014), making $Q_m = 3.65 \times 10^7 - 7.3 \times 10^7$ g m^{-1} yr^{-1}. Thus $V_z = 1.17 - 2.34$ m yr^{-1}. Therefore, it takes approximately 4300–8600 years for a 10 m column of sand to travel 10 km, roughly the scale of some basin-wide aeolian transport corridors (e.g., Edwards, 1993).

5.3.5 *Long-range Atmospheric Connectivity*

Mineral dust (also called mineral aerosols) is produced during aeolian transport when saltating particles strike the surface and provide sufficient kinetic energy to release fine particles (generally < 50 μm) into the air (Gillette, 1974; Gillette, 1977; Marticorena & Bergametti, 1995). These fine particles are suspended in the air stream and can travel extremely large distances (e.g., Swap et al., 1992; Prospero, 1999). Many parts of the world are therefore impacted by sources of desert dust through long-range atmospheric connectivity.

Desert dust has important impacts on the Earth's system. In the atmosphere, it has both direct and indirect effects on radiative forcing (e.g., Charlson & Heintzenberg, 1995; Sokolik & Toon, 1996; Haywood & Boucher, 2000). Deposition of mineral dust onto the ocean significantly impacts marine biogeochemistry by impacting the availability of iron, a sometimes-limiting micronutrient (e.g., Duce & Tindale, 1991; Baker et al., 2003, 2007; Mahowald et al., 2005, 2008, 2009; Okin et al., 2011). Deposition of mineral dust can also significantly impact terrestrial and lacustrine biogeochemistry (e.g., Litaor, 1987; Chadwick et al., 1999; Okin et al., 2004; Ballantyne et al., 2011; Lawrence et al., 2011; Mladenov et al., 2011; Munroe, 2014; Munroe et al., 2021). Dust, deposited on snow from long distances can also have important consequences for the snowmelt and the resultant water available in rivers (e.g., Painter et al., 2007, 2010).

5.4 Functional Connectivity

The concept of functional connectivity is not well developed in the aeolian realm. We can identify three features that control functional connectivity: structural connectivity, wind speed, and wind direction.

Functional connectivity is mainly relevant in areas with nonerodible elements (large rocks, vegetation), as these are typically the objects that break connectivity. Patches of nonerodible but smooth soil surface do not disrupt connectivity so much as reduce the area of a surface that can emit particles.

Particles that are emitted from erodible areas can either jump over smooth, nonerodible surfaces or, if landing on the nonerodible area, can be moved in subsequent hops. Small patches of surface roughness (less than the height of saltation) within an erodible soil can, for a while, disrupt connectivity as they capture saltating particles. Over time, however, the pits and valleys that can trap saltators fill up and the rough patch becomes a smooth, sandy surface that is not a barrier to transport. Furrows within agricultural fields, areas of clodding, or deliberate soil ridging to reduce erosion play this sand-capturing role by providing local depressions that can trap sand thus disrupting connectivity of transport (Fryrear, 1984).

The relevant scale of structural connectivity when considering its control on functional connectivity is the patch sale. During a single event, each of the sand grains (or sand grain-sized aggregates) in the topmost surface layer only makes on the order of one saltation hop that is considerably shorter than one metre (Box 5.1), meaning that for a single particle functional connectivity occurs within a single gap. Nonetheless, it is the patch-scale influence of vegetation on the wind field that controls the amount of transport during an event.

For a patch, the scaled gap size distribution controls total horizontal transport by wind (Okin, 2008) and, therefore, also controls the aggregate functional connectivity. The wind vector interacts with existing structural connectivity during an event in such a way that the portion of the surface that is protected depends on gap size and wind speed (Figure 5.1). Under high wind conditions, less of the surface is protected from the wind. Conversely, the length of functionally connected pathways for transport increases with wind speed, all other things being equal. During an event, these linear pathways, aligned with the direction of the wind, essentially define the functional connectivity (Figure 5.2) and the aggregated transport that occurs along these pathways.

In many cases, the direction of the wind vector, not just its speed, can impact functional connectivity. Vegetation in deserts can be anisotropic which means, for the purposes here, that the gap size distribution is a function of direction. One of the best-known expressions of this is banded vegetation, sometimes called "tiger bush," which results from feedbacks between water flow and vegetation growth on shallow slopes (e.g., Ludwig et al., 1999; Moreno-de las Heras et al., 2011). For water, this anisotropy is striking but somewhat irrelevant because the slope does not change aspect and water always flows downhill. However, the wind can come from any direction, and it is clear that the functional connectivity with respect to aeolian transport would be very different depending on the wind's azimuth.

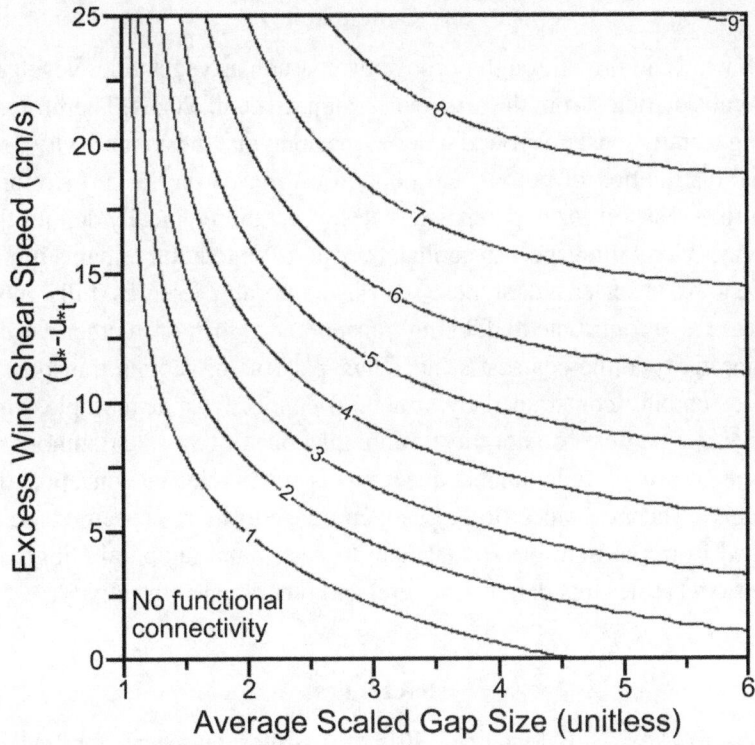

Figure 5.2   Contour plot showing the average length of functionally connected pathways, in units of plant height, against average scaled gap size (i.e., gap size expressed in units of plant height) and the excess wind shear speed (i.e., the difference between the wind speed and the threshold wind speed). This plot was generated using best-fit parameters from Li et al. (2013) for the Shao et al. (1993) flux equation, gap size distributions defined by gamma distributions (as in Okin, 2008), and a maximum scaled gap size for all cases of 20. There is no functional connectivity in the bottom left of the plot meaning that there are no cases where the wind shear speed in the lee of the roughness element exceeds the soil threshold.

Aeolian processes can also result in anisotropic distribution of vegetation, but in contrast to banding due to water transport, the longer gaps are in the direction of flow instead of perpendicular to it (Gillette et al., 2006, McGlynn & Okin, 2006, Zhang et al., 2021b). Functional connectivity of transport on a beach is also subject to both wind direction as well as tide/storm surge (e.g., Nordstrom & Jackson, 1992, Bauer et al., 2009). Tidal elevation and storm surge impact the effective width of the beach, by controlling the area of soil that is too wet to be detached by the wind or is underwater. The direction of the wind with respect to the beach – whether the wind strikes the beach perpendicularly or obliquely – controls the fetch, the distance over which the wind can blow without obstructions.

5.5 Conclusion

Although wind can flow through porous objects such as vegetation, vegetation efficiently scrubs particles from the airstream (Raupach et al., 2001). Therefore, particle transport generally must go around objects, meaning that the connectivity defined by the spatial distribution of porous and nonporous objects on the surface fundamentally controls transport. At gap to patch scales, vegetation typically defines the structural connectivity with respect to aeolian transport. Vegetation remains important at landscape to basin scales but, at these scales, geomorphic features (hills, mountains, valleys, etc.) also contribute to defining transport corridors, or in other words, structural connectivity at the coarsest scale. Thus, patterns of aeolian transport through time are essentially constrained by structural connectivity at multiple, embedded scales. Because particles do not travel more than one or two saltation hops during a single event, however, functional connectivity is only a relevant concept at the finest spatial scales. Thus, consideration of aeolian transport in any specific case must be approached from the multiple spatial (gap to basin) and temporal (single event to longer periods) scales that define structural and functional connectivity.

References

Al-Masrahy, M. A., & N. P. Mountney. 2015. A classification scheme for fluvial–aeolian system interaction in desert-margin settings. *Aeolian Research* 17: 67–88.

Anderson, R. S. 1987. A theoretical model for aeolian impact ripples. *Sedimentology* 34: 943–956.

Anthony, E. J., & T. Aagaard. 2020. The lower shoreface: Morphodynamics and sediment connectivity with the upper shoreface and beach. *Earth-Science Reviews* 210: 103334.

Armbrust, D. V., & A. Retta. 2000. Wind and sandblast damage to growing vegetation. *Annals of Arid Zone* 39: 273–284.

Bagnold, R. A. 1941. *The Physics of Blown Sand and Desert Dunes*. Methuen, New York.

Baker, A. R., S. D. Kelly, K. F. Biswas, M. Witt, & T. D. Jickells. 2003. Atmospheric deposition of nutrients to the Atlantic Ocean. *Geophysical Research Letters* 30: 2296.

Baker, A. R., K. Weston, S. D. Kelly, M. Voss, P. Streu, & J. N. Cape. 2007. Dry and wet deposition of nutrients from the tropical Atlantic atmosphere: Links to primary productivity and nitrogen fixation. *Deep-Sea Research Part I-Oceanographic Research Papers* 54: 1704–1720.

Ballantyne, A. P., J. Brahney, D. Fernandez, C. L. Lawrence, J. Saros, & J. C. Neff. 2011. Biogeochemical response of alpine lakes to a recent increase in dust deposition in the Southwestern, US. *Biogeosciences* 8: 2689–2706.

Bauer, B. O., R. G. D. Davidson-Arnott, P. A. Hesp, S. L. Namikas, J. Ollerhead, & I. J. Walker. 2009. Aeolian sediment transport on a beach: Surface moisture, wind fetch, and mean transport. *Geomorphology* 105: 106–116.

Belly, P. Y. 1964. *Sand Movement by Wind*. Technical Memorandum No. 1. U.S. Army Coastal Engineering Research Center, Washington, DC.

Belnap, J., S. L. Phillips, J. E. Herrick, & J. R. Johansen. 2007. Wind erodibility of soils at Fort Irwin, California (Mojave Desert), USA, before and after trampling disturbance: Implications for land management. *Earth Surface Processes and Landforms* 32: 75–84.

Bogle, R., M. H. Redsteer, & J. Vogel. 2015. Field measurement and analysis of climatic factors affecting dune mobility near Grand Falls on the Navajo Nation, southwestern United States. *Geomorphology* 228: 41–51.

Bradley, E. F., & P. J. Mulhearn. 1983. Development of velocity and shear-stress distributions in the wake of a porous shelter fence. *Journal of Wind Engineering and Industrial Aerodynamics* 15: 145–156.

Brandle, J. R., L. Hodges, & X. H. Zhou. 2004. Windbreaks in North American agricultural systems. Pages 65–78 *in* P. K. R. Nair, M. R. Rao, and L. E. Buck, editors. *New Vistas in Agroforestry: A Compendium for 1st World Congress of Agroforestry, 2004*. Springer Netherlands, Dordrecht.

Chadwick, O. A., L. A. Derry, P. M. Vitousek, B. J. Huebert, & L. O. Hedin. 1999. Changing sources of nutrients during four million years of ecosystem development. *Nature* 397: 491–497.

Charlson, R. J., & J. Heintzenberg. 1995. *Aerosol Forcing of Climate*. Wiley, Chichester, New York.

Cornelis, W. M., & D. Gabriels. 2005. Optimal windbreak design for wind-erosion control. *Journal of Arid Environments* 61: 315–332.

Davidson-Arnott, R. G. D., & M. N. Law. 1990. Seasonal patterns and controls on sediment supply to coastal foredunes, Long Point, Lake Erie. Pages 177–200 *in* K. F. Nordstrom, N. Psuty, & B. Carter, editors. *Coastal Dunes: Form and Process*. Wiley, New York.

de Vries, S., J. S. M. van Thiel de Vries, L. C. van Rijn, S. M. Arens, & R. Ranasinghe. 2014. Aeolian sediment transport in supply limited situations. *Aeolian Research* 12: 75–85.

Dong, Z., P. Lv, Z. Zhang, & J. Lu. 2014. Aeolian transport over a developing transverse dune. *Journal of Arid Land* 6: 243–254.

Draut, A. E. 2012. Effects of river regulation on aeolian landscapes, Colorado River, southwestern USA. *Journal of Geophysical Research: Earth Surface* 117: F02022, doi:10.1029/2011JF002329

Duce, R. A., & N. W. Tindale. 1991. Atmospheric transport of iron and its deposition in the ocean. *Limnology and Oceanography* 36: 1715–1726.

Edwards, S. R. 1993. Luminescence dating of sand from the Kelso Dunes, California. *Geological Society, London, Special Publications* 72: 59.

Fecan, F., B. Marticorena, & G. Bergametti. 1999. Parameterization of the increase of the aeolian erosion threshold wind friction velocity due to soil moisture for arid and semi-arid areas. *Annales Geophysicae-Atmospheres Hydrospheres and Space Sciences* 17: 149–157.

Feng, G., & B. Sharratt. 2007. Validation of WEPS for soil and PM10 loss from agricultural fields within the Columbia Plateau of the United States. *Earth Surface Processes and Landforms* 32: 743–753.

Foken, T. 2006. 50 years of the Monin–Obukhov similarity theory. *Boundary-Layer Meteorology* 119: 431–447.

Fryberger, S. G., & G. Dean. 1979. Dune forms and wind regime. *A study of global sand seas* 1052: 137–169.

Fryrear, D. W. 1984. Soil Ridges-Clods and Wind Erosion. *Transactions of the Asae* 27: 445–448.

Gillette, D. A. 1974. On the production of soil wind erosion aerosols having the potential for long range transport. *Journal des Recherches Atmospheriques* 8: 735–744.

Gillette, D. A. 1977. Fine particulate emissions due to wind erosion. *Transactions of the American Society of Agricultural Engineers* 20: 890–897.

Gillette, D. A. 1988. Threshold friction velocities for dust production for agricultural soils. *Journal of Geophysical Research* 93: 12645–12662.

Gillette, D. A., & A. M. Pitchford. 2004. Sand flux in the northern Chihuahuan desert, New Mexico, USA, and the influence of mesquite-dominated landscapes. *Journal of Geophysical Research-Earth Surface* 109: F04003.

Gillette, D. A., J. E. Herrick, & G. A. Herbert. 2006. Wind characteristics of mesquite streets in the northern Chihuahuan Desert, New Mexico, USA. *Environmental Fluid Mechanics* 6: 241–275.

Gillette, D. A., T. C. Niemeyer, & P. J. Helm. 2001. Supply-limited horizontal sand drift at an ephemerally crusted, unvegetated saline playa. *Journal of Geophysical Research-Atmospheres* 106: 18085–18098.

Gillette, D. A., J. Adams, A. Endo, D. Smith, & R. Kihl. 1980. Threshold velocities for input of soil particles into the air by desert soils. *Journal of Geographical Research* 85: 5621–5630.

Gillies, J. A., J. M. Nield, & W. G. Nickling. 2014. Wind speed and sediment transport recovery in the lee of a vegetated and denuded nebkha within a nebkha dune field. *Aeolian Research* 12: 135–141.

Gillies, J. A., H. Green, G. McCarley-Holder, S. Grimm, C. Howard, N. Barbieri, D. Ono, & T. Schade. 2015. Using solid element roughness to control sand movement: Keeler Dunes, Keeler, California. *Aeolian Research* 18: 35–46.

Greeley, R., & J. D. Iversen. 1985. *Wind as a Geological Process on Earth, Mars, Venus and Titan*. Cambridge University Press, Cambridge.

Hagen, L. J. 2004. Evaluation of the Wind Erosion Prediction System (WEPS) erosion submodel on cropland fields. *Environmental Modelling & Software* 19: 171–176.

Han, G., G. Zhang, & Y. Dong. 2007. A model for the active origin and development of source-bordering dunefields on a semiarid fluvial plain: A case study from the Xiliaohe Plain, Northeast China. *Geomorphology* 86: 512–524.

Haywood, J., & O. Boucher. 2000. Estimates of the direct and indirect radiative forcing due to tropospheric aerosols: A review. *Reviews of Geophysics* 38: 513–543.

Hernández-Calvento, L., D. W. T. Jackson, R. Medina, A. I. Hernández-Cordero, N. Cruz, & S. Requejo. 2014. Downwind effects on an arid dunefield from an evolving urbanised area. *Aeolian Research* 15: 301–309.

Herrick, J. E., J. W. Van Zee, K. M. Havstad, L. M. Burkett, & W. G. Whitford. 2005. *Monitoring Manual for Grasslands Shrublands and Savana Ecosysts*. Volume 1: Quick Start. USDA-ARS Jornada Experimental Range, Las Cruces, New Mexico.

Hesp, P. 2002. Foredunes and blowouts: Initiation, geomorphology and dynamics. *Geomorphology* 48: 245–268.

Houser, C. 2009. Synchronization of transport and supply in beach-dune interaction. *Progress in Physical Geography: Earth and Environment* 33: 733–746.

Houser, C., & S. Mathew. 2011. Alongshore variation in foredune height in response to transport potential and sediment supply: South Padre Island, Texas. *Geomorphology* 125: 62–72.

Iversen, J. D., & B. R. White. 1982. Saltation threshold on Earth, Mars and Venus. *Sedimentology* 29: 111–119.

Iversen, J. D., J. B. Pollack, R. Greeley, & B. R. White. 1976. Saltation threshold on Mars – effect of interparticle force, surface-roughness, and low atmospheric density. *Icarus* 29: 381–393.

Ivester, A. H., & D. S. Leigh. 2003. Riverine dunes on the coastal plain of Georgia, USA. *Geomorphology* 51: 289–311.

Karl, J. W., M. C. Duniway, & T. S. Schrader. 2011. A technique for estimating range-land canopy-gap size distributions from high-resolution digital imagery. *Rangeland Ecology & Management* 65: 196–207.

Kawamura, R. 1951. *Study of Sand Movement by Wind.* University of California, Berkeley, CA.

Kocurek, G., M. Carr, R. Ewing, K. G. Havholm, Y. C. Nagar, & A. K. Singhvi. 2007. White sands dune field, New Mexico: age, dune dynamics and recent accumulations. *Sedimentary Geology* 197: 313–331.

Kocurek, G., & K. G. Havholm. 1993. Chapter 16. Eolian sequence stratigraphy-a conceptual framework. Pages 393–409 *in* H. W. Posamentier and G. P. Allen, editors. *Recent Developments in Siliciclastic Sequence Stratigraphy, AAPG Memoir 58.* The American Association of Petroleum Geologists, Tulsa, OK.

Kocurek, G., & N. Lancaster. 1999. Aeolian system sediment state: Theory and Mojave Desert Kelso dune field example. *Sedimentology* 46: 505–515.

Lancaster, N. 1997. Response of eolian geomorphic systems to minor climate change: Examples from the southern Californian deserts. *Geomorphology* 19: 333–347.

Lawrence, C. R., J. C. Neff, & G. L. Farmer. 2011. The accretion of aeolian dust in soils of the San Juan Mountains, Colorado, USA. *Journal of Geophysical Research* 116: F02013.

Li, J. R., G. S. Okin, & H. E. Epstein. 2009b. Effects of enhanced wind erosion on surface soil texture and characteristics of windblown sediments. *Journal of Geophysical Research-Biogeosciences* 114. G02003. doi:10.1029/2008JG000903

Li, J., G. S. Okin, L. J. Alvarez, & H. E. Epstein. 2009a. Sediment deposition and soil nutrient heterogeneity in two desert grassland ecosystems, southern New Mexico. *Plant and Soil* 319: 67–84.

Li, J., G. S. Okin, L. J. Hartman, & H. E. Epstein. 2007. Quantitative assessment of wind erosion and soil nutrient loss in desert grasslands of southern New Mexico, USA. *Biogeochemistry* 85: 317–332.

Li, J., G. S. Okin, J. Tatarko, N. P. Webb, & J. E. Herrick. 2014. Consistency of wind erosion assessments across land use and land cover types: A critical analysis. *Aeolian Research* 15: 253–260.

Li, J., G. S. Okin, J. E. Herrick, J. Belnap, M. E. Miller, K. Vest, & A. E. Draut. 2013. Evaluation of a new model of aeolian transport in the presence of vegetation. *Journal of Geophysical Research: Earth Surface* 118: 288–306.

Litaor, M. I. 1987. The influence of Eolian dust on the genesis of alpine soils in the front range, Colorado. *Soil Science Society of America Journal* 51: 142–147.

Ludwig, J. A., D. J. Tongway, & S. G. Marsden. 1999. Stripes, strands or stipples: Modelling the influence of three landscape banding patterns on resource capture and productivity in semi-arid woodlands, Australia. *Catena* 37: 257–273.

Ludwig, J. A., G. N. Bastin, V. H. Chewings, R. W. Eager, & A. C. Liedloff. 2007. Leakiness: A new index for monitoring the health of arid and semiarid landscapes using remotely sensed vegetation cover and elevation data. *Ecological Indicators* 7: 442–454.

Ludwig, J. A., R. W. Eager, G. N. Bastin, V. H. Chewings, and A. C. Liedloff. 2002. A leakiness index for assessing landscape function using remote sensing. *Landscape Ecology* 17: 157–171.

Ludwig, J. A., R. W. Eager, A. C. Liedloff, G. N. Bastin, & V. H. Chewings. 2006. A new landscape leakiness index based on remotely sensed ground-cover data. *Ecological Indicators* 6: 327–336.

Ma, R., J. Li, Y. Ma, L. Wei, & Y. Zhang. 2020. A wind tunnel study of the seasonal shelter efficiency of deciduous windbreaks. *Transactions of the ASABE* 63: 913–922.

Mahowald, N. M., A. R. Baker, G. Bergametti, N. Brooks, R. A. Duce, T. D. Jickells, N. Kubilay, J. M. Prospero, & I. Tegen. 2005. Atmospheric global dust cycle and iron inputs to the ocean. *Global Biogeochemical Cycles* 19. GB4025, doi:10.1029/2004GB002402

Mahowald, N., T. D. Jickells, A. R. Baker, P. Artaxo, C. R. Benitez-Nelson, G. Bergametti, T. C. Bond, Y. Chen, D. D. Cohen, B. Herut, N. Kubilay, R. Losno, C. Luo, W. Maenhaut, K. A. McGee, G. S. Okin, R. L. Siefert, & S. Tsukuda. 2008. The global distribution of atmospheric phosphorus deposition and anthropogenic impacts. *Global Biogeochemical Cycles* 22: GB4026.

Mahowald, N. M., S. Engelstaedter, C. Luo, A. Sealy, P. Artaxo, C. R. Benitez-Nelson, S. Bonnet, Y. Chen, P. Y. Chuang, D. D. Cohen, F. Dulac, B. Herut, A. M. Johansen, N. Kubilay, R. Losno, W. Maenhaut, A. Paytan, J. M. Prospero, L. M. Shank, & R. L. Siefert. 2009. Atmospheric iron deposition: Global distribution, variability and human perturbations. *Annual Reviews of Marine Sciences* 1: 245–278.

Marqués, M. A., N. P. Psuty, & R. Rodriguez. 2001. Neglected effects of Eolian dynamics on artificial beach nourishment: The case of Riells, Spain. *Journal of Coastal Research* 17: 694–704.

Marsham, J. H., P. Knippertz, N. S. Dixon, D. J. Parker, & G. M. S. Lister. 2011. The importance of the representation of deep convection for modeled dust-generating winds over West Africa during summer. *Geophysical Research Letters* 38: L16803.

Marticorena, B., & G. Bergametti. 1995. Modeling the atmospheric dust cycle: 1. Design of a soil-derived dust emission scheme. *Journal of Geophysical Research* 100: 16415–16430.

Mayaud, J. R., G. F. S. Wiggs, & R. M. Bailey. 2016. Characterizing turbulent wind flow around dryland vegetation. *Earth Surface Processes and Landforms* 41: 1421–1436.

McGlynn, I. O., & G. S. Okin. 2006. Characterization of shrub distribution using high spatial resolution remote sensing: Ecosystem implications for a former Chihuahuan Desert grassland. *Remote Sensing of Environment* 101: 554–566.

Mladenov, N., R. Sommaruga, R. Morales-Baquero, I. Laurion, L. Camarero, M. C. Diéguez, A. Camacho, A. Delgado, O. Torres, Z. Chen, M. Felip, & I. Reche. 2011. Dust inputs and bacteria influence dissolved organic matter in clear alpine lakes. *Nature Communications* 2: 405.

Moreno-de las Heras, M., P. M. Saco, G. R. Willgoose, & D. J. Tongway. 2011. Assessing landscape structure and pattern fragmentation in semiarid ecosystems using patch-size distributions. *Ecological Applications* 21: 2793–2805.

Muhs, D. R., J. Roskin, H. Tsoar, G. Skipp, J. R. Budahn, A. Sneh, N. Porat, J.-D. Stanley, I. Katra, & D. G. Blumberg. 2013. Origin of the Sinai–Negev erg, Egypt and Israel: Mineralogical and geochemical evidence for the importance of the Nile and sea level history. *Quaternary Science Reviews* 69: 28–48.

Munroe, J. S. 2014. Properties of modern dust accumulating in the Uinta Mountains, Utah, USA, and implications for the regional dust system of the Rocky Mountains. *Earth Surface Processes and Landforms* 39: 1979–1988.

Munroe, J. S., R. McElroy, S. O'Keefe, A. Peters, & L. Wasson. 2021. Holocene records of eolian dust deposition from high-elevation lakes in the Uinta Mountains, Utah, USA. *Journal of Quaternary Science* 36: 66–75.

Nordstrom, K. F., & N. L. Jackson. 1992. Effect of source width and tidal elevation changes on aeolian transport on an estuarine beach. *Sedimentology* 39: 769–778.

Nordstrom, K. F., & S. Hotta. 2004. Wind erosion from cropland in the USA: A review of problems, solutions and prospects. *Geoderma* 121: 157–167.

Okin, G. S. 2008. A new model for wind erosion in the presence of vegetation. *Journal of Geophysical Research-Earth Surface* 113: F02S10.

Okin, G. S., & D. A. Gillette. 2001. Distribution of vegetation in wind-dominated landscapes: Implications for wind erosion modeling and landscape processes. *Journal of Geophysical Research* 106: 9673–9683.

Okin, G. S., & T. H. Painter. 2004. Effect of grain size on remotely sensed spectral reflectance of sandy desert surfaces. *Remote Sensing of Environment* 89: 272–280.

Okin, G. S., B. Murray, & W. H. Schlesinger. 2001a. Degradation of sandy arid shrubland environments: Observations, process modelling, and management implications. *Journal of Arid Environments* 47: 123–144.

Okin, G. S., J. E. Herrick, & D. A. Gillette. 2006. Multiscale controls on and consequences of aeolian processes in landscape change in arid and semiarid environments. *Journal of Arid Environments* 65: 253–275.

Okin, G. S., B. Murray, & W. H. Schlesinger. 2001b. Desertification in an arid shrubland in the southwestern United States: Process modeling and validation. Pages 53–70 in A. Conacher, editor. *Land Degradation: Papers Selected from Contributions to the Sixth meeting of the International Geographical Union's Commission on Land Degradation and Desertification, Perth, Western Australia, 20–28 September 1999.* Kluwer Academic Publishers, Dordrecht.

Okin, G. S., N. Mahowald, O. A. Chadwick, & P. Artaxo. 2004. The impact of desert dust on the biogeochemistry of phosphorus in terrestrial ecosystems. *Global Biogeochemical Cycles* 18: 10.1029/2003GB002145.

Okin, G. S., O. E. Sala, E. R. Vivoni, J. Zhang, & A. Bhattachan. 2018. The interactive role of wind in water in functioning of drylands: What does the future hold? *Bioscience* 68: 670–677.

Okin, G. S., A. J. Parsons, J. Wainwright, J. E. Herrick, B. T. Bestelmeyer, D. P. C. Peters, & E. L. Fredrickson. 2009. Do changes in connectivity explain desertification? *Bioscience* 59: 237–244.

Okin, G. S., A. R. Baker, I. Tegen, N. M. Mahowald, F. J. Dentener, R. A. Duce, J. N. Galloway, K. A. Hunter, M. Kanakidou, N. Kubilay, J. M. Prospero, M. M. Sarin, V. Surapipith, M. Uematsu, & T. Zhuo. 2011. Impacts of atmospheric nutrient deposition on marine productivity: Roles of nitrogen, phosphorus, and iron *Global Biogeochemical Cycles* 25: GB2022.

Page, K. 1971. Riverine source bordering sand dune. *Australian Geographer* 11: 603–605.

Page, K. J., A. J. Dare-Edwards, J. W. Owens, P. S. Frazier, J. Kellett, & D. M. Price. 2001. TL chronology and stratigraphy of riverine source bordering sand dunes near Wagga Wagga, New South Wales, Australia. *Quaternary International* 83–85: 187–193.

Painter, T. H., J. Deems, J. Belnap, A. F. Hamlet, C. C. Landry, & B. Udall. 2010. Response of Colorado River runoff to dust radiative forcing in snow. *Proceedings of the National Academy of Science* 107: 17125–17130.

Painter, T. H., A. P. Barrett, C. Landry, J. Neff, M. P. Cassidy, C. Lawrence, K. E. McBride, & G. L. Farmer. 2007. Impact of disturbed desert soils on duration of mountain snow-cover. *Geophysical Research Letters* 34: L12502.

Prospero, J. M. 1999. Long-term measurements of the transport of African mineral dust to the southeastern United States: Implications for regional air quality. *Journal of Geophysical Research-Atmospheres* 104: 15917–15927.

Psuty, N. 1996. Coastal foredune development and vertical displacement. *Zeitschrift für Geomorphologie. Supplement band* 102: 211–221.

Raupach, M. R. 1992. Drag and drag partition on rough surfaces. *Boundary-Layer Meteorology* 60: 375–395.

Raupach, M. R., D. A. Gillette, & J. F. Leys. 1993. The effect of roughness elements on wind erosion threshold. *Journal of Geophysical Research* 98: 3023–3029.

Raupach, M. R., N. Woods, G. Dorr, J. F. Leys, & H. A. Cleugh. 2001. The entrapment of particles by windbreaks. *Atmospheric Environment* 35: 3373–3383.

Ravi, S., P. D'Odorico, D. D. Breshears, J. P. Field, A. S. Goudie, T. E. Huxman, J. Li, G. S. Okin, R. J. Swap, A. D. Thomas, R. S. Van Pelt, J. J. Whicker, & T. M. Zobeck. 2011. Aeolian processes and the biosphere. *Reviews of Geophysics* 49: RG3001.

Rendell, H. M., M. L. Clarke, A. Warren, & A. Chappell. 2003. The timing of climbing dune formation in southwestern Niger: Fluvio-aeolian interactions and the rôle of sand supply. *Quaternary Science Reviews* 22: 1059–1065.

Sankey, J. B., J. Caster, A. Kasprak, & A. E. East. 2018a. The response of source-bordering aeolian dunefields to sediment-supply changes 2: Controlled floods of the Colorado River in Grand Canyon, Arizona, USA. *Aeolian Research* 32: 154–169.

Sankey, J. B., A. Kasprak, J. Caster, A. E. East, & H. C. Fairley. 2018b. The response of source-bordering aeolian dunefields to sediment-supply changes 1: Effects of wind variability and river-valley morphodynamics. *Aeolian Research* 32: 228–245.

Schlesinger, W. H., J. F. Reynolds, G. L. Cunningham, L. F. Huenneke, W. M. Jarrell, R. A. Virginia, & W. G. Whitford. 1990. Biological feedbacks in global desertification. *Science* 247: 1043–1048.

Shao, Y., M. R. Raupach, & P. J. Findlater. 1993. Effect of saltation bombardment on the entrainment of dust by wind. *Journal of Geophysical Research* 98: 12719–12726.

Sherman, D. J., & W. Lyons. 1994. Beach-state controls on aeolian sand delivery to coastal dunes. *Physical Geography* 15: 381–395.

Sokolik, I. N., & O. B. Toon. 1996. Direct radiative forcing by anthropogenic airborne mineral aerosols. *Nature* 381: 681–683.

Sterk, G., J. Parigiani, E. Cittadini, P. Peters, J. Scholberg, & P. Peri. 2012. Aeolian sediment mass fluxes on a sandy soil in Central Patagonia. *Catena* 95: 112–123.

Swap, R., M. Garstang, S. Greco, R. Talbot, & P. Kallberg. 1992. Saharan dust in the Amazon Basin. *Tellus Series B-Chemical and Physical Meteorology* 44: 133–149.

Thomas, D. S. G., & G. F. S. Wiggs. 2008. Aeolian system responses to global change: Challenges of scale, process and temporal integration. *Earth Surface Processes and Landforms* 33: 1396–1418.

Torshizi, M. R., A. Miri, A. Shahriari, Z. Dong, & R. Davidson-Arnott. 2020. The effectiveness of a multi-row Tamarix windbreak in reducing aeolian erosion and sediment flux, Niatak area, Iran. *Journal of Environmental Management* 265: 110486.

Von Kármán, T. 1931. *Mechanical Similitude and Turbulence*. National Advisory Committee for Aeronautics. https://books.google.co.uk/books?id= ONBCAQAAIAAJ&dq=von+karman&lr=&source=gbs_navlinks_s

Walter, B., C. Gromke, K. C. Leonard, C. Manes, and M. Lehning. 2012. Spatio-temporal surface shear-stress variability in live plant canopies and cube arrays. *Boundary-Layer Meteorology* 143: 337–356.

Webb, N. P., G. S. Okin, & S. Brown. 2014. The effect of roughness elements on wind erosion: The importance of surface shear stress distribution. *Journal of Geophysical Research-Atmospheres* 119: 6066–6084.

Webb, N. P., M. S. Galloza, T. M. Zobeck, & J. E. Herrick. 2016. Threshold wind velocity dynamics as a driver of aeolian sediment mass flux. *Aeolian Research* 20: 45–58.

Webb, N. P., E. Kachergis, S. W. Miller, S. E. McCord, B. T. Bestelmeyer, J. R. Brown, A. Chappell, B. L. Edwards, J. E. Herrick, J. W. Karl, J. F. Leys, L. J. Metz, S. Smarik, J. Tatarko, J. W. Van Zee, & G. Zwicke. 2020. Indicators and benchmarks for wind erosion monitoring, assessment and management. *Ecological Indicators* 110: 105881.

Wolfe, S. A., & W. G. Nickling. 1993. The protective role of sparse vegetation in wind erosion. *Progress in Physical Geography: Earth and Environment* 17: 50–68.

Wolfe, S. A., & W. G. Nickling. 1996. Shear stress partitioning in sparsely vegetated desert canopies. *Earth Surface Processes and Landforms* 21: 607–619.

Wopfner, H., & C. R. Twidale. 1988. Formation and age of desert dunes in the Lake Eyre depocentres in central Australia. *Geologische Rundschau* 77: 815–834.

Yu, L., Z. Lai, & P. An. 2013. OSL chronology and paleoclimatic implications of paleo-dunes in the middle and southwestern Qaidam Basin, Qinghai–Tibetan Plateau. *Sciences in Cold and Arid Regions* 5: 211–219.

Zhang, J., G. S. Okin, B. Zhou, & J. W. Karl. 2021b. UAV-derived imagery for vegetation structure estimation in rangelands: Validation and application. *Ecosphere* 12: e03830.

Zhang, J., W. Guo, B. Zhou, & G. S. Okin. 2021a. Drone-based remote sensing for research on wind erosion in drylands: Possible applications. *Remote Sensing* 13: 283.

6

Glacial Processes

PETER G. KNIGHT

6.1 Perspectives and Context

Discussions of connectivity in geomorphology frequently refer to the idea of a conveyor belt of sediment transfer through a geomorphic system. The conveyor belt is typically interrupted by discontinuities, disturbances and temporary storage points, and Ferguson (1981) described the progress of material through fluvial systems as a 'jerky' conveyor belt. The jerkiness is a measure of the continuity or connectedness of the system, and the interruptions that affect the path from source to sink. In glacier research, the conveyor belt metaphor is well established and can be traced back to the origin of the word 'glacier' itself. Derived from the French word for ice (la glace), the word glacier carries within it, etymologically, not only the idea of ice but also the idea of delivery, provision or supply. Just as a fruiterer supplies fruit, a glacier supplies or delivers ice. A glacier, of course, conveys and delivers much more than just ice. We can think of it as a supply chain or conveyor belt of water, debris, solutes, nutrients, trapped bubbles of ancient atmospheres, microbial communities, meteorites, and even a flight of six Lockheed Lightning P-38 fighters and two B-17 bombers that were lost on the Greenland ice sheet in 1942, buried under snowfall, and entrained into the glacier. One of the planes was subsequently excavated and restored, but the rest are continuing their slow journey through the glacier. Whether, or when, and where they will finally either emerge from the glacier front, be exposed at the surface, buried amongst subglacial debris or flushed out in a meltwater torrent, will depend on the connectivity among the different components of the glacier system.

From a purely glaciological perspective, glaciers involve the accumulation, movement and ablation of ice. In most environments, the ice is associated with water, and water plays a significant role both in glacial processes (glacier dynamics, for example) and in glacial geomorphic processes (such as debris entrainment, transport and deposition). From a geomorphological perspective, we recognise that along with ice there are other components (especially debris) being accumulated, entrained, transported and released, and that the movement of ice and water, as well as being instrumental in the transportation of a debris load, is also instrumental in the processes of erosion, entrainment and deposition that determine the changing composition and volume of that load. Movement of both ice and water is central to the geomorphic connectivity of the glacial system, so in looking at geomorphic connectivity in the context of glacial processes we are interested both in debris or sediment pathways and in the ice and water pathways to which they are related.

Glacial geomorphic processes can be mapped as a network of vertical and longitudinal connections between process domains in the glacier system through which ice, water and debris are transferred or stored. These process domains stretch from subglacial and glacier-surface entrainment sources in continental interiors to depositional sinks in the oceans. Domains can be defined structurally in terms of their position within a flow system from areas of accumulation through to areas of ablation, but the functional or process-related connection of domains is more realistically defined by geographic and temporal patterns of controlling factors such as temperature. The thermal regime within a glacier can vary both spatially and over time, and process domains within glaciers are highly susceptible to thermal control, both through the presence/absence of water and because the dynamic properties of ice itself are temperature dependent. Characteristic geomorphic processes and landforms are associated with different process domains, and with the boundaries between them. For example, drumlin swarms are associated with subglacial environments, whereas terminal moraine ridges are associated with the ice margin. Freezing of subglacial debris to the base of the glacier is associated with a transition zone (geographic or temporal) between warm and cold basal thermal regimes. These geomorphic markers reflect the connectivity of the system, because the supply and removal of debris by ice and water are the fundamental controls on the creation, persistence and destruction of most glacial landforms. As connections between units within the system change over time, sediment delivery changes in response. Connections through and between process domains are subject to interruption and reconnection in response to transient external changes such as climate change and glacier fluctuations. The geography of glacial landforms might therefore be understood in terms of changing connectivity conditions at boundary points within the system. The idea of connectivity can provide a unifying framework for understanding debris transfer and landform creation in the glacier system.

The basic idea of connectivity has long been important in glacier research, but without much explicit reference to connectivity science or use of its terminology. Glacier research has been concerned with debris transport pathways, sediment stores, sediment budgets, and transfers of energy, water and debris through glaciers, because those ideas are fundamental to how glacial geomorphic systems work. There has been some cross-referencing between glacial and non-glacial environments. Etzelmüller et al. (2000) visualised valley glacier sediment transfer systems as pathways through the glacier ending in meltwater evacuation either directly from the glacier bed or indirectly via marginal moraines. They compared how hillslope and fluvial systems were coupled in a glacial setting with how that coupling had been assessed in a mountain environment by Caine (1986). Studies of glacial and proglacial sediment budgets have typically set their conclusions in terms of traditional geomorphic theory such as thresholds or the magnitude and frequency of specific events that make connections between storage points or between sections of the system. For example, Warburton (1990) found that high magnitude, low frequency flood events were dominant in releasing sediment from proglacial storage into distal river systems. We might interpret that as a recognition of sedimentological connectivity between proglacial storage and the distal fluvial system varying with meltwater discharge. Calculating sediment flux through the terminus of the glacier Gepatschferner, Stocker-Waldhuber and Kuhn (2019) found that the highest sediment fluxes from the margin were associated with extreme and episodic high magnitude events such as rockfalls that exceeded normal annual sediment fluxes by orders of magnitude, demonstrating the importance of specific short-term connections along sediment pathways. However, glacier-related research has not generally focused on connectivity theory explicitly, and theoretical treatments of connectivity have not typically focused on glacier systems, concentrating more on terrestrial hillslope and river systems (e.g., Ali & Roy, 2009; Bracken et al., 2013). There is a clear opportunity for glacial geomorphology to engage more with connectivity theory in the way that other areas of geomorphology have done, and for connectivity theory to be applied more explicitly to glacial environments.

There has been some glacier-related work explicitly talking about connectivity, particularly between specific components or units within small parts of the system. For example, Miles et al. (2017) discussed hydrological connectivity between the supraglacial and englacial water systems, and MacDonell et al. (2016) focused on the hydrological connectivity of cryoconite holes on a glacier surface both in terms of their isolation from the atmosphere and surface-water systems and their connection with the englacial drainage network. Collins (1979) used diurnal variations in conductivity in meltwater emerging from the glacier margin to reconstruct subglacial hydrological pathways on the basis of a mixing model that used the time-varying connection between fast-flow and slow-flow pathways to explain changing chemical composition of bulk meltwaters. Slow-flow water pathways beneath a glacier generate more solute-rich meltwaters because of the increased contact time between water and minerals and hence greater opportunity for transfer of material into solution. Small (1987) and Small et al. (1984) discussed connections between the glacier and the ice-marginal moraine in the context of moraine sediment budgets and the controls on delivery of debris from the ice to the moraine, and similarly, Swift et al. (2018) identified connections between basal thrusting in the interior of a glacier and specific sedimentological characteristics of the ice-marginal moraine that demonstrated the interior flow path of the debris. At a smaller scale of observation but again looking at entrainment into the glacier and release into the forefield, Toubes-Rodrigo et al. (2021) looked at how sub-glacially entrained minerals can support microbial ecosystems within basal ice, and how those can, in turn, contribute to soil development in front of the glacier when the basal ice and its load are released from the glacier margin. Lane et al. (2017) similarly looked at connectivity between glaciers and their proglacial areas, focusing on how that connectivity might change as a consequence of glacier retreat. Warburton (1990) is one example of many that have focused even further downstream and have been concerned entirely with proglacial sediment budgets. Many of those, such as Porter et al. (2019) looking at how glacial sediment stores are reworked in the proglacial area, although they are focused beyond the glacier terminus, are dealing with glacier-derived sediments and illustrate one of the fundamental problems in glacial geomorphology: identifying the boundaries of what is 'glacial'. Those examples indicate a common characteristic of the literature: an interest in local connectivity within limited parts of the system, in connectivity between the glacier system and the proglacial systems, and in changes in both glacier-proglacial connectivity and the operation of the proglacial system consequent upon glacier retreat. The example of Collins (1979), and his use of a chemical mixing model to reconstruct subglacial hydrological pathways on the basis of changing composition of bulk meltwater at the glacier margin, illustrates a practical application of a connectivity approach to glacial environments. The characteristics of each component of the glacier system, at each location, depend on connections between components higher up in the system, so an understanding of how that connectivity works can be valuable both for understanding the development of downstream characteristics and also for using downstream characteristics to reconstruct the nature of the higher parts of the system which, in the context of glaciers, may be in the inaccessible and difficult to measure subglacial zone.

6.2 Connecting Source to Sink in Glacier Systems

Our preliminary consideration of connectivity in glaciers has led already to a recognition that glacier systems are very complex and have many components. We are not dealing with a single, simple conveyor belt. As well as thinking longitudinally from the upstream end of a glacier to the margin and out to the proglacial environment, we also have to think vertically from the glacier surface in contact with the atmosphere, through the underlying glacier interior, down into the basal ice layer where the glacier is affected by contact with the bed, and through the sole of the glacier to the bed and the substrate. We need to accommodate the supraglacial, englacial, basal, subglacial, ice-marginal and pro-glacial process domains, and see external connections to time-dependent process environments and landsystems such as the periglacial and paraglacial. If connectivity is an exploration of source-to-sink pathways, the source and the sink in a glacier can be very far apart and separated by a lot of intermediate steps. The transit time for ice through a glacier can be millions of years, and even average values are about 10,000 years, compared with a global average transit time of just weeks for water through river systems. In glaciers, the loading end of the conveyor belt can be a very long way, structurally and functionally, from the delivery end. At the largest scale, source to sink in a glacier system could stretch from the accumulation of snow, atmospheric dust and extra-terrestrial debris high in the interior of a continental ice sheet, to the delivery of that debris into the ocean, potentially millions of years later, from melting icebergs. The 'watershed scale' in this system is potentially planetary, with glaciers mediating between atmosphere and oceans as a major part of the global hydrological cycle. Glaciers play the long game. Part of their slowness is due to the rate of flow of ice and the physical size of ice sheets, with potential flow paths of thousands of miles at literally glacial speeds. Boulton (2006) described ice sheets as being coupled to the earth system across interfaces with the atmosphere, the lithosphere and the ocean. There are more sources, more potential transport pathways and more sinks, than in some other geomorphic systems, and there are deep, long-lasting intermediate storage points such as subglacial lakes, along with short-lived high-magnitude processes such as glacier surges or glacier lake outburst floods that can briefly accelerate sediment transfer and can also make new temporary connections. Connectivity might provide a useful conceptual framework for associating glacial process-environments and landsystems with the conditions of material transfer through the complex, frequently disrupted, glacier system.

Bracken et al. (2015) suggested that 'If we remain bounded by established practices and existing ways of approaching sediment transfer we may not be able to exploit the full potential of the concept of sediment connectivity.' The citation record of that seminal paper is instructive in showing how far glaciologists and glacial geomorphologists have engaged with connectivity theory. Exactly seven years after the first online publication of the paper, it had been cited by 229 subsequent publications, the vast majority of which were concerned with fluvial and hillslope systems. Only one of the citing articles (Swift et al., 2021) focused explicitly on glacial processes. A further ten publications (less than 5%) concerned alpine, proglacial, postglacial or periglacial environments. For example, Altmann et al.

(2020) studied sediment transfer into mountain streams from ancient moraines over extended periods following glacier retreat, and Mancini and Lane (2020) looked at the debuttressing effect of glacier retreat on unstable valley sides. Hassan et al. (2019) were concerned with the legacy of ancient glaciers and their post-glacial impacts on forested mountain catchments. Kummert and Delaloye (2018) looked at sediment transfer from periglacial mountain rock glaciers. Bernhardt et al. (2017) described how deglacial environmental change could be recognised in deep-marine turbidites. These examples illustrate the pervasive impact of glaciers and glaciation far beyond the immediate historical and geographical limits of glaciation. However, that only one of the 229 items citing Bracken et al. (2015) was specifically concerned both with connectivity theory and with current glacial processes and glacial sediment transfer, demonstrates that, despite its potential, and although some of its key components are implicit, connectivity theory is not explicitly at the forefront of great deal of current glacial or glacial geomorphological work. Nevertheless, the ideas that underpin connectivity are prominent in studies of glacial systems, albeit without the shared terminology or contextual framework of literature. For example, Jaeger and Koppes (2016), discussing the role of the cryosphere in source-to-sink systems, focus on the generation of glacigenic sediment by glacial processes, its transfer through the system and its eventual accumulation in the ocean. There is much in common between connectivity literature and glacial literature using more traditional ways of describing sediment fluxes, and the flux of debris from stage to stage that Jaeger and Koppes (2016) discuss could be described in a connectivity framework, but this has not become the usual approach in glacier contexts.

6.3 Debris Pathways in Glaciers

6.3.1 Vertical Connections

Describing glacial sediment production and transfer, Knight (1999) represented a glacier as a vertical system divided into six horizontal layers, and identified what we might now call connections by which debris could move between the layers (Figure 6.1). We can use this vertical model to begin an exploration of connectivity in the glacial sediment transfer system. Following traditional glacier terminology, our six layers are the extraglacial, supraglacial, englacial, basal, subglacial and substrate layers. The top layer is the extraglacial layer above the glacier, including the atmosphere. Below this, the top surface of the glacier itself is the supraglacial layer. Next comes the englacial layer: the interior of the glacier. Below that, where the ice is in contact with the bed or affected by proximity to the bed, is the basal layer. Beneath the base of the glacier is the subglacial layer,

Peter G. Knight

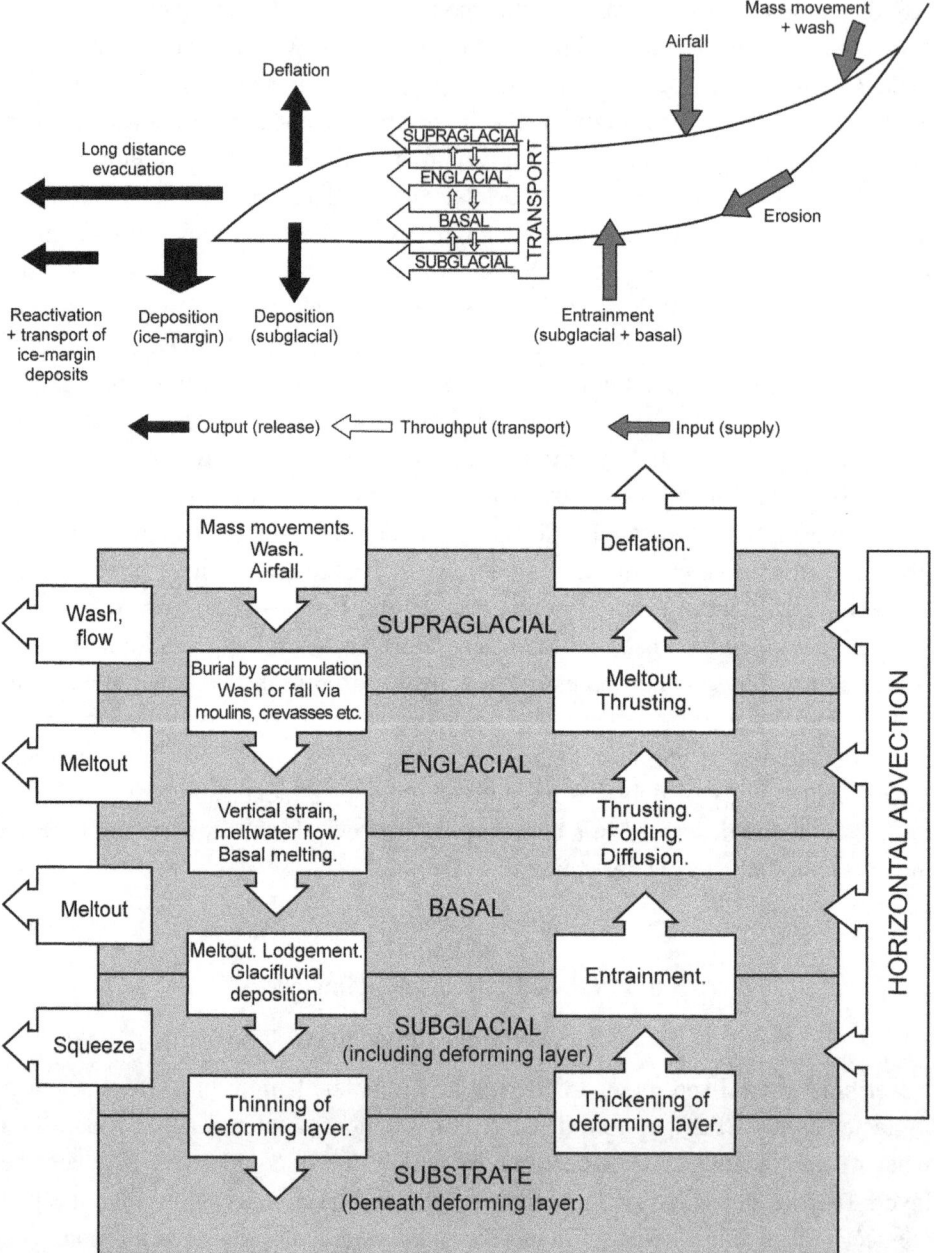

Figure 6.1   Diagrammatic representation of the main components and connections within a glacier sediment transfer system. After Figs. 9.3 and 9.4 in Knight (1999).

which might include deforming subglacial debris that comprises part of the glacier's dynamic system. Finally, below the subglacial layer, the bottom boundary of our system is the substrate: typically we think of this as bedrock or immobile sediment, but where a glacier is floating, then the water (and whatever ground surface is below the water) is effectively this bottom substrate layer.

The top layer of our glacier system is the 'extraglacial' zone above the glacier, which includes the atmosphere and mountains or valley-sides rising above the ice. Debris inputs from this layer onto the glacier surface include meteorites, volcanic ash, dust trapped within snow or deposited direct from the atmosphere, anthropogenic pollutants, rockfalls, landslides and other mass movements, wash of valley-side sediments, and waterborne debris carried onto the surface from ice-marginal lakes or streams.

All of that material derived from the extraglacial layer crosses our first boundary and arrives at the next layer in our sequence, the glacier surface or 'supraglacial' layer. In higher sections of the glacier, material deposited onto the surface is likely to be buried by the accumulation of snow, and delivered downwards to the next layer in our sequence: the 'englacial' layer or glacier interior. In downstream sections of the glacier, in the ablation zone, supraglacial material is unlikely to be incorporated into the glacier by snowfall, as there is typically more ablation than accumulation here, but may penetrate down into the englacial zone by wash or fall into crevasses or moulins. There are therefore several different pathways by which debris from the supraglacial layer can connect downwards into the englacial layer, with different pathways more relevant at different positions along the glacier. Debris can also emerge from the englacial layer upwards onto the glacier surface. If ice is being lost from the surface by ablation processes such as melting or sublimation, the flow path of englacial debris will be upwards relative to the declining surface. As well as emergence by supraglacial melt-out, debris can emerge onto the surface in some cases by thrusting along shear planes under compressive flow within the ice. The lower portions of many glaciers are characterised by extensive coverings of supraglacial debris that has emerged from englacial or basal-layer transport.

A special case of this re-emergence onto the glacier surface of debris derived from the extraglacial zone, after a period of transport in the englacial zone, occurs in the so-called 'blue ice zones' of Antarctica (Bitanja, 1999). These are areas where local ice-flow and meteorological conditions result in enhanced surface ablation by sublimation, and the emergence of englacial ice to the surface, along with its included debris. These zones only comprise about 1% of the Antarctic ice sheet and typically occur close to mountains, where ice flow is held up by the topography, where sublimation and wind erosion are increased by the local microclimate, and where the excess of local ablation over local accumulation results in an upward flow of ice towards the surface. As the catchment area for this ice is the deep interior of the continent, a high proportion of the debris emerging at the surface in these zones is made up of meteorites that accumulated into the ice over long periods and were collected together by the peculiar ice and climate dynamics of the blue ice zones. Meteorites are buried in the accumulation zone and are carried through the ice sheet, but those on a particular trajectory re-emerge and accumulate on the surface in the blue-ice zones. About 66% of all collected meteorite specimens on Earth come from these Antarctic blue ice 'meteorite stranding zones' (Evatt et al., 2016). On flow paths that do not encounter specific topographic conditions, meteorites and other englacial debris will not emerge prematurely at the surface in this way and will continue their longitudinal passage through the englacial zone, but where ice emerges at a blue ice zone, the debris becomes stranded at the ablating surface. If that flight of lost bombers that we mentioned at the start of the chapter had been in Antarctica and followed a trajectory that brought them to the surface in a blue ice zone, they would then remain at the surface, with ice emerging and sublimating from below them, indefinitely, illustrating one of the potential breaks in connectivity in the downflow conveyor belt of glacial transport.

The trajectory of meteorites in Antarctic blue ice zones illustrates another complication in the connectivity between the supraglacial and englacial zones. Evatt et al. (2016) describe how meteorites approaching the upper surface of the ice from below can be warmed by solar radiation penetrating the ice. Meteorites tens of centimetres beneath the surface can be warmed sufficiently to melt the ice with which they are in immediate contact, and sink downwards into the glacier at a rate that matches the upwards flow trajectory of ice to the surface. This could maintain a layer of meteorites below the surface that never emerge into the supraglacial zone but persist in isolated pockets of meltwater. A similar process can affect any debris at or close to the surface if it has a high thermal conductivity. Large rocks simply protect the ice below them from melting, but fine-grained debris or small rocks, especially if dark coloured, can transmit heat that they absorb from the sun to the ice below, and melt their way downwards into the ice surface creating 'cryoconite holes'. MacDonell et al. (2016) explained how cryoconite holes can be an important source and store of water close to cold glacier surfaces and can also provide a microbiological habitat. They also found that cryoconite holes were not always fully isolated from the surrounding drainage system and so, in our terms, provide some potential hydrological connectivity between the supraglacial and englacial layers. Describing near-surface hydraulic conductivity in glaciers Stevens et al. (2018) showed that near-surface ice textures had a major impact on transmissivity of the ice surface and in the rate of transfer of water through the supraglacial system or from the supraglacial to englacial systems.

Material can move downwards from the supraglacial zone by burial into the englacial zone, upwards to the atmosphere by deflation, sublimation, or evaporation, or longitudinally down-glacier by surface wash or by horizontal advection as the ice moves forward. Material buried by subsequent snowfall into the englacial zone makes up the bulk of most glaciers. Where the surface is warm and there is water present, some or all of the surface snow melts to water and refreezes to make ice. This is referred to as superimposed ice and typically does not retain a clear climate signal in its stable isotope composition. Where the surface is cold and dry, snow is compressed and recrystallised into ice without significant melting, and the isotopic composition of the snow is preserved. Also trapped and preserved are bubbles of the atmosphere and atmospheric particulates and solutes. The englacial facies contains a detailed historical record, layer upon layer, of the characteristics of the atmosphere at the date of the transformation of snow to ice at the surface. It also contains an archive of volcanic tephra, industrial pollution and other atmospheric inputs. Alley (2000), describing how ice cores reaching all the way from the surface to the bed are extracted from glaciers, referred to ice cores as 'time machines' allowing us to see what conditions at the glacier surface have been over hundreds of thousands of years. This use of ice cores relies on the connectivity and continuity between the supraglacial and englacial layers, and cannot be employed where that connectivity is broken by local conditions such as surface melting.

In some glaciers, depending on their thermal regime, the englacial layer is characterised by a complex hydrological network that can connect upwards to the supraglacial zone and downwards to the glacier bed. For example, supraglacial water in lakes and streams can penetrate through moulins (glacial sinkholes) or crevasses into the englacial zone and to the bed. Water flows at a macro scale through pipes and channels, but although ice is theoretically permeable along grain boundaries between ice crystals, the vein network is so fine that transmission is extremely limited at that scale.

Addition of material at the glacier surface in the accumulation zone, and its movement outwards towards the margin, results in a downwards trajectory for material in the englacial layer. This is enhanced where melting occurs at the base of the glacier and debris moves downwards through the englacial layer due to this vertical strain, or by meltwater flow through the englacial hydrological system of moulins and channels. In warm-based glaciers, this results in a general transport of debris through the englacial layer downwards to the base where it might be released to the bed or subglacial layer by glacial depositional processes such as meltout or lodgement, or by fluvioglacial deposition.

In many glaciers, an additional layer exists between the englacial layer and the bed: the so-called basal ice layer (Figure 6.2). In this layer, the ice is created or altered by processes operating at the bed, losing some of the characteristics of surface-derived englacial ice and acquiring characteristics peculiar to basal ice (Knight, 1997). The basal ice layer typically includes ice created by the freezing of water at the bed of the glacier. The accretion of this ice can involve entrainment of debris from the bed into the basal layer. Compressive flow can also result in folding

Figure 6.2   Intercalated layers of clean ice (darker) and debris (lighter) in a section of the basal ice layer exposed at the margin of the Greenland ice sheet. Photograph by P.G. Knight.

or thrusting within the basal ice causing a thickening of the basal layer beyond the thickness of accreted ice and raising debris from the bed higher into the glacier. This layer of basally affected ice does not exist in all parts of all glaciers, as it is dependent on specific thermal and dynamic conditions, but where it does exist it can comprise a layer tens of metres in thickness that is significantly richer in debris than the englacial facies above. In glaciers such as ice sheets where there is relatively little supraglacial debris supply from mountain peaks or valley walls, the majority of debris entrainment into the glacier can be from basal sources into this basal ice layer. Processes of debris entrainment include bulk freeze-on of meltwater and debris caused by temperature or pressure changes along flowlines, regelation of ice around basal obstacles, squeezing of deformable subglacial sediment into basal crevasses, folding and thrusting, and flow of fine debris in pressurised water away from the bed into the grain-boundary vein network.

In some glaciers, the base of the glacier rests directly on a substrate of bedrock or rigid sediment. In others, there is an intermediate subglacial layer of deformable sediment that can be mobilised in response to shear stress imposed by the ice.

The degree of mobilisation depends on characteristics of the sediment, including temperature and pore water pressure, that vary over time. In many glaciers, the substrate is mobile for periods when the basal temperature and water pressure are higher, but immobile when temperatures and water pressure are lower. Subglacial deformation can therefore switch on and off over time. This might happen seasonally or might be associated with long-term changes in dynamic conditions. Under deforming bed conditions, the subglacial sediment is in transport with the glacier, so the switching off of the deformation is essentially a depositional event. The thickness of the deforming layer depends on the stress imposed from the glacier and the temperature-related strength of the subglacial sediment, so as well as switching on and off, deformation can change by the thickening or thinning of the deformation layer, which again can be viewed as a kind of erosion (when the layer thickens) and deposition (when the layer thins), and the lower, non-deforming part of the layer can be considered (perhaps temporarily) as part of the lowest layer in our sequence, the substrate.

At the beds of many glaciers, there is also a hydrological system that may extend downwards from the supraglacial and englacial systems, or may be isolated from them. Water can exist at the bed as a film, as a sheet, in continuous channels or in a system of linked cavities. Water can also occur in disconnected water-filled cavities and in subglacial lakes. It is important to note the hydrological connection between glaciers and groundwater. Piotrowski (2006) describes how groundwater recharge beneath glaciers draws not only on water produced at the bed by melting but also, over large areas of some glaciers, on water produced by melting at the surface that has penetrated to the bed via the englacial drainage network. He also observes that groundwater flow beneath glaciers differs from that in other confined aquifers in that the flow is driven by a hydraulic gradient imposed by the overburden of ice, and that some water may be advected horizontally within subglacial sediment being deformed by glacier stress. These points lead us to the importance of considering not only vertical connections between layers in the glacier, but also the longitudinal movement of the 'conveyor belt' that we introduced at the start of the chapter.

The vertical profile of a glacier can be viewed as layers with connections upwards and downwards between each layer, but to make sense of the overall flow of ice, and of water and debris within the ice, we also need to consider the horizontal or longitudinal organisation of glacial processes.

6.3.2 Longitudinal Connections

Reduced to its functional, rather than structural or geographical components, the glacier system can be viewed longitudinally in three sections: inputs (supply, accumulation and entrainment), throughputs (transport or movement) and outputs (ablation, deposition or release). That longitudinal model of processes does not necessarily coincide with the longitudinal geography of the glacier: accumulation and entrainment are not confined to the upper parts of the glacier, nor deposition to the lower parts. If we wish to maintain the conveyor belt model, we need to include people with shovels throwing extra material onto the belt right up until its very end, and people dragging material off the belt and putting it into storage silos at intervals the whole way along!

Inputs to the glacier can be from airfall, mass movement or wash from outside the glacier and from entrainment at the glacier bed. These can occur throughout the geographical length of the glacier. Even though we commonly refer to an up-glacier accumulation zone and a down-glacier ablation zone, this terminology refers to the mass balance of ice. In the accumulation zone, more ice is locally gained than is lost, giving a positive mass balance, whereas in the ablation zone, the local mass balance is negative. The glacier margin is the point where the negative mass balance of the ablation area has entirely removed the material accumulated by the positive mass balance in the accumulation area. In terms of ice, the accumulation area is the dominant input zone and the ablation area is the dominant output zone. However, the same is not necessarily true for debris entrainment and deposition. Debris can be added or entrained throughout the length of the glacier. In many glaciers, especially ice caps and ice sheets, the ice surface in the upper part of the glacier is above the level of the topography, so there is little or no input of debris from valley walls or nunataks. Closer to the margin, however, even down into the ablation zone and to the very edge of the glacier, it is increasingly likely that valley walls and surrounding topography will provide an increasing source area for rockfalls and surface wash. Similarly for entrainment of debris into the base of the glacier, closer to the margin where the ice is thinner and more likely to divide around nunataks or split into valley outlets, the area of glacier-bed contact for debris entrainment is greater (Figure 6.3).

Debris entrainment processes at the bed depend on more than simply the area of glacier in contact with the bed. Processes such as freezing of meltwater and debris to the base are temperature and pressure dependent, so vary both over time (seasonally or in response to longer-term environmental changes) and over distance along a flow line. For example, the basal thermal regime of an ice sheet might include a central cold-based area in the interior, with little liquid water available, surrounded by a zone where the temperature reaches the melting point and the bed has water flowing through a basal drainage network and within permeable subglacial sediments. Closer to the margin, the temperature at the base is typically reduced under thinner ice, and water flowing outwards under cryostatic pressure is likely to freeze, entraining ice and sediment to the glacier bed. That freezing on close to the margin, combined with diminishing ice thickness and hence diminishing

Figure 6.3 Medial moraines on the surface of the Pers Glacier in Switzerland illustrate the down-glacier transport of debris derived from a combination of rockfall from valley walls and entrainment at the glacier bed. Photograph by Erich Vestendaper on Pixabay.

shear stress, can lead to compressive flow, folding, and shearing by which more debris can be entrained from the bed or raised to higher levels in the basal ice. Often, debris emerging at a glacier margin has most recently been entrained (or re-entrained) close to the margin, not in the deep interior of the glacier.

With inputs from earlier and later, and connections upwards and downwards through the vertical structure of a glacier, debris can be in transport within a glacier at all levels. Supraglacially, debris can be carried passively on the ice surface or moved by water across the surface. Englacially, moving ice carries within it debris from the surface that has been buried by accumulation or that has been washed down into the englacial drainage system. The basal layer contains debris raised from the glacier bed. Debris is in transport at the ice-bed interface, and where there is a deformable subglacial sediment layer, debris can be mobile in there as well. The horizontal movement of ice, the flow of water, and the movement of deforming subglacial sediment all contribute to the horizontal debris flux. At the same time, the vertical connections allowing debris to move upwards or downwards through the layers of the glacier's thickness persist.

The conveyor belt's 'jerkiness' is due partly to longitudinal interruptions in transport pathways, some of which also switch on or off vertical connections between different layers. The same variations in basal conditions that lead to changes in entrainment of basal debris also affect deposition of material at the bed throughout the length of the glacier. Changes in temperature, pressure, basal topography and substrate composition can lead to differences in depositional processes along a flowline and over time. Melting of basal ice can release debris by melt-out to the bed. This debris may become part of a deforming subglacial layer or may be plastered down as 'lodgement till' beneath the glacier. Sometimes this might be a permanent resting place until the glacier eventually retreats back to expose the deposited material to postglacial or paraglacial activity, but sometimes, as conditions at the bed change over time, debris may be remobilised or eroded and re-entrained to continue its journey within the ice. That switching between frozen and unfrozen conditions also affects debris transport in the water within the glacier. Generally, water flow through glaciers, and transport of sediment within that water, occurs more quickly than ice flow, so the proportion of flux that occurs in water impacts the overall rate of transport.

Just as the Antarctic blue ice zones provided an example of unusual flow conditions that interrupted the typical pattern of vertical connections within a glacier's profile, there are specific situations that illustrate how the horizontal or longitudinal journey of material through a glacier can be interrupted. These include subglacial lakes and subglacial topographic overdeepenings.

More than 400 subglacial lakes have been identified, the majority beneath the Antarctic ice sheet. The largest is subglacial Lake Vostok. Estimates of its volume average over 5,000 km^3 (Li et al., 2019) making it the 6th largest lake on Earth. Some of the lakes appear to be isolated or disconnected from the rest of the system, and their water is trapped in long-term storage at the bed. Others appear to be hydrologically active: their levels rising and falling in response to flows in and out of the lake. Some, typically those closer to the edge of the ice sheet, appear to be recharged by water derived from the supraglacial zone, and to discharge water into the basal drainage system. Others appear to be connected only to the basal drainage system. Basal water flow is controlled much more by the ice surface slope than by basal topography, and it appears likely that the hydrological connectivity or isolation of subglacial lakes depends on the ice surface and hydraulic gradient, with higher gradients being more likely to drive water into and out from the lakes. Wright et al. (2014) found that substantial and long-lasting water storage in subglacial lakes in the Byrd Glacier catchment only occurred under very low hydraulic gradients. In the ice sheet interior where the surface gradient is very low, basal topography can play an important role but the hydraulic gradient will be sensitive to very small changes in surface topography. Elevation changes as small as 5 m on an ice sheet surface could be sufficient to produce significant changes to subglacial water flow paths (Wright et al., 2008), with consequences for the connectivity of subglacial lakes. Where subglacial lakes are connected to the basal hydrological system, they can form part of long distance transport networks. Wright et al. (2012) proposed that subglacial lakes formed part of a connected 1,000 km pathway for basal water stretching from close to the Antarctic ice divide at Dome C all the way to the ice sheet margin at the coast. Subglacial lakes can therefore be parts of a connected transport system or they can be long-term traps and stores, depending on their topographic and hydraulic conditions.

A potential obstacle to both hydrological and sediment connectivity along the base of a glacier is presented by subglacial topographic overdeepenings (Cook & Swift, 2012; Swift et al., 2021). These are closed topographic basins at the glacier bed, typically tens or hundreds of metres deep and hundreds or thousands of metres long. Overdeepenings are very common, and while some may have a tectonic or other non-glacial origin, most are caused by local variability in erosion rates associated with the geography of glacier bed processes. For example, they may occur where a tributary glacier joins a main glacier trunk, and many glaciers have 'terminal' overdeepenings a short way upglacier from their margin, where erosion rates are typically higher than further inland and higher also than at the very edge of the glacier. These overdeepenings pose an obstruction to the flow of ice, water and sediment because their downstream end offers an adverse slope for the flow to overcome. This adverse slope reduces drainage efficiency, increase basal sediment accumulation, and is associated with pressure-related supercooling that can lead to the freezing of water and debris to the bed even in areas that are not below the melting point.

The formation process for subglacial overdeepenings seems to involve feedbacks between ice, water and sediment. For example, as the overdeepening develops to the point where water flow is reduced by the adverse slope, more erosional debris will remain in the basin rather than being flushed out, a negative feedback limiting further erosion. This negative feedback is enhanced if basal water freezes as a result of the glaciohydraulic supercooling that has been associated with the pressure gradients on the downstream slopes of overdeepenings. The clogging of water pathways by the formation of ice through supercooling can limit sediment transfer through the basal system compared with situations where there is no overdeepening. It is also possible that water from the bed will be forced upwards into the englacial layer to follow higher-level hydrological pathways. Also, it has been suggested that compressive flow associated with both the adverse slope and the freezing on at the bed will cause folding and thrusting of the basal ice, raising basal debris to higher levels. Swift et al. (2018) argue that thrusting is probably the mechanism by which continuity of sediment transfer is maintained through overdeepenings, and subglacial overdeep-enings thus potentially impact both longitudinal and vertical flow paths, and are a potential sink or long-term storage point for subglacial sediment. Swift et al. (2021) propose that overdeepenings are a significant mediating factor in glacial sediment export from glaciated catchments.

The final stage of the glacier conveyor belt is the release or output stage. As with the other functional components of the system, this is not geographically constrained to specific parts of the glacier. Even at the summit of an ice sheet sur-face material can be lost by deflation, sublimation or other processes. Subglacial deposition to the bed or into subglacial lakes can occur at different points along the horizontal pathway through the glacier, and the terminus of a glacier is not the only place where deposition or release of glacial sediment occurs. However, if we are considering a longitudinal pathway through the glacier into the proglacial zone, then the glacier margin is an important transfer point, whether we think of it as a stage within an extended 'glacial' environment or as the boundary marking the end of the glacial zone and its transition into the extraglacial realm. It is also important to remember that glaciers are historical as well as geographical phenom-ena. The end of a glacier can be a point in time as well as a point in space, so a full model of connectivity in glacial environments needs to include not only longitudi-nal changes in the position of an ice margin, but also changes in the very existence of a glacier over time and the release of debris at the bed and surface, not only the terminus, of a collapsing ice sheet.

Material can be transferred out of the glacier front in different ways. Material can be carried away from the margin in icefalls or in icebergs (Figure 6.4). It can be flushed out in water, supraglacially in streams flowing over the ice and subgla-cially in basal meltwater. Material can be released and dumped by melting of ice,

Figure 6.4   Icebergs that calve from water-terminating glaciers can carry debris from the glacier into the proglacial environment. Photograph by Benjamin Alexander on Pixabay.

Figure 6.5 In this image, the edge of the Greenland ice sheet is on the left-hand side, advancing towards an ancient moraine ridge on the right (figure for scale). At this stage in the glacier's progress across the ridge, sediment released from the ice is likely to be trapped between the ice and the moraine. Photograph by P.G. Knight.

or squeezed out in a deforming subglacial sediment layer. Some of this material is immediately transferred away from the margin into the proglacial zone, for example in meltwater streams, or in icebergs at floating margins. However, a great deal of material can be deposited in marginal moraines. Moraines might be viewed as an endpoint for sediment at the edge of the glacier system, or as a storage point at the boundary between the glacial and proglacial systems. The extent to which moraines form, and debris is stored, depends to some extent on the extent of the glacier meltwater system. Meltwater flow is more likely to carry debris away from the glacier margin, so margins with less water and less developed drainage networks have greater potential for moraine formation.

As glacier margins advance and retreat, sediment flux into or through the moraine system varies over time. Knight et al. (2007) described the advance of an ice sheet margin in Greenland across a pre-existing ice-marginal moraine. Initially, with the margin behind the moraine, debris released from the glacier was predominantly held in storage between the ice margin and the moraine ridge ahead of it (Figure 6.5). As the ice advanced across the moraine, it was able to deposit material directly into the proglacial area on the distal side of the moraine. Also, as the glacier overrode the moraine it cannibalised and recycled debris from the moraine, remobilising debris from ice-marginal storage and generating a short-lived peak in debris flux as the debris being released from the ice was supplemented by debris being recycled during the period of moraine overtopping.

Most previous studies of sediment transfer through glacial environments treat the glacial and proglacial environments separately. For example, glacial sediments are treated as an external input in studies of proglacial and paraglacial sediment transfer. Therefore we might treat the glacier margin as the end of our conveyor belt. Whether that is the most useful way to think of glaciers in the context of connectivity is one of the many questions that glacier researchers should consider as they reflect on the value of a connectivity approach to glacial geomorphology. It raises the specific question of exactly where, given the reach of glaciers into the proglacial area, the limits of what Pöppl and Parsons (2018) call a 'geomorphic cell' might be in a glacial context.

6.4 Conclusion: Connectivity and Glacial Geomorphology

Glaciers are sediment transfer systems that at their fullest potential can deliver material from elevations 4 km above sea level in the centre of an ice sheet, along a flow path thousands of km in length, and deposit it into the ocean via calving icebergs. The system is connected vertically with transport pathways both upwards and downwards between supraglacial, englacial and basal layers. It is also connected longitudinally through process environments that are controlled by factors such as temperature and that vary over short and long time scales to generate a dynamic geography of glacial processes. That potential long distance transfer system is one extreme of a range of scales at which glacier systems operate. At the other end of the scale are glaciers just hundreds of metres in length that terminate on land very close to the source area of even the farthest-travelled of their debris load. At all scales, the pathway between initial entrainment and ultimate release can be deflected along alternative routes (englacial, subglacial, etc.) and can be placed into short or long-term stores at many different points along the pathway (subglacial lakes, supraglacial blue ice sublimation zones, marginal moraines, etc.). Even delimiting the boundaries of what we call glacial environments is controversial. Glacier boundaries migrate over time, and glaciers have direct impacts on geomorphic processes geographically beyond the ice margin and long after their historical end point.

Glacier systems are riddled with inputs, outputs, connections, transfers and storage points, and there is a long history of research that considers glaciers and their landscapes in that context. However, it is rare for glaciologists or glacial geomorphologists to talk specifically about connectivity theory, and connectivity has not entered into the language of glacial landscape studies in the way that it has done in other parts of geomorphology. Nevertheless, the examples described in this chapter indicate that there is certainly potential for a much greater and more specific engagement with connectivity theory in glacial geomorphology. However, there are several important questions that will need to be addressed, not least the question of exactly what are the boundaries of the glacier system when we look for internal and external connections.

Pöppl and Parsons (2018) point out that discussion of connectivity within a system requires a recognition of the units or entities between which the connections exist. In other words, we need to define fundamental geomorphic units for which we can identify connections with neighbouring geomorphic units. Pöppl and Parsons propose the 'geomorphic cell' as such a unit, and this might be a useful starting point for engaging with how glacial geomorphology could align itself with the approaches to connectivity used elsewhere. The cell that Pöppl and Parsons envisage, at the largest scale we might imagine it, accommodates the description of a glacier that we used at the start of this chapter. Their cell stretches from the atmosphere at the top to bedrock at the bottom, and it is easy to fit the layers of extraglacial, supraglacial, and so forth, that we used in our earlier description of glaciers, into that model. The cell described by Pöppl and Parsons has internal vertical connections between layers as we described here for glaciers, and also has horizontal connections to adjacent cells. Here we face the problem of deciding on the lateral scale of a geomorphic cell in a glacial context. We might follow Boulton's (2006) assertion that glaciers are connected into the global environmental system via atmosphere, lithosphere and ocean, in which case we could envisage atmosphere and lithosphere as the top and bottom of a cell, ocean as an adjacent downstream cell, and the whole glacier as a single cell within a planetary scale geomorphic model. To work at smaller scales within the glacier system using this same approach, we would need to address the problem of differentiating longitudinally between glacial process cells, or defining horizontal boundaries for the cells. Given the complexity of the glacial environment, this is an exciting prospect for future work.

References

Ali, G. A., & Roy, A. G. (2009). Revisiting hydrologic sampling strategies for an accurate assessment of hydrologic connectivity in humid temperate systems. *Geography Compass*, 3(1), 350–374. https://doi.org/10.1111/j.1749-8198.2008.00180.x

Alley, R. B. (2000). *The Two-Mile Time Machine*. Princeton: Princeton University Press.

Altmann, M., Piermattei, L., Haas, F., Heckmann, T., Fleischer, F., Rom, J., Betz-Nutz, S., Knoflach, B., Müller, S., Ramskogler, K., Pfeiffer, M., Hofmeister, F., Ressl, C., & Becht, M. (2020). Long-term changes of morphodynamics on little ice age lateral moraines and the resulting sediment transfer into mountain streams in the Upper Kauner Valley, Austria. *Water*, 12(12):3375. https://doi.org/10.3390/w12123375

Bernhardt, A., Schwanghart, W., Hebbeln, D., Stuut, J. W., & Strecker, M. R. (2017). Immediate propagation of deglacial environmental change to deep-marine turbidite systems along the Chile convergent margin. *Earth and Planetary Science Letters*, 473, 190–204. https://doi.org/10.1016/j.epsl.2017.05.017

Bitanja, R. (1999). On the glaciological, meteorological and climatological significance of Antarctic blue ice areas. *Reviews of Geophysics*, 37(3), 337–359.

Boulton, G. S. (2006). Glaciers and their coupling with hydraulic and sedimentary processes. In P. G. Knight, ed., *Glacier Science and Environmental Change*. Oxford: Blackwell, pp. 3–22.

Bracken, L. J., Wainwright, J., Ali, G. A., Tetzlaff, D., Smith, M. W., Reaney, S. M., & Roy, A. G. (2013). Concepts of hydrological connectivity: Research approaches, pathways and future agendas. *Earth-Science Reviews*, 119, 17-34. https://doi.org/10.1016/j.earscirev.2013.02.001

Bracken, L. J., Turnbull, L., Wainwright, J., & Bogaart, P. (2015). Sediment connectivity: A framework for understanding sediment transfer at multiple scales. *Earth Surface Processes and Landforms*, 40(2), 177–188.

Caine, N. (1986). Sediment movement and storage on alpine slopes in the Colorado Rocky Mountains. In A. D. Abrahams, ed., *Hillslope Processes*. London: Allen & Unwin, pp. 115–137.

Collins, D. (1979). Quantitative Determination of the Subglacial Hydrology of Two Alpine Glaciers. *Journal of Glaciology*, 23(89), 347–362. https://doi.org/10.3189/S0022143000029956

Cook, S. J., & Swift, D. A. (2012). Subglacial basins: Their origin and importance in glacial systems and landscapes. *Earth Science Reviews*, 115(4), 332–372. http://doi.org/10.1016/j.earscirev.2012.09.009

Etzelmüller, B., Ødegard, R. S., Vatne, G., Mysterud, R. S., Tonning, T., and Sollid, J. L. (2000). Glacier characteristics and sediment transfer system of Longyearbreen and

Larsbreen, western Spitsbergen. *Norsk Geografisk Tidsskrift–Norwegian Journal of Geography*, 54, 157–168. https://doi.org/10.1080/002919500448530

Evatt, G. W., Coughlan, M. J., Joy, K. H., Smedley, A. R. D., Connolly, P. J., & Abrahams, I. D. (2016). A potential hidden layer of meteorites below the ice surface of Antarctica. *Nature Communications*, 7, 10679. https://doi.org/10.1038/ncomms10679

Ferguson, R. (1981). Channel forms and channel changes. In J. Lewin, ed., *British Rivers*. London: Allen & Unwin, pp. 90–125.

Hassan, M. A., Bird, S., Reid, D., Ferrer-Boix, C., Hogan, D., Brardinoni, F., & Chartrand, S. (2019). Variable hillslope-channel coupling and channel characteristics of forested mountain streams in glaciated landscapes. *Earth Surface Processes and Landforms*, 44, 736–751. https://doi.org/10.1002/esp.4527

Jaeger, J. M., & Koppes, M. N. (2016). The role of the cryosphere in source-to-sink systems. *Earth-Science Reviews*, 153, 43–76. https://doi.org/10.1016/j.earscirev.2015.09.011

Knight, P. G. (1997). The basal ice layer of glaciers and ice sheets. *Quaternary Science Reviews*, 16, 975–993. https://doi.org/10.1016/S0277-3791(97)00033-4

Knight, P. G. (1999). *Glaciers*. Cheltenham: Stanley Thornes.

Knight, P. G., Jennings, C. E., Waller, R. I., & Robinson, Z. P. (2007). Changes in ice-margin processes and sediment routing during ice-sheet advance across a marginal moraine. *Geografiska Annaler: Series A, Physical Geography*, 89(3), 203–215. https://doi.org/10.1111/j.1468-0459.2007.00319.x

Kummert, M., & Delaloye, R. (2018). Mapping and quantifying sediment transfer between the front of rapidly moving rock glaciers and torrential gullies. *Geomorphology*, 309, 60–76. https://doi.org/10.1016/j.geomorph.2018.02.021

Lane, S. N., Bakker, M., Gabbud, C., Micheletti, N., & Saugi, G. (2017). Sediment export, transient landscape response and catchment-scale connectivity following rapid climate warming and Alpine glacier recession. *Geomorphology*, 277, 210–227. https://doi.org/10.1016/j.geomorph.2016.02.015

Li, Y., Lu, Y., Zhang, Z., Shi, H., & Xi, H. (2019). Characterizing three-dimensional features of Antarctic subglacial lakes from the inversion of hydraulic potential – Lake Vostok as a case study. *Advances in Polar Science*, 30, 70–75. https://doi.org/10.13679/j.advps.2019.1.00070

MacDonell, S., Sharp, M., & Fitzsimons, S. (2016). Cryoconite hole connectivity on the Wright Lower Glacier, McMurdo Dry Valleys, Antarctica. *Journal of Glaciology*, 62(234), 714–724. https://doi.org/10.1017/jog.2016.62

Mancini, D., & Lane, S. N. (2020). Changes in sediment connectivity following glacial debuttressing in an Alpine valley system. *Geomorphology*, 352, 106987. https://doi.org/10.1016/j.geomorph.2019.106987

Miles, E. S., Steiner, J., Willis, I., Buri, P., Immerzeel, W. W., Chesnokova, A., & Pellicciotti, F. (2017). Pond Dynamics and Supraglacial-Englacial Connectivity on Debris-Covered Lirung Glacier, Nepal. *Frontiers in Earth Science*, 5(69). https://doi.org/10.3389/feart.2017.00069

Piotrowski, J. A. (2006). Groundwater under ice sheets and glaciers. In P. G. Knight, ed., *Glacier Science and Environmental Change*. Oxford: Blackwell.

Pöppl, R. E., & Parsons, A. J. (2018). The geomorphic cell: A basis for studying connectivity. *Earth Surface Processes and Landforms*, 34, 1155–1159. https://doi.org/10.1002/esp.4300

Porter P., Smart M., Irvine-Fynn T. D. L. (2019). Glacial sediment stores and their reworking. In: T. Heckmann & D. Morche eds., *Geomorphology of Proglacial Systems*. Geography of the Physical Environment. Cham: Springer. https://doi.org/10.1007/978-3-319-94184-4_10

Small, R., Beecroft, I., & Stirling, D. (1984). Rates of deposition on Lateral Moraine Embankments, Glacier De Tsidjiore Nouve, Valais, Switzerland. *Journal of Glaciology*, 30(106), 275–281. https://doi.org/10.3189/S0022143000006092

Small, R. J. (1987). Moraine sediment budgets. In A. M. Gurnell & M. J. Clark, eds., *Glacio-Fluvial Sediment Transfer – An Alpine Perspective*. Chichester: John Wiley and Sons, pp. 165–197.

Stevens, I., Irvine-Fynn, T., Porter, P. R., Cook, J., Edwards, A., Smart, M., Moorman, B., Hodson, A., & Mitchell, A. (2018). Near-surface hydraulic conductivity of Northern Hemisphere glaciers. *Hydrological Processes*, 32(7), 850–865. https://doi .org/10.1002/hyp.11439

Stocker-Waldhuber M., Kuhn M. (2019). Closing the balances of ice, water and sediment fluxes through the terminus of gepatschferner. In T. Heckmann and D. Morche, eds. *Geomorphology of Proglacial Systems*. Geography of the Physical Environment. Cham: Springer. https://doi.org/10.1007/978-3-319-94184-4_5

Swift, D. A., Cook, S. J., Graham, D., Midgley, N., Fallick, A. E., Storrar, R., Toubes Rodrigo, M., & Evans, D. (2018). Terminal zone glacial sediment transfer at a temperate overdeepened glacier system. *Quaternary Science Reviews*, 180, 111–131. https://doi.org/10.1016/j.quascirev.2017.11.027

Swift, D. A., Tallentire, G. D., Farinotti, D., Cook, S. J., Higson, W. J., & Bryant, R. G. (2021). The hydrology of glacier-bed overdeepenings: Sediment transport mechanics, drainage system morphology, and geomorphological implications. *Earth Surface Processes and Landforms*, 46, 1–15. https://doi.org/10.1002/esp.5173

Toubes-Rodrigo, M., Potgieter-Vermaak, S., Sen, R., Oddsdottir, E. S, Elliott, D., & Cook, S. (2021). Active microbial ecosystem in glacier basal ice fuelled by iron and silicate comminution-derived hydrogen. *Microbiology Open*, 10(4), e1200. https://doi .org/10.1002/mbo3.1200

Warburton, J. (1990). An alpine proglacial fluvial sediment budget. *Geografiska Annaler Series A, Physical Geography*, 72(3/4), 261–272. https://doi.org/10.2307/521154

Wright, A. P., Siegert, M. J., Le Brocq, A. M., & Gore, D. B. (2008). High sensitivity of subglacial hydrological pathways in Antarctica to small ice-sheet changes. *Geophysical Research Letters*, 35, L17504. https://doi.org/10.1029/2008GL034937

Wright, A. P., Young, D. A., Roberts, J. L., Schroeder, D. M., Bamber, J. L., Dowdeswell, J. A., Young, N. W., Le Brocq, A. M., Warner, R. C., Payne, A. J., Blankenship, D. D., van Ommen, T. D., & Siegert, M. J. (2012). Evidence of a hydrological connection between the ice divide and ice sheet margin in the Aurora Subglacial Basin, East Antarctica. *Journal of Geophysical Research Earth Surface*, 117(F1), F01033. https://doi.org/10.1029/2011JF002066

Wright, A., Young, D., Bamber, J., Dowdeswell, J., Payne, A., Blankenship, D., & Siegert, M. (2014). Subglacial hydrological connectivity within the Byrd Glacier catchment, East Antarctica. *Journal of Glaciology*, 60(220), 345–352. https://doi .org/10.3189/2014JoG13J014

7

Periglacial Processes

MATHILDE BAYENS AND STUART N. LANE

This chapter focuses on periglacial processes in mountain environments, and the effects of connectivity in these environments that result from climate change. 'Periglacial' was first used to describe the angular rock-rubble surfaces that characterize mountain summits formed by the previous action of intense frost (Łoziński, 1909). More generally it is used to describe cold non-glacial environments characterized by frost action (French, 2015), typically close to glacierized regions and so often transitional between glaciated and deglaciated environments. Therefore periglacial regions and processes are particularly affected by contemporary climate change.

Climate warming is leading to reduced snow accumulation, glacier recession (Micheletti & Lane, 2016), permafrost degradation and a subsequent reduction of hillslope stability (Haeberli et al., 2017). These processes may impact sediment supply (Stott & Mount, 2007; Hirschberg et al., 2021) and sediment connectivity significantly (Lane et al., 2017; Mancini & Lane, 2020), and hence sediment flux to catchment outlets. Underlying this changing connectivity are both changes in the landscape itself (e.g., the development of gullies) but also changes in sediment transport capacity because climate change may also modify hydrological regimes and hence sediment mobilization and transport (Stahl et al., 2008; Weber et al., 2010; Lane et al., 2017; Rainato et al., 2018).

Over the next few decades, ongoing global glacier melt and recession (IPCC, 2021) will transform many glaciated mountain environments into recently deglaciated terrain, comprising debris, a developing vegetation cover, lakes and steep slopes with slowly degrading permafrost (Stoffel & Huggel, 2012; Haeberli et al., 2017; Huss et al., 2017; Shugar et al., 2020). This increased area of deglaciated terrain will interact with changing magnitude and frequency of sediment-mobilizing extreme events. During the twentieth century in all regions of Switzerland, as with other Alpine regions, an increase in the number of extreme rainfall events (Schmidli et al., 2002; Schmidli and Frei, 2005), snowmelt and summer runoff in glaciated basins (Zappa & Kan, 2007) has already been observed. These trigger mechanisms can contribute to a changing frequency of sediment mobilization by hillslope processes, such as debris flow events (Rebetez et al., 1997; Hirschberg et al., 2021). With future climate change, it is not only the frequency of those events that is likely to change but also the seasonality: for example of moderate and extreme precipitation events (Brönnimann et al., 2018), and reduction of winter snow accumulation leading to an earlier onset of snow-free conditions (Gobiet & Kotlarski, 2020; IPCC, 2021). Equally, even if the overall frequency of debris flows events remains low, magnitude may increase if larger volumes of sediment are delivered to the channels (Stoffel et al., 2014; Hirschberg et al., 2021). There is emerging evidence that these changes can be seen in increasing sediment yield from mountain basins. Micheletti and Lane (2016) and Lane et al. (2017) report the onset of much high bedload sediment yield from the 1980s in three Swiss glaciated catchments. Costa et al. (2018) reported a similar trend for suspended sediment yield for the Alpine Rhône catchment (5,400 km^2 in area) and it was possible to note a marked increase in deposition rates in the Rhône delta in Lake Geneva as a result (Lane et al., 2019).

What is perhaps less frequently considered in periglacial environments is the role played by connectivity as an influence upon the ways in which the signals of changing climate propagate through river catchments. Sediment transport is not simply a consequence of climate change but, through erosion and deposition, leads to modification of the landscape that in turn influences the ease with which sediment can move through the landscape, that is the degree of connectivity. For instance, glacier recession leads to sidewall debuttressing and local base-level falls which can lead to quite intense erosional processes on sidewalls, notably gully development. Gully development may then extend headwards and dissect the topographic legacies of glacial advance such as lateral moraines which other- wise cause local disconnection (Cossart, 2008) so increasing potential connectivity from upstream to downstream (Lane et al., 2017; Mancini et Lane, 2020).

Connectivity can also evolve in the opposite sense. If increases in connectivity due to gullying translate into increased sediment flux from upstream to downstream, then there may be downstream consequences that lead to disconnection due to sediment deposition. A good example is the formation of alluvial fans at the base of hillslopes, their diffusive surface drainage networks slowing sediment flux to the valley bottom and increasing the tendency to disconnection (Lane et al., 2017; Mancini et Lane, 2020). As a second example, glacial retreat can lead to proglacial lake formation which traps sediment (Geilhausen et al., 2012; Stoffel & Huggel, 2012; Otto, 2019) so decreasing sediment connectivity in glacial catchments (Carrivick & Tweed, 2013; Cossart, 2008; Cossart & Fressard, 2017; Heckmann & Schwanghart, 2013). However, high-magnitude but low-frequency events may lead to localized erosion of lake margins that re-establishes connection, most commonly manifest as glacial lake outburst floods (Marren & Toomath, 2014; Westoby et al., 2014; Worni et al., 2014).

Two important points follow from this introduction: (1) in periglacial environments, connectivity is dynamic, evolving in response to the sediment transport processes it, itself, induces; and (2) connection and disconnection have to be viewed as relative and multi-variate concepts. For most of the time, a landscape is functionally disconnected; sediment does not move. When it does move, then there are locations where it is more likely to be able to move further downstream. This likelihood can be quantified and locations compared in terms of static metrics of (potential) connectivity. This is the basis of an ergodic hypotheses that zones that are more (or less) connected in space are more (or less) frequently connected in time; and it then allows analysis of the spatial patterns of relative structural connectivity as a basis of inferring which zones of a landscape are most able to connect and flux sediment most often. However, because such sediment flux (i.e., functional connectivity) may cause landscape changes that in turn change connection, such static structural representations of connectivity also need to be considered as non-stationary.

In this chapter, we illustrate these points with reference to examples from periglacial zones in deglaciated landscapes. To set this discussion in wider context we consider sediment yield in deglaciating landscapes in relation to what is known as the paraglacial model (Church & Ryder, 1972), and the relationship between this model and evolving connectivity. Using examples, we then: (1) illustrate the spatial variability of the functional connectivity that is typical of the periglacial and proglacial zones of mountain landscapes; (2) consider how the structural connectivity (see Sections 1.2 and 2.4.2) interact with the processes that drive sediment flux; and (3) discuss the ways in which sediment flux (resulting from functional connectivity, see Sections 2.4.2 and 2.4.3) can lead to evolution of the structural connectivity. Our review is illustrated with examples from the Arolla and Ferpecle-Mont Miné Valleys, both located in the Val d'Hérens of Canton Valais, south-west Switzerland.

7.1 Connectivity in the Framework of the Paraglacial Model of Sediment Yield Following Glacier Recession

In a classic paper Church and Ryder (1972) proposed a paraglacial model for how sediment yield evolves in deglaciating landscapes. Whilst it was developed to deal with the interpretation of the legacy of Quaternary deglaciation, it provides a useful frame for thinking about how deglaciated landscapes might evolve over future decades and centuries as we undergo human-induced, rapid climate change in the twenty-first century. The basic assumption of the model is that as a glacier retreats, hillslopes become debuttressed and are oversteepened. These freshly deglaciated zones have significant quantities of poorly-sorted sediment, so providing extensive potential sediment sources especially as they are commonly unvegetated (e.g., Ballantyne, 2002a, 2002b; Carrivick & Heckmann, 2017; Cossart et al., 2018;

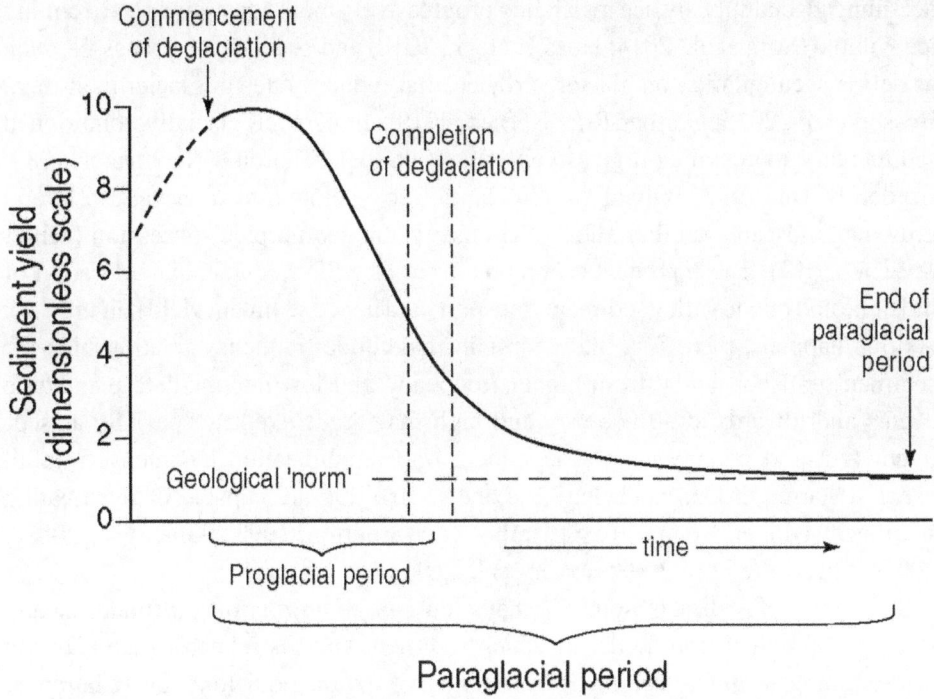

Figure 7.1   Sediment yield following deglaciation and the subsequent paraglacial period (proposed by Church and Ryder (1972) and modified by Ballantyne (2002b)).

Porter et al., 2019; Mancini & Lane, 2020). Whilst there may be landforms in the landscape (e.g., lateral moraines) that can interrupt downslope sediment flux, steep sediment slopes allow for the transfer of considerable quantities of sediment to the valley bottom under the combined action of gravity and water. There, delivered sediment may meet a substantial meltwater subsidy associated with melting glaciers. As sediment transport capacity is a non-linear function of water discharge, with coefficients up to 2–4 (Alley et al., 1997, 2019; Swift et al., 2005; Antoniazza & Lane, 2021) it becomes possible to evacuate an important quantity of valley-delivered sediment (e.g., Beylich et al., 2009, 2017; Bogen, 2010; Evans et al., 2012; Lane et al., 2017). The result is a rapid rise in sediment yield in deglaciating catchments (Figure 7.1; Church & Ryder, 1972; Bratlie, 1994; Ballantyne, 2002a, 2002b; Delmas et al., 2009; Koppes & Montgomery, 2009). Such a trend is confirmed in the very few decadal scale studies of sediment export from deglaciating catchments for both suspended sediment yield (Costa et al., 2018; Lane et al., 2019) and bedload yield (Micheletti & Lane, 2016 Lane et al., 2017).

This pattern is later reversed (Figure 7.1) by two important sets of processes. The first relates to the progressive decline in sediment transport capacity as a result of a declining glacial subsidy, ice melt being progressively more constrained by declining ice volume (Sorg et al., 2014; Huss & Hock, 2018) and/or feedback processes, such as debris accumulation on glacier surfaces, that reduces rates of glacier melt (e.g., Bosson et al., 2015; Gärtner-Roer & Bast, 2019). In parallel, glacially-conditioned sedimentary sources are progressively exhausted (e.g., Church & Slaymaker, 1989; Cruden & Hu, 1993; Ballantyne, 2002a, 2002b; Antoniazza & Lane, 2021) and growing landscape stability allows the onset of biogeomorphic succession (Miller & Lane, 2019). These processes combine to reduce the frequency and duration of hillslope and down-valley sediment transport, and hence sediment yield (Figure 7.1). As this happens, there is a likely shift in magnitude-frequency relationships for sediment transport away from higher frequency and low/intermediate magnitude events and towards low-frequency and high-magnitude events, especially as sediment transport becomes more dependent upon mobilization by intense rainfall events (Marren, 2005; Micheletti & Lane, 2016) that are capable of reactivating stabilized sediment sources (e.g., Harbor & Warburton, 1993; Ballantyne, 2002a, 2002b; Cossart & Fort, 2008; Porter et al., 2019).

The result of sediment source exhaustion can be approximated (under steady state) by an exhaustion model in which sediment yield is related to the addition of remaining available sediment by a negative exponential function (Church & Ryder, 1972; Church & Slaymaker, 1989; Ballantyne, 2002b; Figure 7.1). This function is asymptotic on the primary regional denudation rate of the specific land surface (called the geological 'norm'; Church & Ryder, 1972). It has been suggested that this model is only applicable to the primary paraglacial system in small Alpine (upland) basins characterized by primary sediment transport (Church & Slaymaker, 1989; Harbor & Warburton, 1993). In larger basins, sediment yield is firstly conditioned by the processes typical of the primary paraglacial system but then modified strongly by secondary reworking involving the re-entrainment of sediment stored in primary paraglacial sediment accumulation zones. Such secondary paraglacial reworking of sediment includes, for example, entrenchment of paraglacial alluvial fans, talus cones and paraglacial valleys filled by trunk channels (Church & Ryder, 1972; Jackson Jr. et al., 1982; Brooks, 1994). Secondary paraglacial sediment transport makes the evolution of sediment yield more complex because it potentially influences the sediment budget of trunk streams. Reworked by trunk streams of large catchments, sediment yield rises during the early postglacial period until reaching a peak several millennia after deglaciation whilst paraglacial sediment transport in upland valleys continues to decline (Figure 7.2). After the peak, both primary and secondary sediment sources are progressively depleted (Church & Slaymaker, 1989). Based on a positive relationship between specific sediment yield and catchment area between 10 and ca. 30,000 km^2, the delay of the peak can be explained. The peak is reached earlier (and immediately after

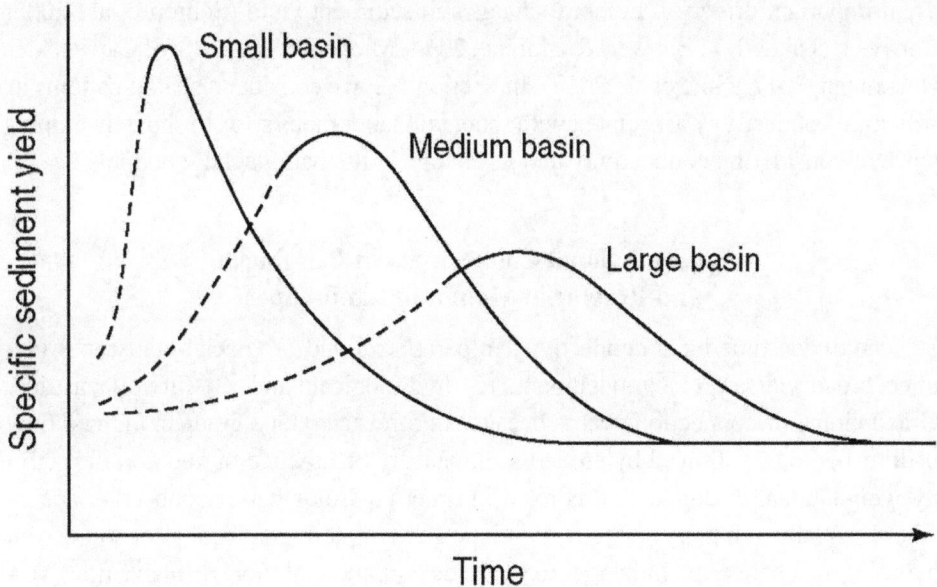

Figure 7.2   The effect of the spatial scale on the temporal sediment yield evolution
as modified by Church and Slaymaker (1989) and presented by/in Ballantyne
(2002b). The time axis spans is approximately 10,000 years.

deglaciation) in smaller than in larger basins. The amplitude of the peak also varies
with catchment size, decreasing as catchment size increases (Figure 7.2) (Harbor
& Warburton, 1993; Ballantyne, 2002b). The relationship between the time delay
of the peak and the catchment size may also reflect the response of fluvial transport, evolution of sediment sources and changes following extreme rainfall events.

Application of the paraglacial model to landscapes that have deglaciated since
the Quaternary has emphasized that it is extreme rainfall events that move sediment of primary and secondary glacigenic origin from steep tributaries to the trunk
valleys, but the downstream sediment travel time can be prolonged by periods of
temporary storage in debris cones, alluvial fans or valley-fills and low-slope river
terraces for example (Ballantyne, 2002b). This temporary storage is likely to be a
feature of sediment flux during the rapid deglaciation we are currently witnessing.
The probability of having such temporary storage areas increases as catchment
size increases. By analogy, larger basins have potentially more sediment available
(due to greater relief and so sediment contributions from slope failure, debris flows
and other hillslope processes) and so also have a higher probability of having more
sources of reworked sediment (Ballantyne, 2002b).

In these analyses, Church and Ryder (1972) (and others) do not specifically consider how connectivity might impact the paraglacial model nor how that connectivity might evolve through time, although this evolution is now well-acknowledged

as an important driver of temporal changes in sediment yield (Schrott et al., 2006; Carrivick et al., 2013; Baewert & Morche, 2014; Micheletti et al., 2015; Carrivick & Heckmann, 2017; Lane et al., 2017). In Section 7.2, we consider a spatial patterns of structural connectivity associated with mountain landscapes so as to show the importance of considering connectivity and its change in the paraglacial model.

7.2 Functional Connectivity in Periglacial and Proglacial Mountain Landscapes

We can divide functional connectivity in periglacial and proglacial landscapes into three broad classes: (1) gravitational; (2) hydrological; and (3) fluvio-lacustrine. Gravitational disconnection occurs because of a reverse-slope gradient along a flow path that is to be followed by eroded sediment. In the absence of geomorphic activity, gravitational disconnection is fixed in time. Hydrological disconnection occurs where, even though there is no reverse-slope gradient, topographically- or soil-driven hydrological changes on hillslopes (e.g., the development of more diffusive flow; vertical infiltration of water) reduce the capacity to transport sediment. Hydrological disconnection is not fixed in time; that is in more extreme rain events, sufficient capacity may still be achieved to move sediment. Fluvio-lacustrine disconnection is similar to hydrological disconnection but relates to situations where topographic slope is low and sediment transport becomes controlled by fluvial or lacustrine processes. Fluvial disconnection is associated with hydraulic processes (see Chapter 2). Hydraulic and lacustrine controls of sediment transport involve thresholds and so the evolution of these controls in space (e.g., the transition from a straight channel to a braided one) or time (e.g., a reduction in discharge) may lead to a change in sediment connectivity.

As with other environments, landforms in Alpine regions play a role in controlling sediment connectivity across different spatial scales. In periglacial and proglacial mountain environments at high altitudes, most landforms are associated with relatively limited vegetation cover and direct anthropogenic influence on topography. Table 7.1 lists examples of such landforms which are characterized according to how they influence sediment connectivity (Heckmann et al., 2019). Table 7.1 also contains brief details about their formation or composition, type of functional connectivity (defined earlier), role in sediment connectivity and examples from the Arolla (Figure 7.3) and Ferpecle-Mont Miné (Figure 7.4) valleys on which the index of connectivity proposed by Cavalli et al. (2013) has been applied.

Landform source defines those zones of a watershed producing and supplying sediment; store/sink landforms are zones of short- or long-term sediment storage; links comprise those landforms able to connect sediment sources to sediment sinks in the sediment cascade; buffers are temporary sediment storage zones that slow sediment flux; and barriers are features that sediment transport pathways.

Table 7.1   *Natural landforms and their role(s) in sediment connectivity in proglacial environments*

Landform	Formation–composition	Type of functional connectivity	Rôle in sediment connectivity	Example from the Arolla or Ferpècle catchment located in the Val d'Hérens in the South Western part of the Swiss Alps
Moraine	Steep or even overhanging slopes commonly covered by a thick mantle of unvegetated **glaciogenic sediment deposits leave by glacier retreat** (Mattson & Gardner, 1991; Ballantyne & Benn, 1994, 1996).	Gravitational process	Barrier due to sediment deposition. Moraines are clear example of a gravitational disconnection formed by past glacial advance.	LIA moraines disconnecting upstream basins from the downstream proglacial margin (Figure 7.3 and 7.4).
Gully system	Sediment **moraine deposits are susceptible to erosion** associated with slope failure, debris flows, snow avalanches, tributary stream activity and surface wash. Debris flows (the dominant agent of sediment reworking in paraglacial environments) may begin the gullying of drift slopes after glacier retreat, a process that may continue to bedrock. Paraglacial gully erosion range from a few millimeters to tens of centimetres per year, and gully systems and cones stabilize within a few decades or centuries (Ballantyne, 2002b).	Gravitational process	Source due to sediment erosion. Barrier if the system is stabilized	These landforms have developed since the Little Ice Age (LIA) and most rapidly since ice retreated through the current proglacial margin since 1970 (Lane et al., 2017) for the Bas and the Haut Glacier d'Arolla systems (Figure 7.3). The retreat of the glacier left a clearly marked **lateral moraines.** This barrier then evolves as the moraine is eroded by gullying (due to ice retreat and notably a base-level fall) and so acts as a sediment source. Gully dissection of the LIA moraine leads to the potential connection of hillslope sediment sources upstream of the moraine (Figure 7.3 location 3).

Table 7.1   (cont.)

Landform	Formation–composition	Type of functional connectivity	Rôle in sediment connectivity	Example from the Arolla or Ferpêcle catchment located in the Val d'Hérens in the South Western part of the Swiss Alps
Scree slope/ cone	**Landform deposits** create in the gullying process from coarse sediments accumulate at the base of the hillslopes. Scree or debris cone is **moderate steep (12–25°) accumulations** of poorly sorted sediment deposited rapidly during repeated debris flows.	Gravitational process	Source, link or buffer depending of the steepness of the landform and so capacity to transport sediments downstream. Buffer cone is temporary sediment storage zones that slow sediment flux.	
Debris/ Alluvial fan	**Alluvial fans (Figure 7.3) form in the gully process with sediment accumulation** at the base of hillslopes. These valley floor deposits are frequently form gentler slopes (generally <15°, often <5°) and commonly involving fluvial sediment sorting (Ballantyne, 2002b).	Gravitational and hydrological process	Fans landforms are diffusive, and so act as a buffer against sediment flux from hillslopes to the valley floor (Lane et al., 2017). This is a form of hydrological disconnection because the **fan** slopes are rarely steep enough to move sediment without the aid of water. As flow of water on the hillslope tends to be diffusive, **fan** development serves as positive feedback, aiding fan growth; but fan progradation across the valley floor may increase the probability that the **fan** reaches the proglacial stream, where sediment transport rates are high. Thus, the initial disconnection associated with **fan** development may be reversed, leading to an increase in connection (Figure 7.3).	Alluvial fan down the Bertol torrent (Figure 7.3)

| Rock glacier | Rock glaciers are composed of rocks with a heterogeneous size distribution which, under a few meters of non-permanently frozen rock debris (i.e., the active layer), are cemented by interstitial ice (Kummert & Delaloye, 2018). **They are generated by creep of frozen ground and convey debris from an upslope area** (source area or rooting zone) towards their front (Delaloye & Morard, 2019). | Gravitational process | Depending on their location with respect to sources of sediment and to the fluvial system, rock-glaciers can act as sediment sources, buffers or barriers.

– sources of sediment, notably at their fronts if located below rock slopes and steep headwater channels (Lugon, 2010; Kummert & Delaloye, 2018; Kummert et al., 2018).

– can be class as buffers rather than barriers if they move, albeit slowly. The deformation of ice drives the downslope movement of a rock glacier. This process means that their connection is gravitational in nature. Active rock glaciers move at low speeds, that is, typically 0.1–2 m yr^{-1} (Delaloye & Morard, 2019). There is now widespread evidence of increased rock glacier flow velocities in regions with retreating glaciers due to climate warming (e.g., Lugon, 2010; Micheletti et al., 2015; Kenner et al., 2018; Kummert et al., 2018) with **destabilized rock glaciers** displaying velocities of up to 10 m per year in the Alps (Delaloye et al., 2007; Lambiel et al., 2008). Such increases will also increase sediment delivery. | In the Arolla catchment, the La Roussette rock glacier (Figure 7.5 location 4, Figure 7.6; Micheletti et al., 2015) recently showed an enhancement of sediment production at the front. It is composed of five lobes and three zones of very different activity: the central depression, the feeding zone and a lateral sedimentary input zone (Figure 7.6). The lobe 3 is the most active with the presence of a front with an average slope of 42°, in contrast to the lobe 1 and 4 which are lightly vegetated (Meyrat, 2018). Nevertheless, the absence of a channel at its front does not allow sediment transfer to the valley floor. This hydrological disconnection shows that climate change does not always lead to enhanced sediment delivery from rock glaciers; it is strongly influenced by connectivity. In this case, the rock glacier is currently a sediment sink which prevents upstream sediment flux to the downstream fluvial system and so the valley bottom (Micheletti et al., 2015) (Figure 7.5 location 4). |

Table 7.1   (cont.)

Landform	Formation–composition	Type of functional connectivity	Rôle in sediment connectivity	Example from the Arolla or Ferpècle catchment located in the Val d'Hérens in the South Western part of the Swiss Alps
			– characterized as barrier when in some cases, boulders are delivered to the front of the rock glacier and then deposited into a narrow channel because of limited water supply. When the sediment leaves the rock glacier, its further transfer may still be limited by sediment transport capacity and so the signal of increased warming on rock glacier sediment flux may not always propagate to downstream alluvial fans and/or the valley bottom	
Protalus Rampart	Description of **discrete talus and finer debris accumulations from rockfalls and avalanches** (Whalley, 2003). These ridges (or ramps) of eroded debris and talus (Figure 7.7) are commonly found in periglacial environments at the base of very steep (60–90°) rocky slopes with sufficient sediment supply and are almost oriented transverse to slopes (Ballantyne & Kirkbride, 1986; Arfstrom, 2003; Shakesby, 2004; Hedding, 2011).	Gravitational process	Buffer or source of sediments coming from rocky slope	

			Store/sink Buffer	
Proglacial plain Valley floor	Over time, river channel of proglacial plain changes are consequences of erosion and deposition processes, which are themselves explained by both exo- or allogenic drivers (e.g., changes in discharge or sediment supply; Lane et al., 1996) and autogenic processes such as diverging decelerating flow and converging accelerating flow and sediment sorting (Ashworth & Ferguson, 1986).	Fluvio-lacustrine process	**Alluvial plains and valley floor zones** illustrate both stores, sinks and buffers. These zones involve fluvio-lacustrine disconnections typically with very low slopes and where, as a result, sediment movement is controlled by transport capacity.	The floodplain of the Ferpècle glacier in the Val d'Herens (Switzerland) illustrates both major types of fluvio-lacustrine disconnections: braided rivers and proglacial lakes (Figure 7.4 and 7.8). More examples: Figure 7.3 location 5 and Figure 7.4 location 7
Lake	**Lake** formation is another **consequence of high glacier meltwater discharges in such environments.** Lakes are formed by glacial retreat, generally located in and may be formed due to over-deepening as a consequence of glacial erosion and/or due to barriers, such as terminal moraines, bedrock, ice or landslide debris. When a glacier retreats, ice eventually decouples from the sandur, forming ponding of meltwater and the potential for storage of sediment within proglacial lake (Syverson, 1998; Schomacker & Kjaer, 2008; Carrivick & Russel, 2013).	Fluvio-lacustrine process	**Store/Sink Buffer** Lakes constitute very effective sediment traps and hence drivers of disconnection as they trap much finer (sand and washload) sediment as well as coarse sediment. Rates of deposition may be such that there is complete lake fill within decades or a few centuries (Liermann et al., 2012; Carrivick & Tweed, 2013) such that the disconnection effect may be temporary. High rates of lacustrine sedimentation have been linked to periods of glacier retreat and following paraglacial reworking of sediment on glacier forelands;	Example of lake formation: Figure 7.4 and 8

Table 7.1 *(cont.)*

Landform	Formation–composition	Type of functional connectivity	Rôle in sediment connectivity	Example from the Arolla or Ferpècle catchment located in the Val d'Hérens in the South Western part of the Swiss Alps
			and attributed to the availability of meltwater and the exposure of fresh and unstable glacigenic deposits to more rapid non-glacial erosion processes (Desloges, 1994; Dirszowsky & Desloges, 1997; Leonard, 1985) (e.g., Figure 7.4 and 7.8b). Whilst some hydraulic connectivity may be conserved by surface (e.g., proglacial lake outflows) and sub-surface (groundwater) pathways, lacustrine disconnection can be very effective. However, even if eventual fill ends this disconnection, such lakes can be breached (Benn et al., 2012) and so rapidly increase sediment connectivity. Some lakes may be temporary, such as where drainage of the side- or trunk valley is blocked by glacier ice or a moraine (Clague, 2000). Possible failure mechanisms of moraine dams include melting of moraine ice cores (Livingstone et al., 2012),	

overtopping and incision triggered by heavy rainstorm or waves generated by avalanche or rockfall and, overtopping by an influx of water caused by rapid drainage of an upstream ice-dammed lake (called jökulhlaup) (Clague, 2000). Especially in ice-marginal lakes, the water depth may also be prone to sudden change due to; (1) opening and closing of outlets by glaciers margin position changes and by changes in glaciers thickness, (2) periodic or episodic outburst floods and (3) recharge events (Livingstone et al., 2012). Rupture or overflow of the impounding material frequently leads to glacial lake outburst floods (e.g., Carling et al., 2002; Carrivick et al., 2004, 2007; Carling, 2009) impacting landscape.

Glacial basins may also be subject to catastrophic outburst floods, that is high-magnitude-low-frequency events (Marren, 2005), that may lead to very significant sediment supply and transfer into storage in proglacial margins (see later). Glacial lake outburst floods are commonly characterized by an exponential increase in discharge (Clague, 2000), able to move huge particles of up to several meters in diameter (Carrivick & Rushmer, 2009).

Table 7.1 *(cont.)*

Landform	Formation–composition	Type of functional connectivity	Rôle in sediment connectivity	Example from the Arolla or Ferpècle catchment located in the Val d'Hérens in the South Western part of the Swiss Alps
			These events are also normally coupled to significant increases in sediment transport capacity. But, it occurs because waves of water commonly move more rapidly than waves of sediment due to the fact that a water wave travels at the mean flow velocity whereas a sediment travels at the near-bed flow velocity. Thus, hydrological-sediment disconnection of is an inherent consequence of this process and leads to extensive deposition on the declining limb of flood hydrographs. Such extreme events may also be important for re-establishing connection between alluvial fans and the river (Baewert & Morche, 2014).	

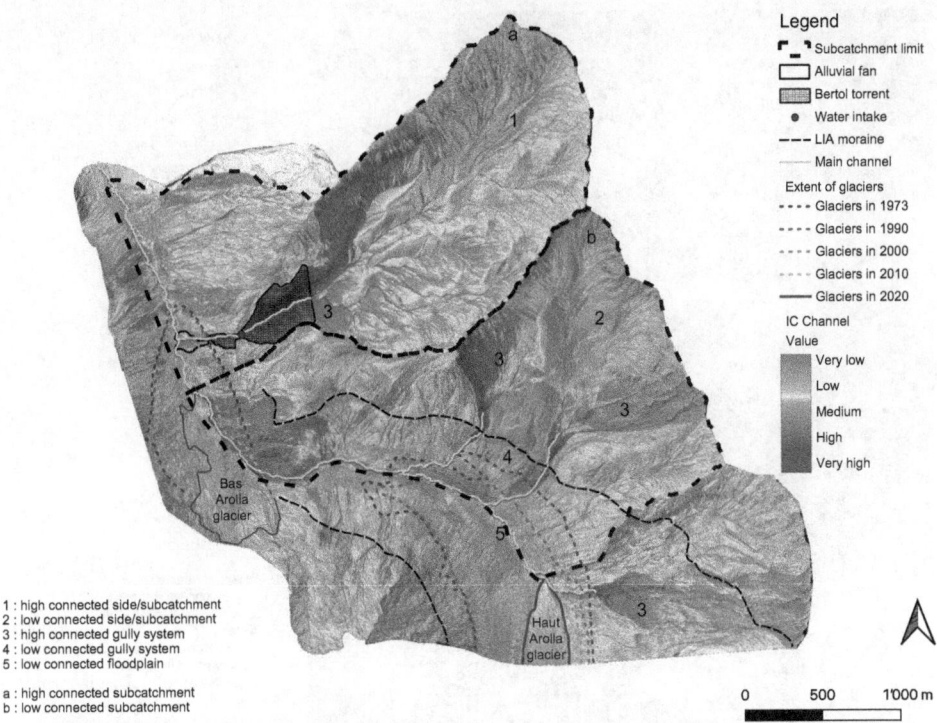

Figure 7.3   Connectivity Index (Cavalli et al., 2013) of a section of the proglacial area of the Bas and Haut Glacier d'Arolla with the surrounding area of the torrent the Bertol relatively well connected. A black and white version of this figure will appear in some formats. For the colour version, refer to the plate section.

In most rivers, for most of the time, sediment transport is limited by flow competence, and so is hydraulically disconnected. An increase in discharge may lead to sediment transport and hence connection, and whilst there is debate as to the extent to which transport is size selective, higher discharges commonly lead to the transport of coarser size fractions. Size-selective transport will cause evolution of the size of sediment on the river bed surface and hence change the discharge required for transport and the conditions that lead to hydraulic connection. Thus, hydraulic disconnection/connection may have a particle size dependence, be dynamic (Carrivick & Rushmer, 2009) and act as an important filter of sediment supplied from upstream via bed material evolution in response to both allogenic changes (e.g., in discharge) and autogenic responses (sediment sorting). Allogenic changes in deglaciating basins are of crucial importance as there may still be strong diurnal discharge fluctuation and seasonal discharge change (Lane et al., 1996; Lane & Nienow, 2019). The magnitude of diurnal discharge cycles, especially late in the meltwater season, may be very high with discharge crossing the critical value required for sediment transport on a daily basis (Perolo et al., 2019). Thus, connection becomes on-off, potentially impacting sediment transport

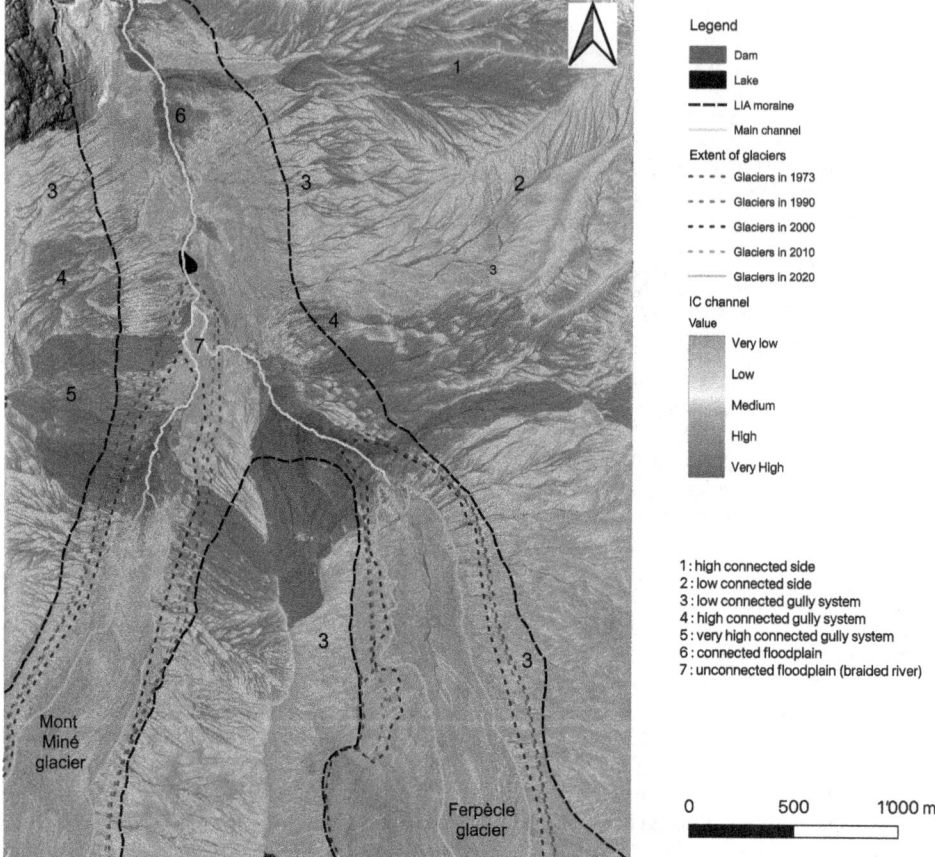

Figure 7.4   Connectivity Index (Cavalli et al., 2013) of the Ferpecle – Mont Miné catchment and joined floodplain. A black and white version of this figure will appear in some formats. For the colour version, refer to the plate section.

both underneath glaciers and in proglacial margins downstream. Discharge variation may also interact with a second allogenic factor, sediment supply which also has a diurnal variation (Lane et al., 1996) although there remain no studies that have coupled the signal of bedload sediment supply from glaciers and surrounding hillslopes to the dynamics of proglacial margins. These diurnal changes may lead to distinct daily cycles of erosion and deposition in proglacial margins and this commonly translates into an autogenic response (cycles of bar formation and destruction, avulsion) and river braiding (Figure 7.8a and 7.8c) (Ashworth & Ferguson, 1986). Bars represent short-term stores for the coarse material and turn rapidly into sources, for example, during floods (Baewert & Morche, 2014). This is clearly a buffering effect in terms of Table 1. However, there may also be larger spatial-scale and longer timescale storage processes. Data suggest that larger-scale patterns of net erosion and net deposition with this erosion-deposition couplet migrate upstream as a glacier retreats (Marren, 2005). Equally, our field

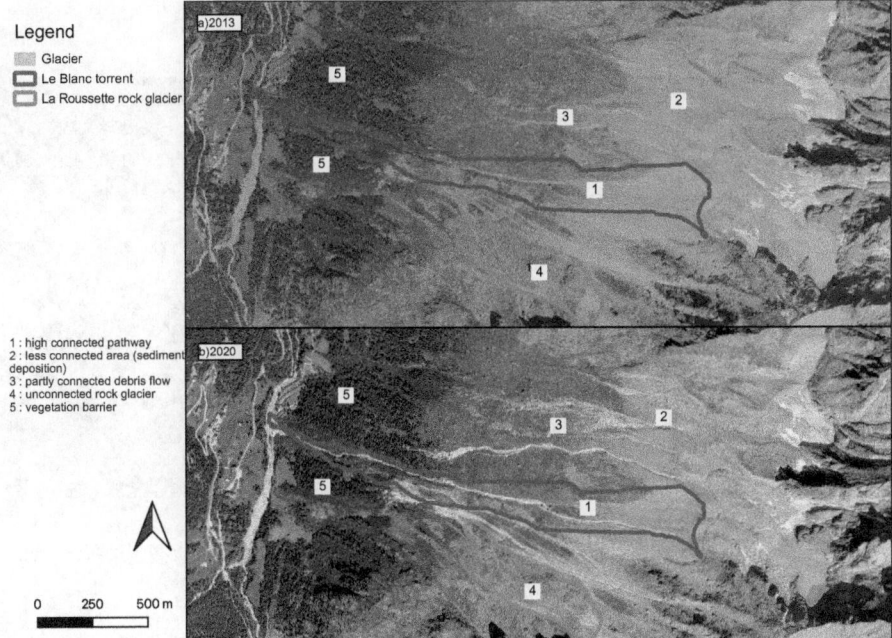

Figure 7.5   Aerial images of the sub-catchment in front of the Arolla village and illustration of relatively high connectivity created by Le Blanc torrent activity in (a) 2013 and (b) 2020. Number 2 shows the location of the La Rousette rock glacier (Micheletti et al., 2015; Meyrat, 2018). A black and white version of this figure will appear in some formats. For the colour version, refer to the plate section.

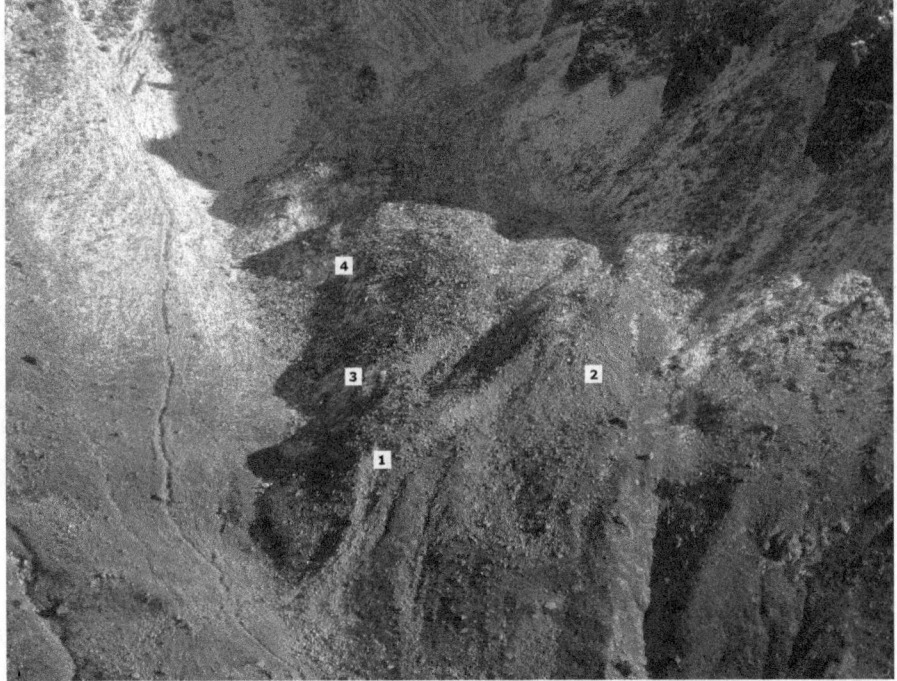

Figure 7.6   Detailed image (taken by Lambiel C.) of the La Rousette rock glacier (Figure 7.5) showing the different lobes (from Meyrat, 2018). A black and white version of this figure will appear in some formats. For the colour version, refer to the plate section.

Figure 7.7   Protalus ramparts (Otemma, Val de Bagnes, Valais Canton, Switzerland) (photo by Bayens M., 2022).

a) Floodplain in 1994 : braided   b) Floodplain in 2004 : lake   c) Floodplain in 2010 : braided

Figure 7.8   Evolution of the join floodplain of Mont Miné and Ferpècle glacier (Val Hèrens) in 1994, 2004 and 2010 (source: SwissTopo). A black and white version of this figure will appear in some formats. For the colour version, refer to the plate section.

observations suggest within a season, subglacial sediment supply exhaustion may cause a transition from net downstream deposition to net downstream erosion but this is a process not yet formally reported.

7.3 Interactions between Structural Connectivity and the Drivers of Sediment Transport

The concept of structural connectivity describes the existing degree of linkage between sediment sources, sinks and storage areas (Cavalli et al., 2013) which controls sediment yield. Sediment storage and flux (and so yield) vary strongly over the short-term (annual to decadal) in Alpine catchments: (1) because functional connectivity is dependent sediment availability and triggering events (Schrott et al., 2006), and (2) due to intermittent storage in valley floors and braidplains (e.g., Warburton, 1990; Orwin and Smart, 2004; Bertoldi et al., 2010; Carrivick et al., 2013). There are many fewer studies of how connectivity evolves in degla- ciating landscapes (exceptions include Borselli et al., 2008; Cavalli et al., 2013; Lane et al., 2017; Mahoney et al., 2018; Mancini & Lane, 2020), not least because the ease of acquiring high-quality digital elevation data of the sort needed for cal- culating connectivity indices has only recently increased. In practice, there are two ways of considering the temporal evolution of structural connectivity. The first is ergodic: in a landscape that is undergoing systematic deglaciation through time, it may be possible to use systematic variations in connectivity in space as a means of inferring its evolution through time. The second involves direct quantification of its evolution using historical topographic data. This is becoming more feasible as time passes and the 1990s revolution in digital topographic data is allowing extraction of high-resolution topographic data from historical data sources, nota- bly aerial imagery. Archival photogrammetric analyses (Micheletti et al., 2015) has now been proven as a tool for quantifying the evolution of connectivity at the multi-decadal timescale (Micheletti & Lane, 2016; Lane et al., 2017).

The potential of the ergodic approach is reflected in the application of the sed- iment connectivity index (Cavalli et al., 2013) to the proglacial area of the Haut Glacier d'Arolla (Figure 7.3). Stream and hillslope torrents are now relatively well connected to the outlet of the catchment, compared to surrounding, upstream (and glaciated) areas (Figure 7.3). Proglacial rockwalls and slope subsystems are highly coupled due to mass movement activity such as rockfalls (Schrott et al., 2006), gully systems and debris flows (Ballantyne, 2002b). As a response to LIA gla- cier retreat, paraglacial hillslopes failures have begun to form, and create such mass movements (Blair, 1994; Curry et al., 2006; Hugenholtz et al., 2008; Curry et al., 2009). Debris flows and gullying can have significant impact on sediment delivery and transport processes downstream between hillslopes and glaciers, and hillslopes and rivers (Cossart, 2008). Here, spatially systematic changes in the connectivity suggested by static indices can be seen for gully systems in the Arolla (Figure 7.3 location 3 and 4) and Ferpecle-Mont Miné (Figure 7.4 location 3, 4 and 5) case studies. Close to the Haut Glacier d'Arolla (Figure 7.3 location 3 and 4) there was no evidence of gully infilling suggesting that these gullies are still in a phase of incision. This incision occurred in parallel with ice melt and at a faster rate at the end of the deglaciation period. Without this incision, it is likely that the LIA moraine would probably remain as a sediment sink (Bosson et al., 2015), disconnecting the upper basins from the proglacial area of the Haut Glacier d'Arolla (Lane et al., 2017; Figure 7.3). Figure 7.9 shows an enlarged portion of the

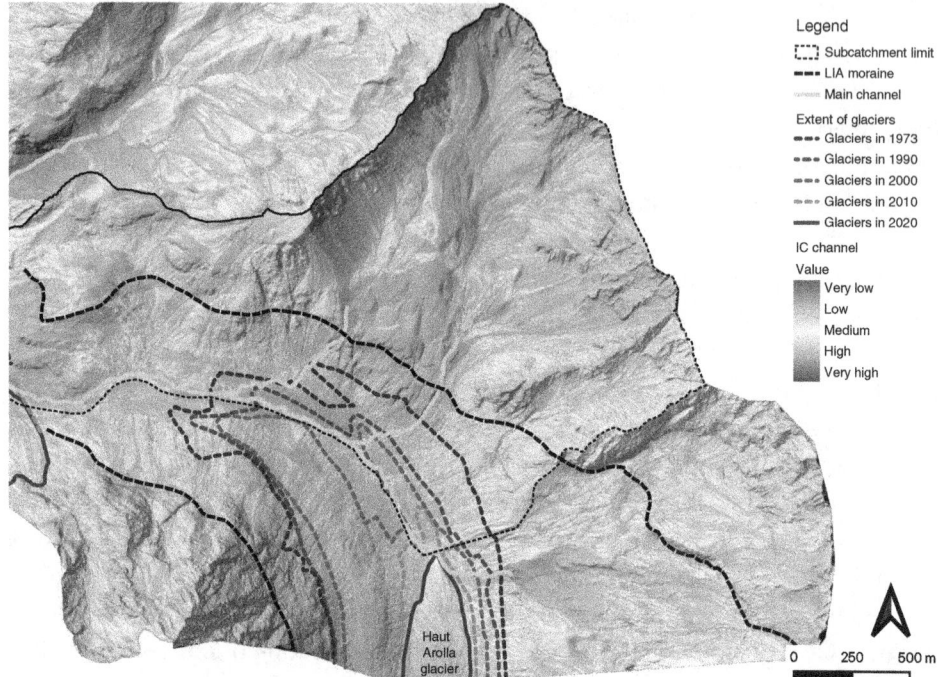

Legend

⌐⌐⌐ Subcatchment limit
━ ━ ━ LIA moraine
╌╌╌ Main channel

Extent of glaciers
━•• Glaciers in 1973
━•• Glaciers in 1990
━•• Glaciers in 2000
━•• Glaciers in 2010
━ Glaciers in 2020

IC channel
Value
Very low
Low
Medium
High
Very high

0   250   500 m

Figure 7.9   Connectivity index (Cavalli et al., 2013) of the surrounding proglacial area of the Haut Glacier d'Arolla with the historical extents of the glacier in 1973, 1990, 2000, 2010 and 2020. A black and white version of this figure will appear in some formats. For the colour version, refer to the plate section.

connectivity analysis for the Haut Glacier d'Arolla shown in Figure 7.3. Neither side of the proglacial floodplain shows a progressive increase of connectivity from upstream to downstream. Rather, the landscape has evolved from well connected and poorly connected hillslope sediment supply (Figure 7.3, location 2, 3 and 4). Thus, there is no clear evidence of a systematic evolution of connectivity through time suggesting that local influences on the development of connection following glacier recession are important.

The alternative approach is to use historical digital elevation data to analyse the evolution of connectivity through time (e.g., Lane et al., 2017; Mancini & Lane, 2020). The main challenge to these analyses is obtaining elevation data with sufficient precision to allow connectivity to be estimated. The lower the precision of the data, the more likely that disconnection is an artificial consequence of noise in topographic data. The traditional hydrological solution to this problem is to force hydrological connection through drainage basins by filling all sinks along a flow path but this also removes all gravity-driven disconnection. Lane et al. (2017) and Maninci and Lane (2020) dealt with this by treating the level of sink filling probabilistically and comparing the level of fill required to connect a flow path to the likely Digital elevation model noise. A greater fill than the likely DEM noise is more likely to be associated with a true geomorphologically driven disconnection. For both the Haut Glacier d'Arolla (Lane et al., 2017) and the Glacier d'Otemma (Mancini & Lane, 2020) it was possible to identify a systematic increase in hydrological connection following from gully incision, limited only by the thickness of sediment that had accumulated on valley sidewalls.

7.3.1 Vegetation and Functional Connectivity

Static analyses rarely consider the role of vegetation in sediment connectivity. Yet, vegetation cover can constitute a sediment buffer, or even a barrier that cannot be circumvented by flows (Meyer et al., 1995; Bochet et al., 2000). It needs emphasis because a longer timescale response of the landscape to deglaciation is the development of vegetation (see Miller & Lane, 2019). Vegetation development may significantly increase energy losses (Lee et al., 2000), encourage subsurface rather than overland flow and so trap sediments from eroded zones upslope (e.g., Dabney et al., 1999; Cammeraat & Imeson, 1999). However, the effectiveness of vegetation barriers in sediment trapping changes through time. After extreme rainfall events, certain vegetation barriers may be destroyed. That said, the dynamics of vegetation development impacts on sediment retention remain uncertain and poorly understood (Rey, 2004). There is evidence of potential importance. Figure 7.5 shows a clear effect of vegetation on the connection of debris flows to the valley bottom. A series of small debris flows associated with breaches of glacial moraine upstream entered a forest zone due to an extreme rainfall event in August 2019. There was a clear differential effect in the distance downslope that the debris flows transferred through the forest. Only one connected with the valley bottom, the Le Blanc torrent, and this was the only one that flowed down a zone that had previously been impacted by snow avalanches (Figure 7.5; 'Swisstopo', n.d.). Three important points follow. First, vegetation development may lead to significant sediment disconnection but this may require fully mature forest to develop which may take decades to centuries in deglaciated landscapes (Miller & Lane, 2019). Second, other natural hazards, in this case, snow avalanches, may impact this effect and the probability of connection is a synergistic consequence. Third, the timescales of vegetation development are commonly long such that it is possible that a second extreme event occurs before mature forest has recovered. The state of connectivity after the 2019 event makes it more probable that a future event of the same size (or even a smaller size) can connect with the valley bottom. Thus, connectivity changes in one event may have important legacy effects that lead to non-linear response. As yet, there are very few studies of the evolution of connectivity during individual events, and this is a topic that merits research.

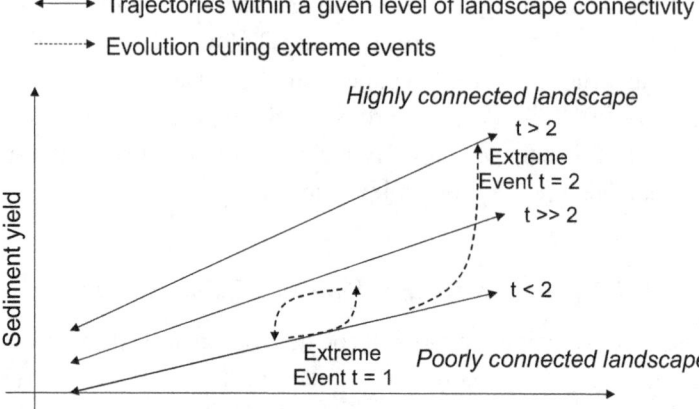

Figure 7.10   Conceptualization of dynamically evolving connectivity in the context of drivers of sediment yield.

A key point emerges. The transition from a well connected to a poorly connected landscape due to external forcing may not be continuous. Much of the literature review has emphasized how extreme events (e.g., rainfall-driven; glacial lake outburst floods) can lead to an increase in connectivity. Thus, it is possible for a given basin to remain in or return to a highly connected state. This was illustrated in Figure 7.5 where the legacy of an extreme storm event in August 2019 leaves the landscape in a more connected state and more sensitive to future extreme events. Figure 7.10 conceptualizes these higher frequency changes in terms of the relationship between an external driver (e.g., discharge) and sediment yield. In the poorly connected landscape, there is a relatively slower increase of sediment yield as a function of external drivers due to poor connection associated with buffering landforms. An extreme event at time $t = 1$, is not sufficient to cause significant landscape-scale change in connectivity. A bigger event at time $t = 2$ leads to a major landscape reorganization to have a significantly higher degree of connectivity, such that there is now a much higher sensitivity of sediment yield to extreme forcing and the landscape is a weakened buffer. Through time, processes such as vegetation development may lead to the recovery of the buffering capacity, landscape-scale recovery of connectivity occurs and the yield-forcing curve evolves back to the poorly connected state. This is very poorly studied in Alpine landscapes. Figure 7.10 emphasizes that connectivity may be one of the major reasons why a given degree of external forcing can lead to substantially different sediment yields (e.g., scatter in discharge–sediment transport relations). It also emphasizes the point that connectivity may have substantial landscape-scale legacy effects (Mancini & Lane, 2020).

7.3.2 Human-induced Impacts on Connectivity in Deglaciating Landscapes

The discussion thus far has assumed that connection and disconnection in deglaciating environments can be attributed entirely to natural processes as modified by humans via changing climate. However, human activities can play a direct role in Alpine periglacial environments. A primary driver of this process is water management for hydropower. How this impacts connectivity depends on the kind of infrastructure that is used and its management. The most extreme example is dams designed to cause water storage and that may interrupt downstream sediment flux for all or most of the time. The presence of melting glaciers and high precipitation rates due to relief, as well as low population densities, has made Alpine environments important in water storage for hydropower. Dams (Figure 7.4) can directly trap all sediment sizes in water reservoirs (Vörösmarty et al., 2003; Syvitski & Milliman, 2007). In the absence of flushing flows events, dams are characterized as permanent sediment storage (Stutenbecker et al., 2018). Their importance at the landscape scale depends on where the dam is built within the river basin with respect to hillslope sediment sources (Lane et al., 2022). If sediment is completely stored behind the dam, and the dam is not flushed, processes that influence connectivity in the river downstream of the dam may also change. For instance, reduced sediment flux to downstream may cause a river to incise (e.g., Galay, 1983; Williams & Wolman, 1984; Smith & Mohrig, 2017). This incision may disconnect the river from its floodplain (Kroes & Hupp, 2010; Renshaw et al., 2014). Reduced sediment supply and incision may cause the river to evolve from multi-thread towards being more single thread (Galay, 1983; Kondolf & Swanson, 1993; Draut et al., 2011) which eases the extent to which sediment transport can be transferred downstream and so increases within hydraulic connectivity. However, these processes may be countered by the extent to which the dam is flushed, a process that maintains sediment flux to downstream, at least partially. The need to flush increases as the size of the dam storage (and hence the time required to fill it) decreases. The smallest kinds of infrastructure in Alpine environments involve water offtakes that store very little water, as water is then transferred laterally to storage or directly for hydropower production (Gabbud & Lane, 2016). Such infrastructure may need to be emptied multiple times per day in glaciated basins with very high rates of sediment supply (Gabbud et al., 2019). Whenever water is retained or transferred, significant hydraulic disconnection can result; the water that drives sediment transport capacity is removed. If sediment connectivity is maintained, this has the opposite effect on rivers to those downstream of dams that are not flushed. Reduced transport capacity plus maintained supply commonly leads to significant downstream sediment deposition which may encourage the development of river braiding (Ashworth & Ferguson, 1986; Bertoldi et al., 2010; Bakker et al., 2019) and a buffering effect in connectivity terms. Significant downstream sediment flux still occurs (Bakker et al., 2018), notably of finer sediment (Costa et al., 2018), and this may be sufficient to counter the widely held and erroneous perception that Alpine dams have led to reduced sediment supply to major sinks in downstream lakes (Lane et al., 2019). Two important points need emphasis. First, one of the key characteristics of human-induced changes in connectivity is that they can change not only due to environmental changes but also due to management practices. Thus, Bakker et al. (2018) reported how offtake management was influenced by the hydrological capacity of transfer tunnels and decisions around health and security, and that as these changed, so the extent of hydraulic connectivity changed. Excessive glacier melt coupled to a decision to flush more frequently during the night increased hydraulic connectivity and this was sufficient to reverse the tendency to deposition in the river downstream of the offtakes studied. Second, hydropower infrastructure is not always installed in the main river but also in steep lateral tributaries. The effects of this infrastructure are less well studied but also clear. In the Arolla catchment flushing events associated with a water intake led to significant gully formation (Figure 7.3) and increased hydrological connection between the hillslope and the main river floodplain (Figure 7.3 location 3).

Hydropower is not the only infrastructure to impact connectivity in Alpine deglaciating environments. As the consequences of sediment delivery to valley bottoms can be serious in terms of flood risk and damage to infrastructure, sediment delivery from tributaries may be significantly reduced via check dams and similar infrastructure. For decades, humans have developed check dams as strategies in order to reduce damage related to mountain streams and debris flows activity. Different types of check dams exist and vary in their impact on morphological processes. Check-dam functions are bed stabilization, hillslopes consolidation, decreasing slope, retention and sediment transport regulation. To fulfil their need(s) and reduce specific natural risks or damage, check dams can have different shapes and positions within the watersheds. This kind of infrastructure tends to have several functions at the same time. For example, to reduce sediment incision and catastrophic consequences due to landslide movement, dams may be used to decrease energy availability and hence the erosive power of debris flows upstream (Piton et al., 2017). Then, if the aim of the dam is to regulate sediment transport in carrying out their function, check dams have a tendency to change channel morphology (Chiu et al., 2021), to create sediment barriers or blankets and so reduce sediment connectivity potentially substantially (Piton et al., 2017).

7.4 Connectivity in Deglaciating Landscapes: Its Role in Longer Term and Large-scale Sediment Transport

The earlier sections have considered the controls on sediment connectivity in deglaciating Alpine landscapes; their evolution through time in response to sediment transport, the very process that sediment connectivity regulates; and the role of human activities and vegetation development as shorter- and longer-term drivers of connectivity respectively. To draw this review together, this section attempts to revisit the classic paraglacial model of Church and Ryder (1972) to consider what these processes mean at the landscape-scale and in the context of landscape response over the following decades to centuries and longer. The dominant view of the sediment yield evolution at the catchment-scale is that the rate of sediment transport at any time after deglaciation is a function of the percentage of remaining available sediment for transport, and so can be modelled by an exponential decline through time (Ballantyne, 2002b).

This review shows how the evolution of connectivity could contribute to this pattern, even if as yet there are very few quantitative studies of how it does so. Given the altitude of Alpine catchments, it is highly likely that long-term post-glacial weathering rates are unable to keep up with erosion, which is why the exhaustion is asymptotic on long-term geological controls. However, it is quite likely that the negative exponential in sediment yield hides substantial variability in yield through time and between river basins, even adjacent river basins.

The temporal evolution is illustrated well for the Arolla catchment shown in Figure 7.5. Close to the Arolla village, the well connected torrent Blanc (Figure 7.5 location 1) is surrounded by gully and debris flows systems (Figure 7.5 location 3). Between 2013 (Figure 7.5a) and 2020 (Figure 7.5b) the activity and sediment production from those processes improved, leading to a potential increase of hydrological connectivity on hillslopes. However, because of mature forest between these zones and the floodplain, sediment was readily trapped (Figure 7.5 location 5) and the increase of hydrological connectivity did not translate into catchment-average sediment yield. Between 2013 (Figure 7.5a) and 2020 (Figure 7.5 b), an increase in sediment sources is clear above the moraine barrier and the high hydrological connectivity of the Blanc torrent is still well observed (Figure 7.5 location 1). Secondly, upstream and closer to the Bas and Haut d'Arolla Glaciers, glacier retreat has resulted in two sub-catchments. The first one (Figure 7.3, sub-catchment a) contains the Bertol torrent and has been impacted for longer by glacial retreat than the second one (Figure 7.3, sub-catchment b). Due to a longer time since deglaciation, the Bertol torrent involves a well developed fluvial, debris-flow and gully system, sub-catchment a (Figure 7.3) with improved connectivity over time. In contrast, sub-catchment b (Figure 7.3) is still adjusting to the effects of glacier retreat. There, proglacial sediment supply remains present and has not reached its maximum peak of sediment production. Therefore, the connectivity of the sub-catchment a (Figure 7.3) is actually higher than sub-catchment b (Figure 7.3) and sediment from sub-catchment a (Figure 7.3) is more readily transferred to the floodplain and the sub-catchment outlet (Figure 7.3). This evidence suggests that at any one period of time, a deglaciating landscape is likely to have a spatial variation of where in the Church and Ryder (1972) model an individual basin is to be found, as basins pass from being initially well connected to being ultimately poorly connected. Ultimately, this should control the rate of exponential decline to long-term non-glacial weathering rates. However, this transition from well connected to poorly connected may not be continuous because of extreme events for example as illustrated by Figure 7.10.

References

Alley, R. B., Cuffey, K. M. & Zoet, L. K. (2019). Glacial erosion: Status and outlook. *Annals of Glaciology* 60 (80): 1–13. https://doi.org/10.1017/aog.2019.38.

Alley, R. B., Cuffey, K. M., Evenson, E. B., Strasser, J. C., Lawson, D. E. & Larson, G. J. (1997). How glaciers entrain and transport basal sediment: Physical constraints. *Quaternary Science Reviews* 16 (9): 1017–1038. https://doi.org/10.1016/S0277-3791(97)00034-6.

Anselmetti, F. S., Bühler, R., Finger, D., Girardclos, S., Lancini, A., Rellstab, C. & Sturm, M. (2007). Effects of Alpine hydropower dams on particle transport and lacustrine sedimentation. *Aquatic Sciences* 69 (2): 179–198. https://doi.org/10.1007/s00027-007-0875-4.

Antoniazza, G. & Lane, S. N. (2021). Sediment yield over glacial cycles: A conceptual model. *Progress in Physical Geography: Earth and Environment*, 45(6), 842–865. https://doi.org/10.1177/0309133321997292.

Arfstrom, J. D. (2003). Protalus ramparts and transverse ridge moraines on Mars: Indicators of surface ice depositional processes. In *34th Annual Lunar and Planetary Science Conference*, March 17–21, 2003, League City, TX, abstract no.1050.

Ashmore, P. (1993). Contemporary erosion of the Canadian landscape. *Progress in Physical Geography* 17: 190–204.

Ashworth, P. J. & Ferguson, R. I. (1986). Interrelationships of channel processes, changes and sediments in a proglacial braided river. *Geografiska Annaler: Series A, Physical Geography* 68 (4): 361–371. https://doi.org/10.1080/04353676.1986.11880186.

Baewert, H. & Morche, D. (2014). Coarse sediment dynamics in a proglacial fluvial system (Fagge River, Tyrol). *Geomorphology* 218 (August): 88–97. https://doi.org/10.1016/j.geomorph.2013.10.021.

Bakker, M., Costa, A., Silva, T. A. Stutenbecker, L., Girarclos, S., Loizeau, J.-L., Molnar, P., Schlunegger, F. & Lane, S. N. (2018). Combined flow abstraction and climate change impacts on an aggrading Alpine river. *Water Resources Research* 54, 223–242.

Bakker, M., Antoniazza, G., Odermatt, E. & Lane, S. N. (2019). Morphological Response of an Alpine Braided Reach to Sediment-Laden Flow Events. *Journal of Geophysical Research: Earth Surface* 124 (5): 1310–1328. https://doi.org/10.1029/2018JF004811.

Ballantyne, C. K. (2002a). A General Model of Paraglacial Landscape Response. *The Holocene* 12 (3): 371–376. https://doi.org/10.1191/0959683602hl553fa.

Ballantyne, C. K. (2002b). Paraglacial Geomorphology. *Quaternary Science Reviews* 21 (18–19): 1935–2017. https://doi.org/10.1016/S0277-3791(02)00005-7.

Ballantyne, C. K. & Benn, D. I. (1994). Paraglacial slope adjustment and resedimenfation following recent glacier retreat, Fåbergstølsdalen, Norway. *Arctic and Alpine Research* 26: 255–269.

Ballantyne, C. K. & Benn, D. I. (1996). Paraglacial slope adjustment during recent deglaciation and its implications for slope evolution in formerly glaciated environments. Anderson, M. G. and Brooks, S., Editors, *Advances in Hillslope Processes* 2: 1173–1195.

Ballantyne, Colin K. & Kirkbride, M. P. (1986). The characteristics and significance of some lateglacial protalus ramparts in upland Britain. *Earth Surface Processes and Landforms* 11 (6): 659–671. https://doi.org/10.1002/esp.3290110609.

Benn, D. I., Bolch, T., Hands, K., Gulley, J., Luckman, A., Nicholson, L. I., Quincey, D., Thompson, S., Toumi, R. & Wiseman, S. (2012). Response of debris-covered glaciers in the mount everest region to recent warming, and implications for outburst flood hazards. *Earth-Science Reviews* 114 (1–2): 156–174. https://doi.org/10.1016/j.earscirev.2012.03.008.

Bertoldi, W., Zanoni, L. & Tubino, M. (2010). Assessment of morphological changes induced by flow and flood pulses in a gravel bed braided river: The Tagliamento River (Italy). *Geomorphology* 114 (3): 348–360. https://doi.org/10.1016/j.geomorph.2009.07.017.

Beylich, A. A., Laute, K., Liermann, S., Hansen, L., Burki, V., Vatne, G., Fredin, O., Gintz, D. & Berthling, I (2009). Subrecent sediment dynamics and sediment budget of the braided sandur system at Sandane, Erdalen (Nordfjord, Western Norway). *Norsk Geografisk Tidsskrift – Norwegian Journal of Geography* 63 (2): 123–131. https://doi.org/10.1080/00291950902907934.

Beylich, A. A., Laute, K. & Storms, J. E. A. (2017). Contemporary suspended sediment dynamics within two partly glacierized mountain drainage basins in western Norway (Erdalen and Bødalen, Inner Nordfjord). *Geomorphology* 287 (June): 126–143. https://doi.org/10.1016/j.geomorph.2015.12.013.

Blair, R. W. (1994). Moraine and valley wall collapse due to rapid deglaciation in mount cook national park, New Zealand. *Mountain Research and Development* 14 (4): 347. https://doi.org/10.2307/3673731.

Bochet, E., Poesen, J. & Rubio, J. L. (2000). Mound development as an interaction of individual plants with soil, water erosion and sedimentation processes on slopes. *Earth Surface Processes and Landforms* 25: 847–867.

Bogen, J. (2010). Sediment dynamics of glacier-fed rivers. *IAHS-AISH Publication*, 181–188.

Borselli, L., Cassi, P., & Torri, D. (2008). Prolegomena to Sediment and Flow Connectivity in the Landscape: A GIS and Field Numerical Assessment. *CATENA* 75 (3): 268–77. https://doi.org/10.1016/j.catena.2008.07.006.

Bosson, J.-B., Deline, P., Bodin, X., Schoeneich, P., Baron, L., Gardent, M. & Lambiel C. (2015). The influence of ground ice distribution on geomorphic dynamics since the little ice age in proglacial areas of two cirque glacier systems: Influence of ground ice on geomorphic dynamics in proglacial areas. *Earth Surface Processes and Landforms* 40 (5): 666–680. https://doi.org/10.1002/esp.3666.

Bratlie, B. (1994). *Senkvartære Sedimenter Og Glasialhistorie i Van Keulenfjorden, Svalbard.* Master's Thesis, Universitetet i Oslo, Norway.

Brönnimann, S., Rajczak, J., Fischer, E. M., Raible, C. C., Rohrer, M. & Schär, C. (2018). Changing seasonality of moderate and extreme precipitation events in the alps. *Natural Hazards and Earth System Sciences* 18 (7): 2047–2056. https://doi.org/10.5194/nhess-18-2047-2018.

Brooks, G. R. (1994). The fluvial reworking of late Pleistocene drift, Squamish river drainage basin, southwestern British Colombia. *Géographie Physique et Quaternaire* 48 (1): 51–68. https://doi.org/10.7202/032972ar.

Cammeraat, L. H. & Imeson, A. C. (1999). The evolution and significance of soil-vegetation patterns following land abandonment and fire in Spain. *CATENA* 37: 107–127.

Carling, P. A. (2009). Morphology, sedimentology and palaeohydraulic significance of large gravel dunes, Altai Mountains, Siberia. *Sedimentology* 43 (4): 647–664. https://doi.org/10.1111/j.1365-3091.1996.tb02184.x.

Carling, P. A., Kirkbride, A. D., Parnachov, S., Borodavko, P. S. & Berger, G. W. (2002). Late quaternary catastrophic flooding in the Altai mountains of south–central Siberia: A synoptic overview and an introduction to flood deposit sedimentology. In *Flood and Megaflood Processes and Deposits*, edited by I. Peter Martini, Victor R. Baker, and Guillermina Garzn, 17–35. Oxford, UK: Blackwell Publishing Ltd. https://doi .org/10.1002/9781444304299.ch2.

Carrivick, J. L., Geilhausen, M., Warburton, J., Dickson, N. E., Carver, S. J., Evans, A. J. & Brown, L. E. (2013). Contemporary geomorphological activity throughout the proglacial area of an alpine catchment. *Geomorphology* 188 (April): 83–95. https:// doi.org/10.1016/j.geomorph.2012.03.029.

Carrivick, J. L. & Heckmann, T. (2017). Short-term geomorphological evolution of proglacial systems. *Geomorphology* 287 (June): 3–28. https://doi.org/10.1016/j .geomorph.2017.01.037.

Carrivick, J. L., Pringle, J. K., Russell, A. J. & Cassidy, N. J. (2007). GPR-derived sedimentary architecture and stratigraphy of outburst flood sedimentation within a bedrock valley system, Hraundalur, Iceland. *Journal of Environmental and Engineering Geophysics* 12 (1): 127–143. https://doi.org/10.2113/JEEG12.1.127.

Carrivick, J. L. & Rushmer, E. L. (2009). Inter- and intra-catchment variations in proglacial geomorphology: An example from Franz Josef Glacier and Fox Glacier, New Zealand. *Arctic, Antarctic, and Alpine Research* 41 (1): 18–36. https://doi .org/10.1657/1523-0430-41.1.18.

Carrivick, J. L. & Russel, A. J. (2013). Glaciofluvial landforms of deposition. In Elias, S. A. (ed.), *The Encyclopedia of Quaternary Science*. Amsterdam: Elsevier, 2: 6–17.

Carrivick, J. L., Russell, A. J., Tweed, F. S. & Twigg, D. (2004). Palaeohydrology and sedimentary impacts of Jökulhlaups from Kverkfjöll, Iceland. *Sedimentary Geology* 172 (1–2): 19–40. https://doi.org/10.1016/j.sedgeo.2004.07.005.

Carrivick, J. L. & Tweed, F. S. (2013). Proglacial lakes: Character, behaviour and geological importance. *Quaternary Science Reviews* 78 (October): 34–52. https://doi .org/10.1016/j.quascirev.2013.07.028.

Cavalli, M., Trevisani, S., Comiti, F. & Marchi, L. (2013). Geomorphometric assessment of spatial sediment connectivity in small alpine catchments. *Geomorphology* 188 (April): 31–41. https://doi.org/10.1016/j.geomorph.2012.05.007.

Chiarle, M., Iannotti, S., Mortara, G. & Deline,P. (2007). Recent debris flow occurrences associated with glaciers in the Alps. *Global and Planetary Change* 56 (1–2): 123–136. https://doi.org/10.1016/j.gloplacha.2006.07.003.

Chiu, Y.-F., Tfwala, S. S., Hsu, Y.-C., Chiu, Y.-Y., Lee, C.-Y. & Chen, S.-C. (2021). Upstream morphological effects of a sequential check dam adjustment process. *Earth Surface Processes and Landforms* 46 (13): 2527–2539. https://doi.org/10.1002/esp.5178.

Church, M. & Ryder, J. M. (1972). Paraglacial sedimentation: A consideration of fluvial processes conditioned by glaciation. *Geological Society of America Bulletin* 83 (10): 3059. https://doi.org/10.1130/0016-7606(1972)83[3059:PSACOF]2.0.CO;2.

Church, M. & Slaymaker, O. (1989). Disequilibrium of holocene sediment yield in glaciated British Columbia. *Nature* 337 (6206): 452–454. https://doi.org/10.1038/337452a0.

Clague, J. (2000). A Review of Catastrophic Drainage of Moraine-Dammed Lakes in British Columbia. *Quaternary Science Reviews* 19 (17–18): 1763–1783. https://doi .org/10.1016/S0277-3791(00)00090-1.

Cossart, É. (2008). Landform Connectivity and Waves of Negative Feedbacks during the Paraglacial Period, a Case Study: The Tabuc Subcatchment since the End of the Little Ice Age (Massif Des Écrins, France). *Géomorphologie: Relief, Processus, Environnement* 14 (4): 249–260. https://doi.org/10.4000/geomorphologie.7430.

Cossart, E. & Fort, M. (2008). Sediment release and storage in early deglaciated areas: Towards an application of the exhaustion model from the case of Massif Des Écrins (French Alps) since the Little Ice Age. *Norsk Geografisk Tidsskrift – Norwegian Journal of Geography* 62 (2): 115–131. https://doi.org/10.1080/00291950802095145.

Cossart, É. & Fressard, M. (2017). Assessment of structural sediment connectivity within catchments: Insights from graph theory. *Earth Surface Dynamics* 5 (2): 253–268. https://doi.org/10.5194/esurf-5-253-2017.

Cossart, É., Viel, V., Lissak, C., Reulier, R., Fressard, M. & Delahaye, D. (2018). How might sediment connectivity change in space and time? *Land Degradation & Development* 29 (8): 2595–2613. https://doi.org/10.1002/ldr.3022.

Costa, A., Anghileri, D. & Molnar, P. (2018). Hydroclimatic control on suspended sediment dynamics of a regulated alpine catchment: A conceptual approach. *Hydrology and Earth System Sciences* 22 (6): 3421–3434. https://doi.org/10.5194/hess-22-3421-2018.

Costa, A., Molnar, P., Stutenbecker, L., Bakker, M., Silva, T. A., Schlunegger, F., Lane, S. N., Loizeau, J.-L. & Girardclos, S. (2018). Temperature signal in suspended sediment export from an alpine catchment. *Hydrology and Earth System Sciences* 22 (1): 509–528. https://doi.org/10.5194/hess-22-509-2018.

Coulthard, T. J. & Van de Wiel, M. J. (2013). Climate, tectonics or morphology: What signals can we see in drainage basin sediment yields? *Earth Surface Dynamics* 1 (1): 13–27. https://doi.org/10.5194/esurf-1-13-2013.

Cruden, D. M. & Hu, X. Q. (1993). Exhaustion and steady state models for predicting landslide hazards in the canadian rocky mountains. *Geomorphology* 8 (4): 279–285. https://doi.org/10.1016/0169-555X(93)90024-V.

Curry, A. M., Cleasby, V. & Zukowskyj, P. (2006). Paraglacial response of steep, sediment-mantled slopes to post-"Little Ice Age" glacier recession in the central Swiss Alps. *Journal of Quaternary Science* 21 (3): 211–225. https://doi.org/10.1002/jqs.954.

Curry, A. M., Sands, T. B. & Porter, P. R. (2009). Geotechnical Controls on a Steep Lateral Moraine Undergoing Paraglacial Slope Adjustment. *Geological Society, London, Special Publications* 320 (1): 181–197. https://doi.org/10.1144/SP320.12.

Dabney, S.M, Liu, Z., Lane, M., Douglas, J., Zhu, J. & Flanagan, D. C. (1999). Landscape benching from tillage erosion between grass hedges1 paper presented at international symposium on tillage translocation and tillage erosion held in conjunction with the 52nd annual conference of the soil and water conservation society, Toronto, Canada, 24–25 July 1997.1. *Soil and Tillage Research* 51 (3–4): 219–231. https://doi.org/10.1016/S0167-1987(99)00039-2.

Delaloye, R. & Morard, S. (2019). *Géomorphologie Périglaciaire*. Géographie Dép. Géosciences, Université de Fribourg – GG.02262–263.

Delaloye, R., Strozzi, T., Lambiel, C., Perruchoud, E. & Raetzo, H. (2007). *Landslide-like development of rockglaciers detected with ERS-1/2 SAR Interfero-Metry*. Presented at the Proceedings of the Frascati, Italy, FRINGE 2007 Work-shop, 26–30 November 2007 (ESA SP-649, February 2008).

Delmas, M., Calvet, M. & Gunnell, Y. (2009). Variability of quaternary glacial erosion rates – a global perspective with special reference to the eastern Pyrenees. *Quaternary Science Reviews* 28 (5–6): 484–498. https://doi.org/10.1016/j.quascirev.2008.11.006.

Desloges, J. R. (1994). Varve deposition and sediment yield record at three small lakes of the southern Canadian Cordillera. *Arctic and Alpine Research* 26: 130–140.

Dirszowsky, R. W. & Desloges, J. R. (1997). Glaciolacustrine sediments and neoglacial history of the Chephren Lake Basin, Banff National Park, Alberta. *Geographie Physique et Quaternaire* 51: 41–53.

Draut, A. E., Logan, J. B. & Mastin, M. C., (2011). Channel evolution on the dammed Elwha River, Washington, USA. *Geomorphology*, 127, 71–87

Dunne, T. & Leopold, L. B. (1978). *Water in Environmental Planning*. San Francisco. CA: W. H. Freeman and Co., 818. https://doi.org/10.1002/esp.3290040322.

Evans, D. J. A., Hiemstra, J. F., Boston, C. M., Leighton, I., Cofaigh, C. Ó. & Rea, B. R. (2012). Till stratigraphy and sedimentology at the margins of terrestrially terminating ice streams: Case study of the western Canadian prairies and high plains. *Quaternary Science Reviews* 46 (July): 80–125. https://doi.org/10.1016/j.quascirev.2012.04.028.

Fatichi, S., Rimkus, S., Burlando, P., Bordoy, R. & Molnar, P. (2015). High-resolution distributed analysis of climate and anthropogenic changes on the hydrology of an alpine catchment. *Journal of Hydrology* 525 (June): 362–382. https://doi.org/10.1016/j.jhydrol.2015.03.036.

French, H. (2015). *Periglacial Environments*. Oxford University Press. https://doi.org/10.1093/obo/9780199363445-0038.

Gabbud, C., Robinson, C. & Lane, S. N., (2019). Summer is in winter: disturbance-driven shifts in macroinvertebrate communities following hydroelectric power exploitation. *Science of the Total Environment*, 650, 2164–2180.

Gabbud, C. & Lane, S. N. (2016). Ecosystem impacts of alpine water intakes for hydropower: The challenge of sediment management. *WIREs Water* 3 (1): 41–61. https://doi.org/10.1002/wat2.1124.

Galay, V. J. (1983). Causes of river bed degradation. *Water Resources Research*, 19(5), 1057–1090.

Gärtner-Roer, I. & Bast, A. (2019). (Ground) ice in the proglacial zone. In Tobias Heckmann & David Morche (eds.), *Geomorphology of Proglacial Systems*. 85–98. Cham: Springer International Publishing, https://doi.org/10.1007/978-3-319-94184-4_6.

Geilhausen, M., Otto, J.-C., Morche, D. & Schrott, L. (2012). Decadal sediment yield from an alpine proglacial zone inferred from reservoir sedimentation (Pasterze, Hohe Tauern, Austria). In *Erosion and Sediment Yields in the Changing Environment*, 161–172.

Gobiet, A. & Kotlarski, S. (2020). *Future Climate Change in the European Alps*. Oxford Research Encyclopedia of Climate Science, Oxford University Press. https://doi.org/10.1093/acrefore/9780190228620.013.767.

Gurnell, A. M. (1983). Downstream channel adjustments in response to water abstraction for hydro-electric power generation from alpine glacial melt-water streams. *The Geographical Journal* 149 (3): 342. https://doi.org/10.2307/634009.

Haeberli, W., Schaub, Y. & Huggel, C. (2017). Increasing risks related to landslides from degrading permafrost into new lakes in de-glaciating mountain ranges. *Geomorphology* 293 (September): 405–417. https://doi.org/10.1016/j.geomorph.2016.02.009.

Harbor, J. & Warburton, J. (1993). Relative rates of glacial and nonglacial erosion in alpine environments. *Arctic and Alpine Research* 25 (1): 1-7. https://doi.org/10.2307/1551473.

Harvey, A M. (2001). *Coupling between Hillslopes and Channels in Upland Fluvial Systems: Implications for Landscape Sensitivity*, Illustrated from the Howgill Fells, Northwest England, 26.

Heckmann, T., Cavalli, M. & Marchi, L. (2019). Sediment Connectivity in Proglacial Areas. In Heckmann, T. & Morche, D. (eds) *Geomorphology of Proglacial Systems – Landform and Sediment Dynamics in Recently Deglaciated Alpine Landscapes*, 271–87. (Geography of the Physical Environment), Cham, Schweiz: Springer.

Heckmann, T. & Morche, D. (2019). Geomorphology of proglacial systems: landform and sediment dynamics in recently deglaciated alpine landscapes. In Heckmann, T. & Morche, D. (eds.), *Geography of the Physical Environment*. Cham: Springer International Publishing, https://doi.org/10.1007/978-3-319-94184-4.

Heckmann, T. & Schwanghart, W. (2013). *Geomorphic Coupling and Sediment Connectivity in an Alpine Catchment – Exploring Sediment Cascades Using Graph Theory*, 15.

Hedding, D. W. (2011). Pronival rampart and protalus rampart: A review of terminology. *Journal of Glaciology* 57 (206): 1179–1180. https://doi.org/10.3189/002214311798843241.

Hirschberg, J., Fatichi, S., Bennett, G. L., McArdell, B. W., Peleg N., Lane, S. N., Schlunegger, F. & Molnar, P. (2021). Climate change impacts on sediment yield and debris-flow activity in an alpine catchment. *Journal of Geophysical Research: Earth Surface* 126 (1). https://doi.org/10.1029/2020JF005739.

Hugenholtz, C. H., Moorman, B. J., Barlow, J. & Wainstein, P. A. (2008). Large-scale moraine deformation at the Athabasca glacier, Jasper National Park, Alberta, Canada. *Landslides* 5 (3): 251–260. https://doi.org/10.1007/s10346-008-0116-5.

Huss, M., Bookhagen, B., Huggel, C., Jacobsen, D., Bradley, R. S., Clague, J. J., Vuille, M. & al. (2017). Toward mountains without permanent snow and ice: Mountains without permanent snow and ice. *Earth's Future* 5 (5): 418–435. https://doi.org/10.1002/2016EF000514.

Huss, M. & Hock, R. (2018). Global-scale hydrological response to future glacier mass loss. *Nature Climate Change* 8 (2): 135–140. https://doi.org/10.1038/s41558-017-0049-x.

IPCC. (2021). *Climate Change 2021: The Physical Science Basis. Contribution of Working Group I to the Sixth Assessment Report of the Intergovernmental Panel on Climate Change* [Masson-Delmotte, V., P. Zhai, A. Pirani, S. L. Connors, C. Péan, S. Berger, N. Caud, Y. Chen, L. Goldfarb, M. I. Gomis, M. Huang, K. Leitzell, E. Lonnoy, J. B. R. Matthews, T. K. Maycock, T. Waterfield, O. Yelekçi, R. Yu, and B. Zhou (Eds.)]. Cambridge University Press. In Press.

Jackson, L. E., MacDonald, G. M. & Wilson, M. C. (1982). Paraglacial origin for terraced river sediments in bow valley, Alberta. *Canadian Journal of Earth Sciences* 19 (12): 2219–2231. https://doi.org/10.1139/e82-196.

Kenner, R., Phillips, M., Limpach, P., Beutel, J., & Hiller, M. (2018). Monitoring mass movements using georeferenced time-lapse photography_ Ritigraben Rock Glacier, Western Swiss Alps. *Cold Regions Science and Technology*, 145, 127–134. https://doi.org/10.1016/j.coldregions.2017.10.018.

Kondolf, G. M. & Swanson, M. L. (1993). Channel adjustments to reservoir construction and instream gravel mining, Stony Creek, California. *Environmental Geology and Water Science*, 21, 256–269.

Koppes, M. N. & Montgomery, D. R. (2009). The relative efficacy of fluvial and glacial erosion over modern to orogenic timescales. *Nature Geoscience* 2 (9): 644–647. https://doi.org/10.1038/ngeo616.

Kroes, D. E. & Hupp, C. R. (2010). The effect of channelization on floodplain sediment deposition and subsidence along the Pocomoke River, Maryland. *Journal of the American Water Resources Association*, 46 686–699

Kummert, M. & Delaloye, R. (2018). Mapping and quantifying sediment transfer between the front of rapidly moving rock glaciers and torrential gullies. *Geomorphology* 309 (May): 60–76. https://doi.org/10.1016/j.geomorph.2018.02.021.

Kummert, M., Delaloye, R. & Braillard, L. (2018). Erosion and sediment transfer processes at the front of rapidly moving rock glaciers: Systematic observations with automatic cameras in the western Swiss Alps. *Permafrost and Periglacial Processes* 29 (1): 21–33. https://doi.org/10.1002/ppp.1960.

Lambiel, C., Delaloye, R., Strozzi, T., Lugon, R. & Raetzo, R. (2008). ERS InSAR for Detecting the Rock Glacier Activity. Presented at the Kane, D. L. & K. M. Hinkel (eds), *Proceedings of the 9th International Conference on Permafrost*, June 29–July 3, 2008, Fairbanks, Alaska 2: 1019–1024.

Lane, S., Bakker, M., Balin, D., Lovis, B. & Regamey, B. (2014). Climate and Human Forcing of Alpine River Flow. In Anton Schleiss, Giovanni de Cesare, Mario Franca, & Michael Pfister (eds.), *River Flow 2014*, 7–15. CRC Press, Lausanne. https://doi .org/10.1201/b17133-5.

Lane, S. N., Bakker, M., Costa, A., Girardclos, S., Loizeau, J.-L., Molnar, P., Silva, T., Stutenbecker, L. & Schlunegger, F. (2019). Making stratigraphy in the Anthropocene: Climate change impacts and economic conditions controlling the supply of sediment to Lake Geneva. *Scientific Reports* 9 (1): 8904. https://doi.org/10.1038/ s41598-019-44914-9.

Lane, S. N., Gaillet, T. & Goldenschue, L. (2022). Restoring morphodynamics downstream from Alpine dams: development of a geomorphological version of the serial discontinuity concept. *Geomorphology*, 402. https://doi.org/10.1016/j.geomorph.2022.108131.

Lane, S. N. & Nienow, P. W. (2019). Decadal-scale climate forcing of alpine glacial hydrological systems. *Water Resources Research*, 55: 2478–2492. https://doi .org/10.1029/2018WR024206.

Lane, S. N., Richards, K. S. & Chandler, J. H. (1996). Discharge and sediment supply controls on erosion and deposition in a dynamic alluvial channel. *Geomorphology* 1C5 (1): 1–15. https://doi.org/10.1016/0169-555X(95)00113-J.

Lane, S. N., Bakker, M. Gabbud, C., Micheletti, N. & Saugy, J.-N. (2017). Sediment export, transient landscape response and catchment-scale connectivity following rapid climate warming and alpine glacier recession. *Geomorphology* 277 (January): 210–227. https://doi.org/10.1016/j.geomorph.2016.02.015.

Lee, K.-H., Isenhart, T. M., Schultz, R. C. & Mickelson, S. K. (2000). Multispecies riparian buffers trap sediment and nutrients during rainfall simulations. *Journal of Environmental Quality* 29 (4): 1200–1205. https://doi.org/10.2134/jeq2000.004724 25002900040025x.

Leonard, E. M. (1985). Glaciological and climatic controls on lake sedimentation, canadian rocky mountains. *Zeitschrift Fur. Gletscherkunde Und Glazialgeologie* 21: 35–42.

Liermann, S., Beylich, A. A. & Welden, A. (2012). Contemporary suspended sediment transfer and accumulation processes in the small proglacial Sætrevatnetsub-catchment, Bødalen, western Norway. *Geomorphology*, 167: 91–101.

Livingstone, S. J., Clark, C. D., Piotrowski, J. A., Tranter, M., Bentley, M. J., Hodson, A., Swift, D. A. & Woodward, J. (2012). Theoretical framework and diagnostic criteria for the identification of Palaeo-subglacial lakes. *Quaternary Science Reviews* 53 (October): 88–110. https://doi.org/10.1016/j.quascirev.2012.08.010.

Łozinzki, W. von. (1909). Über Die Mechanische Verwitterung Der Sandsteine Im Gemässigten Klima. *Bulletin International de l'Academie Des Sciences de Cracovie*, 1: 1–25. Classe des Sciences Mathematiques et Naturelles.

Lugon, R. (2010). Rock-glacier dynamics and magnitude–frequency relations of debris flows in a high-elevation watershed: Ritigraben, Swiss Alps. *Global and Planetary Change*, 73 (3–4): 202–210, ISSN 0921-8181, https://doi.org/10.1016/j .gloplacha.2010.06.004.

Mahoney, D. T., Fox, J. F. & Al Aamery, N. (2018). Watershed erosion modeling using the probability of sediment connectivity in a gently rolling system. *Journal of Hydrology* 561 (June): 862–883. https://doi.org/10.1016/j.jhydrol.2018.04.034.

Mancini, D. & Lane, S. N. (2020). Changes in sediment connectivity following glacial debuttressing in an alpine valley system. *Geomorphology* 352 (March): 106987. https://doi.org/10.1016/j.geomorph.2019.106987.

Marren, P. M. (2005). Magnitude and frequency in proglacial rivers: A geomorphological and sedimentological perspective. *Earth-Science Reviews* 70 (3–4): 203–251. https:// doi.org/10.1016/j.earscirev.2004.12.002.

Marren, P. M. & Toomath, S. C. (2014). Channel pattern of proglacial rivers: Topographic forcing due to glacier retreat: Channel pattern of proglacial rivers. *Earth Surface Processes and Landforms* 39 (7): 943–951. https://doi.org/10.1002/esp.3545.

Mattson, L. E. & Gardner, J. S. (1991). Mass wasting on valley-side ice-cored moraines, boundary glacier, Alberta, Canada. *Geografiska Annaler: Series A, Physical Geography* 73 (3–4): 123–128. https://doi.org/10.1080/04353676.1991.11880337.

Meyer, L. D., Dabney, S. M. & Harmon, W. C. (1995). Sediment-trapping effectiveness of stiff-grass hedges. *Transactions of the ASAE* 38 (3): 809–815. https://doi.org/10.13031/2013.27895.

Meyrat, R. (2018). *L'utilisation de la photogrammétrie Structure from Motion pour le suivi des glaciers rocheux*. (mémoire de license non publié). Université de Lausanne, Faculté des géosciences et de l'environnement, Suisse.

Micheletti, N., Lambiel, C. & Lane, S. N. (2015). Investigating decadal-scale geomorphic dynamics in an Alpine mountain setting. *Journal of Geophysical Research: Earth Surface* 120 (10): 2155–2175. https://doi.org/10.1002/2015JF003656.

Micheletti, N. & Lane, S. N. (2016). Water yield and sediment export in small, partially glaciated Alpine watersheds in a warming climate: Climate change control in small Alpine watersheds. *Water Resources Research* 52 (6): 4924–4943. https://doi.org/10.1002/2016WR018774.

Milan, D. J. (2012). Geomorphic impact and system recovery following an extreme flood in an upland stream: Thinhope Burn, Northern England, UK. *Geomorphology* 138 (1): 319–328. https://doi.org/10.1016/j.geomorph.2011.09.017.

Miller, H. R. & Lane, S. N. (2019). Biogeomorphic feedbacks and the ecosystem engineering of recently deglaciated terrain. *Progress in Physical Geography: Earth and Environment* 43 (1): 24–45. https://doi.org/10.1177/0309133318816536.

Orwin, J. F. & Smart, C. C. (2004). Short-term spatial and temporal patterns of suspended sediment transfer in proglacial channels, small river glacier, Canada. *Hydrological Processes* 18 (9): 1521–1542. https://doi.org/10.1002/hyp.1402.

Otto, J.-C. (2019). Proglacial lakes in high mountain environments. In Tobias Heckmann & David Morche (eds.), *Geomorphology of Proglacial Systems*. Cham: Springer International Publishing, 231–247. https://doi.org/10.1007/978-3-319-94184-4_14.

Petts, G. E. & Bickerton, M. A. (1994). Influence of water abstraction on the macroinvertebrate community gradient within a glacial stream system: La Borgne d'Arolla, Valais, Switzerland. *Freshwater Biology* 32 (2): 375–386. https://doi.org/10.1111/j.1365-2427.1994.tb01133.x.

Perolo, P., Bakker, M., Gabbud, C., Moradi, G., Rennie, C. & Lane, S. N. (2019). Subglacial sediment production and snout marginal ice uplift during the late ablation season of a temperate valley glacier. *Earth Surface Processes and Landforms*, 44, 1117-1136.

Piton, G., Simon C., Recking, A., Tacnet, J. M., Liébault, F., Kuss, D., Quefféléan, Y. & Marco, O. (2017). Why do we build check dams in alpine streams? an historical perspective from the French experience: A review of the subtle knowledge of 19th century torrent-control-engineers. *Earth Surface Processes and Landforms* 42 (1): 91–108. https://doi.org/10.1002/esp.3967.

Porter, P. R., Smart, M. J. & Irvine-Fynn, T. D. L. (2019). Glacial sediment stores and their reworking. In Tobias Heckmann & David Morche, (eds.), *Geomorphology of Proglacial Systems*. Cham: Springer International Publishing, 157–176. https://link.springer.com/chapter/10.1007/978-3-319-94184-4_10.

Rainato, R., Picco, L., Cavalli, M., Mao, L., Neverman, A. J. & Tarolli, P. (2018). Coupling climate conditions, sediment sources and sediment transport in an Alpine basin: Climate, sediment sources and sediment transport in an Alpine basin. *Land Degradation & Development* 29 (4): 1154–1166. https://doi.org/10.1002/ldr.2813.

Ravanel, L. & Deline, P. (2011). Climate influence on rockfalls in high-Alpine steep Rockwalls: The north side of the Aiguilles de Chamonix (Mont Blanc massif) since the end of the "Little Ice Age". *The Holocene* 21 (2): 357–365. https://doi .org/10.1177/0959683610374887.

Rebetez, M., Lugon, R. & Baeriswyl, P.-A. (1997). Climatic change and debris flows in high mountain regions: The case study of the Ritigraben Torrent (Swiss Alps). In Henry F. Diaz, Martin Beniston, & Raymond S. Bradley, (eds.), *Climatic Change at High Elevation Sites.* Dordrecht: Springer, 139–157. https://doi.org/10.1007/ 978-94-015-8905-5_8.

Renshaw, C. E., Abengoza, K., Magilligan, F. J., Dade, W. B. & Landis, J. D. (2014). Impact of flow regulation on near-channel floodplain sedimentation. *Geomorphology*, 205, 120–127.

Rey, F. (2004). Effectiveness of vegetation barriers for Marly sediment trapping. *Earth Surface Processes and Landforms* 29 (9): 1161–1169. https://doi.org/10.1002/esp.1108.

Schmidli, J. & Frei, C. (2005). Trends of heavy precipitation and wet and dry spells in Switzerland during the 20th century. *International Journal of Climatology* 25 (6): 753–771. https://doi.org/10.1002/joc.1179.

Schmidli, J., Schmutz, C., Frei, C., Wanner, H. & Schär, C. (2002). Mesoscale precipitation variability in the region of the European Alps during the 20th century: Alpine precipitation variability. *International Journal of Climatology* 22 (9): 1049–1074. https://doi.org/10.1002/joc.769.

Schomacker, A. & Kjaer, K. H. (2008). Quantification of dead-ice melting in ice-cored moraines at the high-arctic glacier Holmstr Mbreen, Svalbard. *In Boreas* 37:211–225.

Schrott, L., Götz, J., Geilhausen, M. & Morche, D. (2006). Spatial and temporal variability of sediment transfer and storage in an Alpine basin (Reintal Valley, Bavarian Alps, Gemany). *Geographica Helvetica* 61 (3): 191–200. https://doi.org/10.5194/gh-61-191-2006.

Shakesby, R. A. (2004). Protalus ramparts. *In Encyclopedia of Geomorphology* 1: 813–814.

Shugar, D. H., Burr, A., Haritashya, U. K., Kargel, J. S., Watson, C. S., Kennedy, M. C., Bevington, A. R., Betts, R. A., Harrison, S. & Strattman, K. (2020). Rapid worldwide growth of glacial lakes since 1990. *Nature Climate Change* 10 (10): 939–945. https:// doi.org/10.1038/s41558-020-0855-4.

Sloan, J., Miller, J. R. & Lancaster, N. (2001). Response and recovery of the Eel River, California, and its tributaries to floods in 1955, 1964, and 1997. *Geomorphology* 36 (3–4): 129–154. https://doi.org/10.1016/S0169-555X(00)00037-4.

Smith, V. B. & Mohrig, D. (2017). Geomorphic signature of a dammed Sandy River: The lower trinity river downstream of Livingston dam in Texas, USA. *Geomorphology*, 297, 122–136

Sorg, A., Mosello, B., Shalpykova, G., Allan, A., Clarvis, M. H. & Stoffel, M. (2014). Coping with changing water resources: The case of the Syr Darya river basin in Central Asia. *Environmental Science & Policy* 43 (November): 68–77. https://doi .org/10.1016/j.envsci.2013.11.003.

Stahl, K., Moore, R. D., Shea, J. M., Hutchinson, D. & Cannon, A. J. (2008). Coupled modelling of glacier and streamflow response to future climate scenarios: modelling of glacier and streamflow. *Water Resources Research* 44 (2): 1-13. https://doi .org/10.1029/2007WR005956.

Stoffel, M. & Huggel, C. (2012). Effects of climate change on mass movements in mountain environments. *Progress in Physical Geography: Earth and Environment* 36 (3): 421–439. https://doi.org/10.1177/0309133312441010.

Stoffel, M., Mendlik, T., Schneuwly-Bollschweiler M. & Gobiet, A. (2014). Possible impacts of climate change on debris-flow activity in the Swiss Alps. *Climatic Change* 122 (1–2): 141–155. https://doi.org/10.1007/s10584-013-0993-z.

Stott, T. & Mount, N. (2007). Alpine proglacial suspended sediment dynamics in warm and cool ablation seasons: implications for global warming. *Journal of Hydrology* 332 (3–4): 259–270. https://doi.org/10.1016/j.jhydrol.2006.07.001.

Sturm, M. (1986). Formation of a strandline during the 1984 Jokulhlaup of strandline lake. *Artic* 39: 267–269.

Stutenbecker, L., Delunel, R., Schlunegger, F., Silva, T.A., Šegvić, B., Girardclos, S. Bakker, M. & et al. (2018). Reduced sediment supply in a fast eroding landscape? A multi-proxy sediment budget of the Upper Rhône Basin, Central Alps. *Sedimentary Geology* 375 (November): 105–19. https://doi.org/10.1016/j.sedgeo.2017.12.013.

Surian, N., Righini, M., Lucía, A., Nardi, L., Amponsah, W., Benvenuti, M., Borga, M. & al. (2016). Channel response to extreme floods: insights on controlling factors from six mountain rivers in northern Apennines, Italy. *Geomorphology* 272 (November): 78–91. https://doi.org/10.1016/j.geomorph.2016.02.002.

Swift, D. A., Nienow, P. W. & Hoey, T. B. (2005). Basal sediment evacuation by subglacial meltwater: Suspended sediment transport from Haut Glacier d'Arolla, Switzerland. *Earth Surface Processes and Landforms* 30 (7): 867–883. https://doi.org/10.1002/esp.1197.

Swisstopo, Office fédérale de topographie, ed. (2004). *DHM25: The Digital Height Model of Switzerland.*

Syverson, K. M. (1998). Sediment record of short-lived ice-contact lakes, Burroughs glacier, Alaska. *In Boreas* 27:44–54.

Syvitski, J. P. M. & Milliman, J. D. (2007). Geology, geography, and humans battle for dominance over the delivery of fluvial sediment to the coastal ocean. *The Journal of Geology* 115 (1): 1–19. https://doi.org/10.1086/509246.

Vörösmarty, C. J, Meybeck, M., Fekete, B., Sharma, K., Green, P. & Syvitski, J. P. M. (2003). Anthropogenic sediment retention: Major global impact from registered river impoundments. *Global and Planetary Change* 39 (1–2): 169–190. https://doi.org/10.1016/S0921-8181(03)00023-7.

Warburton, J. (1990). An Alpine proglacial fluvial sediment budget. *Geografiska Annaler: Series A, Physical Geography* 72 (3–4): 261–272. https://doi.org/10.1080/04353676.1990.11880322.

Weber, M., Braun, L., Mauser, W. & Prasch, M. (2010). Contribution of rain, snow and ice-melt in the upper Danube discharge today and in the future. *Supplementi di Geografia Fisica e Dinamica Quaternaria* 33: 221–230.

Westoby, M. J., Glasser, N. F., Brasington, J., Hambrey, M. J., Quincey, D. J. & Reynolds, J. M. (2014). Modelling outburst floods from moraine-dammed glacial lakes. *Earth-Science Reviews* 134 (July): 137–159. https://doi.org/10.1016/j.earscirev.2014.03.009.

Whalley, W. B. (2003). Rock glaciers and protalus landforms: Analogous forms and ice sources on Earth and Mars. *Journal of Geophysical Research* 108 (E4): 8032. https://doi.org/10.1029/2002JE001864.

Williams, G. P. & Wolman, M. G. (1984). *Downstream effects of dams on alluvial rivers.* US Geological Survey Professional Paper 1286, 83 pp.

Winsemann, J., Brandes, C. & Polom, U. (2011). Response of a proglacial delta to rapid high-amplitude lake-level change: An integration of outcrop data and high-resolution shear wave seismics. *Basin Research* 23 (1): 22–52. https://doi.org/10.1111/j.1365-2117.2010.00465.x.

Worni, R., Huggel, C., Clague, J. J., Schaub, Y. & Stoffel, M. (2014). Coupling glacial lake impact, dam breach, and flood processes: A modeling perspective. *Geomorphology* 224 (November): 161–176. https://doi.org/10.1016/j.geomorph.2014.06.031.

Zappa, M. & Kan, C. (2007). Extreme heat and runoff extremes in the Swiss Alps. *Natural Hazards and Earth System Sciences* 7 (3): 375–389. https://doi.org/10.5194/nhess-7-375-2007.

8

Coastal and Deltaic Environments

PAOLA PASSALACQUA AND DAVID MOHRIG

8.1 Introduction

The coastal environment constitutes the triple point interface between land, sea, and atmosphere. As such, the system that shapes this environment arguably possesses the broadest ranges in types and styles of connectivity of any one of the geomorphic environments presented in this book. In fact, the breath is sufficiently great that all aspects cannot be covered in a single chapter, so this contribution is not intended to be exhaustive, rather it is intended to highlight important styles of connectivity in the coastal environment, with particular focus on coastal rivers and river deltas. We will highlight examples of structural connectivity that are driven by physical adjacency of disparate topographic elements (Bracken et al., 2013), functional connectivity set by the array of transport processes controlling the magnitude and directionality of fluxes across the landscape (Bracken et al., 2013), and process connectivity that highlights and identifies how all of the variables interact to define the system's state (Ruddell and Kumar, 2009a). Process connectivity becomes particularly important and helpful in systems affected by nonlocal variables, a common state for coastal environments where, for example, information from the sea is commonly propagated tens to hundreds of kilometres inland through channel networks (Nittrouer et al., 2012) and atmospheric disturbances spawned hundreds of kilometres offshore can significantly modify coastline topography and ecology (Bevington et al., 2017; Olliver et al., 2020).

The coastal zone can be divided into a set of environments that from upstream to downstream comprise (1) coastal rivers and floodplains, (2) deltas and estuaries, (3) wetlands and tidal flats, (4) bays and lagoons, and (5) beach and barrier systems. In this chapter, we will focus on the first two environments as they are responsible for the conveyance of fluxes of water, solids, and solutes to the shore and thus for building the coast. In particular, we will highlight mechanisms of connectivity that are associated with and regulate the exchange of fluxes across sub-environments and their associated time scales (Aalto et al., 2008; Allison et al., 1998; Anderson et al., 2014; Cahoon et al., 2011; Hiatt & Passalacqua, 2015; Mertes, 1997; Wright et al., 2018); for example, the exchange between coastal rivers and their flood-plains or between the deltaic channel network and islands (landmasses surrounded by channels), and the role of plants in modulating connectivity and being affected by connectivity. Our discussions of connectivity in these sub-environments are not intended to be comprehensive, but rather will focus on generalizable relations and properties that enhance our understanding of how the coastal environment functions and how these functions compare to those acting in the other environments that make up Earth's landscape. The exclusion of other coastal environments in this discussion is not indicative of their importance or community interest in these systems, but instead simply is a concession to the brevity of this chapter. Coastal rivers and river deltas are well posed for a connectivity analysis given the complex and dynamic nature of river networks and the presence of sub-environments whose connectivity or dis-connectivity controls the health and well-being of coastal landforms (Passalacqua, 2017; Tessler et al., 2015). Defining connectivity mechanisms in these landforms can help perform a similar analysis in other coastal environments.

We will consider the landward extent of the coastal environment as being defined by the most upstream locations in rivers and streams that tidal plus sea level information are translated to, while its seaward extent is the shoreline. Both of these boundaries are dynamic in the sense that their positions vary over time. The instantaneous upstream position shifts due to the competition between the local river discharge and sea level elevation, while shoreline position varies with the combined magnitudes of uprush from breaking waves, tides, and offshore- and onshore-directed winds. These boundaries can also be characterized using seasonal attributes for the defining variable, and in fact continue adjusting at centennial and millennial scales with changing climate and associated sea level (Anderson et al., 2014; Geleynse et al., 2015). It is therefore necessary to state explicitly the timescale being considered when discussing connectivity within the coastal environment. As mentioned earlier, here we will emphasize a range of scales that spans from the scales associated with flux exchange between sub-environments, thus days and weeks, to the decadal scales relevant for coastal management and its impact on people and man-made infrastructure. How the different types of connectivity within the coastal environment can be used in management practices is the focus of Chapter 15.

Landscapes arise from relatively small differences in extremely large fluxes of solids and solutes across Earth's surface. Movements of these solids and solutes are connected with fluid motion that in the coastal zone is predominantly fresh and seawater, though air also plays an important role in the wind-blown transport of sediment, particularly sand, which constructs coastal dune fields that are recognized as adding considerable resilience to the immediate shoreline. The solids are primarily grains produced by the weathering of sedimentary, igneous, and metamorphic rocks, but locally biogenically produced carbonate grains may dominate, and detrital organic material, while representing a small fraction of deposited and eroded substratum, has an outsized role in the carbon cycle and Earth's climate. In most cases, changes in topography arise from imbalances in the flux of solids across Earth's surface that either cause sediment to accumulate and topography to grow, or promote erosion of the substratum and elevation loss. Whichever is the case, this new topography regulates subsequent patterns of fluid and solids flow, which in turn modifies later topography. Explicit study of co-evolving patterns in solids transport, fluid transport, and topography is a branch of geomorphology known as morphodynamics that was first codified within the coastal system. The morphodynamic approach was originally defined by Wright and Thom (1977), who applied it to studies of beach systems. They saw this approach being applied at all-time scales from instantaneous to geologic and across a wide range of space scales. Here we adopt a morphodynamic approach to study connectivity in the coastal environment. As a result, when evaluating connectivity within the coastal system, we will independently consider the flow of fluid and solids through its sub-environments (Pearson et al., 2020). We argue that considering separately the flow of the two phases within the coastal zone at their relevant spatial and temporal scales is necessary to characterize accurately this system's structural, functional, and process connectivity.

8.2 Definition of Coastal Rivers and River Deltas

The coastal segment of a river system contains the length of river channel upstream from the coastline having flow that responds to the elevation of standing water in the receiving marine basin. Upstream of this reach, river flow approximates normal flow, the case where changes in channel discharge are accomplished through changes in depth and therefore water-surface elevation, with no change in the downstream water-surface slope. Moving downstream into the coastal segment, commonly referred to as the backwater zone, changes in river discharge become increasingly tied to changes in downstream water-surface slope and commensurate adjustments to flow velocity with lesser change in flow depth. These differences in both the ranges of water-surface elevations and flow velocities tied to the same change in river discharge affect connectivity in transport of water, solutes, and solids throughout coastal river systems.

A delta is a geomorphic feature that forms where a river enters a standing body of water (Elliot, 1986; Bhattacharya, 2006). The system appears as a discrete protuberance on the shoreline caused by the accumulation of sediment supplied by a feeder at a rate faster than that at which the sediment can be redistributed by basinal processes (tides, waves). In addition to providing a suite of important environmental services, these systems are hot spots for energy and food production and, most importantly, host hundreds of millions of people (Edmonds et al., 2020). Deltas are characterized by a wide range of spatial and temporal scales: as such, approaches to study these systems vary and provide complementary information to each other. From the quantification of delta structure (e.g., the channel network, island shapes) to dynamics over time (e.g., avulsions, branching) and in response to changing forcing (e.g., sea level rise, subsidence). The past decade has seen the application of high fidelity numerical models and also the development of reduced complexity approaches to modelling deltas. Physical experiments continue to provide access to scaled analogues of real systems. Mathematical approaches such as graph and network theory and information theory are starting to become common as they provide mathematical ways to deal with the spatial and temporal scales involved as well as the numerous variables that represent and regulate fluxes at different times and different locations in the systems. The use of these new techniques relates to the concept of metrics and the quest for ways of characterizing and comparing deltas and validating the results of numerical and physical experiments.

8.2.1 Relevant Space and Time Scales

The importance of scale in any analysis of a natural system is widely recognized for at least two reasons pertaining to connectivity. First, the size of a geomorphic feature and timescale over which measurable change in its shape can be detected are correlated. Bigger features tend to adjust over longer time scales than smaller features. Second, the dominant process or processes acting on a geomorphic feature change with the increasing time and space scales being investigated (Cowell & Thom, 1994; Yuill et al., 2009). Every riverine and deltaic system is constructed out of and can be divided into a set of sub-environments whose boundaries are defined by change in elevation and/or slope of its surface topography. We use these sub-environments to define the minimum time and space scales considered in the contribution. Our minimum time scale is associated with measurable change within a single sub-environment or across the boundaries of adjacent sub-environments. Channels are the dominant topographic elements of rivers and deltas and we will use their dimensions to focus our evaluation; spatial scales being considered here are equal to and greater than the median depth or relief associated with channels constituting the primary network for the coastal system. In practice, this minimum channel dimension restricts our discussion of distances exceeding 1 metre and areas exceeding 1 square metre. Changes in connectivity that lead to measurable topographic adjustments have been shown to occur at daily time scales, days to months associated with individual storm and flood events, seasonal/annual time scales, and decadal scales associated with climate oscillations. While longer scales are also important in coastal rivers and deltas, we will focus on the four scales just described because they are particularly relevant to society, its communities, and infrastructure.

A second set of connected time and space scales are those associated with equilibrium topography and form. Both are important in coastal settings where the net accumulation and redistribution of sediment builds the landscapes. The timescale for equilibrium topography can be estimated from successive land-surface maps and the time interval separating them. Using such a data set, Wagner et al. (2017) defined an invariant equilibrium elevation for Wax Lake Delta (WLD) and then estimated a timescale of 16 years for any subaerial topography on the delta to reach its equilibrium elevation. This timescale, associated with a prograding-aggrading landform ~8 km in length, separates shorter-term topographic changes arising because of system connectivity, from the maintenance of topography via a balance between connectivity driven sedimentation and the longer-term sink of land-surface subsidence plus rising sea level. Connectivity plays an equally important role in preserving the geometries of channels, the most important forms in the coastal systems discussed here. Analysing lidar-derived digital elevation models, Mason and Mohrig (2019) documented how independent motions of the right- and left-banks of a coastal meandering river are compensated over timescales of less than 10 years. Differential migration of river banks along channel bends counterbalances past bank migrations so that a statistically steady state channel width is maintained. Land construction and maintenance of surface elevation and topography are all linked to connectivity within systems at timescales relevant to engineering projects and motivates its examination here.

8.3 Structural Connectivity

8.3.1 Definition of Structural Connectivity in River Deltas and Coastal Rivers

Structural connectivity pertains to the existence of pathways that put in communication sub-environments within a landscape (Figure 8.1). As described throughout this book, structural connectivity does not imply the existence of a flux, but rather the existence of a path that can convey fluxes of water, solutes, and solids, either at all times or under specific conditions. The simplest way to think about structural connectivity in the coastal environment is the river delta network in itself as a set of primary and secondary channels that convey fluxes to the shore and control their

(a)   (b)

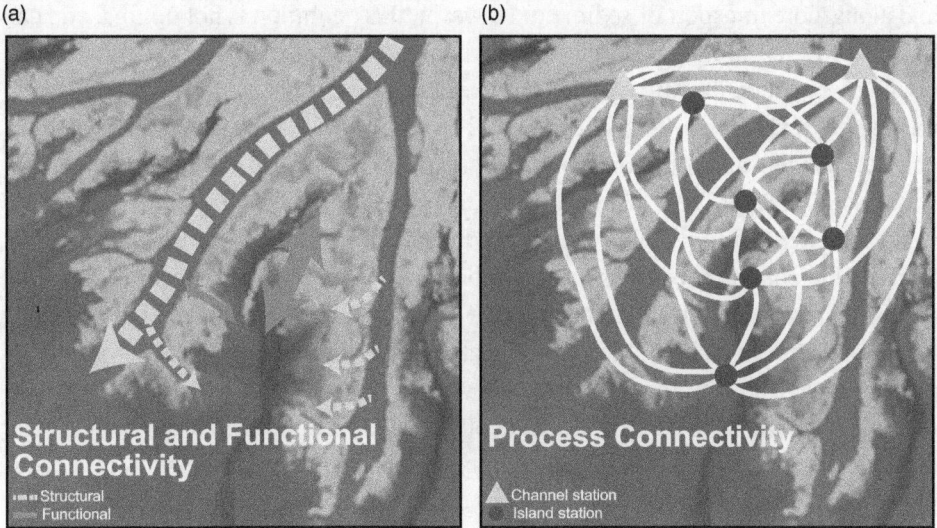

Figure 8.1   Illustration of structural and functional (a) and process connectivity
(b) elements in river deltas. Figure reproduced from Passalacqua (2017). A black
and white version of this figure will appear in some formats. For the colour version,
refer to the plate section.

distribution to the delta plain. Looking at a delta system, it is natural to be captured by
the channel network; often complex (but not always), the network channels are the
main arteries of the system used to transport water, solids, and solutes from the apex
to the shore. This view of the delta system from a hydrodynamic perspective stems
from the similar behavior of their upstream counterparts, tributary river networks.

While channels and networks of channels are the most obvious components in the
structural connectivity of coastal rivers and deltas, another style of connectivity that
is highlighted by a lack of or minimal channelization is also important in these systems. These unchannelized elements are the river banks that circumscribe the aforementioned channels, as well as the shoreline, which is a style of bank defining the
interface between the terrestrial and marine realms. Channel banks commonly take
the form of local topographic highs that separate channel fluxes from the overbank
sub-environments, except during larger transport events when the transfer occurs
along extended segments of bank line rather than through discrete channels. Not
surprisingly, the frequency and magnitudes of these transfers depend on bank line
topography relative to channelized water-surface elevations, which coevolve through
morphodynamic feedbacks explored in Section 8.4 on functional connectivity.

It is common to define a shoreline as the interface between subaerial topography, or
land, and standing water of the greater marine environment. This definition is straightforward to employ at locations where the wave climate and land pattern of the coast
are sufficient to generate beaches and/or wave-cut platforms via the onshore, offshore,

and alongshore transport of sediment. However, this definition is not optimal and can be argued to fail in many coastal river and deltaic settings, because the connectivity boundary is not defined by a water surface elevation, but rather by a rapid change in the degree of lateral confinement of flow. The central point here is that the mouths of coastal channels are commonly entirely subaqueous, yet still dominate routing for water, solids, and solutes. Where studied, these subaqueous sections of deltaic channels can extend multiple kilometers from the positions of their most distal subaerial banks. This confining of transport by subaqueous topography has also been found along the oceanward edge of both wetlands and the islands separating deltaic distributary channels. The topographic troughs that develop in these interdistributary settings and large wetlands are the product of drainage from a largely subaqueous nourishment area (Shaw et al., 2016). Subaqueous channelization of both distributary and tributary-like nourishing flows lead to a second representation for the shoreline that is not defined to a level of standing water, but rather to the loss of all guiding topographic confinement for flows emanating from the terrestrial realm.

8.3.2 Information on Coastal Rivers Processes from Structural Connectivity

The structural connectivity between a meandering coastal river and its adjacent floodplain or wetland is a function of the topography occurring along its bank line and its elevation relative to the water surface in the river. Bank lines are composed of levees, scroll bars, counterpoint bars, mouths of floodplain channels, cutoff river bends, and inactive terraces. These bank types have different characteristic distributions of elevation; tracking which ones occur where is important as each results in different degrees of water, solids, and solutes exchange during the same flood. The inner banks of bends tend to host point bar/scroll bar complexes (Wolman & Leopold, 1957; Bridge, 2009), while natural levees are most common on outer banks of bends, but are also reported flanking both banks of straight channel segments (Wolman & Leopold, 1957; Brierley et al., 1997). Using greater than 1,000 measurements of scroll bar versus levee-crest topography from the coastal Trinity River, Mason and Mohrig (2019) demonstrated that inner-bank scroll bar elevations were systematically lower than outer-bank levees. Lower still are the elevations associated with the accretionary ridges of counterpoint bars, the concave bank deposits that can form adjacent to and downstream of point bars (Hickin, 1979; Page & Nanson, 1982; Smith et al., 2009; Sylvester et al., 2021). Also present at relatively low elevations are the mouths of floodplain channels, as well as tie channels connecting oxbow lakes to the active river (Rowland et al., 2005, 2009; Day et al., 2008). The resulting array of relative bank line elevations leads to connectivity between a river and its overbank across a range of discharges (Mertes, 1997; Czuba, 2019). Tull et al. (2022) have recently shown that floodplain channels, counterpoint bar surfaces, and abandoned river bends all contribute to coastal river-floodplain connectivity at less than bank-full discharges. What controls the distribution of these elements along bank lines is addressed in Section 8.4 on functional connectivity.

8.3.3 Information on River Delta Processes from Delta Network Structure

8.3.3.1 Surface Processes

The structure of the delta network results from the hydrodynamic and morphodynamic processes acting on the system. Taking the Ganges-Brahmaputra-Meghna Delta as example, its complex and heterogeneous network results from the interplay of large fluvial and tidal forcing. The network structure of deltas in itself carries information about processes; thus an analysis of structural connectivity in delta networks can lead to inference of the processes acting on the system. For example, in the GBMD a statistical analysis of island geometry (area, boundary roughness, etc.) highlights scaling breaks and transitions in statistical behaviour that mark corresponding transitions between tidal, fluvial, and inactive portions of the system (Passalacqua et al., 2013). These characteristics inform the mechanisms by which fluxes are partitioned over the delta plain. The so-called inactive part of the system, for example, is separated from the main rivers and fed only by narrow channels; a clear example of how distance from the rivers and characteristics of local pathways are reflected into the delivery of fluxes to a region. While a structural analysis may not be enough to quantify fully flux partitioning over a delta, network-based approaches that use river width for flux estimation have been shown in a few systems to deliver good estimates of flux partitioning at least to the first-order (Dong et al., 2020; Hariharan et al. 2022). The additional advantage of a structural analysis of delta networks is that it can be used to design simpler numerical models that focus only on parts of the system and leverage the structural similarity of subareas to expand the results to a larger spatial extent. The work of Perignon et al. (2020) on the GBMD, for example, showed that with a machine learning approach, various clusters of islands of similar characteristics can be identified as a result of the processes acting on the system. This type of classification can be leveraged to focus high fidelity modelling experiments in islands representative of each cluster and then extrapolate the results to the region of the system that behaves similarly.

Another way in which structural connectivity is informative is in the presence of secondary channels as connectors of rivers and islands. As discussed in Section 8.4, the behaviour of water-surface elevation at various islands in the WLD in coastal Louisiana can be distinguished based on the presence or lack of secondary channels and the topographic characteristics of channel banks. This mechanism of structural connectivity of main channels and island interiors via secondary channels results in Pintail Island being more hydrologically disconnected than downstream islands, such as Greg and Mike Islands, where large flux exchange between rivers and islands is measured also at discharges less than median values (Hiatt & Passalacqua, 2015). This is an important point that will be discussed in Section 8.4 on functional connectivity; connectivity becomes more frequent moving toward the coast with respect to the bank-full/flood behaviour that commonly characterizes upstream river systems.

8.3.3.2 Subsurface Architecture

While the surface network has inherent structural complexity and a vast range of scales that matter from an eco-geomorphic perspective, the subsurface, meaning the stratigraphic record resulting from the time evolution of the system, presents an additional set of challenges and complexity. Studying the subsurface structure is a three-dimensional (3D) problem, different from the mostly 2D nature of surface networks, as fluids are transported not only on the horizontal plane but also at different depths. Additionally, access to 3D information of the subsurface is much more limited; while cores exist in many systems, coring is expensive and not possible anywhere one would like information.

Studying the subsurface is important for a variety of reasons: from a geologic perspective, the subsurface records some of the past dynamics in the system. Those features are still imprinted in the record, thus, give us access to the past of the system and how it may have responded to previous changes in forcing. The subsurface is also the medium in which water, contaminants, and energy resources are contained and travel underground. In many delta systems, the presence of naturally produced pollutants, such as arsenic, is ubiquitous in groundwater and access to clean water for the millions of people living on coastal areas is often a challenge (Michael & Voss, 2008). The analysis of the subsurface then becomes crucial not only for uncovering past dynamics, but also for predicting the future response of the system to changing forcing and for directing management solutions that favor the sustainability of delta systems, their ecosystem services, and the people living on them.

In response to the need for subsurface information and the challenges outlined earlier, the past decade has seen an increased development of approaches to analyze or infer subsurface structure based on image processing and statistics. Multiple point geostatistics, for example, have been used for quantifying uncertainty based on a large set of images (Scheidt et al., 2015), and approaches that mix reduced complexity modelling and observations have been used for probabilistic reconstruction of aquifer structure (van Dijk et al., 2016). Modelling and data analysis have also been used to understand how much of the subsurface distribution and properties of sand bodies can be inferred from surface networks. An important concept in the analysis of subsurface connectivity is that of percolation; while mostly studied in a theoretical sense, percolation thresholds mark an abrupt transition from sparse geobodies to fully connected geobodies that can lead to the transport of water and contaminants over large distances (the whole domain). Thus, knowledge of percolation zones in the subsurface can inform on contaminant presence and transport and be accounted for in management strategies.

8.4 Functional Connectivity

To first-order, coastal landscapes can be lumped into three primary sub-environments, channels, topographically unconfined overbank, and shorelines. Commonly two scales of channels are present, one that sets the maximum depth of scouring within the system and carries significant water discharge across the entire range of water-surface elevations and a second grouping of channels that connects the larger channels to the relatively unconfined topographic surfaces. Coastal river systems are dominated by a single primary channel, while deltaic systems may have a network of these large channels. Both systems have secondary channels that act to connect primary channels to floodplain, wetlands, and/or island tops. A spectrum of physical and ecological mechanisms governs the flux directions and magnitudes both within and between these three very broadly defined sub-environments. These mechanisms define the functional connectivity for coastal river systems and deltas (Figure 8.1). In this section, we will begin by exploring functional connectivity within the primary, larger channels and then move onto a discussion of the connectivity between primary channels and adjacent unchannelized surfaces. This connectivity between primary channels and floodplain, wetlands, and/or island tops must include evaluation of depositional and erosional processes that are tied to construction and maintenance of levees and secondary channels, as well as the role of plants in trapping versus buffering exchanges across this interface. Our evaluation of functional connectivity in these coastal systems will wrap up with a discussion of transport states leading to conserving versus non-conserving deltaic shorelines.

8.4.1 Functional Connectivity Associated with the Primary Channels

The hydraulics connected to open-channel flow approaching a standing body of water have been thoroughly developed in engineering textbooks for more than 60 years (Chow, 1959). What is less developed are the consequences of this flow with gradually varying depth on the transport of sediment through these channels. Because of this uncertainty, the morphodynamic feedbacks between fluid transport and solids transport and topography that are responsible for building channelized coastal environments remain incompletely understood, in spite of the fact that they have been qualitatively identified for a comparable length of time (Lane, 1957). An appropriate first step in accessing how upland flows are connected to shorelines involves estimating the segment of a channelized system affected by the downstream, coastal boundary condition. The length scale for this backwater channel segment (L_b) can be derived from evaluation of the full momentum equation for turbulent open-channel flow and is set by a characteristic flow depth (H) and bed slope (S): $L_b = HS^{-1}$ (Paola & Mohrig, 1996). For a steep (10^{-3}), shallow (1 m) channel this distance is relatively short (1 km), while for a gentle (10^{-4}), deeper (10 m) channel this distance grows (100 km). For the Mississippi River, for example, this length is roughly 500 km (Nittrouer et al., 2011, 2012). Downstream connectivity in the flow of water and bed-material sediment are remarkably different within this backwater reach. For example, while the Mississippi River channel is adjusted in form to ensure continuity in downstream water flux for all possible discharges, continuity in bed-material flux is broken for all but the flood-magnitude discharges. At low water discharges, sediment fluxes decrease downstream by greater than an order of magnitude (Nittrouer et al., 2012). This spatial convergence in the sediment flux induces temporary sediment storage within an upstream portion of the backwater segment of river channel. This temporarily stored sediment appears to be evacuated and transported to the shoreline during floods when measured sand fluxes in the downstream backwater zone are two orders of magnitude greater than those at low flow (Nittrouer et al., 2011). Interestingly, this two order of magnitude increase in sediment flux was connected to only a 3.4 times increase in water flux. Relatively small changes in water flux produce significantly greater changes in bed-material sediment flux; at low values for water flux even sand-bedded coastal channels fall below the threshold of initial motion for sand. This behaviour points to a punctuated connectivity in the flow of bed-material sediment from upstream of the backwater zone to the coast at timescales most commonly ranging from weeks to months, associated with river floods.

The downstream channel deepening leading to punctuated flow of bed material, commonly sand, through the backwater zones of perennial coastal rivers is also spatially connected to a systematic downstream change in river-bend kinematics. Rates of lateral migration drop significantly moving toward the coast and this reduction affects what the floodplain or wetland adjacent to the channel is composed of, which in turn affects connectivity between the channel and overbank sub-environments. In particular, Hassenruck-Gudipati et al (2022) has shown that the overbank area constructed of counterpoint bar or eddy bar deposits (Sylvester et al., 2021) is reduced, as are the number of channel bends that have grown to the point of cutoff, and through this process have been transferred from the active channel to overbank sub-environments. Tull et al., 2022 has demonstrated that these two styles of low-lying regions positioned along a river's bank line, act as corridors for exchanges between the river channel and overbank sub-environments during flooding. Finally, this downstream reduction in the bend migration rates imparts a first-order structure to the shallow subsurface deposits of meandering coastal rivers controlling the time-integrated width of higher permeability channel-filling or channel-belt deposits approaching the coastline. Fernandes et al. (2016) has documented up to a twenty-fold reduction through the backwater zone as the channel-belt drops from 20-times the characteristic channel width to only as single channel width at the coast.

The process of river meandering itself has embedded in it a connectivity between channel banks that warrants discussion. While line models have been quite success-ful at exploring the evolution of meandering channels and channel bends (Howard & Knutson, 1984; Ikeda et al., 1981; Sun et al., 1996), there is no physical reason justifying tandem motion of the inner and outer banks of channels with equal dis-placement vectors. Realizing this deficiency, many recent models for meandering channels have strived to allow banks to move independently (Asahi et al., 2013; Darby et al., 2002; Eke et al., 2014). Recent work on the coastal Trinity River confirms the independent behaviour in bank mobility, producing large changes in cross-sectional area and cross-section shape (Mason & Mohrig, 2018). In addition, these changes to any particular bend set up its later evolution because compensating amounts of outer bank erosion, inner bank deposition, and/or point bar erosion have been shown to take place during successive floods (Mason & Mohrig, 2019) so that on about the decadal time scale a statistically steady channel shape and width are main-tained. Sediment from the inner and outer sidewalls of river bends affects not only local riverine sediment discharge and budgets (Aalto et al., 2008; Lauer & Parker, 2008a, b; Swanson et al., 2008), but also sediment exchange between the river and its local floodplain, which in turn affects local floodplain sedimentation and connec-tivity. It is now clear that these motions of channel sidewalls should not be expected to be constant throughout an inundation event (Darby et al., 2002; Hooke, 1979; Janes et al., 2017). It is also clear that these compensating sidewall adjustments that maintain a dynamic channel width are also observed in straighter channels of deltaic distributary networks (Shaw et al., 2013). Floodplain processes and the connectivity of rivers and their floodplain are further discussed in the next section.

8.4.2 *Functional Connectivity Associated with Secondary Channels and Floodplains*

According to a traditional view of delta systems, channel transport and expansion into the islands (floodplains) happens only in flooding conditions, when islands are supposed to get inundated and nutrients and sediments be delivered to the island interiors. This view supports the classification of a system like the WLD as "river-dominated" given that the riverine input from the Atchafalaya River (feeder) is much larger than waves or tides, leading to flooding being mainly controlled by river discharge. There is a clear difference, however, in the hydraulics of upstream river networks and of downstream deltaic systems; deltas have hydraulically mild slopes, which means that their normal flow tends to be subcritical, thus characterized by Froude numbers less than one. Since the Froude number is the ratio of the flow velocity to the celerity of surface waves, a Froude number less than one implies that the denominator dominates the flow behavior with flows being controlled by downstream conditions, that is, the water level at the bay or outlet. From classic open channel flow, then, it is to be expected that discharge, which is an upstream control, would play a relatively small role in the hydrodynamics of delta system. Why would then inundation be controlled exclusively by discharge? As a functional and process connectivity analysis can highlight, it is in fact not the case.

An analysis of remotely sensed images from Landsat missions at WLD (Geleynse et al., 2015) showed that, while high discharges were associated with high water level, many instances of inundation through time could not be explained with high discharge magnitude. It was in fact wind from the South the main factor associated with delta inundation. The role of wind fronts in modifying significantly water level had already been quantified by others (Walker & Hammack, 2000). Direct quantification of the ubiquitous nature of overbank flow in delta systems came with a connectivity analysis through the collection of observations at the WLD (Hiatt & Passalacqua, 2015). The authors studied a bifurcation by collecting discharge measurements along two channels, Godwall Pass and Main Pass. Regardless of tidal conditions, the authors measured consistently a decrease in channel discharge of up to 54%. The channels and the islands of WLD were thus found to be hydrologically connected even during the mean to low flow conditions that characterized the field campaign. This finding led to the notion that both channels and islands are part of the hydrological network of a delta and should be studied in conjunction rather than in compartments; this concept is the foundation of connectivity as mass and energy transfer through sub-environments within a landscape.

As is the case with deltaic channels and adjacent islands, connectivity between coastal rivers and adjacent floodplains/wetlands is dominated by a summation of depositional and erosion processes associated with the construction of bank line levees and the secondary channels traversing these levees. These two forms of topography dominate the distributions of bank line elevation because bank line elements associated with laterally migrating channel bends, in particular scroll bars, swaths of counterpoint bar deposits and cutoff channel bends commonly disappear downstream as migration rates drop with distance into the backwater zone. The connectivity of water and sediment fluxes between a channel and its overbank surface are of first-order importance because these fluxes control the patterns and amounts of deposition and erosion that modify bank line elevations, which affect in turn future connectivity (Aalto et al., 2008; Bevington & Twilley, 2018; Dean et al., 2016; Howard, 1992; Kesel et al., 1974; Pierce & King, 2008). Regional, high-resolution time-lapse topographic data available from successive airborne geophysical surveys are documenting time and space variations in surface elevation associated with this morphodynamic feedback showing that they are more dynamic than model treatments to date. In particular, airborne lidar surveys are leading the way in defining topographic change associated with functional connectivity in coastal river systems and deltas.

Construction of levees along the bank lines of channels is the result of a focused, net deposition of sediment that was suspended within the uppermost fraction of the channelized water column and then either advected or mixed into slower moving water covering the adjacent overbank surface (Pizzuto, 1987; Mertes, 1997; Asselman & Middelkoop, 1995; Ferguson & Brierley, 1999; Adams et al., 2004; Cazanacli & Smith, 1998; Smith & Perez-Arlucea, 2008; Smith et al., 2009). Once transferred into a water column above a floodplain, wetland, or island surface, it is most commonly assumed that this suspended sediment simply settles onto the bed where the degree of re-entrainment as bedload or reworking by later flows is minimal. As a result of this assumption, numerical models describing levee construction have not included the processes of bedload transport, instead, restricting this mode of sediment transport to erosional crevasse channels that cut across levees (Slingerland & Smith, 1998; Nienhuis et al., 2018). Accurately characterizing the transport of sediment across levees becomes important when evaluating the origin of the secondary channels that traverse levees, but for the moment we will continue to explore the constructional levee properties.

The idealized ramp-like topography that characterizes levees has been shown to develop naturally under transport where sediment supply or concentration decreases with distance from the channel because sediment is being transported from the flow onto the aggrading levee surface (Pizzuto, 1987; Asselman & Middelkoop, 1995; Törnqvist & Bridge, 2002). In the end-member case where sediment settles onto the levee and is not re-entrained into flowing water, the structureless beds of sediment are characterized by the systematic and selective removal of the coarsest grains with distance away from the channel edge (Baitis, 2008). For this end-member case at WLD (Wagner et al., 2017), it is not surprising that analysis of time-lapse topography clearly shows that the interior areas on levees are aggrading slower than areas nearer the channels. Interestingly, the same data sets demonstrate that in spite of differences in bed-aggradation rate, all terrestrial surfaces are aggrading toward a similar, relatively high, stable, elevation. The identification of a high, stable elevation is consistent with higher elevations receiving less mineral sediment due to the vertical sorting of sediment within the water column and decreased frequency of inundation (Pethick, 1993; Marani et al., 2010). As we shall discuss later in this section, it is also consistent with greater baffling of these higher level surfaces from inundating flows by vegetation. We interpret this tie between levee elevation and vegetation as producing the positive correlation between elevation and percent organic matter measured in the top 10 cm of levee deposits by Bevington and Twilley (2018), as well as overall carbon sequestration at WLD (Shields et al., 2017).

Secondary or floodplain channels connect to a primary distributary channel or river at its bank line and therefore define local topographic lows that allow over-bank inundation at less than bank-full conditions. These secondary channels serve as conduits for exchanging water, solids, and solutes between primary channels and adjacent floodplains, wetlands, or island tops. Transport directions in secondary channels have been seen to change sign in both deltaic and riverine systems (Hiatt & Passalacqua, 2015, 2017; Day et al., 2008; Rowland et al., 2009; Tull et al., 2022) as a function of differences in water-surface elevation between the relatively unconfined surface and primary channel that are connected by the secondary conduit. Evolution and long-term stability of secondary channels is primarily set by the movement of sediment. Imbalances in sediment flux act to aggrade and/or degrade the bed, sidewalls, and banks, changing the cross-sectional form and length of secondary channels over time. Particular attention has been paid to a specific style of secondary conduits, crevasse channels, because they are relatively large and, under the right conditions, can grow to the point of capturing the entire discharge of the connected primary channel, causing it to avulse (Fisk, 1952; Smith et al., 1989). Initiated by flows that overtop the banks and levees of primary channels, crevasse channels generate most of their form via progressive erosion of pre-existing bank and levee deposits. The study by Yuill et al. (2016) of the West Bay diversion on the lower Mississippi River provides an illustrative analogue for the initial phase of crevasse development, during which erosional processes connected to a divergence in sediment flux led to increasing water and sediment discharge within the channel. This sediment was transported outward onto the less confined surface where decelerating flow led to a sediment-flux convergence and accumulation of a crevasse-splay deposit (Millard et al., 2017; North & Davidson, 2012). In the West Bay case, splay growth and associated reductions in water surface slope began to stabilize the channel within five years of its formation. Sometimes these channels persist for centuries, constructing large crevasse splays and new land, but they can also heal quickly. Necessary conditions for stability of crevasse channels have been explored numerically by Slingerland and Smith (1998) and Nienhuis et al. (2018). Finally, it is worth noting that high-resolution bare-earth models derived from airborne lidar surveys of floodplains are revealing a broader spectrum in size, shape, and location of secondary channels than previously has been recognized and incorporated into models of channel-overbank connectivity (David et al., 2017; Hassenruck-Gudipati, 2021). We expect that this increase in channel diversity will be accompanied by an increase in models necessary to accurately capture their formational processes.

8.4.3 The Role of Vegetation

Vegetation has been known to play an important role in geomorphic systems, including in shaping river networks (Melton, 1957). Recent numerical modeling studies have quantified the role of vegetation in the hydro-morphodynamic evolution of delta systems. Vegetation helps trapping sediment (Nardin & Edmonds, 2014) and is the main control on the hydrological connectivity of channels and islands (Hiatt & Passalacqua, 2017) among discharge fluctuations, topography, and surface roughness. The presence of vegetation, in fact, determines the amount of water flow "leaked" from channels to islands, leading, in particular, to hydrological dis-connectivity with high island roughness compared to channel roughness. This behaviour is also reflected in the reduction in process connections among island water levels and external forcings in the presence of island vegetation (Sendrowski & Passalacqua, 2017) (to be discussed in Section 8.5).

Spatial variability in vegetation cover can be seen as a structural connectivity element, resulting from functional connectivity elements in the form of fluxes of water and sediment causing topographic changes which then affect plant type and distribution. In modelling, the spatial distribution of vegetation is often not accounted for as vegetation patches are often "subgrid" elements in commonly employed model resolutions. Full two-way eco-geomorphic coupling is rarely done. These gaps in research practice are due to the challenges encountered in fully coupling the system as wide ranges of spatial and temporal scales are involved. Plants have been shown to facilitate in certain cases sediment trapping, regulated by characteristics such as plant spacing and stem density. At the same time, plants organize in patches and they can affect morphodynamics at system scale, thus calling for modelling from particle scale all the way to system scale. The range of temporal scales is also quite large; from storm events, to seasonal changes, to morphodynamic feedback over years and more.

Wright et al. (2018) analyzed the role of vegetation patches with numerical modelling and random distributions of vegetation patches of different size. The simulation results show the existence of a threshold of vegetation cover above which roughness can effectively be considered uniform. Below this threshold, the existence of preferential flow paths (channels with velocity large enough for transporting sediment) leads to a larger flow from channels to islands than would be expected under a scenario of uniform roughness. Natural systems, such as the

WLD, are characterized by seasonality in vegetation cover; these numerical results suggest that during the winter months, when vegetation dies off, island surface roughness could be modeled as uniform. The same result would apply for summer and fall months when biomass is at its peak, when again surface roughness could be assigned a uniform and higher value. During the times of vegetation development, acknowledging the actual spatial distribution and size of vegetation patches, may lead to very different predictions on the behaviour of flows from those obtained under uniform roughness conditions. Remote sensing techniques are expected to play a major role by allowing the quantification of vegetation characteristics at high resolution, including density and patch size, shape, and overall spatial and temporal distributions of vegetation in coastal and fluvial environments.

As we have already discussed in the context of fluid connectivity, vegetation also affects the flow of sediment between primary channels and neighbouring unconfined surfaces. Resulting spatial changes in sediment flux act to evolve the topography at the interfaces between these sub-environments. While sparse vegetation can act to locally accelerate flow and increase turbulence, thereby limiting sedimentation and even eroding the bed (Larsen, 2019; Temmerman et al., 2007; Yamasaki et al., 2019; Yang et al., 2016; Zong & Nepf, 2010), it is far more common for protruding vegetation to decelerate flows, inducing deposition (Christiansen et al., 2000; Fagherazzi et al., 2012; Kirwan & Murray, 2007). This correlation between vegetation and sedimentation can be so strong that in coastal settings such as tidal salt marshes, an increase in vegetation biomass is modeled as always favoring sediment deposition (Kirwan et al., 2010; Kirwan & Mengonigal, 2013). Deltaic systems appear more interesting, with increasing vegetation density increasing sediment trapping and vertical surface accretion up to a point beyond which the presence of vegetation reduces sediment transport across the banks of major channels, creating a buffering effect that promotes greater transport of solids to the coastline (Nardin et al., 2016; Olliver et al., 2020). While unambiguously identifying the causal links between vegetation growth and surface elevation remains difficult, remarkable correlations between vegetation communities and local elevation are suggestive and commonly connected with the emergence of topography of deltaic island-tops and floodplains (Carle et al., 2015; Kleinhans et al., 2018; Smith et al., 2020). In addition, indirect evidence for buffering is found in the analysis of time-lapse lidar surveys by Wagner et al. (2017) where neutral to slightly reduced island elevations were bordered by highest density vegetation. The contributions of supratidal, intertidal, and shallow subtidal plants to sediment trapping and buffering of island tops likely plays a significant, but poorly understood, role in decoupling the growth rates of subaerial land and subaqueous platform at WLD so that the ratio of subaerial land to total area is decreasing as the delta matures (Shaw et al., 2018).

Temporal and spatial changes in community and density of vegetation fringing the banks of channels also affects connectivity within riverine and deltaic systems by producing temporal and spatial changes in bank strength connected with rooting (Abernethy & Rutherfurd, 2001; Lauzon & Murray, 2018; Simon & Collison, 2002). In particular, increase in effective strength of vegetated banks significantly reduces the lateral mobility of channels and the reworking of the coastal surface and its topography by migrating channels. The role of plants in affecting channel form and kinematics has received a great deal of attention because of interests in comparing Earth's modern landscapes to those of an earlier Earth, prior to the advent of terrestrial plants (Gibling & Davies, 2012), as well as unvegetated landscapes of other planets and moons (Dietrich & Perron, 2006). Ielphi and Lapôtre (2020) conclude that the presence of vegetated banks accounts for an order of magnitude slowdown in river migration rates. A similar behaviour has been measured for the distributary channels of the prograding WLD where the linked successions of surface elevation and vegetation classes lead to a downstream increase in channel mobility (Shaw et al., 2013). As a result, the older, higher island of the proximal delta are more stable than the younger, lower islands of the distal delta. Shaw et al. (2013) identified the occurrence of Salix nigra (Black Willow) as correlating with at least an order of magnitude reduction in both lateral and downstream motion of island edges and their linked distributary channels. The emergence of new vegetation communities as surfaces aggrade due to the addition of sediment and subsurface organic mass (Carle et al., 2015; Bevington et al., 2017), can act to stabilize the network delivering water, solids, and solutes to the coastline.

Up until this point, our discussion of vegetation has focused on its role in governing the flow of water and sediment within coastal settings. While clearly important, these feedbacks fail to describe the complete bio-morphodynamic system, because they do not explicitly account for the adverse effects of sedimentation on the plants themselves. Hot spots in sedimentation can become hot spots in plant mortality within coastal environments. Using experimental mesocosms within a backbarrier marsh, Walters and Kirwan (2016) demonstrated that burial of Spartina alterniflora by more than 60 cm of overwash sand led to plant mortality, while new plant growth was suppressed in the 30–60 cm sand addition treatments. A similar set of thresholds has been identified within coastal dune systems where grasses and seedlings buried at different depths in sand under greenhouse or field conditions yielded no emergence, leading to death at sand thicknesses roughly exceeding 60 cm (Maun, 1998). In the coastal channel-margin environment, Hassenruck-Gudipati (2021) found that zones where levee sedimentation exceeded 60 cm became hot spots in plant mortality, affecting vegetation roughness over multiple years and floods, further influencing levee sedimentation and the evolution of topography along river banks. Determining exactly how functional connectivity tunes the vegetation-sedimentation feedbacks governing land-surface accretion is particularly important as the greater coastal system responds to increasing rates of sea level rise.

At the broadest scale, sedimentation constructing deltas is driven by spatial reduction in velocities as channelized and relatively unchannelized deltaic flows reach the open coast. A central question regarding connectivity in sediment flux is how leaky are these coastal landforms. Do they retain nearly all of the sediment delivered from upland rivers and is there a grain-size dependence on retention? The answer to the former question is no, roughly two-thirds of the sediment delivered to deltas is not captured there, moving on into the fully marine realm (Allison et al., 1998; Bobrovitskaya et al., 1996; Goodbred & Kuehl, 1998; Kim et al., 2009). The answer to the latter question is yes, with a majority of the sediment bypassing deltas being 100 microns and smaller in nominal diameter (Shaw & Mohrig, 2014b; Odezulu et al., 2021). This very fine sand and mud has small enough settling velocities to be regularly suspended by wind-driven water waves and advected tens to hundreds of kilometers along and offshore where it can later accumulate as shoreline deposits (Draut et al., 2005) and continental shelf mud blankets (Weight et al., 2011), or pass across the continental shelf-slope break into a deeper marine setting (Mullenbach et al., 2004). What is preferentially retained within deltas is the sediment fraction consisting of grains greater than 100 microns in diameter. This particularly holds true for deltas dominated by riverine- and tidally driven transport processes where, for example, the tips of distributary channels on WLD define the positions where all mineral grains that are fine sand or larger have been scrubbed via selective deposition from the transport system (Shaw & Mohrig, 2014a, b). On the other hand, coarser sediment is not conserved within deltas exposed to wave climates that produce and drive appreciable alongshore and/or cross-shore sediment fluxes through remobilization of delta-front deposits. A particularly well-documented example of delta erosion via waves is tied to the engineered repositioning of the mouth of the Brazos River, Texas, USA, in 1929 (Anderson et al., 2014; Rodriguez et al., 2000). This engineered avulsion cut off sediment delivery to the old Brazos River delta and without this sediment influx, wave-driven alongshore sediment transport removed tens of square kilometres of delta within a decade. Sediment from the old delta contributed to particularly rapid growth of a new delta at the new river mouth situated about ten kilometres down drift. For these cases, where waves rework deposited sediment to produce beaches and/or low relief sea scarps, the shoreline separating subaerial from subaqueous topography serves as a reasonable demarcation of delta extent. The same cannot be said of deltas dominated by riverine and tidal processes. In these cases, adopting the shoreline excludes most or all of the sediment laid down in constructing the subaqueous delta platforms upon which subaerial wetlands and islands can grow

(Cahoon et al., 2011; Shaw et al., 2018), as well as the subaqueous extensions of distributary channels that can persist seaward for multiple kilometres (Shaw & Mohrig, 2014a).

8.5 Process Connectivity

Information theory statistics have been proposed to quantify couplings among hydrological variables (Ruddell & Kumar, 2009a, b) and expanded to multiple variables spatially distributed for the quantification of process connectivity in deltaic systems (Passalacqua, 2017; Figure 8.1). Metrics based on Shannon entropy, such as Mutual Information and Transfer Entropy (Schreiber, 2000), can be used to quantify the information transfer from one variable (source) to another (sink) (Ruddell & Kumar, 2009a, b) or from multiple sources to one sink (Goodwell & Kumar, 2017a, b), based on the probability density function of the variables. These metrics allow the quantification of couplings over time and at different spatial locations in strength, directionality, and statistical significance without limitations due to the commonly observed nonlinearity of these relationships (which would be problematic for classic correlation metrics such as the Pearson linear correlation coefficient).

In a study of process connectivity at WLD, Sendrowski and Passalacqua (2017) quantified the relationships between water surface elevation, river discharge, wind speed and direction, and tides from observations collected over a period of three months. The work highlighted that process connectivity varies through time due to the effect of discharge, wind, and tides acting at different scales on the water level. Using a moving window of 10 days (related to the length of wind fronts in this area), the authors showed the effect of wind in transferring information to water level fluctuations over short timescales. This work also showed the effect of vegetation as water level measured in island interiors became essentially disconnected from external forcing once vegetation was developed, an example, as discussed in Sections 8.3 and 8.4, of how structural connectivity, in the form of secondary channels and bank topography, affects functional and process connectivity.

An analysis of process connectivity can also be used to compare numerical model results to observations and identify which processes are well represented or missing from a numerical model. This operation can be seen as a form of validation of modelling results beyond the typically employed comparisons of modeled and observed time series of a given variable at sensor locations. Most importantly, this type of comparison ensures that the model gets the right answer for the right reason. In the application of Sendrowski et al. (2018), the process network of water level, discharge, wind, and tides obtained from field observations was compared to that obtained from Delft3D numerical modelling. The locations monitored in the field and in the model were the same islands and channels at WLD. The comparison showed that the model could easily capture the behaviour of water transport through channels, but linkages between water level in island interiors and forcings were missing, thus highlighting that the mechanisms of connectivity between channels and islands were not accurately represented in the numerical model.

The environmental variables and transport processes affecting the flow of surface water through coastal river systems and deltas are the same ones governing sediment flux in these systems. However, exploration of process connectivity for the flow of solids is less developed than that for water. This is because transport of solids is focused at the bases of channelized and unchannelized flows making direct measurements difficult and large transport events are commonly associated with storms and floods that further exacerbate direct measurements of sediment flux. As a result, process connectivity has primarily been explored using numerical models or through pre- and post-event correlations and inferences. Numerical models have particularly explored the combined roles of floods, tides, and vegetation on island-top sedimentation (Nardin & Edmonds, 2014; Olliver et al., 2020), while pre- and post-event measurements have focused on the relative roles of floods, hurricanes, and cold fronts in changing island-top and marsh elevations (Bevington et al., 2017). Exploring teleconnections between climate and coastal sediment flux is still in its infancy, but studies such as that by Fraticelli (2006) are quite promising. Her analysis revealing a punctuated growth history for the Brazos River delta highlighted the significance of successive La Nina-to-El Nino cycles in delivering bursts in sediment flux to the coastline. It also highlighted a potential for extreme non-locality to be part of the coastal record.

8.6 Current Issues in Coasts and Deltas Related to Connectivity

Many factors are putting deltas at increasing risk, including sea level rise, extreme weather events, and natural and anthropogenic subsidence. Climate model predictions (Hirabayashi et al., 2013) show that many of the major rivers in the world will experience more frequent events possibly leading to increasing inundation, including the Ganges–Brahmaputra–Meghna Delta. For other systems, for example, the Mississippi River, the discharge magnitude currently exceeded with probability 1 over 100 will become rarer, possibly leading to land loss. Holistic approaches to risk analysis (Tessler et al., 2015) have shown how changes in economic resources could put at increasing risk deltas located in wealthy countries, where resources to manage and limit risk are currently available but are not guaranteed in future scenarios of increasing cost of energy resources. These factors and numerous hurricanes and storm events, including Hurricane Katrina in coastal Louisiana in

August 2005, have motivated research in support of coastal restoration interventions, particularly in the form of nature-based engineering (Giosan et al., 2014; Temmerman et al., 2015), where systems are designed and managed to enable natural processes and the system's inherent capability of growing land. As further discussed in Chapter 15, connectivity principles such as those presented in this chapter, can be leveraged in the assessment and restoration of the environmental services that characterize coastal systems.

References

Aalto, R., Lauer, J. W. & Dietrich, W. E., 2008. Spatial and temporal dynamics of sediment accumulation and exchange along Strickland River floodplains (Papua New Guinea) over decadal-to-centennial timescales. *Journal of Geophysical Research: Earth Surface*, 113(F1), F01S04. doi:10.1029/2006JF000627.

Abernethy, B. & Rutherfurd, I. D., 2001. The distribution and strength of riparian tree roots in relation to riverbank reinforcement. *Hydrological Processes*, 15(1), 63–79.

Adams, P. N., Slingerland, R. L. & Smith, N. D., 2004. Variations in natural levee morphology in anastomosed channel flood plain complexes. *Geomorphology*, 61(1–2), 127–142.

Allison, M. A., Kuehl, S. A., Martin, T. C. & Hassan, A., 1998. Importance of floodplain sedimentation for river sediment budgets and terrigenous input to the oceans: Insights from the Brahmaputra-Jamuna River. *Geology*, 26(2), 175–178.

Anderson, J. B., Wallace, D. J., Simms, A. R., Rodriguez, A. B. & Milliken, K. T., 2014. Variable response of coastal environments of the northwestern Gulf of Mexico to sea-level rise and climate change: Implications for future change. *Marine Geology*, 352, 348–366.

Asahi, K., Shimizu, Y., Nelson, J. & Parker, G., 2013. Numerical simulation of river meandering with self-evolving banks. *Journal of Geophysical Research: Earth Surface*, 118, 2208–2229. doi:10.1002/jgrf.20150.

Asselman, N. E. & Middelkoop, H., 1995. Floodplain sedimentation: Quantities, patterns and processes. *Earth Surface Processes and Landforms*, 20(6), 481–499.

Baitis, E., 2008. *Grain sizes of recent siliciclastic deposits in Wax Lake Delta, Louisiana.* Thesis (B.S.), The University of Texas at Austin, 26 p.

Bevington, A. E., Twilley, R. R., Sasser, C. E. & Holm Jr, G. O., 2017. Contribution of river floods, hurricanes, and cold fronts to elevation change in a deltaic floodplain, northern Gulf of Mexico, USA. *Estuarine, Coastal and Shelf Science*, 191, 188–200.

Bevington, A. E. & Twilley, R. R., 2018. Island edge morphodynamics along a chronosequence in a prograding deltaic floodplain wetland. *Journal of Coastal Research*, 34(4), 806–817.

Bhattacharya, J. P., 2006. *Deltas, in Facies Models Revisited, SEPM Society for Sedimentary Geology.* doi:10.2110/pec.06.84.0237.

Bobrovitskaya, N. M., Zubkova, C. & Meade, R. H., 1996. Discharges and yields of suspended sediment in the Ob' and Yenisey Rivers of Siberia. In *Erosion and Sediment Yield: Global and Regional Perspectives*, eds. D.E. Walling & B.W. Webb, 115–123. Wallingford, UK: International Association of Hydrological Sciences Press.

Bracken, L. J., Wainwright, J., Ali, G. A., Tetzlaff, D., Smith, M. W., Reaney, S. M., & Roy, A. G., 2013. Concepts of hydrological connectivity: Research approaches, pathways and future agendas. *Earth-Science Reviews*, 119. https://doi.org/10.1016/j.earscirev.2013.02.001.

Bridge, J. S., 2009. *Rivers and Floodplains: Forms, Processes, and Sedimentary Record.* John Wiley & Sons, Chichester.

Brierley, G. J., Ferguson, R. J. & Woolfe, K. J., 1997. What is a fluvial levee? *Sedimentary Geology,* 114, 1–9. doi:10.1016/S0037-0738(97)00114-0.

Cahoon, D. R., White, D. A. & Lynch, J. C., 2011. Sediment infilling and wetland formation dynamics in an active crevasse splay of the Mississippi River delta. *Geomorphology,* 131(3–4), 57–68.

Carle, M. V., Sasser, C. E. & Roberts, H. H., 2015. Accretion and vegetation community change in the Wax Lake Delta following the historic 2011 Mississippi River flood. *Journal of Coastal Research,* 31(3), 569–587.

Cazanacli, D. & Smith, N. D., 1998. A study of morphology and texture of natural levees – Cumberland Marshes, Saskatchewan, Canada. *Geomorphology,* 25(1–2), 43–55.

Chow, V. T., 1959. *Open-Channel Hydraulics.* New York, McGraw-Hill, 680 p.

Christiansen, T., Wiberg, P. L. & Milligan, T. G., 2000. Flow and sediment transport on a tidal salt marsh surface. *Estuarine, Coastal and Shelf Science,* 50(3), 315–331.

Cowell, P. J. & Thom, B. G., 1994. Morphodynamics of coastal evolution. In *Coastal Evolution: Late Quaternary Shoreline Morphodynamics,* (eds.) Carter, R. W. G., Woodroffe, C. D., 33–86, Cambridge: Cambridge University Press.

Czuba, J. A., David, S. R., Edmonds, D. A. & Ward, A. S., 2019. Dynamics of surface-water connectivity in a low-gradient meandering river floodplain. *Water Resources Research,* 55, 1849–1870, doi:10.1029/2018WR023527.

Darby, S. E., Alabyan, A. M. & Van de Wiel, M. J. 2008b. Numerical simulation of bank erosion and channel migration in meandering rivers. *Water Resources Research,* 38, 1163, doi:10.1029/2001WR000602, 2002.

David, S. R., Edmonds, D. A. & Letsinger, S. L., 2017. Controls on the occurrence and prevalence of floodplain channels in meandering rivers: Controls on floodplain channels in Meandering Rivers. *Earth Surface Processes and Landforms,* 42(3), 460–472. https://doi.org/10.1002/esp.4002.

Day, G., Dietrich, W. E., Rowland, J. C. & Marshall, A., 2008. The depositional web on the floodplain of the Fly River, Papua New Guinea. *Journal of Geophysical Research: Earth Surface,* 113, 1–19. doi:10.1029/2006JF000622.

Dean, D. J., D. J. Topping, J. C. Schmidt, R. E. Griffiths & T. A. Sabol, 2016. Sediment supply versus local hydraulic controls on sediment transport and storage in a river with large sediment loads. *Journal of Geophysical Research: Earth Surface,* 121, 82–110. doi:10.1002/2015JF003436.

Dietrich, W. E. & Perron, J. T., 2006. The search for a topographic signature of life. *Nature,* 439(7075), 411–418.

Dong, T. Y., McElroy, J. A. N. B., Il'icheva, E., Pavlov, M., Ma, H., Moodie, A. J. & Moreido, V. M., 2020. Predicting water and sediment partitioning in a delta channel network under varying discharge conditions. *Water Resources Research,* 56(11), p.e2020WR027199. https://doi.org/10.1029/2020WR027199.

Draut, A. E., Kineke, G. C., Velasco, D. W., Allison, M. A. & Prime, R. J., 2005. Influence of the Atchafalaya River on recent evolution of the chenier-plain inner continental shelf, northern Gulf of Mexico. *Continental Shelf Research,* 25(1), 91–112.

Edmonds, D. A. L., Caldwell, R., Brondizio, E. S. & Mo Siani, S., 2020. Coastal flooding will disproportionately impact people on river deltas. *Nature Communications,* 11, 1–8, https://doi.org/10.1038/s41467-020-18531-4.

Eke, E., Parker, G. & Shimizu, Y. 2014. Numerical modeling of erosional and depositional bank processes in migrating river bends with self-formed width: morphodynamics of bar push and bank pull. *Journal of Geophysical Research: Earth Surface*, 119, 1455–1483, doi:10.1002/2013JF003020.

Elliot, T., 1986. Deltas. In *Sedimentary Environments: Processes, Facies, and Stratigraphy*, (ed.) Reading, H. G., 113–154. Blackwell Scientific Publication, Oxford.

Fagherazzi, S., Kirwan, M. L., Mudd, S. M., Guntenspergen, G. R., Temmerman, S., D'Alpaos, A., et al. 2012. Numerical models of salt marsh evolution: Ecological, geomorphic, and climatic factors. *Reviews of Geophysics*, 50, RG1002. https://doi.org/10.1029/2011RG000359.

Ferguson, R. J. & Brierley, G. J., 1999. Levee morphology and sedimentology along the lower Tuross River, south-eastern Australia. *Sedimentology*, 46(4), 627–648.

Fernandes, A. M., Tornqvist, T. E., Straub, K. M. & Mohrig D., 2016. Connecting the backwater hydraulics of coastal rivers to fluvio-deltaic sedimentology and stratigraphy. *Geology*, 44 (12), 979–982. doi:10.1130/G37965.1.

Fisk, H. N. 1952. *Geological Investigations of the Atchafalaya Basin and Problem of Mississippi River Diversion*. Vicksburg, MS: US Army Corps of Engineers.

Fraticelli, C. M., 2006. Climate forcing in a wave-dominated delta: The effects of drought–flood cycles on delta progradation. *Journal of Sedimentary Research*, 76(9), 1067–1076.

Geleynse, N., M. Hiatt, H. Sangireddy & P. Passalacqua, 2015. Identifying environmental controls on the shoreline of a natural river delta, *Journal of Geophysical Research Earth Surface*, 120, 877–893, doi:10.1002/2014JF003408.

Gibling, M. R. & Davies, N. S., 2012. Palaeozoic landscapes shaped by plant evolution. *Nature Geoscience*, 5(2), 99–105.

Giosan, L., J. Syvitski, S. Constantinescu & J. Day, 2014. Climate change: Protect the world's deltas. *Nature* 516, 31–33.

Goodbred S. L. & Kuehl S. A., 1998. Floodplain processes in the Bengal Basin and the storage of Ganges-Brahmaputra river sediment: An accretion study using 137Cs and 210Pb geochronology. *Sedimentary Geology*, 121, 239–258.

Goodwell, A. E. & Kumar, P., 2017a. Temporal information partitioning: Characterizing synergy, uniqueness, and redundancy in interacting environmental variables. *Water Resources Research*, 53(7), 5920–5942. doi:10.1002/2016WR020216.

Goodwell, A. E. & Kumar, P., 2017b. Temporal information partitioning networks (TIPNets): A process network approach to infer ecohydrologic shifts. *Water Resources Research*, 53(7), 5899–5919. doi:10.1002/2016WR020218.

Hariharan, J., Piliouras, A., Schwenk, J., & Passalacqua, P., 2022. Width-based discharge partitioning in distributary networks: How right we are, *Geophysical Research Letters*, 49(14), e2022GL097897. doi:10.1029/2022GL097897.

Hassenruck-Gudipati, H. J., 2021. *Understanding Fluvial Topography: Morphodynamic Processes That Build River Levees and Cut Terraces*. The University of Texas at Austin, Dissertation, 115 p.

Hassenruck-Gudipati, H.J., Passalacqua, P. and Mohrig, D., 2022. Natural levees increase in prevalence in the backwater zone: Coastal Trinity River, Texas, USA. Geology, 50(9), 1068–1072.

Hiatt, M. & P. Passalacqua, 2015. Hydrological connectivity in river deltas: The first-order importance of channel-island exchange, *Water Resources Research*, 51, 2264–2282, doi:10.1002/2014WR016149.

Hiatt, M. & P. Passalacqua, 2017. What controls the transition from confined to unconfined flow? Analysis of hydraulics in a coastal river delta, *Journal of Hydraulic Engineering*, 143, p. 6, doi:10.1061/(ASCE)HY.1943-7900.0001309.

Hickin, E. J., 1979. Concave-bank benches on the Squamish River, British Columbia, Canada. *Canadian Journal of Earth Sciences*, 16, 200–203, doi:10.1139/e79-018.

Hirabayashi, Y., R. Mahendran, S. Koirala, L. Konoshima, D. Yamazaki, S. Watanabe, H. Kim & S. Kanae, 2013. Global flood risk under climate change. *Nature Climate Change*, 3, 816–821. doi:10.1038/NCLIMATE1911.

Hooke, J. M. 1979. An analysis of the processes of river bank erosion. *Journal of Hydrology* 42, 39–62, ISSN 0022-1694, https://doi.org/10.1016/0022-1694(79)90005-2.

Howard, A., 1992. Modeling channel migration and floodplain sedimentation in meandering streams. In *Lowland Floodplain Rivers: Geomorphological Perspectives*, (eds.) P. A. Carlingand & G. E. Petts, pp. 1–41, Hoboken NJ: John Wiley & Sons Ltd.

Howard, A. D. & T. R. Knutson, 1984. Sufficient conditions for river meandering: A simulation approach. *Water Resources Research*, 20, 1659–1667. doi:10.1029/WR020i011p01659.

Ielpi, A. & Lapôtre, M. G., 2020. A tenfold slowdown in river meander migration driven by plant life. *Nature Geoscience*, 13(1), 82–86.

Ikeda, S., G. Parker & K. Sawai, 1981. Bend theory of river meanders: Part 1. Linear development. *Journal of Fluid Mechanics*, 112, 363–377. doi:10.1017/S0022112081000451.

Janes, V. J. J., Nicholas, A. P., Collins, A. L. et al. 2017. Analysis of fundamental physical factors influencing channel bank erosion: results for contrasting catchments in England and Wales. *Environmental and Earth Science*, 76, 307, https://doi.org/10.1007/s12665-017-6593-x.

Kesel, R. H., K. C. Dunne, R. C. McDonald, K. R. Allison & B. E. Spicer, 1974. Lateral erosion and overbank deposition on the Mississippi River in Louisiana caused by 1973 flooding. *Geology*, 2(9), 461–464. doi:10.1130/0091-7613(1974)2h461:LEAODOi2.0.CO;2.

Kim, W., Mohrig, D., Twilley, R., Paola, C. & Parker, G., 2009. Is it feasible to build new land in the Mississippi River delta?: EOS. *Transactions, American Geophysical Union*, 90(42), 373–384.

Kirwan, M. L., Guntenspergen, G. R., D'Alpaos, A., Morris, J. T., Mudd, S. M. & Temmerman, S., 2010. Limits on the adaptability of coastal marshes to rising sea level. *Geophysical Research Letters*, 37, L23401. doi:10.1029/2010GL045489.

Kirwan, M. L. & Megonigal, J. P., 2013. Tidal wetland stability in the face of human impacts and sea-level rise. *Nature*, 504(7478), 53–60.

Kirwan, M. L. & Murray, A. B. 2007. A coupled geomorphic and ecological model of tidal marsh evolution. *Proceedings of the National Academy of Sciences*, 104(15), 6118–6122.

Kleinhans, M. G., de Vries, B., Braat, L. & van Oorschot, M., 2018. Living landscapes: Muddy and vegetated floodplain effects on fluvial pattern in an incised river. *Earth Surface Processes and Landforms*, 43(14), 2948–2963. https://doi.org/10.1002/esp.4437.

Lane, E. W., 1957. *A Study of the Shape of Channels Formed by Natural Streams Flowing in Erodible Material*. Missouri River Division Sediment Series Report 9: Omaha, Nebraska, U.S. Army Corps of Engineers, pp. 1–106.

Larsen, L. G., 2019. Multiscale flow-vegetation-sediment feedbacks in low-gradient landscapes. *Geomorphology*.

Lauer, J.W. & Parker, G. 2008a. Net local removal of floodplain sediment by river meander migration. *Geomorphology*, 96, 123–149, ISSN 0169-555X, https://doi.org/10.1016/j.geomorph.2007.08.003.

Lauer, J. W. & Parker, G. 2008b, Modeling framework for sediment deposition, storage, and evacuation in the floodplain of a meandering river: Theory. *Water Resources Research*, 44, W04425, doi:10.1029/2006WR005528.

Lauzon, R. & Murray, A. B., 2018. Comparing the cohesive effects of mud and vegetation on delta evolution. *Geophysical Research Letters*, 45(19), 10–437.

Marani, M., D'Alpaos, A., Lanzoni, S., Carniello, L. & Rinaldo, A., 2010. The importance of being coupled: Stable states and catastrophic shifts in tidal biomorphodynamics. *Journal of Geophysical Research: Earth Surface*, 115, F04004, doi:10.1029/2009JF001600.

Mason, J. & Mohrig, D., 2019. Differential bank migration and the maintenance of channel width in meandering river bends. *Geology*. doi:10.1130/G46651.1.

Mason, J. & Mohrig, D., 2019. Scroll bars are inner bank levees along meandering river bends. *Earth Surface Processes and Landforms*. doi:10.1002/esp.4690.

Mason, J. & Mohrig, D., 2018. Using time-lapse lidar to quantify river bend evolution on the meandering coastal Trinity River, Texas, USA. *Journal of Geophysical Research – Earth Surface*, 123(5), 1133–1144, https://doi.org/10.1029/2017JF004492.

Maun, M. A., 1998. Adaptations of plants to burial in coastal sand dunes. *Canadian Journal of Botany*, 76(5), 713–738.

Melton, M. A., 1957. *An Analysis of the Relations Among Elements of Climate, Surface Properties, and Geomorphology*. Technical Report 11 Office of Naval Research Department of Geology, Columbia University.

Mertes, L. A., 1997. Documentation and significance of the perirheic zone on inundated floodplains. *Water Resources Research*, 33(7), 1749–1762.

Michael, H. A. & Voss, C. I., 2008. Evaluation of the sustainability of deep groundwater as an arsenic-safe resource in the Bengal Basin. *Proceedings of the National academy of Sciences*, 8531–8536, doi:10.1073/pnas.0710477105.

Millard, C., Hajek, E. & Edmonds, D. A., 2017. Evaluating controls on crevasse-splay size: Implications for floodplain-basin filling. *Journal of Sedimentary Research*, 87(7), 722–739.

Mullenbach, B. L., Nittrouer, C. A., Puig, P. & Orange, D. L., 2004. Sediment deposition in a modern submarine canyon: Eel Canyon, northern California. *Marine Geology*, 211(1–2), 101–119.

Nardin, W. & Edmonds, D. A., 2014. Optimum vegetation height and density for inorganic sedimentation in deltaic marshes. *Nature Geoscience*, 7(10), 722–726.

Nardin, W., Edmonds, D. A. & Fagherazzi, S., 2016. Influence of vegetation on spatial patterns of sediment deposition in deltaic islands during flood. *Advances in Water Resources*, 93, 236–248.

Nienhuis, J. H., Törnqvist, T. E. & Esposito, C. R., 2018. Crevasse Splays Versus Avulsions: A Recipe for Land Building with Levee Breaches. *Geophysical Research Letters*, 45, 4058–4067, doi:10.1029/2018GL077933.

Nittrouer, J. A., Mohrig, D., Allison, M. A. & Peyret, A.-P., 2011. The Lowermost Mississippi River: A Mixed Bedrock-Alluvial Channel: *Sedimentology*, 58, 1914–1934, doi:10.1111/j.1365-3091.2011.01245.x.

Nittrouer, J. A., Mohrig, D. & Allison, M. A., 2011. Punctuated sand transport in the lowermost Mississippi River. *Journal of Geophysical Research-Earth Surface*, 116, p.F04025, doi:10.1029/2011JF002026.

Nittrouer, J. A., Shaw, J., Lamb, M. P. & Mohrig, D., 2012. Spatial and temporal trends for water-flow velocity and bed material sediment transport in the lower Mississippi River. *Geological Society of America Bulletin*, 124, 400–414, doi: 10.1130/B30497.1.

North, C. P. & Davidson, S. K., 2012. Unconfined alluvial flow processes: Recognition and interpretation of their deposits, and the significance for palaeogeographic reconstruction. *Earth-Science Reviews*, 111(1–2), 199–223.

Odezulu, C. I., Swanson, T. & Anderson, J. B., 2021. Holocene progradation and retrogradation of the Central Texas Coast regulated by alongshore and cross-shore sediment flux variability. *The Depositional Record*, 7(1), 77–92.

Olliver, E. A., Edmonds, D. A. & Shaw, J. B. 2020. Influence of floods, tides, and vegetation on sediment retention in Wax Lake Delta, Louisiana, USA. *Journal of
Geophysical Research: Earth Surface*, 125, e2019JF005316. https://doi.org/
10.1029/2019JF005316.

Page, K. & Nanson, G., 1982. Concave-bank benches and associated floodplain formation.
Earth Surface Processes and Landforms, 7, 529–543, doi:10.1002/esp.3290070603.

Paola, C. & Mohrig, D., 1996. Paleohydraulics revisited: Paleoslope estimation in coarse-
grained braided rivers. *Basin Research*, 8, 243–254, doi:10.1046/j.1365-2117
.1996.00253.x.

Passalacqua, P., S. Lanzoni, C. Paola & A. Rinaldo, 2013. Geomorphic signatures of
deltaic processes and vegetation: The Ganges-Brahmaputra-Jamuna case study.
Journal of Geophysical Research Earth Surface, 118(3), 1838–1849, doi:10.1002/
jgrf.20128.

Passalacqua, P., 2017. The *Delta Connectome*: A network-based framework for studying
connectivity in river deltas. *Geomorphology*, 277, 50–62, doi:10.1016/j.geomorph
.2016.04.001.

Pearson, S. G., van Prooijen, B. C., Elias, E. P., Vitousek, S. & Wang, Z. B., 2020.
Sediment connectivity: A framework for analyzing coastal sediment transport pathways. *Journal of Geophysical Research: Earth Surface*, 125(10), e2020JF005595.

Perignon, M., J. Adams, I. Overeem & P. Passalacqua, 2020. Dominant process zones in a
mixed fluvial-tidal delta are morphologically distinct. *eSurf*, 8, 809–824, https://doi
.org/10.5194/esurf-8-809-2020.

Pethick, J., 1993. Shoreline adjustments and coastal management: Physical and biological
processes under accelerated sea-level rise. *Geographical Journal*, 159, 162–168.

Pierce A. R. & S. L. King, 2008. Spatial dynamics of overbank sedimentation in floodplain
systems. *Geomorphology*, 100, 256–268. doi:10.1016/j.geomorph.2007.12.008.

Pizzuto, J. E., 1987. Sediment diffusion during overbank flows. *Sedimentology*, 34, 301–
317. doi:10.1111/j.1365-3091.1987.tb00779.x.

Rodriguez, A. B., Hamilton, M. D. & Anderson, J. B., 2000. Facies and evolution of the
modern Brazos Delta, Texas: Wave versus flood influence. *Journal of Sedimentary
Research*, 70(2), 283–295.

Rowland, J. C., Dietrich, W. E., Day, G. & Parker, G., 2009. Formation and maintenance
of single-thread tie channels entering floodplain lakes: Observations from three
diverse river systems. *Journal of Geophysical Research*, 114, F02013, doi:10.1029/
2008JF001073.

Rowland, J. C., Lepper, K., Dietrich, W. E., Wilson, C. J. & Sheldon, R., 2005. Tie channel sedimentation rates, oxbow formation age and channel migration rate from opti-
cally stimulated luminescence (OSL) analysis of floodplain deposits. *Earth Surface
Processes and Landforms*, 30, 1161–1179, doi:10.1002/esp.1268.

Ruddell, B. L. & P. Kumar, 2009a. Ecohydrologic process networks: 1. Identification.
Water Resources Research, 45, p. W03419, doi:10.1029/2008WR007279.

Ruddell, B. L. & P. Kumar, 2009b. Ecohydrologic process networks: 2.
Analysis and characterization. *Water Resources Research*, 45, W03420,
doi:10.1029/2008WR007280.

Scheidt, C., A. Fernandes, C. Paola & J. Caers, 2015. Can geostatistical models represent
natures variability? An analysis using flume experiments. *Petroleum Geostatistics*.
doi:10.3997/2214-4609.201413624.

Schreiber, T., 2000. Measuring information transfer. *Physical Review Letters*, 85, 461–464.
doi:10.1103/PhysRevLett.85.461.

Sendrowski, A. & P. Passalacqua, 2017. Process connectivity in a naturally prograding river
delta. *Water Resources Research*, 53, 3, 1841–1863. doi:10.1002/2016WR019768.

Sendrowski, A., K. Sadid, E. Meselhe, R. W. Wagner, D. Mohrig & P. Passalacqua, 2018. Transfer entropy as a tool for hydrodynamic model validation. *Entropy*, 20, 58. doi:10.3390/e20010058.

Shaw, J. B., Estep, J. D., Whaling, A. R., Sanks, K. M. & Edmonds, D. A., 2018. Measuring subaqueous progradation of the Wax Lake Delta with a model of flow direction divergence. *Earth Surface Dynamics*, 6(4), 1155–1168.

Shaw, J. B., Mohrig, D. & Wagner, R. W., 2016. Flow patterns and morphology of a prograding river delta. *Journal of Geophysical Research – Earth Surface*, 121(2), 372–391. doi:10.1002/2015JF003570.

Shaw, J. B. & Mohrig, D., 2014a. The importance of erosion in distributary channel network growth, Wax Lake Delta, Louisiana, USA. *Geology*, 42(1), 31–34.

Shaw, J. B. & Mohrig, D., 2014b. *Supplemental material: The importance of erosion in distributary channel network growth, Wax Lake Delta, Louisiana, USA.* GSA DATA REPOSITORY 2014008.

Shaw, J. B., Mohrig, D. & Whitman, S. K., 2013. The morphology and evolution of channels on the Wax Lake Delta, Louisiana, USA. *Journal of Geophysical Research – Earth Surface*, 118, 1562–1584. doi:10.1002/jgrf.20123.

Simon, A. & Collison, A. J., 2002. Quantifying the mechanical and hydrologic effects of riparian vegetation on streambank stability. *Earth surface processes and landforms*, 27(5), 527–546.

Smith, N. D., Cross, T. A., Dufficy, J. P. & Clough, S. R., 1989. Anatomy of an avulsion. *Sedimentology*, 36(1), 1–23. https://doi.org/10.1111/j.1365-3091.1989.tb00817.x.

Smith, D. G., Hubbard, S. M., Lecki, D. A. & Fustic, M., 2009. Counter point bar deposits: Lithofacies and reservoir significance in the meandering modern Peace River and ancient McMurray Formation, Alberta, Canada. *Sedimentology*, 56, 1655–1669, doi:10.1111/j.1365-3091.2009.01050.x.

Smith, B. C., Moffett, K. B. & Mohrig, D., 2020. Short-term ecogeomorphic evolution of a fluvial delta from hindcasting intertidal marsh-top elevations (HIME). *Remote Sensing*, 12, 1517, doi:10.3390/rs12091517.

Smith, V., Mason, J. & Mohrig, D., 2020. Reach-scale changes in channel geometry and dynamics due to the coastal backwater effect: The lower Trinity River, Texas. *Earth Surface Processes and Landforms*, 45(3), 565–573, doi:10.1002/esp.4754.

Smith, N. D. & Pérez-Arlucea, M., 2008. Natural levee deposition during the 2005 flood of the Saskatchewan River. *Geomorphology*, 101(4), 583–594.

Shields, M. R., Bianchi, T. S., Mohrig, D., Hutchings, J., Kenney, W. F., Kolker, A. S. & Curtis, J. H., 2017. Carbon storage in the Mississippi River Delta enhanced by ecosystem engineering. *Nature Geoscience*, 10(11), doi:10.1038/NGEO3044.

Slingerland, R. & Smith, N. D., 1998. Necessary conditions for a meandering-river avulsion. *Geology*, 26(5), 435–438.

Sun, T., P. Meakin, T. Jossang & K. Schwarz, 1996. A simulation model for meandering rivers. *Water Resources Research*, 32, 2937–2954. doi: 10.1029/96WR00998.

Swanson, K. M., Watson, E., Aalto, R., Lauer, J.W., Bera, M.T., Marshall, A., Taylor, M.P., Apte, S.C. & Dietrich, W.E., 2008. Sediment load and floodplain deposition rates: Comparison of the Fly and Strickland rivers, Papua New Guinea. *Journal of Geophysical Research*, 113, F01S03, doi:10.1029/2006JF000623.

Sylvester, Z., Durkin, P. R., Hubbard, S. M. & Mohrig, D., 2021. Autogenic translation and counter-point-bar deposition in meandering rivers. *Geological Society of America Bulletin*, https://doi.org/10.1130/B35829.1.

Temmerman, S. & Kirwan, M. L., 2015. Building landwith a rising sea. *Science*, 349, 588–589. http://dx.doi.org/10.1126/science.aac8312.

Temmerman, S., Bouma, T. J., Van de Koppel, J., Van der Wal, D., De Vries, M. B. & Herman, P. M. J., 2007. Vegetation causes channel erosion in a tidal landscape. *Geology*, 35(7), 631–634.

Tessler, Z. D., C. J. Vörösmarty, M. Grossberg, I. Gladkova, H. Aizenman, J. P. M. Syvitski & E. Foufoula-Georgiou, 2015. Profiling risk and sustainability in coastal deltas of the world. *Science*, 349, 638–643. doi:10.1126/science.aab3574.

Törnqvist, T. E. & Bridge, J. S., 2002. Spatial variation of overbank aggradation rate and its influence on avulsion frequency. *Sedimentology*, 49(5), 891–905.

Tull, N., Passalacqua, P., Hassenruck-Gudipati, H., Rahman S., Wright K., Hariharan J., & Mohrig, D., 2022. Bidirectional river-floodplain connectivity during combined pluvial-fluvial events. *Water Resources Research*, 58(3), e2021WR030492, https://doi.org/10.1029/2021WR030492.

van Dijk, W. M., A. L. Densmore, R. Sinha, A. Singh, & V. R. Voller, 2016. Reduced-complexity probabilistic reconstruction of alluvial aquifer stratigraphy, and application to sedimentary fans in northwestern India. *Journal of Hydrology*, 541, 1241–1257, https://doi.org/10.1016/j.jhydrol.2016.08.028.

Wagner, R. W., Lague, D., Mohrig, D., Passalacqua, P., Shaw, J. & Moffett, K., 2017. Elevation change and stability on a prograding delta. *Geophysical Research Letters*, 44(4), 1786–1794, doi:10.1002/2016GL072070.

Walker, N. D. & A. B. Hammack, 2000. Impacts of winter storms on circulation and sediment transport: Atchafalaya-Vermilion Bay Region, Louisiana, U.S.A. *Journal of Coastal Research*, 16(4), 996–1010.

Walters, D. C. & Kirwan, M. L., 2016. Optimal hurricane overwash thickness for maximizing marsh resilience to sea level rise. *Ecology and Evolution*, 6(9), 2948–2956.

Weight, R. W., Anderson, J. B. & Fernandez, R., 2011. Rapid mud accumulation on the central Texas shelf linked to climate change and sea-level rise. *Journal of Sedimentary Research*, 81(10), 743–764.

Wolman, M. G. & Leopold, L. B., 1957. *River flood plains: Some observations on their formation* (No. 282-C, 87–109). US Government Printing Office.

Wright, K., M. Hiatt & P. Passalacqua, 2018. Hydrological connectivity in vegetated river deltas: The importance of patchiness below a threshold. *Geophysical Research Letters*, 45, 10416–10427, https://doi.org/10.1029/2018GL079183.

Wright, L. D. & Thom, B. G., 1977. Coastal depositional landforms: A morphodynamic approach. *Progress in Physical Geography*, 1(3), 412–459.

Yamasaki, T. N., de Lima, P. H., Silva, D. F., Cristiane, G. D. A., Janzen, J. G. & Nepf, H. M., 2019. From patch to channel scale: The evolution of emergent vegetation in a channel. *Advances in Water Resources*, 129, 131–145.

Yang, J. Q., Chung, H., & Nepf, H. M., 2016. The onset of sediment transport in vegetated channels predicted by turbulent kinetic energy. *Geophysical Research Letters*, 43(21), 11–261.

Yuill, B., Lavoie, D. & Reed, D. J., 2009. Understanding subsidence processes in coastal Louisiana. *Journal of Coastal Research*, 10054, 23–36.

Yuill, B. T., Khadka, A. K., Pereira, J., Allison, M. A. & Meselhe, E. A., 2016. Morphodynamics of the erosional phase of crevasse-splay evolution and implications for river sediment diversion function. *Geomorphology*, 259, 12–29.

Zong, L. & Nepf, H., 2010. Flow and deposition in and around a finite patch of vegetation. *Geomorphology*, 116(3–4), 363–372.

Part III

Quantifying Connectivity in Geomorphology

9

Measuring Connectivity

Methodologies for Assessing Connectivity
as a Property of Soils and Landscapes

SASKIA D. KEESSTRA, ARTEMI CERDÀ AND RICHARD E. BRAZIER

9.1 Introduction

Whilst the term 'connectivity' in hydrological and sediment-based research is well known, it is neither used consistently in the existing literature, nor is it clear from that literature, that the connectivity of a landscape, or part of a landscape can be measured (e.g., Keesstra et al., 2018; Parsons et al., 2015). However, it is argued that understanding how source areas of water or sediment are connected to receiving surface waters and deposition areas, is an essential step towards improvement of land management to mitigate flooding, soil erosion and water quality problems. The first part of this chapter, therefore, briefly explores the differences between structural and functional, or process-based, connectivity, specifically with reference to the movement of water and sediment through a landscape to enable thinking about ways to measure and monitor connectivity.

We and others argue that most existing studies do not actually measure connectivity (Turnbull et al., 2018). Instead, they address only part of the story, as they quantify the runoff, or sediment delivery with no information about the source areas, and the fate of the detached soil particles and water molecules. Existing work may describe structural change in a landscape (López-Vicente et al., 2021a), which can perhaps elucidate the potential for connectivity to occur, or indeed the emergent spatial properties of an eco- or landscape system, but it does not quantify the connectivity of a system (defined here as any part of the landscape that is connected) in a process-based manner through time. Moreover, the degree of connectivity in a landscape is dependent on the rainfall properties, land use and soil properties and conditions. All these parameters change over time.

Alternatively, a great deal of work describes fluxes of water and sediment at (sometimes multiple) points in a landscape and infers connectivity of the system via analysis of time series data; from rainfall peak to hydrograph peak or start of sediment flux until peak sediment flux within an event (Keesstra et al., 2019;

Edokpa et al., 2022). Landforms such as rills and gullies also shed light onto the connectivity of a landscape (Rodrigo-Comino et al., 2020). Such data are doubtless useful to understand catchment function, but, alone, they do not provide any information that is able to quantify (e.g.,) how well connected sediment sources are to the outlets of the catchments from which they originate.

Finally, there are many examples of water and, particularly, sediment tracing studies, that attempt to link, either directly or indirectly, water or sediment sources with their sinks (which might more usefully be termed temporary stores). Whilst direct tracing techniques have great potential to describe how quickly material travels over measurable distances, and therefore how connected the system may be, such approaches are rarely used to answer this type of question. In addition, indirect tracing techniques, often termed fingerprinting approaches (in the case of sediments), tend to focus on the source apportionment of material that has left a catchment. Such experiments do not actually explain whether the material that is eroded from a source (hillslope, gully or channel bank) has left the catchment, thus not proving whether connectivity has indeed occurred at all (Keesstra et al., 2018; Turnbull et al., 2018).

This chapter aims to give an overview of the current state of the art when it comes to techniques and methodologies that aim to measure connectivity. In addition, we give an overview of a wish list of (improved) methodologies that are needed to be able to truly assess connectivity, perhaps by combining the structural, functional and tracing approaches that are summarised earlier.

9.2 Basic Concepts to Embed Connectivity Measurement Techniques

To bring the general concept of connectivity down to measurable units, requires to discuss the ways in which empirical measurements of each component of a landscape can be brought together to quantify connectivity. For this, it is important to evaluate, first of all, the diagram introduced by Keesstra et al. (2018); adapted here in Figure 9.1. These authors introduced the concept of stocks and flows in the landscape. For this they used the wording landscape phases and fluxes, these words have some relation with the wording of structural and functional connectivity but are not exactly the same. Their approach allows for focusing on landscapes at a snapshot in time and the fluxes that go through those landscapes at those times. This, in turn, allows for selecting measuring options for each of these elements in different settings and scales. Earlier studies (e.g., Bracken et al., 2013, 2015) argue that connectivity comprises both *structural* (or potential) and *functional* (or process-based) elements and that studies trying to measure connectivity should embrace the notion of interaction (ideally connection) between these elements, in order to describe the connectivity of a system meaningfully.

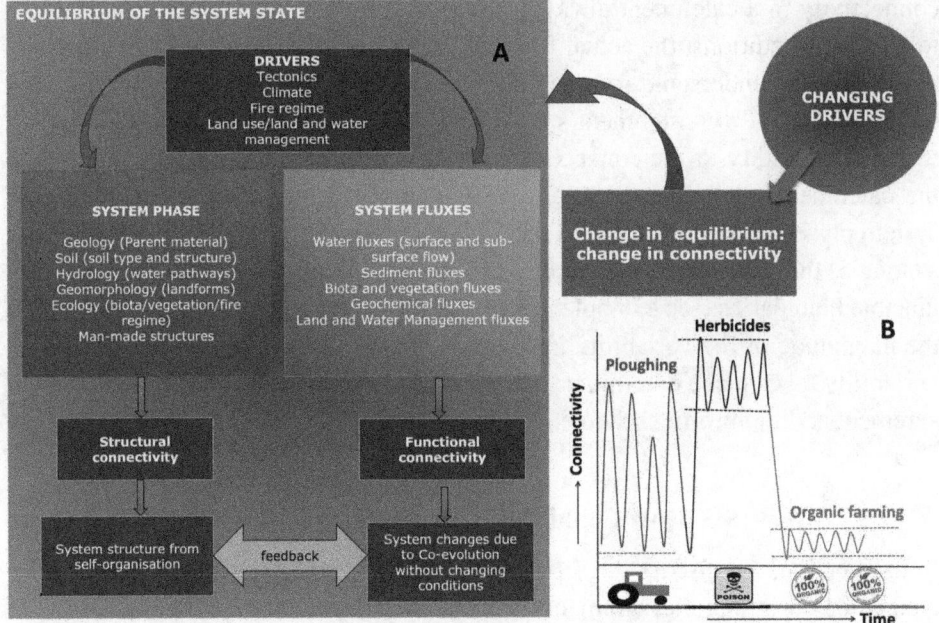

Figure 9.1 (A) Diagram indicating the use of stocks (phases) and flows (fluxes) in the landscape and their drivers (adapted from Keesstra et al., 2018). (B) example of three dynamic equilibria in an agricultural landscape which was first ploughed (high variability of connectivity, but rough ploughed soil reduces connectivity), then under herbicides without ploughing (low infiltration capacity due to soil compaction and removal of vegetation by herbicides) and finally under organic farming without ploughing (low connectivity due to year around vegetation cover and high infiltration capacity of the soil) (cf. Keesstra et al., 2016). A black and white version of this figure will appear in some formats. For the colour version, refer to the plate section.

Figure 9.1A also shows the feedbacks between the way we consider system phases (structural connectivity representing the potential connectivity of a landscape) and the fluxes that actually pass through the landscape (functional connectivity). As long as the conditions of the drivers do not change, the landscape is in dynamic equilibrium (cf. Chorley & Kennedy, 1971). In this conceptual framework, it is argued that, over time, when the landscape system is at equilibrium (Saco & Moreno-De Las Heras, 2013; Keesstra et al., 2018), the system fluxes will remain within certain limits due to the variability in the system phase. This condition means that the system phase may change; but over a longer period of time, the system exhibits the same range of phases. This idea is similar to theories such as channel-forming discharge in fluvial geomorphology and hydrology (Wolman & Miller, 1960; Doyle et al., 2007) where extreme events determine the boundary conditions of such systems.

An example to show how the insights of the diagram in Figure 9.1A can be used to determine the monitoring scheme is shown in Figure 9.1B. The functional connectivity in a catchment always varies, even if all drivers are stable. But due to weather conditions, the actual fluxes through the landscape vary from one day to the next. To understand the natural range of the functional connectivity, a good understanding of the catchment's functioning is needed. The time needed to capture the variability of the connectivity varies depending on the conditions within the catchment (geology, climatic, management, geomorphology, etc. blue box, system phase in Figure 9.1A) and the drivers (top purple box in Figure 9.1A). But as long as they are stable, the amplitude of the variation will not vary beyond predictable boundaries (see example in Figure 9.1B). Therefore, the understanding of the magnitude of the variability in connectivity, and how often the most extreme variability takes place determines the length and organisation of the needed measurement and monitoring scheme.

9.3 How Could We Measure Connectivity?

In this section, we discuss the different ways that connectivity should be considered when designing measurements and monitoring plans:

– Structure. First, we measure structure, that is, the potential for a landscape or ecosystem to be connected, or the emergence of form from processes that have connected the landscape. Topographical surveys are relevant to measure structural connectivity.
– Fluxes. Secondly, we also need to measure fluxes (water and sediment driven, mainly). These measurements, alongside change to the structure (multiple measurements of the structure through time), allow us to quantify feedback and thresholds which are intrinsic to the understanding of connectivity. Run-off discharge and sediment concentration are the most used measurements.

Within the disciplines of hydrology, geomorphology and ecology much progress has been made to measure/monitor/observe different aspects or properties of the landscape and ecosystem that relate to and are often described as connectivity. The following section reviews such approaches, under the headings of structural and functional connectivity and discusses the capacity of these measurements to support a more holistic understanding of connectivity.

9.3.1 *Structure*

The degree to which parts of a landscape or ecosystem are linked (structural connectivity) has developed as a concept quite strongly over the past 25 years (see the conceptual representation of structural connectivity and its evolution through time due to an event introducing a change due to a flux event in Figure 9.2). Examples from ecology, such as the work of With and King, (1997) illustrate the importance of structure in controlling the spatial dependence of a system and how tipping points can be reached beyond which habitats become fragmented or disconnected. The pioneer work done in Ecology about connectivity inspired the geomorphology and hydrology investigations.

In geomorphology, landscape form (or structure) is commonly (and has for some time been) measured and interpreted to understand dominant processes that may have influenced its geomorphic evolution (Brunsden & Thornes, 1979). Measurements of structural connectivity are nowadays demonstrated via a wide range of surveying techniques, which might include data capture using LiDAR (Light Detection and Ranging)(Rodrigo Comino et al., 2018), traditional surveying (Hooke & Souza, 2021), terrestrial laser scanning (Martinez-Agirre et al., 2020), conventional photogrammetry (Mouyen et al., 2020) or 2-D digital photogrammetry approaches (Puttock et al., 2013) or widely used 3-D Structure from Motion (SfM) photogrammetry (Cucchiaro et al., 2018; Forsmoo et al., 2018). The combination of all these methodologies and techniques allows quantifying the potential connectivity of a landscape (for example, the topographic form that controls the pathways of runoff of the next rainfall event), or the structures that have emerged from the dominant processes that have shaped that landscape during the last competent event. Yet most measurements of structure (and inferences of structural connectivity) are only taken once (Lopez-Vicente et al., 2021a) and therefore only describe one snapshot in time. Some workers employ repeat surveys (Walter et al., 2018), which move some way towards describing the evolution of structural connectivity, but still do not measure the way in which changes in structure came about, or indeed the way in which connectivity changes within rainfall/run-off event. Recent advances in survey methodologies and techniques, platforms, sensors and post-processing data workflows and tools (Priddy et al., 2019) have clearly provided a revolution for geomorphologists. We can now rapidly obtain frequent measurements of structure across larger spatial extents, so arguably we must now bring the power of these tools together to confront the true measurements of connectivity.

9.3.2 Fluxes

Taking an ecological perspective, Tischendorf et al. (2001), consider landscape connectivity to be related to the movement of biota, between resources, or habitats. Such a view considers the landscape or ecosystem structure to be somewhat static, playing a supporting role to the more obviously dynamic 'life' within. This emphasis on the functional connectivity of/in landscape-/ecosystems has been well explored in the behavioural ecologies of both plants and animals, whereby movement *is* connectivity, that can be measured in terms of distances travelled, frequencies of journey or migration between patches, within a fragmented, but largely unchanging system (Petit & Burel, 1998). Analogous research is regularly undertaken in hydrology, where flow of water through a catchment is measured, perhaps at multiple locations downstream, to measure the lag times between rainfall and the onset or peak of flow, again within an assumption that the catchment properties do not change, or are static during the measurement period (Abotalib et al., 2021). Similarly, the geomorphic literature is replete with examples of measurements of 'sediment delivery' typically based upon suspended sediment fluxes at a catchment outlet scale and rates of erosion within that catchment quantified at a significantly smaller scale (Burt & Allison, 2010; Heckmann & Vericat, 2018). Changes in sediment delivery ratios are taken to mean changes in connectivity within a catchment, between the source and the outlet (Fryirs, 2013; Lopez-Vicente et al., 2021b). Setting aside the false relationship between measurements at different scales which cannot be reconciled (Parsons et al., 2006), it is clear that such measurements only focus on a representation of sediment flux, and do not consider how hillslope/channel/catchment structure may be changing in response to (or in support of) sediment dynamics. Thus, that past events have a strong effect on the current sediment delivery as was also shown in the study by Masselink et al. (2017) in Navarra, Spain. Here they showed that during large events the sediment is moved from the hillslope to the channel, and during smaller events the channel is slowly cleaned of the deposited sediment. Luk et al., (1993) mention that the sequence of events also determines sediment fluxes on the plot scale. These complex interactions make it even more important to develop measurement and monitoring techniques that are independent of time and place.

9.4 Measuring Connectivity at Multiple Spatial Scales

The connectivity of a system at any given time can only be inferred from measurable variables and parameters. In Section 9.2 we have presented a series of key variables to analyse the structure of a landscape, to study fluxes of water and sediment, and to identify sources, sinks and pathways.

When assessing connectivity of sediment and water on the hillslope scale we identify four main measurement scales to understand connectivity: particle (mm), aggregate (cm), plot (m) and slope (m-km). A nested approach can help to understand how those four scale levels affect the system as a whole and how they interact. Some variables, such as soil organic matter (mm, particle level), grain size and soil organic matter (cm, aggregate scale), vegetation (m, plot scale) and vegetation distribution and morphology (m-km plot scale) are interlinked when looking through scales, and many of the variables are dependent on connectivity. In turn this means that when connectivity increases many soil and watershed properties

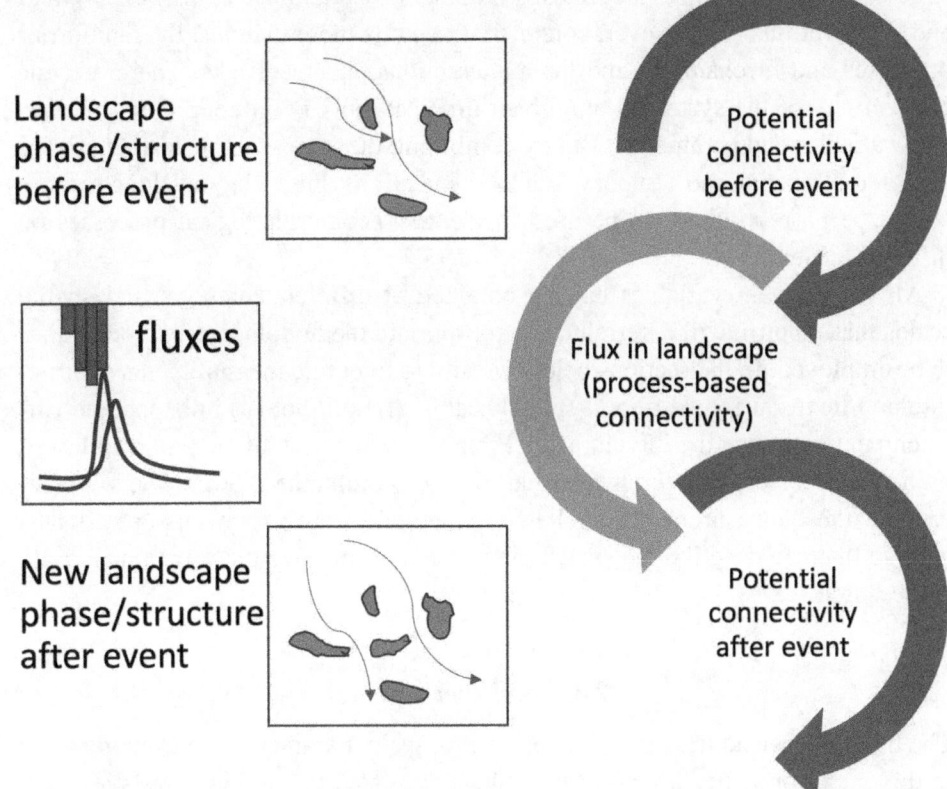

Figure 9.2 Conceptual framework of the links and feedbacks between system phases (structural connectivity, representing the potential connectivity) and fluxes (functional connectivity). A black and white version of this figure will appear in some formats. For the colour version, refer to the plate section.

can change, and humans have modified all landscape and ecosystems, profoundly in many cases, and typically to increase connectivity (through landscape drainage).

As explained in general in Figure 9.1, it is necessary to make the structural and functional connectivity measurable units. In Figure 9.2 this concept is elaborated. Any particular system is evaluated from one snapshot/phase to the next. This image gives you the potential connectivity at a specific moment in time. Next, the fluxes (water and sediment fluxes) go through the landscape and adapt it during an event. The resulting landscape can then be measured again. The most suitable methods to measure the phases depend on the spatial scale of interest. In the case of the fluxes, also the temporal scale of the process that is causing the flux is of importance when deciding the best methods.

However, in most studies, connectivity is used conceptually and is not directly measured, it is inferred from variables we can measure. We can measure the structure of the landscape (landforms, roughness, vegetation, etc.) as well as water and sediment fluxes. However, connectivity entails more than just the landforms, processes and mechanisms and their interactions and feedbacks. Therefore, the connectivity of the system at any given time can only be inferred from measurable variables and parameters. But by combining these measurable characteristics at different spatial and temporal scales connectivity might be developed into a landscape property that can be used to identify geomorphological processes on different scales.

All the relevant variables can be obtained at different temporal and spatial scales that, in turn, will determine the appropriate methodologies and techniques to be employed. In the sections below we aim to integrate measurements of structure and fluxes at four different spatial scales: (i) soil/plot (ii) hillslope and (iii) catchment including the fluvial reach. First, we define each of the scales followed by a review of the measurements that need to be obtained. Secondly, we present how these measurements can be combined and what parameters or variables emerge that can describe the system response and improve the way connectivity is measured.

9.4.1 Soil/Plot Scale

The first scale we address is the pedon-to-plot scale: Parameters we want to assess in this scale for static, structural variables are related to soil characteristics such as texture, soil organic matter, structure and bulk density of the soil. Typically, these characteristics are measured in the field or in the laboratory. When zooming out the soil particles can form aggregates that have properties such as aggregate stability, soil biota and soil moisture; also these variables are usually measured and monitored with field and laboratory monitoring Table 9.1.

When considering the plot scale, soil properties influencing the infiltration and with that the run-off potential, are of specific importance. Both naturally occurring soil properties such as soil water repellency (Tessler et al., 2013), crusting, soil structure, aggregate stability as well as the influence of fire (Cerdà & Doerr, 2008; Keesstra et al., 2014), post-fire management (Smith et al., 2012; Lopez-Vicente et al., 2021a, b) and agricultural land management (Prats et al., 2013) and man-made structures (Mekonnen et al., 2017; Llena et al., 2019) on these properties influence the connectivity at the plot scale.

At the aggregate scale, we see additional processes such as soil aggregate dispersion, splash erosion, which have their own monitoring options (Fernandez-Raga et al., 2017). When moving to the plot scale processes such as infiltration, sheet flow, groundwater recharge, interception, erosion and deposition, soil transpiration are added to the list of fluxes. These variables can be measured in multiple ways

Table 9.1 *Phase and flux parameters and methodologies for the soil to plot scale*

Measurement scale	Structure		Fluxes	
	Variables	Methods	Variables	Methods
Particle	Soil characteristics such as texture, SOM structure, bulk density, water repellency	Lab, field sampling	Soil moisture changes Swelling/shrinking Physical particles (bacteria, viruses, DNA, synthetic)	Soil physics lab, field sampling
Aggregate	Aggregate stability Biota Soil moisture	Lab Continuous monitoring (SMC)	Dispersion Splash erosion Physical particles (bacteria, viruses, DNA, synthetic)	Lab, field sampling Spectroscopy
Plot	Vegetation Stoniness Roughness	SfM OBIA Photogrammetry TLS + methodologies above	Infiltration Sheet flow Through flow Piping Groundwater recharge Interception Soil transpiration Sheet erosion Deposition Tillage erosion	Rainfall simulation Infiltration measurements Tracing Fingerprinting Oxygen isotopes TLS Silt fences Change detection Sediment traps Dyes, RFID plus the earlier mentioned techniques

such as via rainfall simulators, infiltration measurements, tracers and fingerprinting and many more.

At the plot scale, which is often used to assess soil parameters for modelling (Bezak et al., 2021; Tekwa et al., 2021; Sharma et al., 2022; Barrena-González et al., 2020), a whole new set of methodologies comes into play such as SfM, Object Based Image Analysis (OBIA) and Terrestrial Laser Scanners (TLS). On the soil scale the fluxes are related to soil physical processes such as soil moisture changes and swelling and shrinking of soils. But also changes and movement of physical particles, such as bacteria, viruses, DNA, and so on which can be measured only in the laboratory after field sampling.

9.4.2 Hillslope Scale

The processes that occur on the hillslope scale become gradually more complex. The processes described earlier in the soil pedon-to-plot scale also occur within this scale and influence the processes that dominate at this scale. However, regardless of the increased complexity, specific processes can be measured at the hillslope scale. For structural connectivity, we can measure vegetation distribution, landforms and relief, human structures such as terraces, dams, roads and land management interventions (Rodrigo-Comino et al., 2020; Cerdà et al., 2021). All these elements on the hillslope form the base for the potential transfer paths of fluxes. The way this can be measured is ever more complex. In the past, we had to rely on field mapping or aerial photography, but nowadays, deploying drones, and high-resolution remote sensing from satellites allows multiple options for surface assessments. One relatively new, but low-cost, low-technology methods is ISUM (Improved Stock Unearthing Measurement; Rodrigo-Comino & Cerdà, 2018) which uses biomarkers to assess the change in soil level since planting of the crop (such as vines or oranges). The fluxes on this scale are related to flows of water and sediment forming rills, gullies and mass movements. These fluxes can be measured with tracers such as rare earth elements and magnetic tracers (Lizaga et al., 2019), which may also enable understanding of the linkage between the source and sink of material (Table 9.2).

In previous research, the parameters influencing connectivity of the sediments and water at the hillslope scale have been studied: the patchy distribution of the vegetation, the changes in vegetation cover, the influence of rock fragments cover, organic matter; soil water repellency and agricultural land management impacts (Yan and Gao, 2021; Novara et al., 2019; Turski et al., 2022). Change in vegetation cover is the key factor of the run-off generation and the flow of sediments and water at the hillslope scale (Turnbull et al., 2010). The patchy distribution of the vegetation was shown a key factor that determines the runoff at pedon scale as most of the rainfall infiltrates into the soil when vegetation is present. Some researchers found that there is a contrasted response of the vegetated patches and the bare patches (Kröpft et al., 2013; Palacio et al., 2014; Saco et al., 2020). This patchy distribution of the flows causes most of the sediments and water to be trapped in the vegetated patches, and results in the disconnection

Table 9.2 *Phase and flux parameters and methodologies for the hillslope scale*

Measurement scale	Structure		Fluxes	
	Variables	Methods	Variables	Methods
Hillslope	Vegetation distribution Geomorphology Landforms Relief Transfer paths Human structures Effects of land management	Drone based imagery, Geomorphological mapping + methodologies from earlier ISUM	Concentrated flow Rill erosion Gully erosion Mass movements Radionuclides, Dyes, plus the earlier mentioned techniques	LIDAR UAVs Sediment traps Weirs/flumes Real time discharge monitoring of water and sediment

of the flows. Consequently, mapping is necessary to foresee the path and the discharge of the runoff (Shosany et al., 2012; Puttock et al., 2013). Also the presence of rock fragments on the soil surface impacts the water and sediment fluxes in a similar way (Hikel et al., 2013; Li et al., 2022). Recent work has shown that the type of plants, not just the biomass, can affect the run-off generation (Cerdà et al., 2021).

At the hillslope scale, land management becomes more important to take into account, which will also impact the measurement technique that will be chosen as the resulting characteristics of the structural properties influence the connectivity both on the plot as well as on the whole hillslope (see Table 9.2).

9.4.3 Reach Scale

The reach scale (the channel) has a specific set of measurement techniques that do not apply to the other scales. This separate set of techniques is also reflected by the research community that is working on the reach scale that is relatively disconnected from research community studying the other scales discussed here. The techniques identified for structure comprise methodologies to assess bed and bank form and erodibility that involve old and new school techniques to assess small-scale landscape forms; TLS/SfM/OBIA, but also geomorphological mapping. The fluxes for the sediment transport in the channel (bedload, bank erosion and suspended sediment) can be assessed with a multitude of techniques such as ion tracing, bedload tracing, hysteresis analysis and physio-chemical sensors

Table 9.3 *Phase and flux parameters and methodologies for the catchment scale (including the fluvial reach)*

Measurement scale	Structure		Fluxes	
	Variables	Methods	Variables	Methods
Reach	bed form bank form/ erodibility	TLS/SfM/OBIA geomorphological surveys	bank erosion, Bedload transport, discharge Suspended sediment	Ions, bedload tracing Hysteresis Physio-chemical sensors
Catchment	Vegetation distribution Geomorphology Landforms Relief Transfer paths Human structures Effects of land management	Satellite + methodologies earlier	Catchment scale sediment and water discharge	Real time discharge monitoring of water and sediment

(see Powell et al., 2007; Ferguson, 2008; Miller et al., 2021; Oyewumi et al., 2022) (Table 9.3).

9.4.4 Catchment Scale

Measurement of sediment and water connectivity at the river catchment scale is essential to provide a means of clarifying the link between upstream erosion and downstream sediment yield and the role of sediment storage across different temporal scales and to determine how precipitation inputs are either retained within the catchment or transferred downstream to the basin outlet. Elaborating sediment and water budgets may be the best way to approximate connectivity determination but achieving spatial and temporal representativeness presents methodological challenges (Bekele & Gemi, 2021; Berihun et al., 2022) (Table 9.3).

For this scale different techniques become appropriate, such as satellite imagery and catchment scale sediment and water discharge measurements, alongside tracing studies that attribute previous location of material that is then observed to be transported out of catchment. The latter tracing techniques are important to connect to the smaller scales as these types of measurements are usually available. Satellite imagery is often available free of charge, and discharge measurements are undertaken by governmental institutions and often available free of charge or for a limited amount of money. Possibly tracing studies at catchment scales would link these two sets of freely available information.

9.4.5 Nested Approaches

Following the nested approach in the methodologies assessment, the variables that describe the state of a system (Potential structural connectivity) seem to be all valid for the scales that are smaller than the scale studied (Koch et al., 2021; Richards et al., 2021). When the next step is made towards measuring the fluxes (functional connectivity) the methodologies available are in many cases not suitable to be transferred to a different scale, especially when moving from the plot to the hillslope scale (and even more so when upscaling to the combined area of all hillslopes in a catchment). The available techniques for plot scale such as rainfall simulation are just not feasible to undertake over a larger area due to practical or economic reasons. However, emerging methodologies such as high temporal and spatial resolution imagery and sediment or pollutant tracers (see Old et al., 2012, for example) could provide a method to cover, in some instances, the scale gap between the plot and hillslope scale.

We especially think that there is much to gain in methodology development by combining these emerging methodologies with well established, more traditional methodologies (such as geomorphological mapping, sediment collection with sediment traps and by sampling suspended sediment). Examples of methodologies that could potentially be linked are real time monitoring of soil moisture on the point scale, sediment tracing with for instance rare earth oxides or fallout radionuclides (Masselink et al., 2017) and high-resolution imagery for high precision digital elevation models (DEMs). These DEMs could also be used when taken at different time steps, so that changes in landforms and erosion features can be detected using object-based image analysis. The difference of DEMs method is however unsuitable for especially agricultural areas where changes because of ploughing and soil compaction are often higher than changes because of erosion.

Running combined method experiments, such as those described earlier, in parallel and independently from each other can serve as a means to gain better insights into the processes we are studying by explaining the measured differences. Such a combined approach will have a twofold benefit: first, it will show the capabilities of certain techniques to capture the process of interest, and secondly it will give insights into the connectivity of hillslopes when bringing the measurement techniques together.

Table 9.4. presents the combined list of different key variables (combining Tables 9.1, 9.2 and 9.3) alongside a list of some of the specific methodologies and techniques. These have been grouped by the three measurement components presented in Figure 9.1.

Table 9.4 *Measuring connectivity from soil to the catchment scale*

Measurement Scale	Structure		Fluxes	
	Variables	Methods	Variables	Methods
Particle	Texture SOM	Lab, field sampling	Soil moisture changes Swelling/ shrinking Physical particles (bacteria, viruses, DNA, synthetic)	Soil physics lab, field sampling
Aggregate	Aggregate stability Biota Soil moisture	Lab Continuous monitoring (SMC)	Dispersion Splash erosion Physical particles (bacteria, viruses, DNA, synthetic)	Lab, field sampling Spectroscopy
Plot	Vegetation Stoniness Roughness	SfM OBIA Photogrammetry RADAR Aerial photography TLS + methodologies earlier	Infiltration Sheet flow Through flow Piping Groundwater recharge Interception Soil transpiration Sheet erosion Deposition Tillage erosion Dyes, RFID plus the earlier mentioned techniques	Rainfall simulation Infiltration measurements Tracing Fingerprinting Oxygen isotopes TLS Silt fences Change detection Sediment traps
Hillslope	Vegetation distribution Geomorphology Landforms Relief Transfer paths Human structures Effects of land management	Drone based imagery, Geomorphological mapping + methodologies from above	Concentrated flow Rill erosion Gully erosion Mass movements Radionuclides, Dyes, plus the earlier mentioned techniques	LIDAR UAVs Sediment traps Weirs/flumes Real time discharge monitoring of water and sediment

Table 9.4 *(cont.)*

Measurement Scale	Structure		Fluxes	
	Variables	Methods	Variables	Methods
Reach	bed form bank form/ erodibility	TLS/SfM/OBIA geomorphological surveys	bank erosion, Bedload transport, discharge Suspended sediment	Ions, bedload tracing Hysteresis Physio-chemical sensors
Catchment	Vegetation distribution Geomorphology Landforms Relief Transfer paths Human structures Effects of land management	Satellite + methodologies above	Catchment scale sediment and water discharge	Real time discharge monitoring of water and sediment

Example of the integration of variables and methodologies of the three different measurement components. The arrows indicated variables that can be upscaled form one of the measurement scale to the other.

9.5 Examples of Implementation of Measuring the Concept of Connectivity

9.5.1 Example 1: A Catchment Scale Measuring and Monitoring Scheme for Connectivity using a Nested Approach

To explain a possible way to approach the catchment-scale measurement and monitoring scheme we take as an example a meso-scale (i.e., 100–1,000 km^2) catchment. Measurement of connectivity is based on a nested approach (Figure 9.3). First, fluxes are measured by following a nested-catchment approach, with hydro-sedimentological gauging stations at different stream-order tributaries until reaching the mainstream. Gauging stations measure, in a continuous way and with the most up to date methodologies and highest time resolution (e.g., five-minute): (i) groundwater level, (ii) stream discharge, (iii) stream water conductivity, and (iv) stream suspended sediment and bedload transport; in parallel, automatic water sampling is set up both for routine time-integrated sampling and high resolution monitoring during flood events, as well as enabling the potential for fingerprinting of sediments to be performed, if desirable. Where available, historical hydro-sedimentological data records of water and sediment fluxes will also be used. Meteorological stations

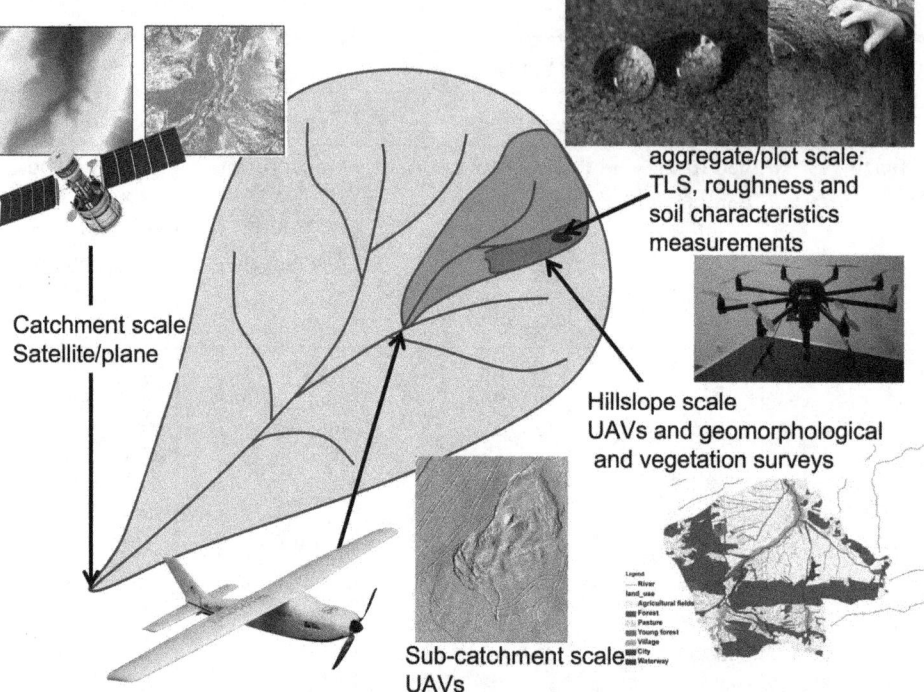

Figure 9.3   Set of possible methodologies for different scales within a catchment (adapted from Keesstra et al., 2018). A black and white version of this figure will appear in some formats. For the colour version, refer to the plate section.

would be required to be adequately distributed across the catchment to quantify drivers and input through measurements of: (i) rainfall, (ii) wind, (iii) evapotranspiration, (iv) moisture, and (v) temperature, though notably improvements to rainfall radar might enable remote quantification of rainfall intensity at five-minute time-steps (Zhang et al., 2019). Secondly, characteristics of landscape structure would be measured mainly by combining different remote sensing techniques (Smith & Vericat, 2014). DEMs would be elaborated from LiDAR and SfM surveys. Past aerial data and information about vegetation, soil and land-uses would be obtained from satellite, aerial photographs and hyperspectral surveys. Thirdly, water tracing (Soulsby et al., 2015) and sediment fingerprinting (Walling, 2013) would be carried out to determine sources, pathways and residence time within the catchment. Water tracing is based on: (i) the characterization of the different water sources (by soil water and groundwater sampling) (Heidbüchel et al., 2012); and (ii) the determination of the run-off processes at the event scale by separating hydrographs (e.g., SEC, stable isotopes, solutes) (Klaus & McDonnell, 2013). Sediment fingerprinting follows the same structure: (i) the sediment sources are characterized separating the surface (by landscape units approach based on type zones defined using landscape structure data) from the subsurface (by field recognizance, e.g., gullies, eroding banks) (e.g., Blake et al., 2009; Owens et al., 2012); (ii) determination of sediment fingerprints of material in transit at the event scale by suspended sediment sampling (e.g., traps, bed sampling, automatic sampling) (e.g., Smith & Blake, 2014), and (iii) establishing the long-term residence time of the sediment within the catchment by combined modelling and tracing methodologies (Evrard et al., 2015) and the sediment of the floodplain and terraces (i.e., >100 yr; e.g., Lecce & Pavlowsky, 2014; Kalantari et al., 2019). Finally, after measuring fluxes at nested scales, evaluating the structural changes of the basin and applying the water tracing and sediment fingerprinting, hydrological and sedimentological data will be integrated permitting establishment of water and sediment budgets across representative spatial (i.e., from the bare slope, through the river channel and up to large-scale river catchments) and temporal scales (i.e., starting at the hourly daily scales, passing from the seasonal to the annual scales, and finally reaching the long-term/centurial/millennial time scales) that demonstrate linkage between key landscape units.

9.5.2 *Example 2: Measuring Connectivity of Soil Carbon Erosion*

Managing Soil Organic Carbon (SOC) stocks and storage capacity requires prevention of losses through physical erosion (Garcia-Ruiz et al., 2017). Research on SOC erosion focuses on processes at point or plot scale, where there is negative feedback between SOC stocks and erosion susceptibility with low C content. Lower SOC generally correlates with disintegration of soil aggregates and loss of soil structure, which reduces the physical stability of the soil profile (Chenu & Le Bissonnais, 2000), which is usually measured following experiments such as explained in Table 9.1. Accelerated erosion (0.05–50 mm yr^{-1} loss of surface soil depth) can occur during or immediately after physical interventions such as tillage (Cerdà et al., 2020). Physical erosion along hillslopes can lead to greenhouse-gas emissions from deeper SOC that is exposed as layers erode from the upper slopes or to C sequestration sinks in the depositional environments. Studies show a range of +1 Pg yr^{-1} of additional sequestration to −1 Pg yr^{-1} additional C release (Smith, 2004; see Figure 9.4) with predictions that soil erosion is a source of atmospheric carbon (Lugato et al., 2018; Smith et al., 2020). These kinds of numbers are often derived from modelling studies as they are difficult to measure directly. However, measurements on hillslope scale could be done using methodologies as explained in Tables 9.2 and 9.3. Especially remote sensing techniques (from drones to satellites) are useful to assess changes in topography as well as deriving the actual reduction of organic matter in the top soil layer (e.g., Angelopoulou, et al., 2019; Andries et al., 2021). It is important to note that this is only possible in areas where there the soil is bare at least for part of the year, when the measurements are done.

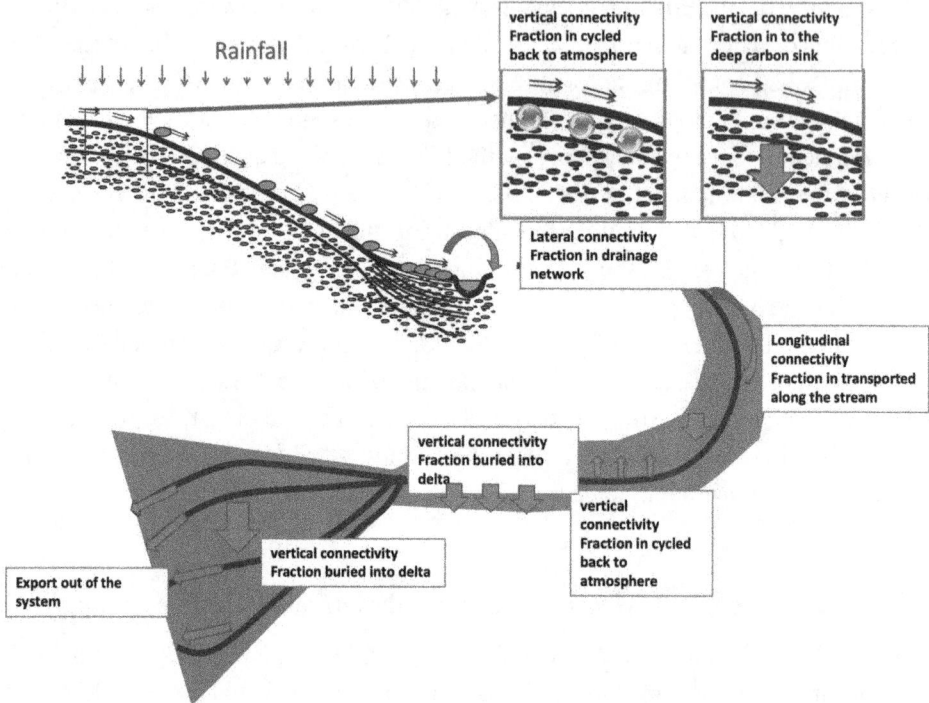

Figure 9.4   Conceptual figure demonstrating the Connectivity principle. Small-scale soil processes are connected and integrated across watersheds. Transmission and multiplication of these processes is transported along stream and river networks before with soil constituents (e.g., soil organic matter or soil aggregates) are exported from the system. A black and white version of this figure will appear in some formats. For the colour version, refer to the plate section.

Geomorphological connectivity can be used as an approach to identify where it is necessary to quantify the transport and fate of eroded SOC (Bracken et al., 2015; Keesstra et al., 2018). Key principles identify the mass transport linkages that impact carbon cycling at a range of scales across the landscape. The linkages include (i) vertical connectivity of soil profiles that expose deeper *C* to the atmosphere at the point of erosion or that bury and preserve a fraction of the carbon in deposition environments at the toe of eroding slopes, (ii) lateral connectivity that releases a fraction of the transported *C* to the surface water drainage networks where (iii) vertical connectivity during surface water can releases greenhouse gasses (GHGs) to the atmosphere as dissolved organic matter is oxidised in the water column or buried particle-bound SOC through sedimentation where it is preserved (Fryirs et al., 2007; see Figure 9.4). This oxidation and burial occur during river transport, in deltas and on the coastal marine shelf. The use of geomorphological knowledge is essential to be able to determine where measurement and monitoring equipment, such as mentioned in Table 9.4, can best be installed.

The exact selection of the most adequate measurement methods depends on the local conditions in the field, depending on climate, relief, vegetation and soil type.

Negative feedback is a major concern, whereby the reduced fertility of eroding agricultural land can result in lower net primary production, reduced root mass to stabilise soil against erosion (Vannoppen et al., 2015), and reduced plant carbon input below ground (Sombrero & de Benito; 2010), thus reducing soil carbon sequestration (Chirinda et al., 2014). Studying the connectivity of carbon cycle dynamics at the landscape scale will improve the understanding of the mitigation potential of agricultural carbon loss to erosion. For this purpose, small-scale measurements determining soil strength can be included, as well as bank erosion stability measurements (surveys) and connectivity potential using remote sensing of the hillslopes may be considered in the monitoring plan.

9.6 Conclusions and a Wish List of Methodologies to Measure and Monitor Connectivity

Figure 9.5 shows a diagrammatic representation of the key components that may be used to measure connectivity. For each component we have identified a series of key variables alongside a series of techniques that may be used to measure each variable.

For each landscape phase/structure scale and associated fluxes, there are a set of measurement techniques available. However, the connection between the different sections in terms of time and scale is still largely an open question.

The hypothetical experimental designs in the examples (Section 9.5 of this chapter) show that the theoretical base is there, but the connection of the different elements in the measurement scheme is still vague. In Figure 9.5 we list the key variables and main techniques at different scales, highlighting areas where our understanding of the system is most developed and also where the greatest confidence lies in the measurement approaches that could be combined to quantify connectivity.

However, a key goal must be to improve our ability to quantify dynamics in any system, as this will determine the minimum length of a monitoring campaign needed to fully assess the response of a system to a given flux (see Figure 9.1). This is especially true for dryland environments, with larger temporal differences in terms of fluxes of sediment and water due to erratic rainfall and temporal differences in vegetation cover. Especially, areas where wildfire and subsequent post-fire management treatments have been implemented create a rapidly changing environment, for which the connectivity is not easy to capture (Lopez-Vicente et al., 2020).

The revolution in photogrammetry and the acquisition of such imagery with drones and even readily available smart phones, has brought enormous opportunities

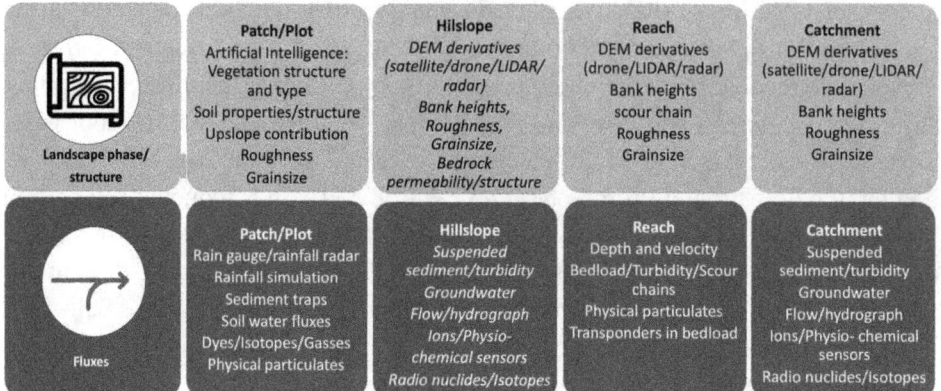

Figure 9.5   Measuring connectivity: measurement components, key variables and techniques.

for connectivity research and will continue to be explored in more detail in the near future. Where post-processing of such data was only previously possible for a few specialists and very laborious, these type of techniques are now available for many researchers.

Regardless of all these developments there is still a large gap between studies done on hillslope processes and studies done in channel environments. The processes on the hillslope are different from the in-channel processes and typically are measured by different researchers whereas, in reality, the two are (at times well) connected (Masselink et al., 2017; Keesstra et al., 2019).

We conclude that there is great potential to measure, or perhaps more appropriately, quantify connectivity of water and sediment across a wide range of scales and landscape and ecosystems. However, we also argue that currently very few, if any, studies actually do this well. We argue that in part this is because few researchers formulate explicit questions about connectivity at the heart of their research, preferring to experiment within traditional scales of interest; hillslopes, channels and catchments, using measurement techniques that, in isolation, do not bridge gaps across scales in a coherent manner.

This chapter has given an overview of the difficulties in measuring and monitoring connectivity. Most studies do not actually measure connectivity, but only the potential of a landscape to allow connectivity to occur; or the connectivity that occurs at a given moment. This fact shows the two opportunities that exist to make it possible to monitor connectivity: assess the potential connectivity and the water and sediment fluxes through those landscapes. These components finally may result in the desired knowledge on the connectivity of the research area. In this chapter we identify three spatial levels of connectivity: soil, hillslopes and catchments. In addition, to be able to measure and monitor connectivity the stocks

Figure 3.5 An almost fully connected pathway delivering sediment from near the top of the hillslope to a point adjacent to a river channel (photo by the author).

Figure 3.6 A landscape of disconnected mass movements in which sediment-gravity processes move material only part way down the hillslopes (photo by the author).

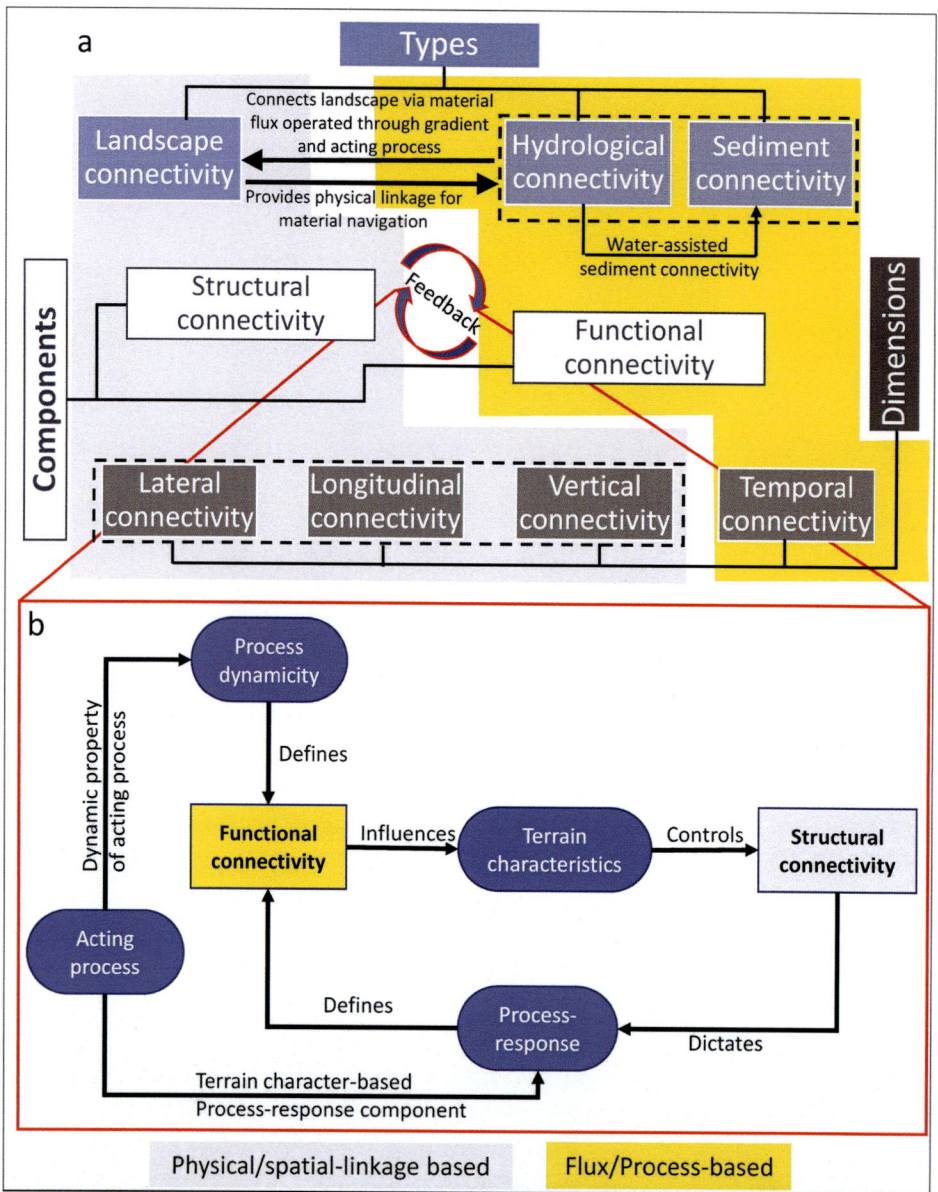

Figure 4.1 (a) The interdependency and interactions between different factors of geomorphic connectivity. (b) the feedback system among components of connectivity and their driving factors. Singh et al. (2021)

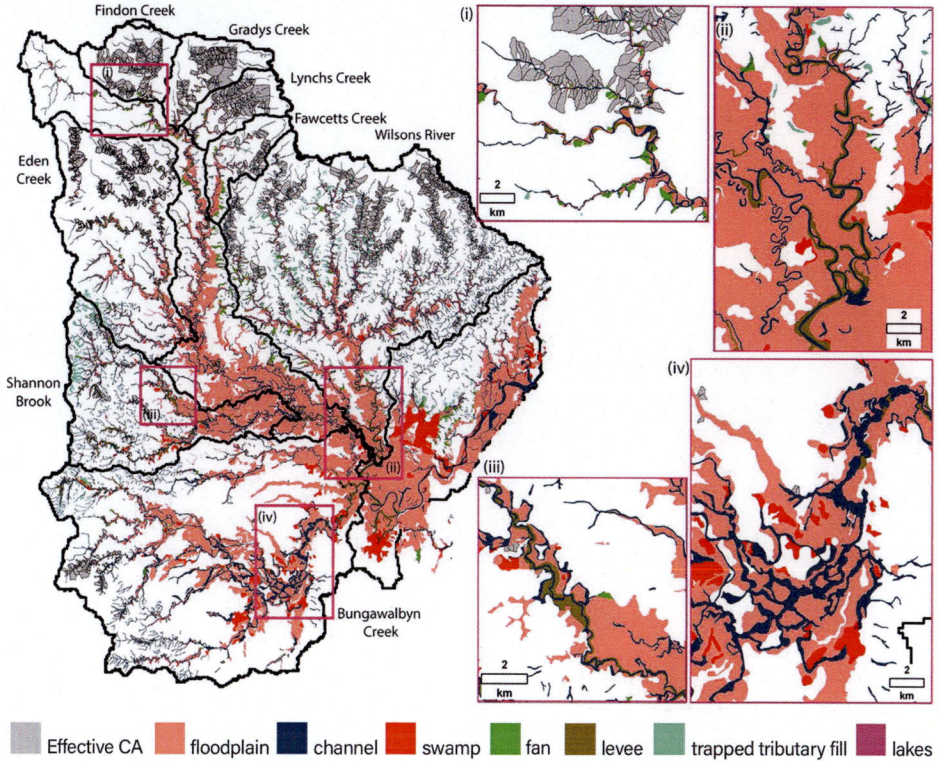

Figure 4.2   Spatial distribution of the ECA and buffers in the Richmond catchment (Khan et al., 2021).

b)

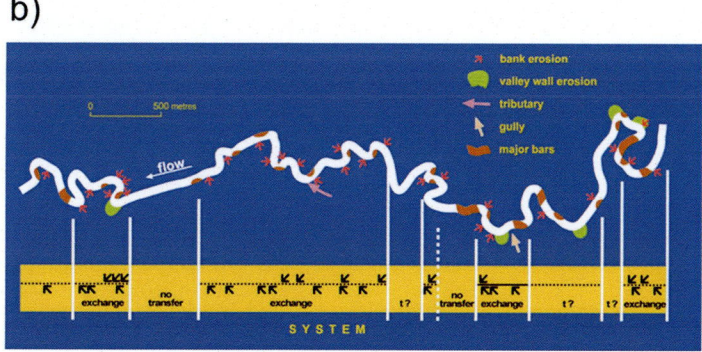

Figure 4.3   (b) connectivity mapping of the River Dane channel, NW England (after Hooke, 2003).

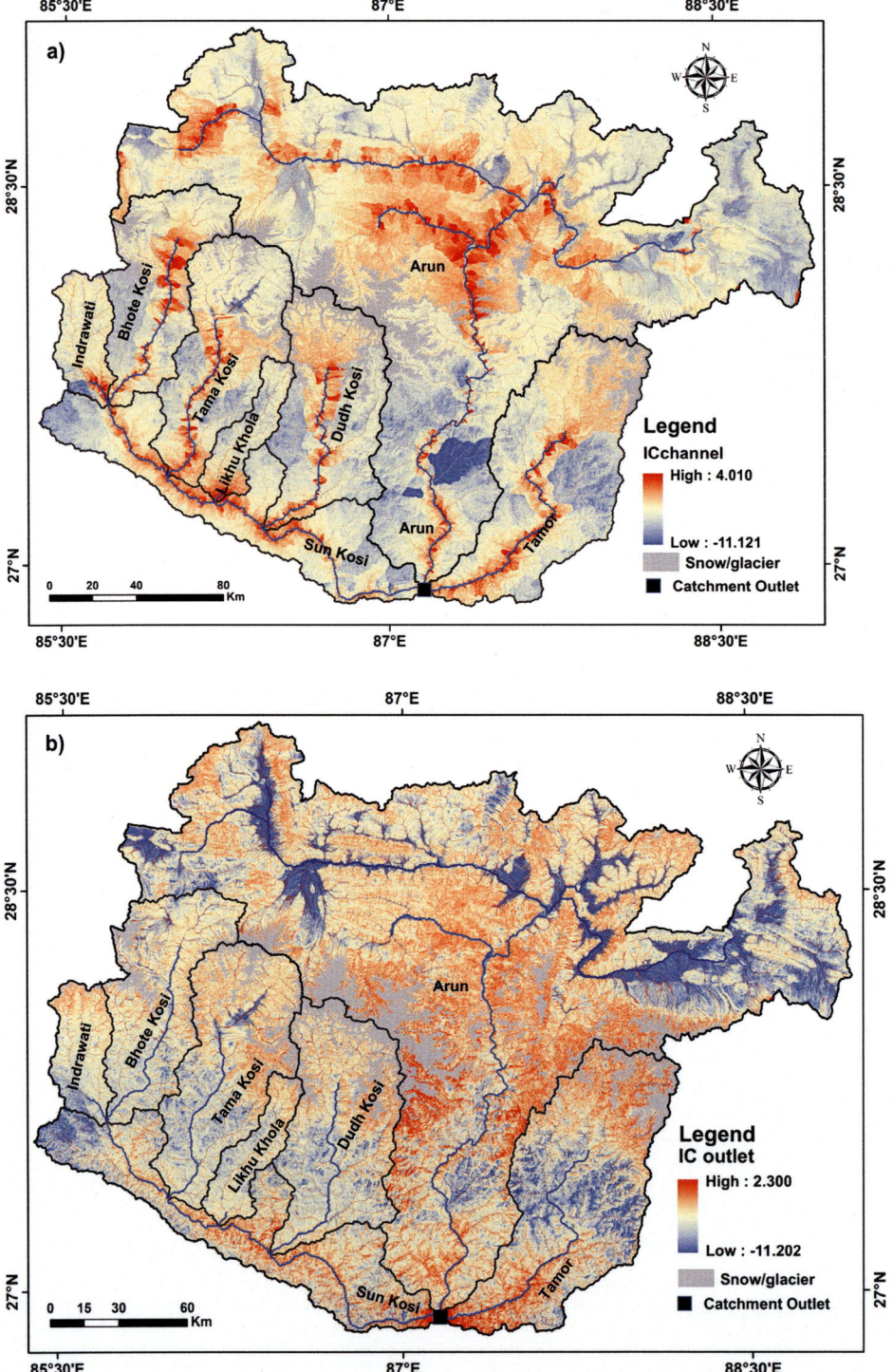

Figure 4.4   (a) IC_channel map of upper Kosi basin, (b) IC_outlet map of upper Kosi basin (after Mishra et al., 2019).

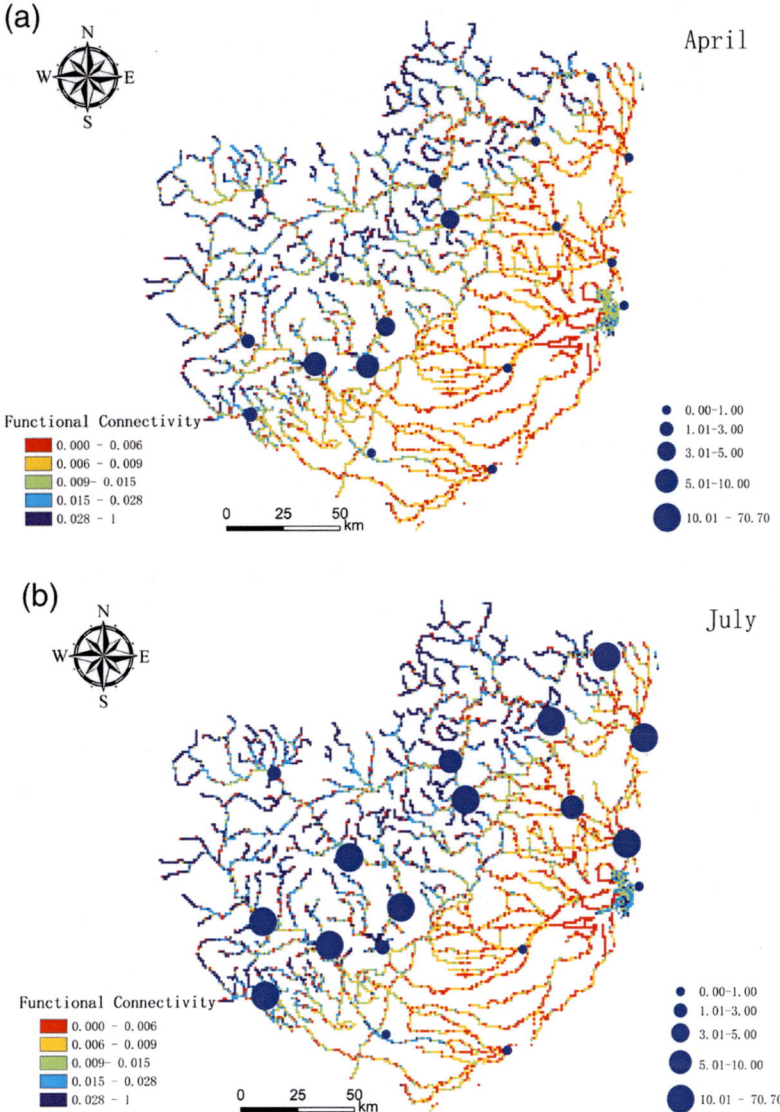

Figure 4.5   Spatial pattern of water and sediment functional connectivity and discharge measured at hydrological stations in the Baiyangdian Basin in (a) April, (b) July and (c) December 2016 (after Liu et al., 2020).

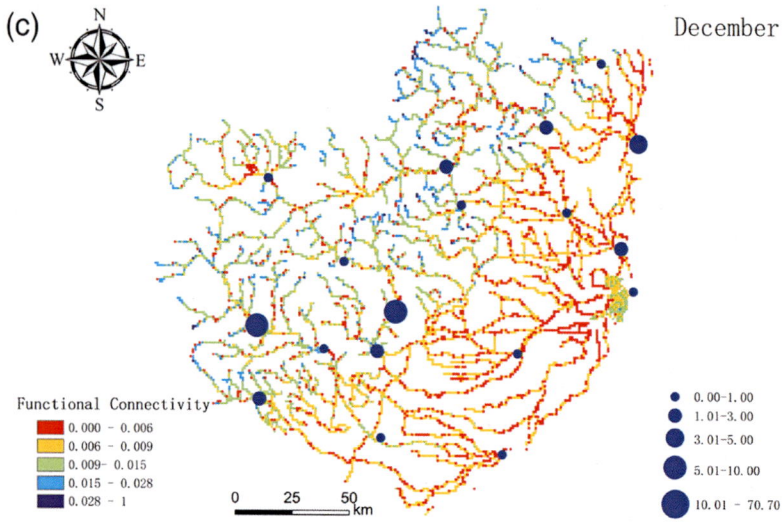

(c)

December

Functional Connectivity
- 0.000 – 0.006
- 0.006 – 0.009
- 0.009 – 0.015
- 0.015 – 0.028
- 0.028 – 1

0   25   50
km

- 0.00–1.00
- 1.01–3.00
- 3.01–5.00
- 5.01–10.00
- 10.01 – 70.70

Figure 4.5   (cont.)

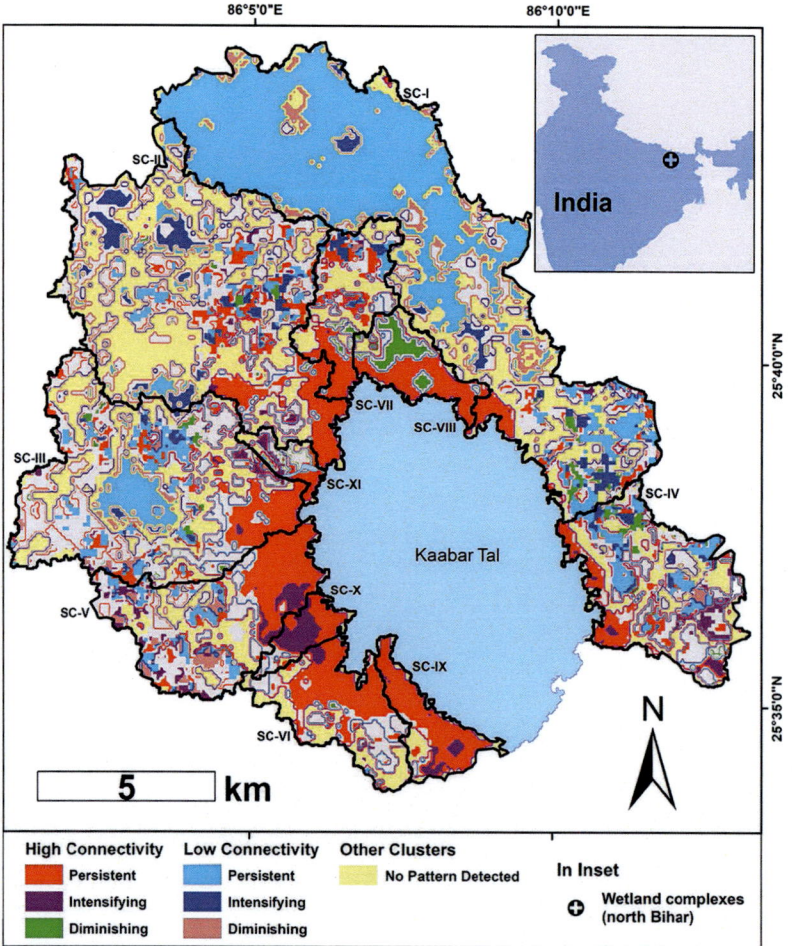

Figure 4.7   The dynamicity of the wetland-catchment connectivity as a function of land-use/land-cover (LULC) over approximately three decades for a wetland complex in north Bihar, Kaabar Tal, for the post-monsoon season (modified after Singh & Sinha, 2019) (Singh et al., 2021).

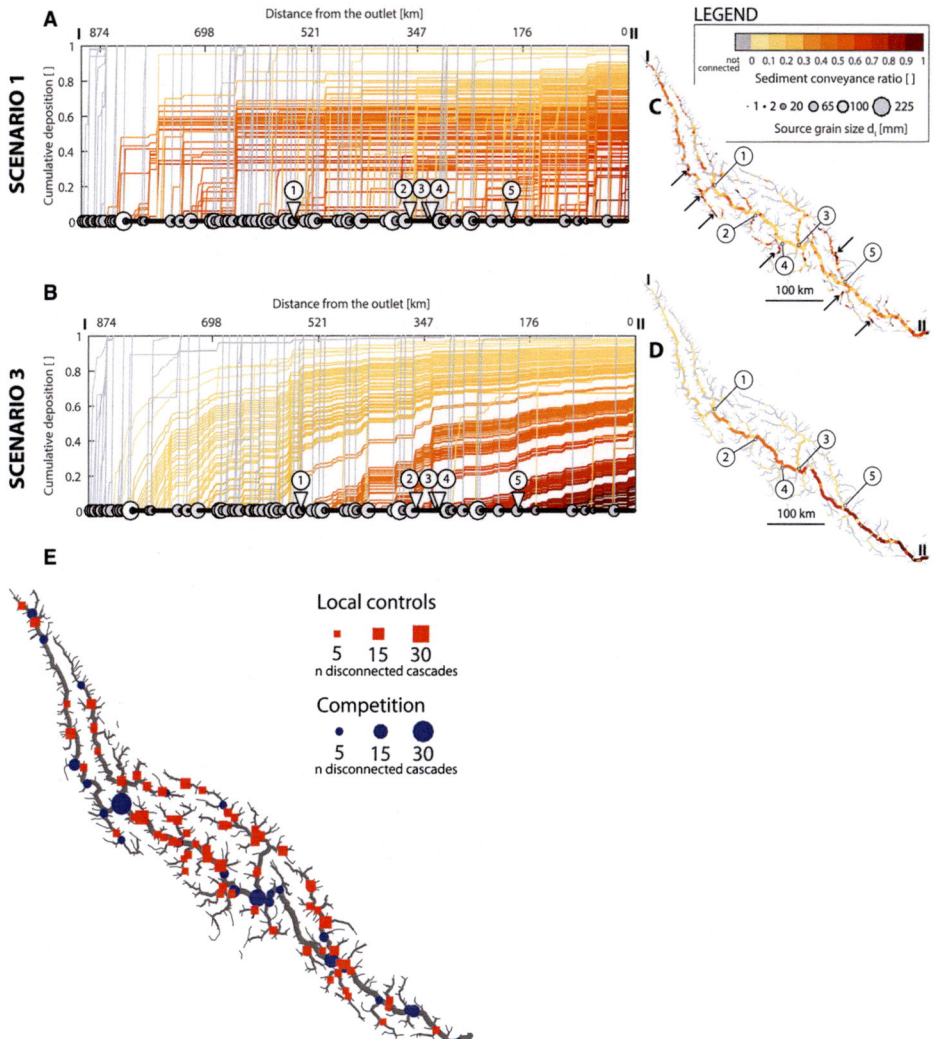

Figure 4.8   Network connectivity for two scenarios in the River Mekong system. (A) and (B) show deposition trajectories (y-axis), respectively of the sediment conveyance ratio along the main stem of the Da River. Dots indicate the source grain size d1 of each cascade (identical between scenarios). Numbers and triangles indicate the location of major tributaries. (C) and (D) are the sediment conveyance ratio mapped on the network throughout the river basin. Arrows in Figure 4.8C indicate some hotspots of sediment recruitment. (E) Hotspots of disconnectivity for scenario 3. Red squares and blue dots indicate edges where multiple cascades are interrupted either due to local hydromorphologic controls or competition. The marker size indicates the number of interrupted cascades. (Schmitt et al., 2016)

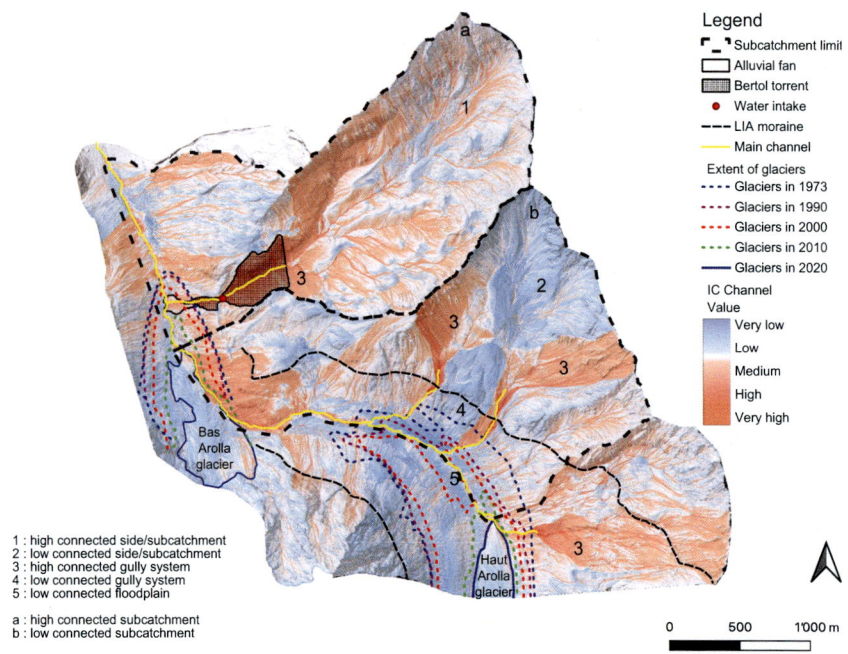

Legend

⌐_⌐ Subcatchment limit
▢ Alluvial fan
▨ Bertol torrent
● Water intake
--- LIA moraine
— Main channel

Extent of glaciers
····· Glaciers in 1973
····· Glaciers in 1990
····· Glaciers in 2000
····· Glaciers in 2010
—— Glaciers in 2020

IC Channel
Value
Very low
Low
Medium
High
Very high

1 : high connected side/subcatchment
2 : low connected side/subcatchment
3 : high connected gully system
4 : low connected gully system
5 : low connected floodplain

a : high connected subcatchment
b : low connected subcatchment

0   500   1'000 m

Figure 7.3   Connectivity Index (Cavalli et al., 2013) of a section of the proglacial area of the Bas and Haut Glacier d'Arolla with the surrounding area of the torrent the Bertol relatively well connected.

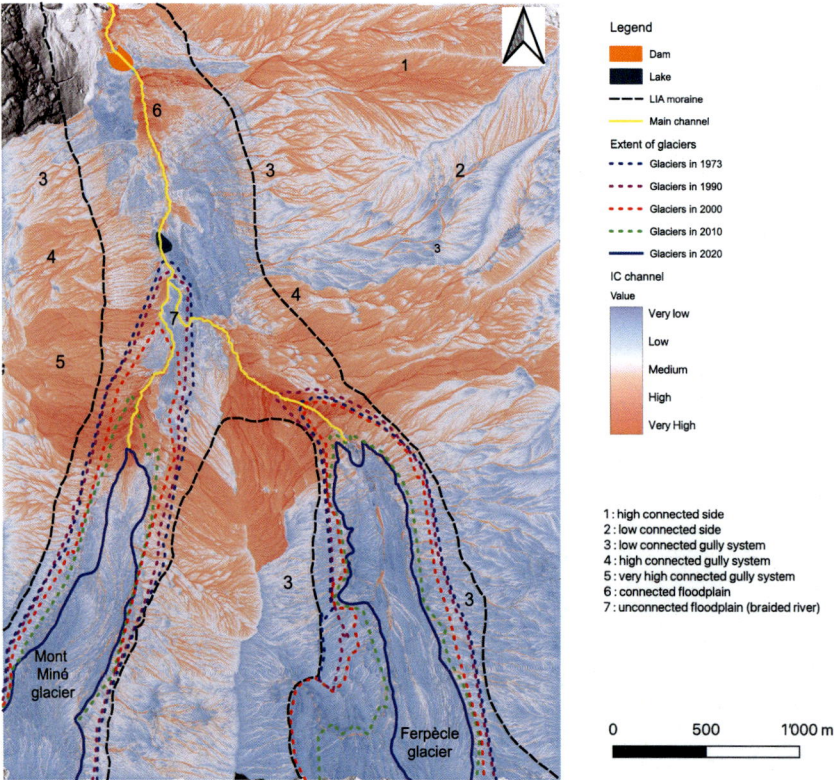

Legend

▧ Dam
▪ Lake
--- LIA moraine
— Main channel

Extent of glaciers
····· Glaciers in 1973
····· Glaciers in 1990
····· Glaciers in 2000
····· Glaciers in 2010
—— Glaciers in 2020

IC channel
Value
Very low
Low
Medium
High
Very High

1 : high connected side
2 : low connected side
3 : low connected gully system
4 : high connected gully system
5 : very high connected gully system
6 : connected floodplain
7 : unconnected floodplain (braided river)

0   500   1'000 m

Figure 7.4   Connectivity Index (Cavalli et al., 2013) of the Ferpecle – Mont Miné catchment and joined floodplain.

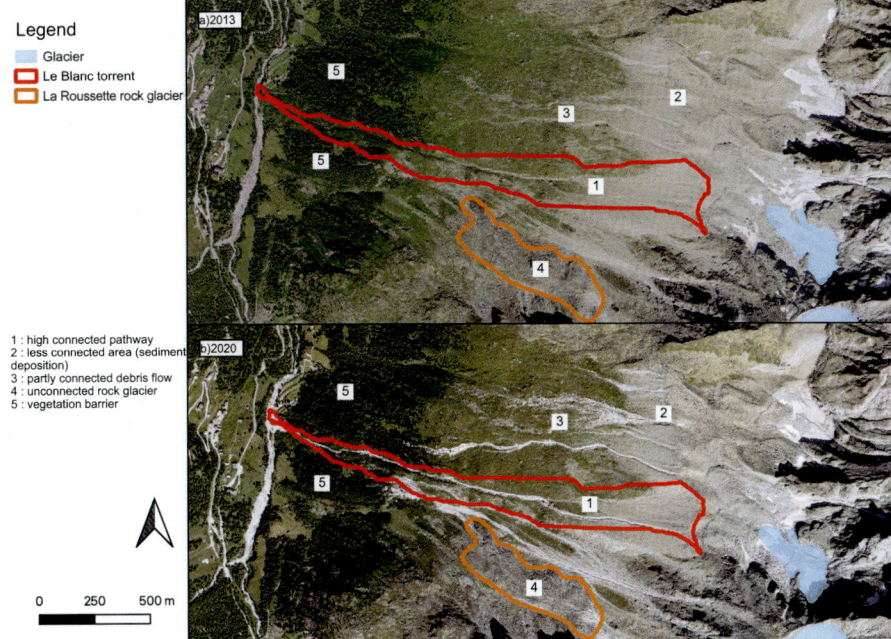

Figure 7.5 Aerial images of the sub-catchment in front of the Arolla village and illustration of relatively high connectivity created by Le Blanc torrent activity in (a) 2013 and (b) 2020. Number 2 shows the location of the La Rousette rock glacier (Micheletti et al., 2015; Meyrat 2018).

Figure 7.6 Detailed image (taken by Lambiel C.) of the La Rousette rock glacier (Figure 7.5) showing the different lobes (from Meyrat, 2018).

a) Floodplain in 1994 : braided   b) Floodplain in 2004 : lake   c) Floodplain in 2010 : braided

Figure 7.8   Evolution of the join floodplain of Mont Miné and Ferpècle glacier (Val Hèrens) in 1994, 2004 and 2010 (source: SwissTopo).

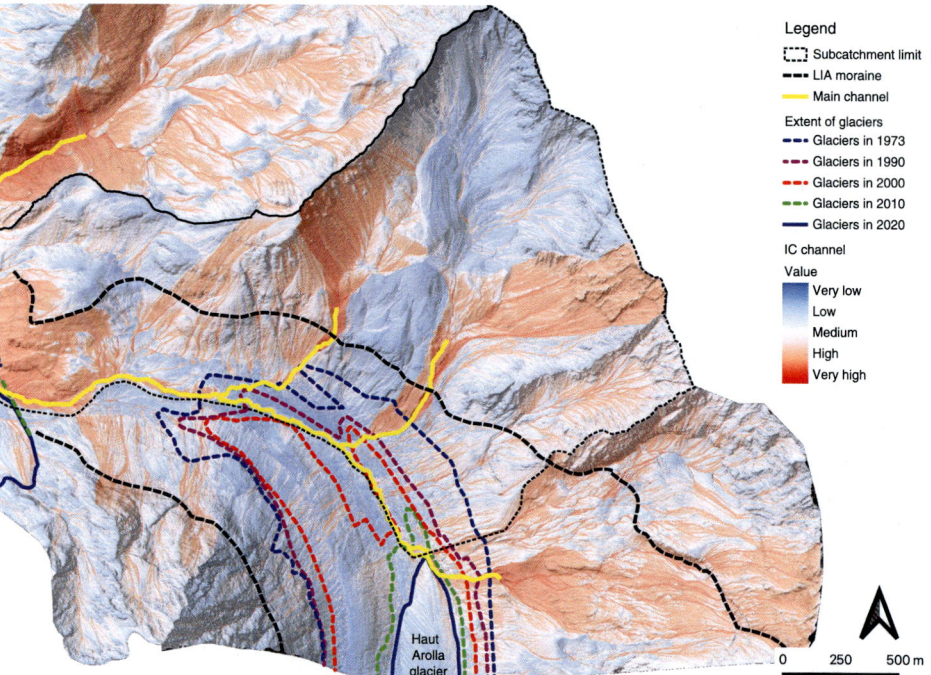

Figure 7.9   Connectivity index (Cavalli et al., 2013) of the surrounding proglacial area of the Haut Glacier d'Arolla with the historical extents of the glacier in 1973, 1990, 2000, 2010 and 2020.

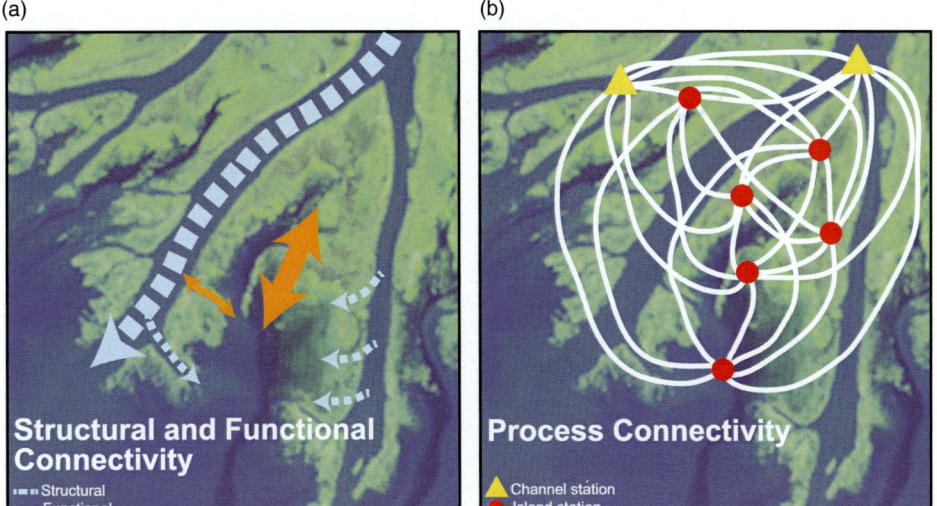

Figure 8.1 Illustration of structural and functional (a) and process connectivity (b) elements in river deltas. Figure reproduced from Passalacqua (2017).

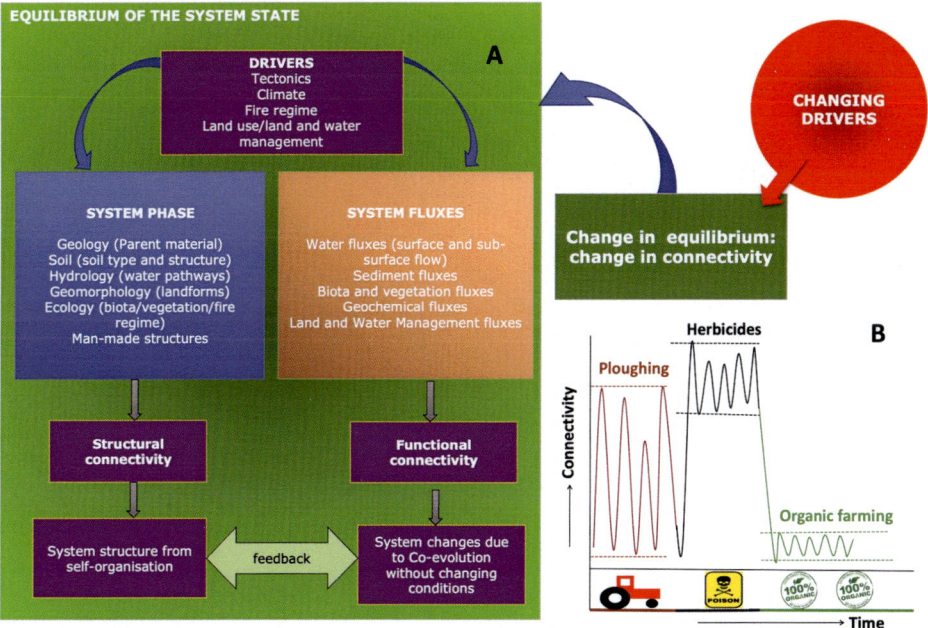

Figure 9.1 (A) Diagram indicating the use of stocks (phases) and flows (fluxes) in the landscape and their drivers (adapted from Keesstra et al., 2018). (B) example of three dynamic equilibria in an agricultural landscape which was first ploughed (high variability of connectivity, but rough ploughed soil reduces connectivity), then under herbicides without ploughing (low infiltration capacity due to soil compaction and removal of vegetation by herbicides) and finally under organic farming without ploughing (low connectivity due to year around vegetation cover and high infiltration capacity of the soil) (cf. Keesstra et al., 2016).

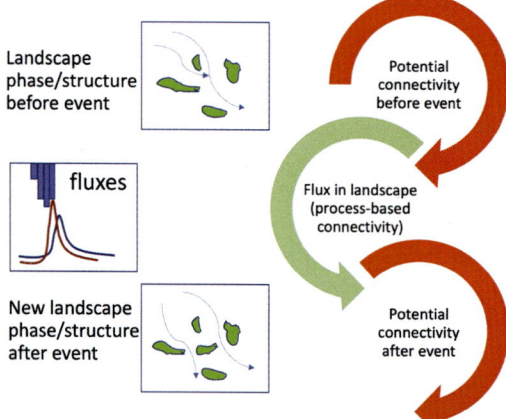

Figure 9.2   Conceptual framework of the links and feedbacks between system phases (structural connectivity, representing the potential connectivity) and fluxes (functional connectivity).

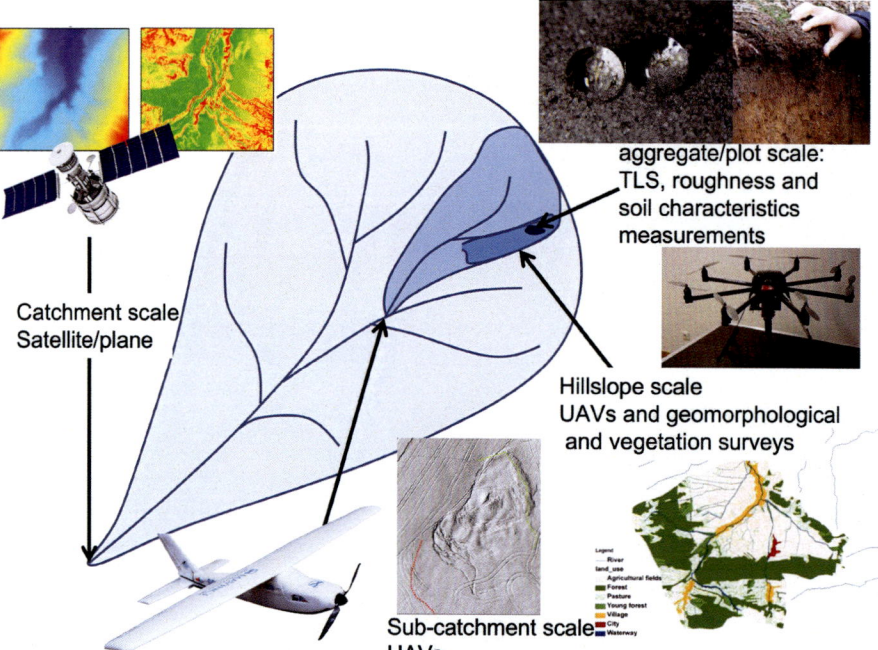

Figure 9.3   Set of possible methodologies for different scales within a catchment (adapted from Keesstra et al., 2018).

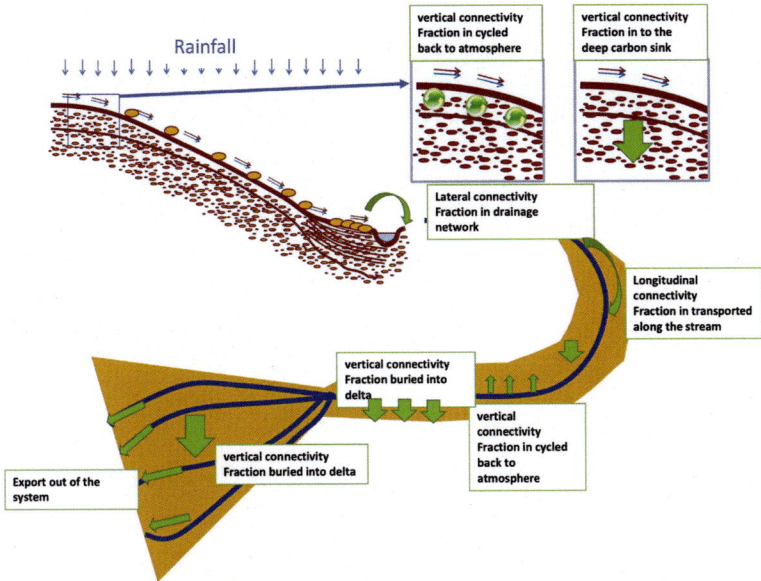

Figure 9.4 Conceptual figure demonstrating the Connectivity principle. Small-scale soil processes are connected and integrated across watersheds. Transmission and multiplication of these processes is transported along stream and river networks before with soil constituents (e.g., soil organic matter or soil aggregates) are exported from the system.

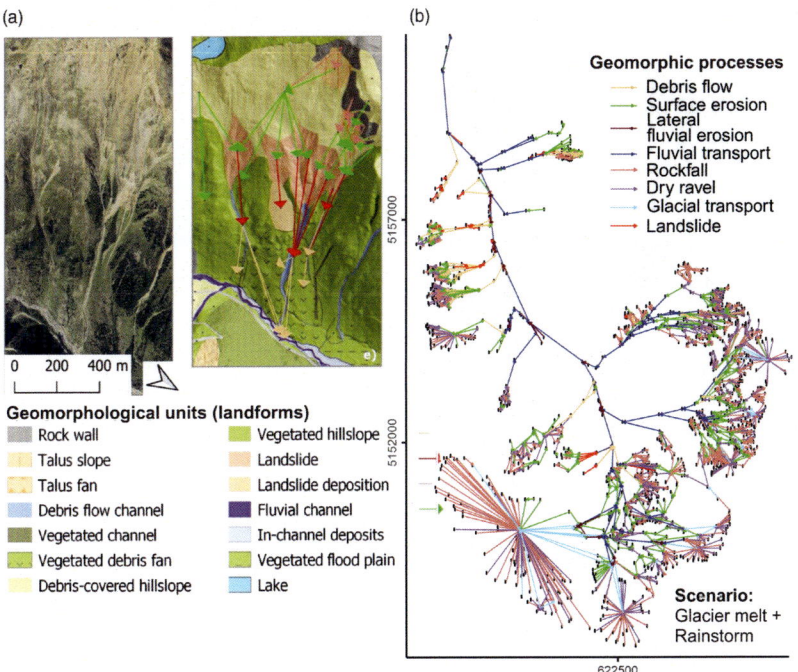

Figure 10.1 Landforms depicted on a geomorphological map (a) as the fundamental unit of a sediment connectivity assessment. On the map (b), landforms are represented by nodes and linked by edges that represent geomorphic processes under a 'glacier melt + rainstorm' scenario. Figure put together from original figures in Buter et al. (2022).

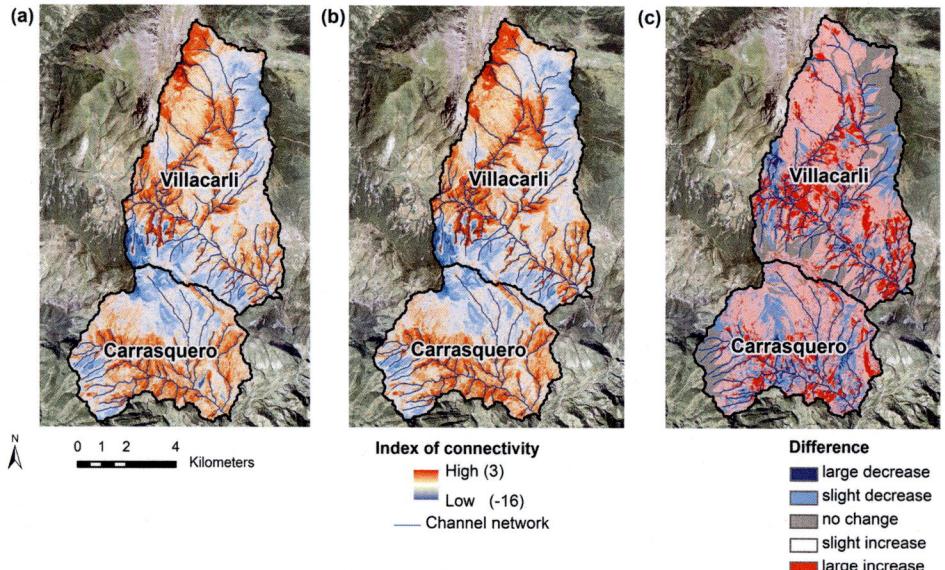

Figure 10.2   Multitemporal map of the IC for a catchment in the Spanish Pyrenees. Panel (a) reflects the situation in April, (b) in August and (c) shows the change in IC between the two points of time. Figure taken from Foerster et al. (2014).

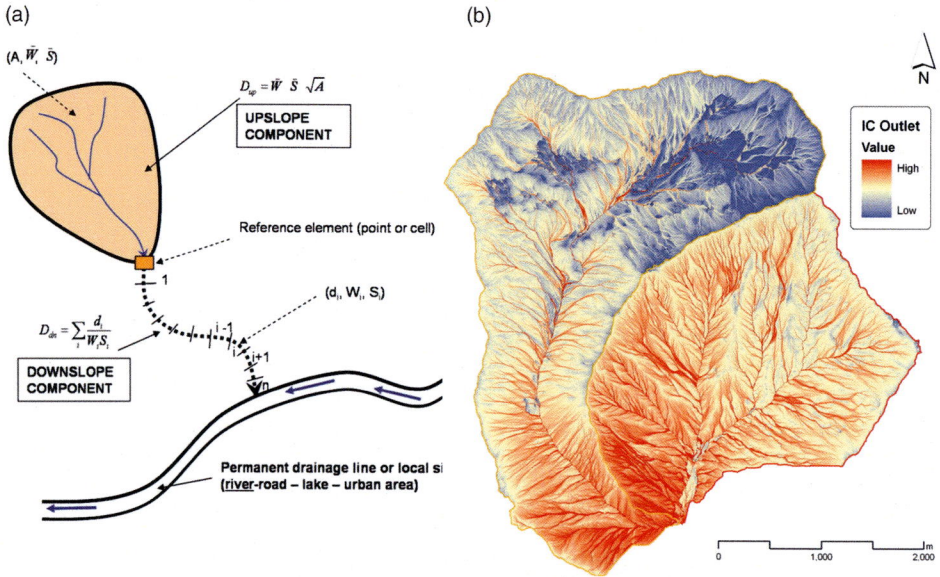

Figure 10.4   (a) Conceptual diagram explaining the computation of IC for each raster cell; the upslope component refers to the contributing area, the downslope component refers to the flow pathway until the proximal user-specified target is reached. (b) IC map (resolution: 1 m, target = catchment outlet) of the Gadria and Strimm catchments, Val Venosta, South Tyrol, Italy. Figures taken from Cavalli et al. (2013).

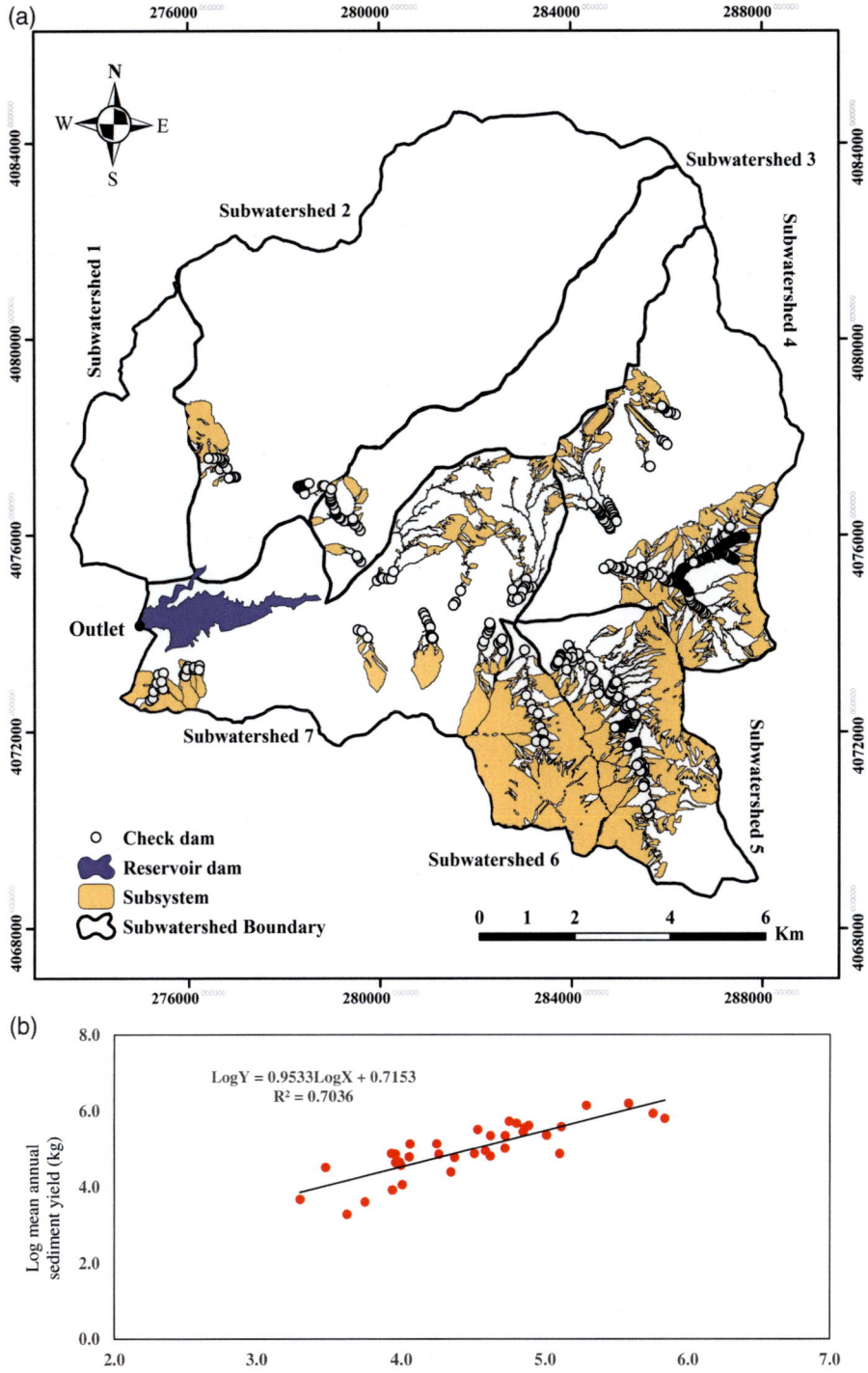

Figure 10.5 (a) Map of the Taham watershed, Zanjan province, NW Iran, with check dams and their corresponding SCA delineated on a DEM. (b) Regression analysis of log-transformed mean annual sediment yield on log-transformed SCA size. Figures taken from Najafi et al. (2021b).

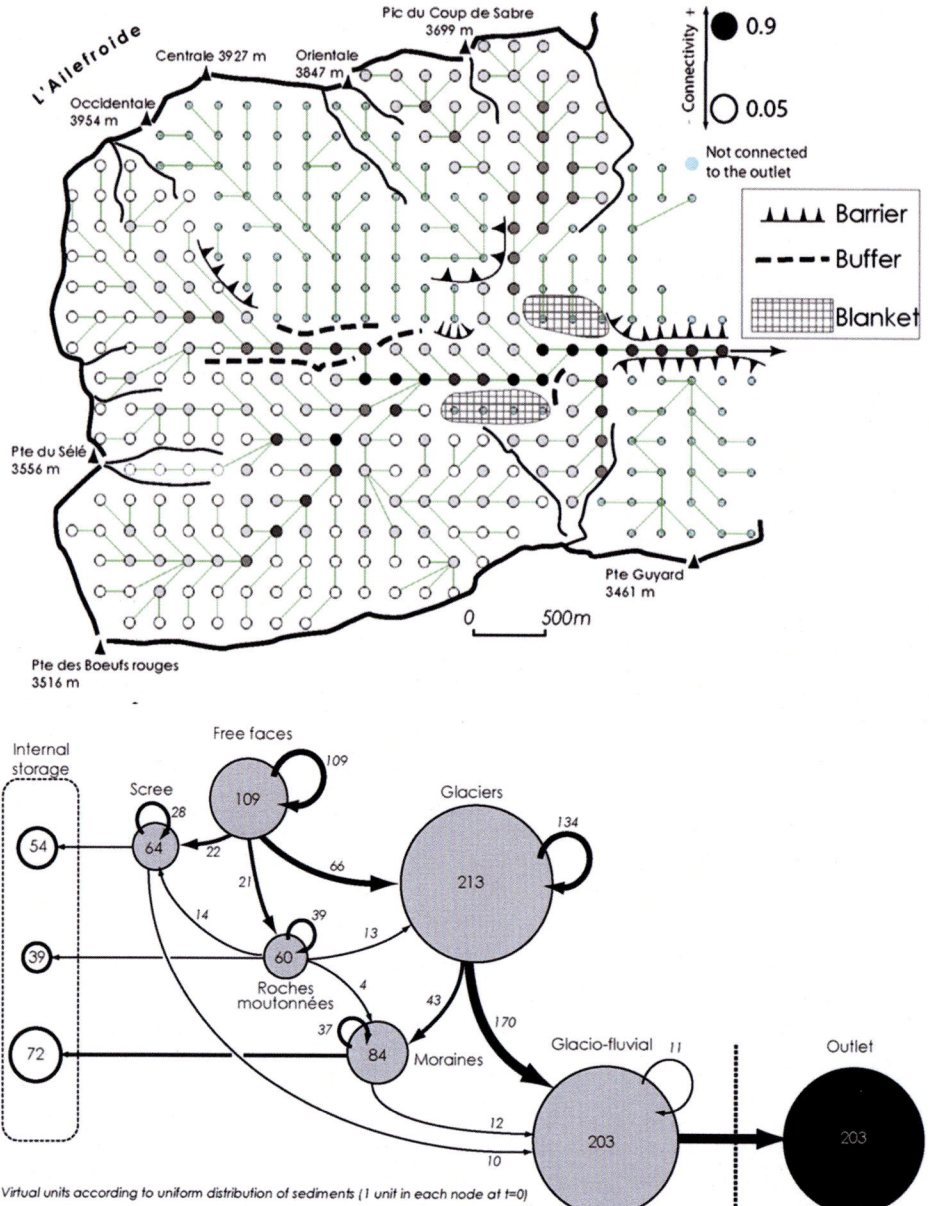

Figure 10.6 Connectivity assessment of the Celse-Nière catchment, French Alps, under current conditions (the original study also shows scenarios with a decoupling and a re-coupling of previously decoupled landscape units). Top panel shows the network of sediment pathways according to flow directions, and disrupted by buffers, barriers and blankets according to a geomorphological map. The nodes are coloured according to their connectivity index value. Bottom panel shows simulated rates of sediment transfer between system compartments, assuming initial (t = 0) uniform distribution of sediment across nodes. Figure taken from Cossart and Fressard (2017; selected parts of their Figure 7).

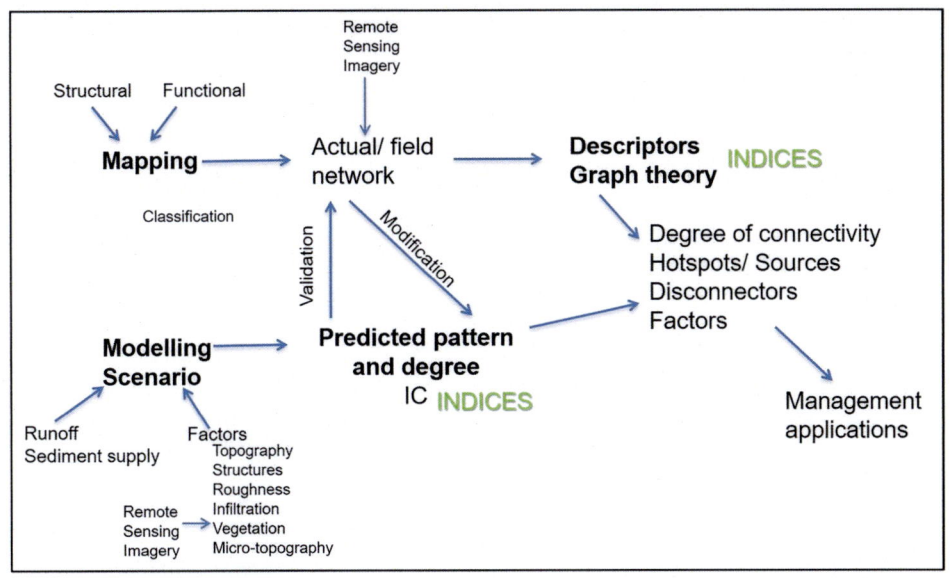

Figure 10.7 Classification of connectivity index approaches and applications proposed by Hooke and Souza (2021).

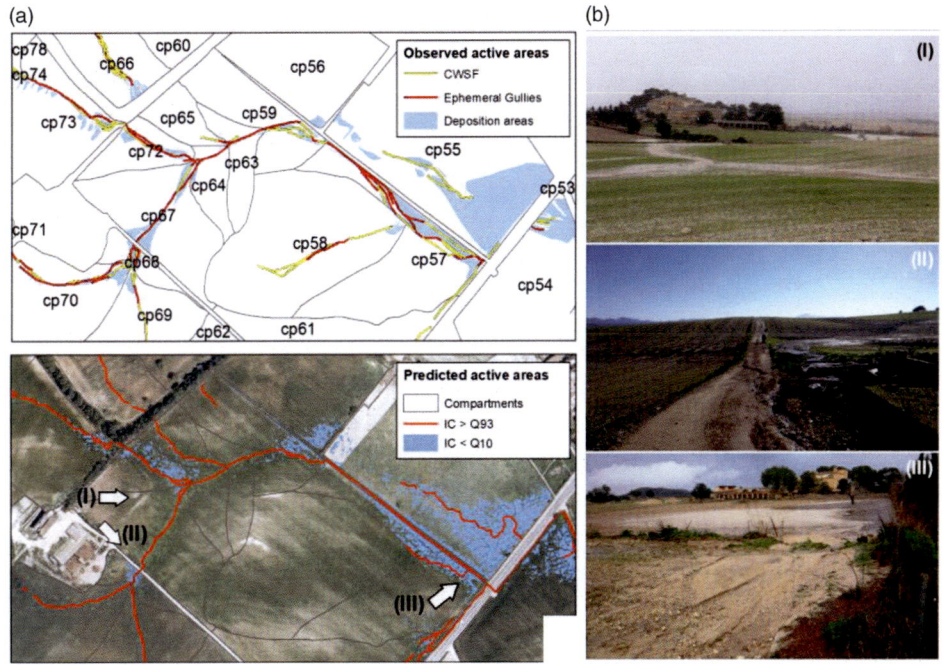

Figure 10.8 (a) Ephemeral gullies, deposition areas and concentrated water and sediment flows mapped from aerial photos after a rainstorm event at Can Revull catchment, Mallorca, Spain. (b) Aerial image with high- and low-IC areas highlighted and the locations of photos I, II and III (right). Figure taken and rearranged from Calsamiglia et al. (2020).

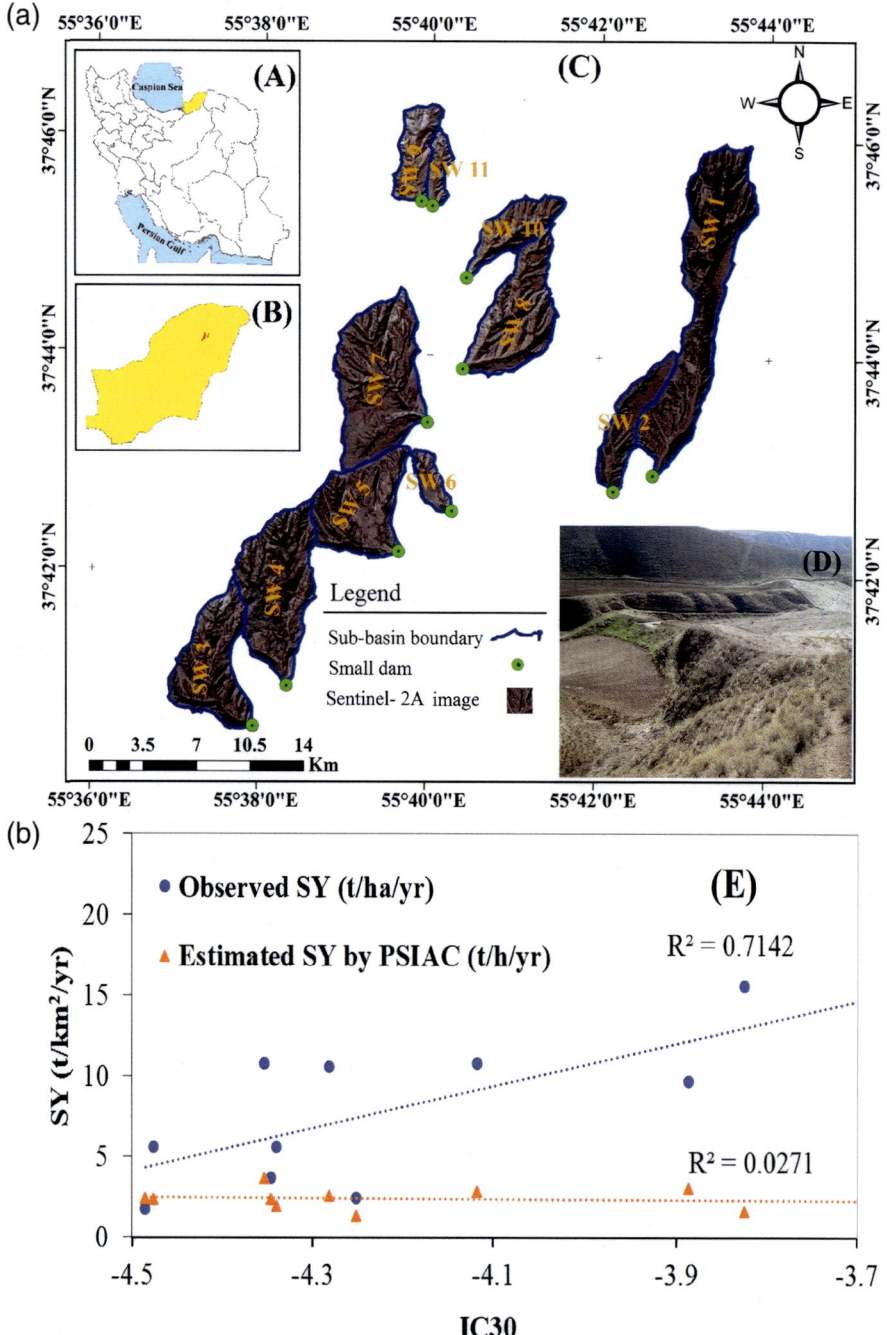

Figure 10.9   (a) Map of eleven check dams and their contributing areas in the Shour-Dareh basin, Golestan province, N. Iran. (b) Regression analysis of mean annual specific sediment yield (SSY) from these catchments on their mean IC (blue circles). The estimated SSY by PSIAC refers to an empirical, semi-quantitative method to assess sediment production. Figures taken from Arabkhedri et al. (2021).

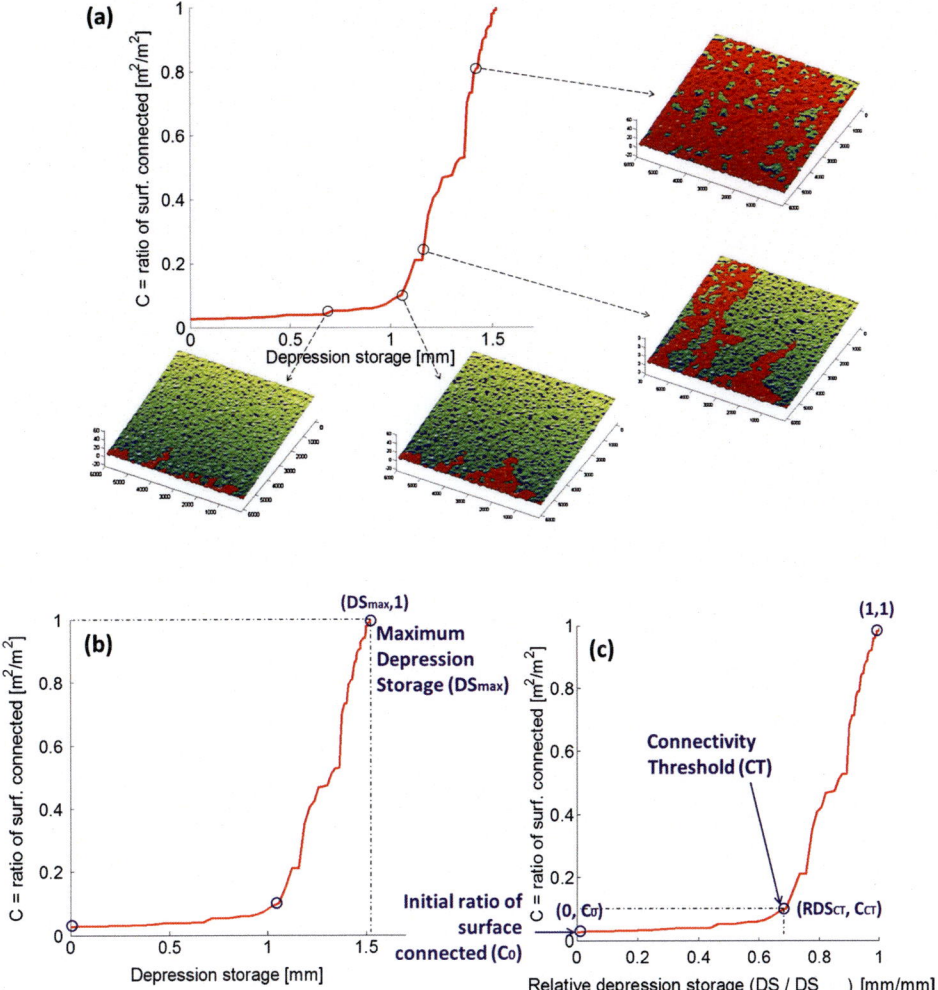

Figure 11.2   Definition of the RSC function (from Peñuela et al. (2015)). (a) As the depression storage fills, the relative proportion of the surface becomes connected to the outflow edge. (b) shows the RSC function; and (c) the normalized RSC function. The connectivity threshold is the inflexion point in the RSC function, and together with the initially connected surface a maximum depression storage, these three parameters can be used to estimate sub-grid patterns of flow for models of connectivity.

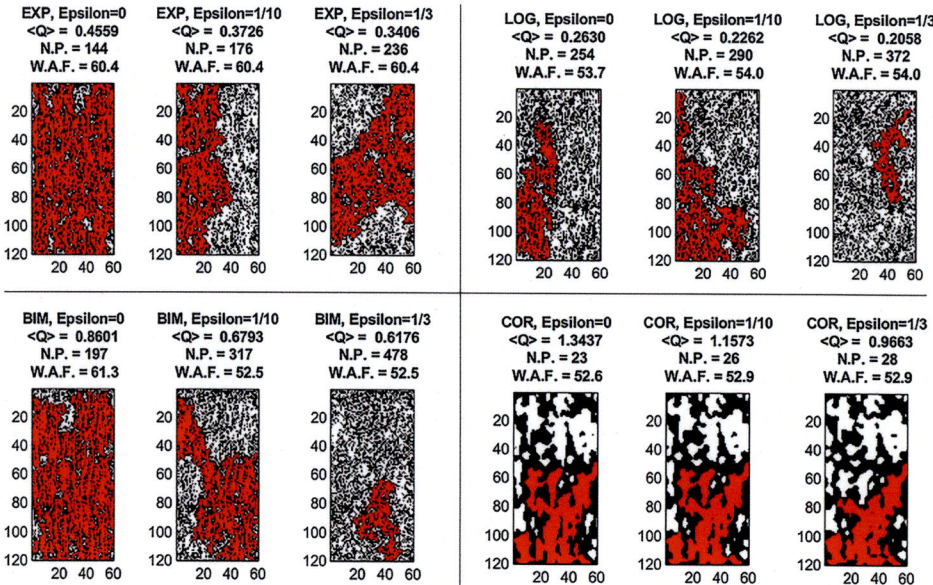

Figure 11.3 Simulations by Harel & Mouche (2014) comparing how uncorrelated and correlated infiltrability fields affect the runoff from a rectangular surface (120 cells in the downslope direction from the top to the bottom of the plots by 60 cells across the slope) The three uncorrelated surfaces are defined by exponential (EXP), log-normal (LOG) and bimodal (BIM) distribution functions. The correlated (COR) surfaces are generated using an isotropic multi-Gaussian field. The Epsilon values specify the amount of cross-slope flow in the downslope direction: Epsilon = 0 means that all flow goes into the immediately adjacent downslope cell; Epsilon = 1/3 means that flow is equally distributed between the immediately adjacent downslope cell and the two lateral cells in the downslope direction. <Q> is the mean runoff flow rate, N.P. is the number of runoff patterns and W.A.F. is the wet area fraction, defined as the number of wet pixels divided by the total number of pixels.

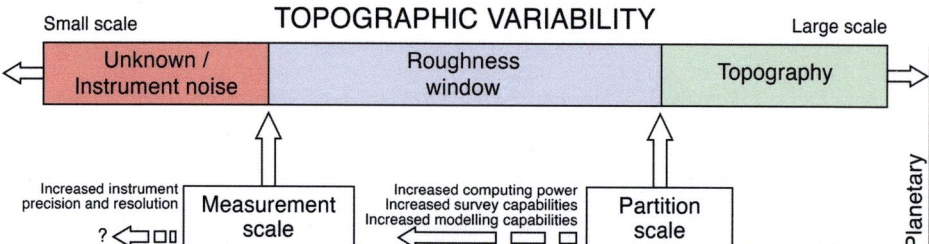

Figure 11.4 Definition of how information about topographic variability is dependent on the scale at which measurement is possible (and may thus change as technology improves) and the scale between which macroscale topography is represented, as opposed to surface roughness (Smith, 2014). Different model applications or changes in technology may also cause this distinction of the 'partition scale' to vary.

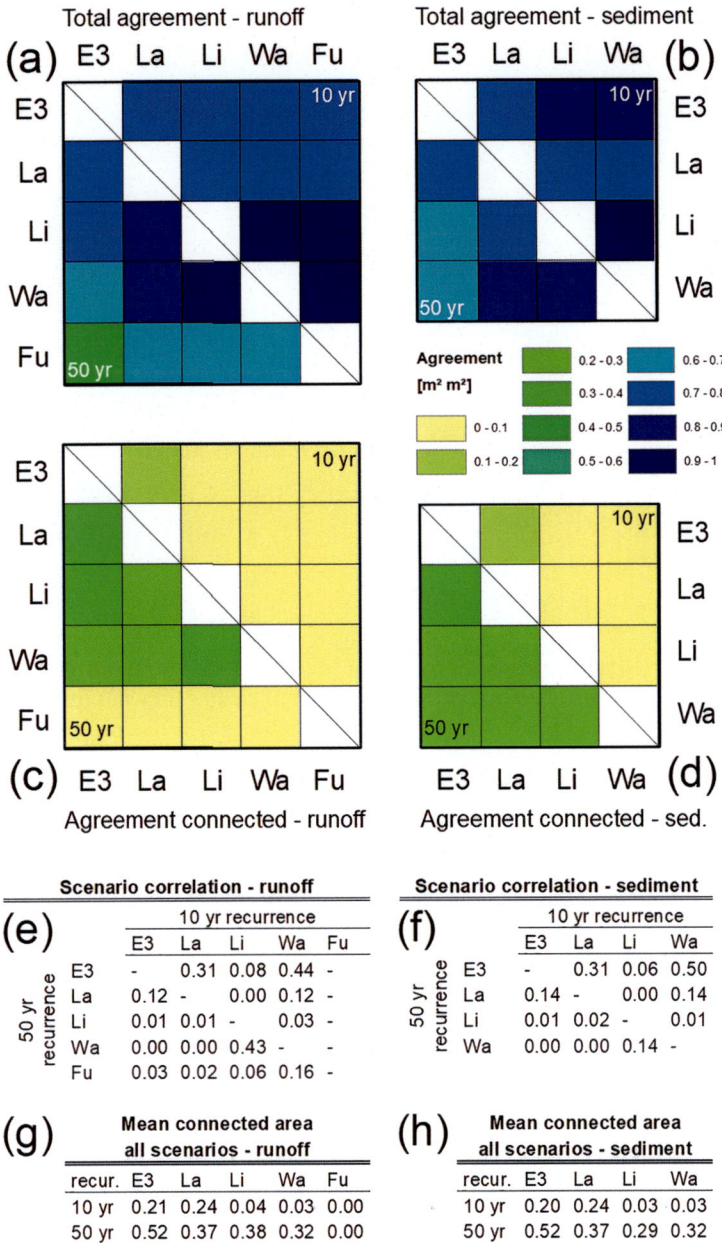

Figure 11.6   Agreement between the five different runoff and erosion models evaluated in terms of their connectivity characteristics by Baartman et al. (2020). The extent to which the models agree regarding the total area (connected and unconnected) are shown for (a) runoff and (b) sediment flux. For the connected areas only, model agreement is shown in (c) for runoff (c) and (d) for sediment flux. Colours indicate the mean area of agreement averaged over all scenarios. Tables (e) and (f) show the correlation in total connected area between model pairs in different scenarios for runoff (e) and sediment (f). (g) and (h) show the connected area, averaged over all scenarios, for runoff (g) and sediment (h). E3: Erosion3D, LA: LandSoil, Li: OpenLISEM, Wa: Watershed, Fu: FullSWOF_2D.

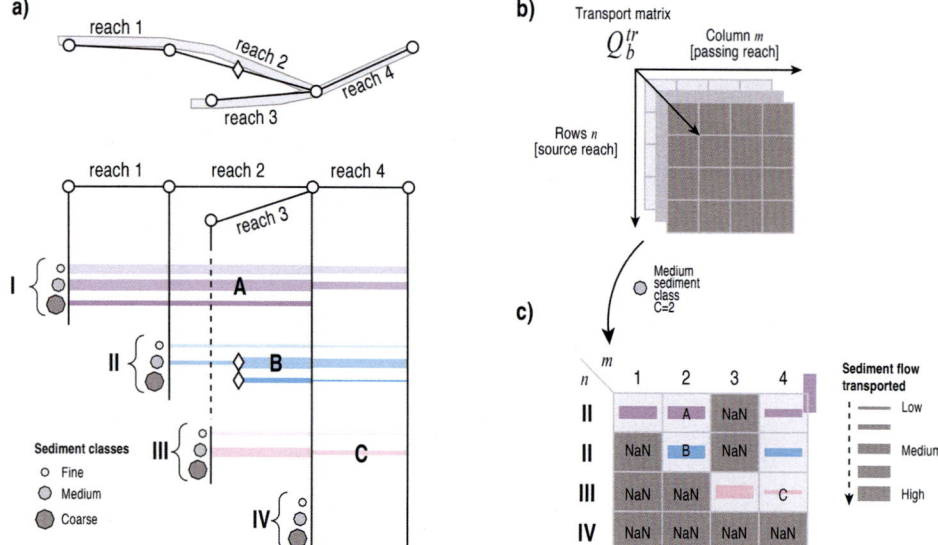

Figure 11.7 Visual representation of the basis of the CASCADE model (Tangi et al., 2019). (a) shows how a simplified network of channel reaches might produce multiple cascades of sediment and how each cascade is made up of different particle sizes (simplified here as fine, medium and coarse grained). In (b), the model data structure is shown as a three-dimensional matrix, showing each reach of the system as the matrix columns, the source reach of each cascade as the matrix rows, repeated for each size class of sediment. An example of the outcome of sediment-flux calculations is shown in (c). The values of *m* are the reach numbers in Arabic numerals, and *n* are the source reaches in Roman numerals. The cells A, B and C show how the calculations relate back to the catchment structure in (a).

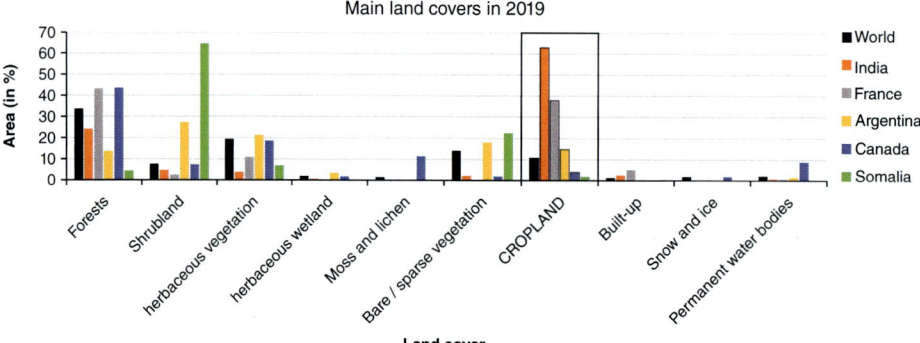

Figure 12.1 Main land covers in the world and in selected countries (Data source: Copernicus, Land monitoring service, Global Land Cover).

Figure 12.2 Spatial and human-decision factors affecting connectivity processes in woody crops. (A) Linear framework plantation in steep slopes at the Mosel Valley (Germany); (B) Irregular framework plantation in steep slopes at the Montes de Málaga (Spain); (C) "Undisturbed" hillslopes cultivated with conventional soil managed olive orchards in the Jaén province (Spain); (D) Agricultural terraces stopping flow connectivity processes on steep slopes in volcanic soils at Canary Islands (Spain); (E) Spontaneous vegetation cover growing along the interrow areas in steep slope vineyards in the Saar Valley (Germany); and, (F) Application of straw mulch in organic farming vineyards plantations in Valencia province (Spain). All pictures were taken by J. Rodrigo-Comino.

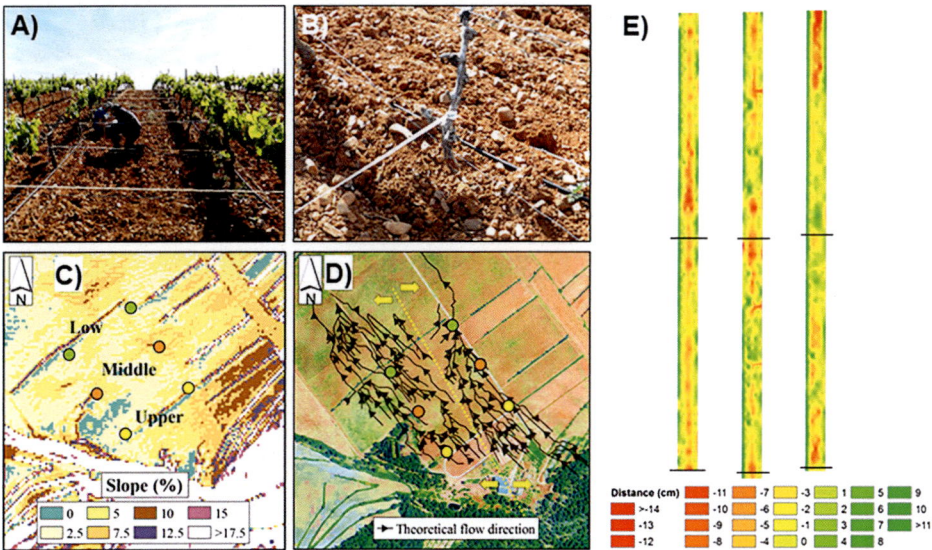

Figure 12.3 Example of ISUM applications and maps to assess connectivity processes. (A) Field campaign measuring cross sectional transects; (B) Detection of the graft union on the vine; (C) Localization and inclination map of three experimental plots at different hillslope positions; (D) Using the tool flow direction of ArcGIS 10.5 (ESRI, USA) to contextualize the resulting ISUM map; and, (E) Examples of ISUM maps. All images were taken/generated by J. Rodrigo-Comino.

Figure 12.4   Connectivity processes activated on bare surfaces in olive (A) and almond (B) orchards. All pictures were taken by J. Rodrigo-Comino.

Figure 12.5   Climate and human-decision factors affecting connectivity processes in alpine grasslands. (A) Fragmented meadow patches; (B) Sediment transport in erosion gullies; (C) Artificial planting practices in extremely degraded grasslands. All pictures were taken by G.L. Wu and Y.-F. Liu.

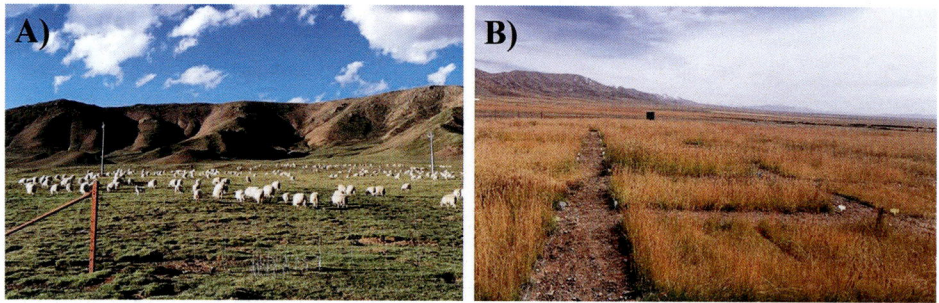

Figure 12.6   Example of managed practices in grazing grasslands. (A) fencing management; (B) reseeding in extremely degraded grasslands. All pictures were taken by G.L. Wu and Y.-F. Liu.

Figure 12.7   Traditional elevated drystone wall in abandoned fields in Huesca province, Spanish Pyrenees (A). Steep viticulture in terraces in Orense province, DO Ribeira Sacra, NW Spain (B). Ponding in a cereal field in the central Ebro river valley in Zaragoza province, NE Spain (C). Dis-connectivity of a mud-flow generated in a cereal field (right part of the picture) and the adjacent vineyard (left part of the picture) due to the presence of elevated narrow buffer strips of grass and an unpaved trail; picture taken near Barbastro, Huesca province (D). Large check dam built in an abandoned agro-ecosystem to reduce the sediment yield; picture taken in Albacete province, SE Spain (E). Buffer strips of forest between the cultivated hillslopes and the Duero river; picture taken near Pinhão, northern Portugal (F). All pictures were taken by M. López-Vicente.

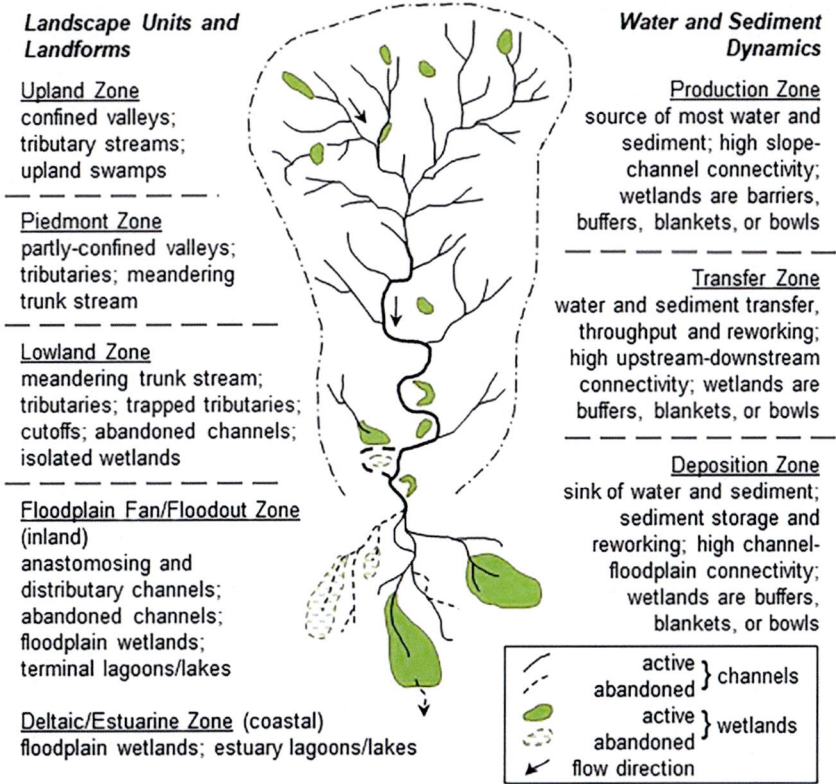

Landscape Units and Landforms

Upland Zone
confined valleys; tributary streams; upland swamps

Piedmont Zone
partly-confined valleys; tributaries; meandering trunk stream

Lowland Zone
meandering trunk stream; tributaries; trapped tributaries; cutoffs; abandoned channels; isolated wetlands

Floodplain Fan/Floodout Zone (inland)
anastomosing and distributary channels; abandoned channels; floodplain wetlands; terminal lagoons/lakes

Deltaic/Estuarine Zone (coastal)
floodplain wetlands; estuary lagoons/lakes

Water and Sediment Dynamics

Production Zone
source of most water and sediment; high slope-channel connectivity; wetlands are barriers, buffers, blankets, or bowls

Transfer Zone
water and sediment transfer, throughput and reworking; high upstream-downstream connectivity; wetlands are buffers, blankets, or bowls

Deposition Zone
sink of water and sediment; sediment storage and reworking; high channel-floodplain connectivity; wetlands are buffers, blankets, or bowls

active } channels
abandoned
active } wetlands
abandoned
flow direction

Figure 13.1   The role of wetlands as disconnectors in landscape units and sediment process zones in catchments.

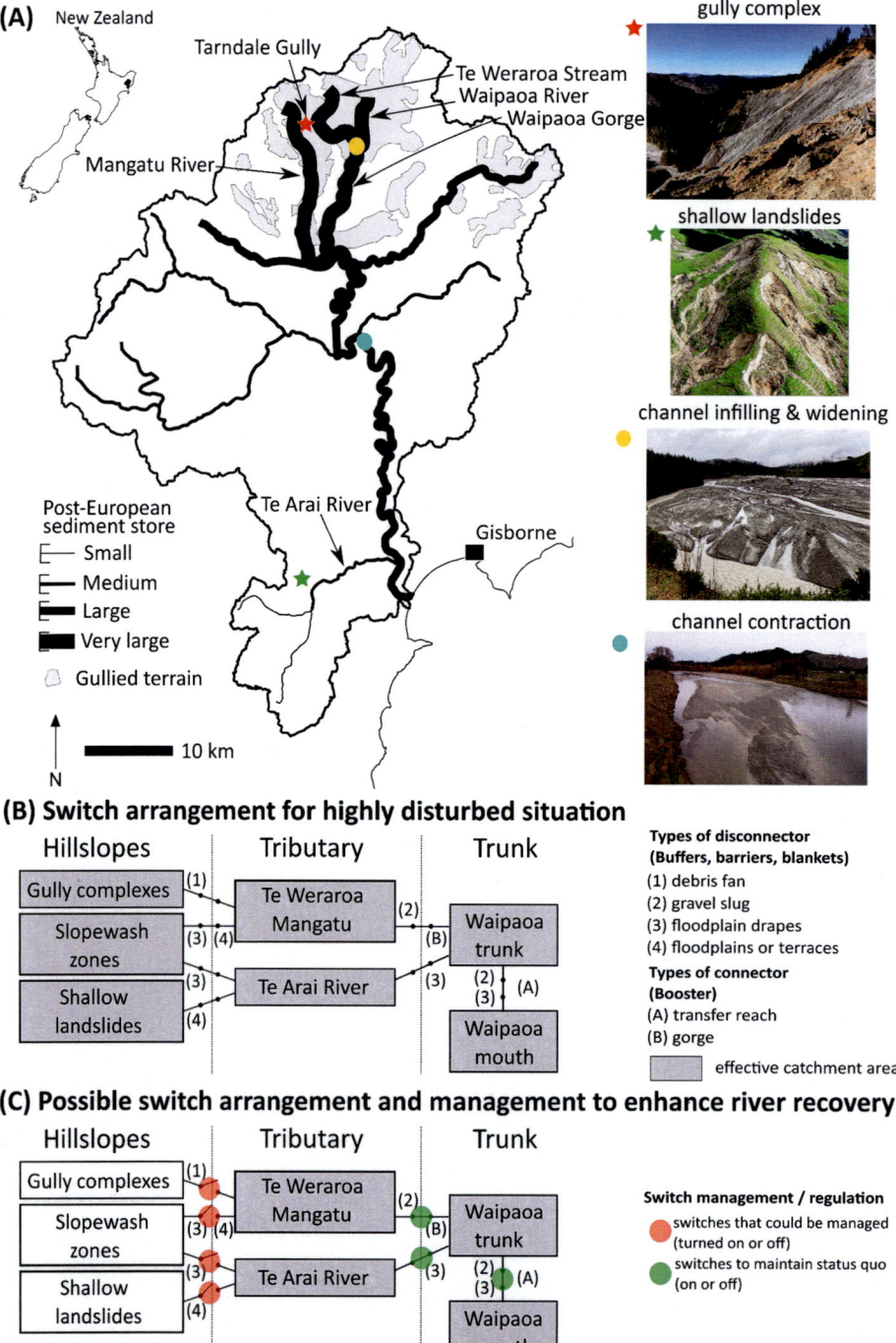

(A) New Zealand

Tarndale Gully

Te Weraroa Stream
Waipaoa River
Waipaoa Gorge

Mangatu River

gully complex

shallow landslides

channel infilling & widening

channel contraction

Post-European
sediment store

⊢— Small
⊢— Medium
⊢— Large
⊢— Very large

⬭ Gullied terrain

Te Arai River

Gisborne

10 km

N

(B) Switch arrangement for highly disturbed situation

Hillslopes	Tributary	Trunk
Gully complexes	Te Weraroa Mangatu	Waipaoa trunk
Slopewash zones		
Shallow landslides	Te Arai River	Waipaoa mouth

(1)
(2)
(B)
(3) (4)
(3)
(3)
(4)
(2)
(3)
(A)

Types of disconnector
(Buffers, barriers, blankets)
(1) debris fan
(2) gravel slug
(3) floodplain drapes
(4) floodplains or terraces

Types of connector
(Booster)
(A) transfer reach
(B) gorge

effective catchment area

(C) Possible switch arrangement and management to enhance river recovery

Hillslopes	Tributary	Trunk
Gully complexes	Te Weraroa Mangatu	Waipaoa trunk
Slopewash zones		
Shallow landslides	Te Arai River	Waipaoa mouth

(1)
(2)
(B)
(3) (4)
(3)
(3)
(4)
(2)
(3)
(A)

Switch management / regulation
● switches that could be managed (turned on or off)
● switches to maintain status quo (on or off)

Figure 13.2 Regulating switches to manage sediment (dis)connectivity in the Waipaoa catchment. (B, C) modified from Fryirs et al. (2007). Landslides photo: Mike Crozier.

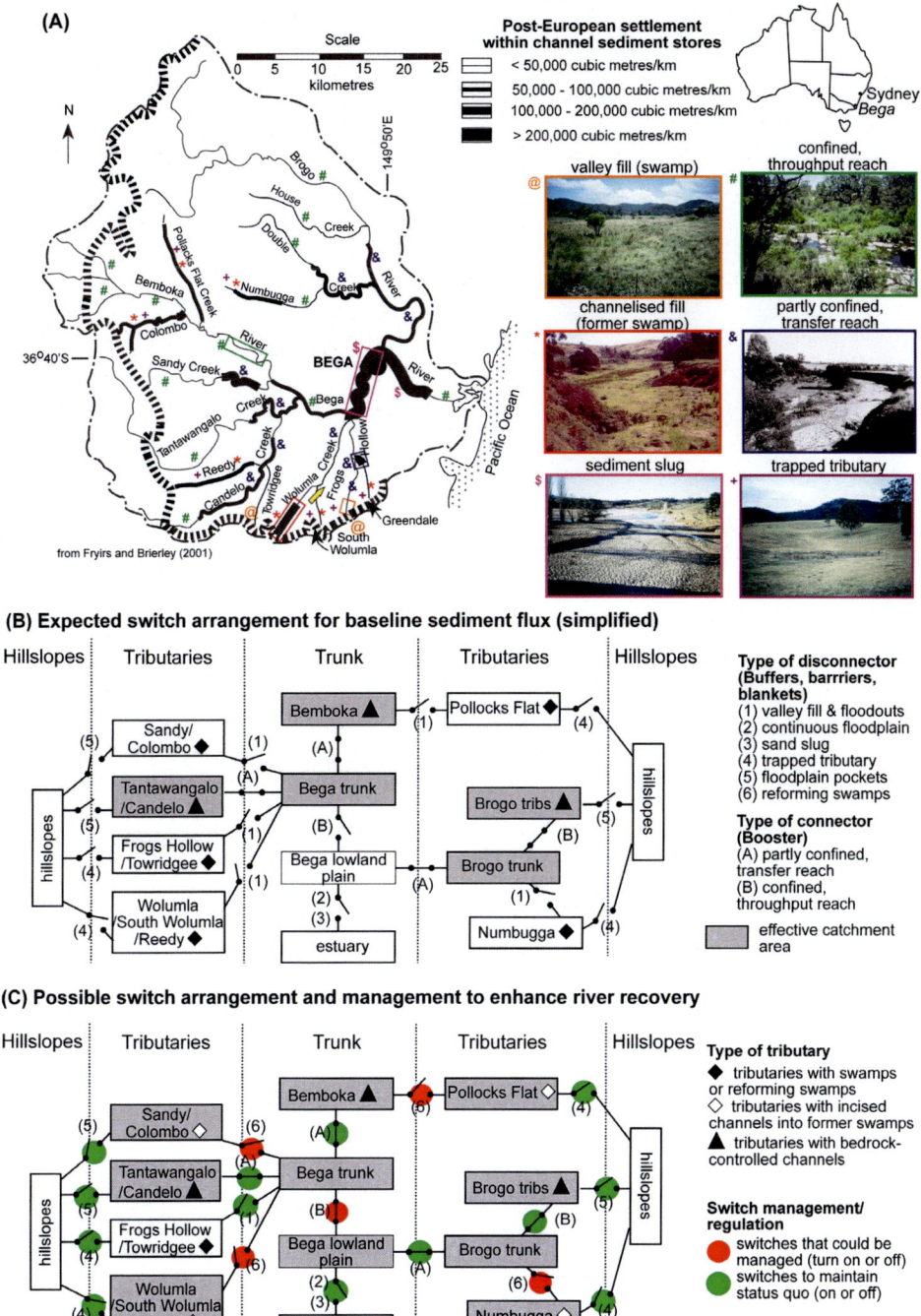

Figure 13.3 Regulating switches to manage sediment (dis)connectivity in the Bega catchment. (A) modified from Fryirs and Brierley (2001) and Pöppl et al. (2020). (B, C) modified from Fryirs et al. (2007).

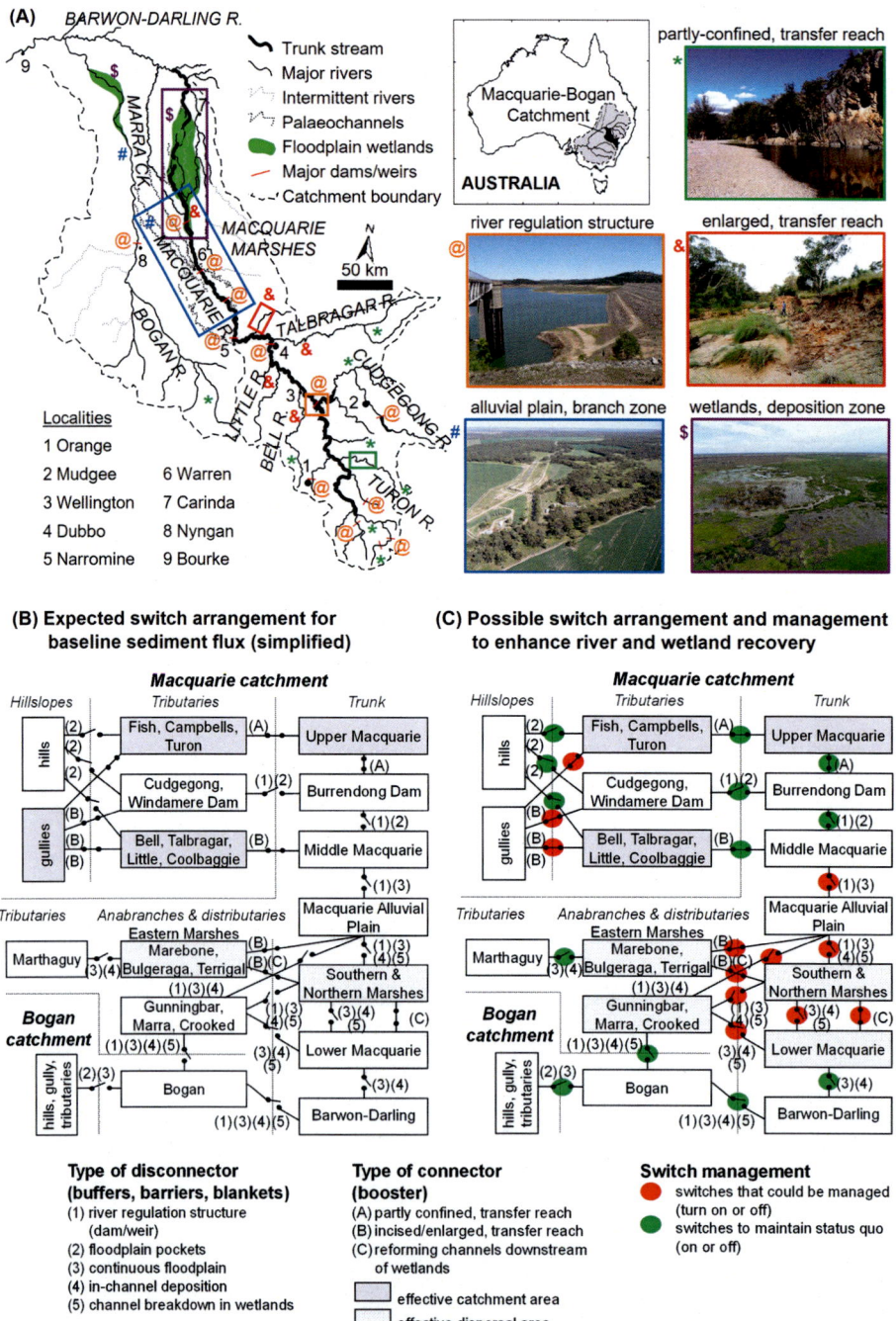

Figure 13.4   Regulating switches to manage sediment (dis)connectivity in the Macquarie catchment. (A) catchment map modified from Ralph and Hesse (2010), sediment budget data from DeRose et al. (2003).

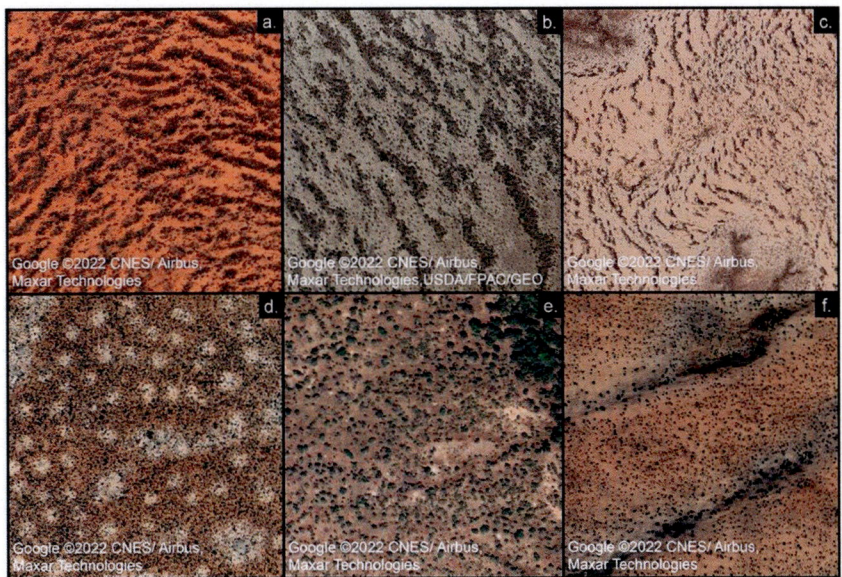

Figure 14.1 Examples of vegetation patterns in drylands around the world. (a) Northern Territory, Australia (Google Earth^{TM} 23°29'6.80"S, 133°51'46.02"E), (b) Texas, USA (Google Earth^{TM} 30°48'44.78"N, 103°23'28.41"W), (c) Baja California, Mexico (Google Earth^{TM} 26°52'43.87"N, 112°51'56.27"W), (d) Kursin, Kenya (Google Earth^{TM} 1° 9'3.10"N, 40°22'21.49"E), (e) W Benin Niger National Park, Benin (Google Earth^{TM} 12° 0'32.05"N, 2°59'36.25"E), (f) Nara, Mali (Google Earth^{TM} 15° 6'17.70"N, 8°10'48.37"W).

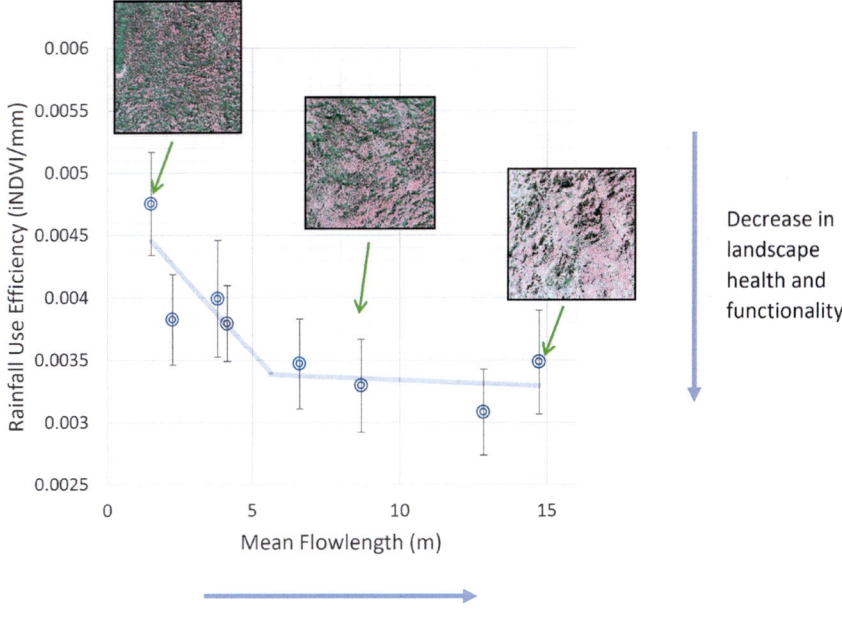

Figure 14.2 Response of landscape health and functionality (quantified using rainfall use efficiency over a period of 13 years) to changes in hydrologic connectivity (quantified using mean flow length) for Mulga dryland sites in Australia with different levels of degradation (adapted from Saco et al., 2020).

Figure 14.3 Experimental restoration in Lake Mere, Australia. (a) pre-experimental conditions, (b) brush piles set in the field with initial vegetation growth, and (c) established vegetation (photo credit David J. Tongway, CSIRO).

Figure 14.4 Connectivity modifiers (ConMOd) installed in a degraded grassland plot in Canyonlands National Park (July 26, 2016) (adapted from USGS, public domain).

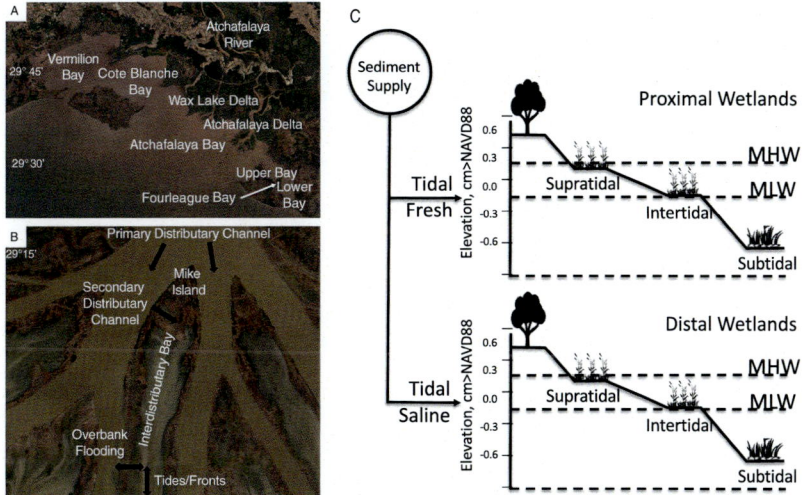

Figure 15.1 (A) Map of the large delta estuaries of the Atchafalaya River that include Fourleague Bay, Atchafalaya Bay, Cote Blanche Bay and Vermilion Bay (Landsat Image, Google Earth, 2021). (B) Morphologic features of coastal deltaic floodplain that defines the connectivity between primary and secondary channels with delta islands that include wetlands defined by hydrogeomorphic zones and interdistributary bay that is coupled to Gulf of Mexico by tides and fronts. (C) Diagrams of marsh platforms illustrating the hydrogeomorphic zones in the proximal and distal wetlands of an active coastal basin, such as described for the Atchafalaya River Delta Estuaries. (Figure from Twilley et al., 2019).

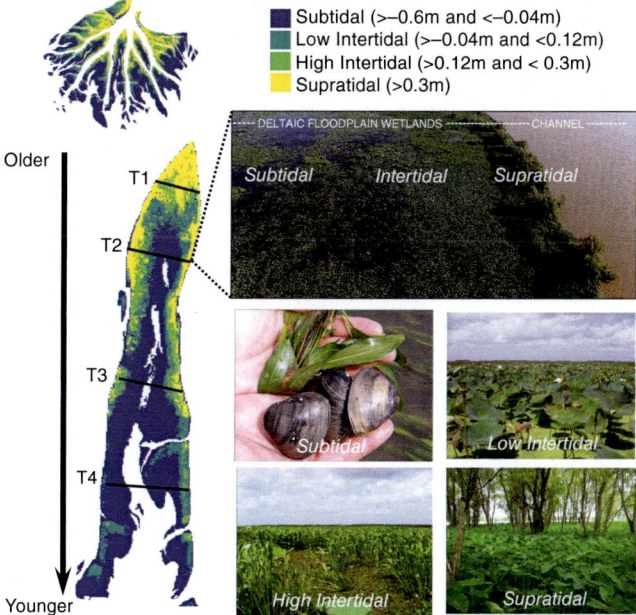

Figure 15.2 Map of Mike Island in WLD, Louisiana, showing the distribution of hydrogeomorphic zones based on elevation records from USGS Atchafalaya 2 project LiDAR Survey 2012 digital elevation model (4 m resolution). The black lines across Mike Island delineate four chronosequence transects (from younger to older: T4–T1) mainly defined by the distance to apex of Mike Island and the characterization of cross-sectional morphology (Figure from Bevington & Twilley, 2018). Three hydrogeomorphic zones (subtidal, intertidal and supratidal) are distinguished by sediment surface elevation relative to MHW and MLW.

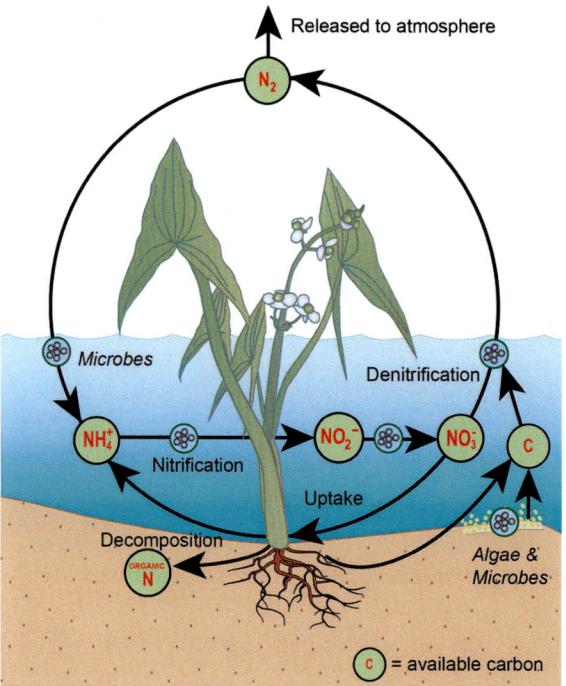

Figure 15.3   A simplified illustration of the nitrogen and phosphorus cycles in a wetland Treatment Wetlands'; images from IAN, University of Maryland (modified from Kadlec & Knight (1996), 'Treatment Wetlands'; images from IAN, University of Maryland).

(a)

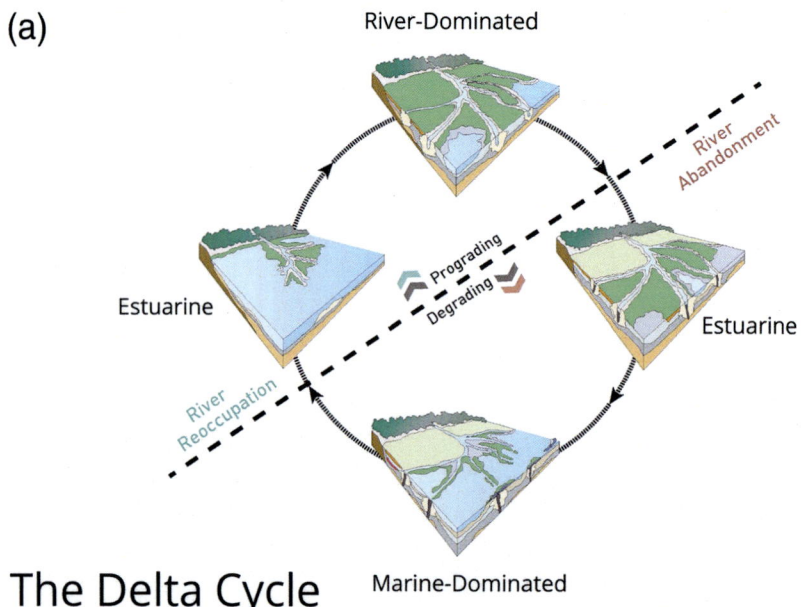

Figure 15.4   (a) Landscape changes of delta cycle associated with river occupation (prograding active delta) compared to stages with river abandonment (degrading inactive delta) modified from Penland et al. (1988).

and flow within every spatial level has been introduced to allow for the identification of available techniques to actually assess connectivity at the given scale. However, in this chapter, we also identified key questions that need answering to make measuring connectivity on different scales reliable and useful. For this purpose, the following research questions for measuring and monitoring connectivity show the most urgent gaps in our knowledge:

– Which are the most useful techniques to measure connectivity at (and across) each scale?
– If we need different measurement techniques for different land uses, how can we standardise results to support comparison of connectivity in different landscape settings?
– How can we compare measurement techniques designed for different time and space realms?
– How can we bring together our measures of the emergent properties of connectivity with tracing approaches that get us close to quantifying how connected a hillslope/river reach/catchment is?
– How can we precisely quantify the distribution of travel distances of parcels of water or particles of sediment that could be defined at any scale of interest, for any cell, or for any timestep?
– How can we compare such a distribution across all spatial scales and all temporal scales, to show how the distribution changes and therefore indicate how connectivity changes with scale of interest, or land management interventions?
– How can we include after hydrological connectivity and sediment connectivity also the connectivity of carbon stocks and flows?
– How can we measure and monitor other substances in landscapes such as pollutants and metals?
– What is the role of gully erosion in catchment scale erosion and how can it be incorporated in connectivity research?
– How can post-processing become available for all environmental research, with emphasis on hydrological and sediment connectivity?
– How can we connect the hydrological and sediment dynamics on the hillslope with the dynamics in the channel?

By solving these measurement issues in connectivity research, an enhanced understanding of the connectivity process will follow that in turn will improve our understanding of the geomorphological genesis. The improved understanding will then enable research to implement nature-based, long-term and economically viable solutions to the problems that society encounters due to anthropogenic changes in connectivity.

Acknowledgements

We would like to thank the participants of the WG2 of COST Action CONNECTEUR: Connecting European Connectivity Research (ES1306) for their contributions to the development of the ideas in this chapter and the good times we spend together.

References

Abotalib, A. Z., Heggy, E., El Bastawesy, M., Ismail, E., Gad, A., & Attwa, M. (2021). Groundwater mounding: A diagnostic feature for mapping aquifer connectivity in hyper-arid deserts. *Science of The Total Environment*, 801, 149760.

Andries, A., Morse, S., Murphy, R. J., Lynch, J., Mota, B., & Woolliams, E. R. (2021). Can current earth observation technologies provide useful information on soil organic carbon stocks for environmental land management policy? *Sustainability*, 13(21), 12074.

Angelopoulou, T., Tziolas, N., Balafoutis, A., Zalidis, G., & Bochtis, D. (2019). Remote sensing techniques for soil organic carbon estimation: A review. *Remote Sensing*, 11(6), 676.

Barrena-González, J., Rodrigo-Comino, J., Gyasi-Agyei, Y., Pulido, M., & Cerdá, A. (2020). Applying the RUSLE and ISUM in the Tierra de Barros Vineyards (Extremadura, Spain) to Estimate Soil Mobilisation Rates. *Land*, 9(3), 93.

Bekele, B., & Gemi, Y. (2021). Soil erosion risk and sediment yield assessment with universal soil loss equation and GIS: In Dijo watershed, Rift valley Basin of Ethiopia. *Modeling Earth Systems and Environment*, 7(1), 273–291.

Berihun, M. L., Tsunekawa, A., Haregeweyn, N., Tsubo, M., Fenta, A. A., Ebabu, K., Sultan D. & Dile, Y. T. (2022). Reduced runoff and sediment loss under alternative land capability-based land use and management options in a sub-humid watershed of Ethiopia. *Journal of Hydrology: Regional Studies*, 40, 100998.

Bezak, N., Mikoš, M., Borrelli, P., Alewell, C., Alvarez, P., Anache, J. A. A., Baartman J et al. (2021). Soil erosion modelling: A bibliometric analysis. *Environmental Research*, 197, 111087.

Blake, W. H., Wallbrink, P. J., Wilkinson, S. N., Humphreys, G. S., Doerr, S. H., Shakesby, R. A., & Tomkins, K. M. (2009). Deriving hillslope sediment budgets in wildfire-affected forests using fallout radionuclide tracers. *Geomorphology*, 104(3–4), 105–116.

Bracken, L. J., Wainwright, J., Ali, G. A., Tetzlaff, D., Smith, M. W., Reaney, S. M. & Roy, A. G. (2013). Concepts of hydrological connectivity: Research approaches, pathways and future agendas. *Earth-Science Reviews*, 119, 17–34. https://doi.org/10.1016/j.earscirev.2013.02.001

Bracken, L. J., Turnbull, L., Wainwright, J., & Bogaart, P. (2015). Sediment connectivity: A framework for understanding sediment transfer at multiple scales. *Earth Surface Processes and Landforms*, 40, 177–188. https://doi.org/10.1002/esp.3635

Brunsden, D., & Thornes, J. B. (1979). Landscape sensitivity and change. *Transactions of the Institute of British Geographers*, 4, 463–484.

Burt, T., & Allison, R. J. (eds.). (2010). *Sediment Cascades: An Integrated Approach*. John Wiley & Sons.

Cerdà, A. & Doerr, S. H. (2008). The effect of ash and needle cover on surface runoff and erosion in the immediate post-fire period. *Catena*, 74, 256–263.

Cerdà, A., Rodrigo-Comino, J., Yakupoğlu, T., Dindaroğlu, T., Terol, E., Mora-Navarro, G., Arabameri, A., Radziemska, M., Novara, A., Kavian, A., Vaverková, M. D.,

Abd-Elmabod, S. K., Hammad, H. M. & Daliakopoulos, I. N. (2020). Tillage versus no-Tillage. Soil properties and hydrology in an organic persimmon farm in Eastern Iberian Peninsula. *Water* 12, 1539. https://doi.org/10.3390/w12061539

Cerdà, A., Franch-Pardo, I., Novara, A., Sannigrahi, S. & Rodrigo-Comino, J. (2021). Examining the effectiveness of catch crops as a nature-based solution to mitigate surface soil and water losses as an environmental regional concern. *Earth Systems and Environment*. https://doi.org/10.1007/s41748-021-00284-9.

Cerdà, A., Lucas-Borja, M. E., Franch-Pardo, I., Úbeda, X., Novara, A., López-Vicente, M., Popović, Z. & Pulido, M. (2021). The role of plant species on runoff and soil erosion in a Mediterranean shrubland. *Science of The Total Environment*, 799, 149218.

Chenu, C., Le Bissonnais, Y., & Arrouays, D. (2000). Organic matter influence on clay wettability and soil aggregate stability. *Soil Science Society of America Journal*, 64(4), 1479–1486.

Chirinda, N., Roncossek, S. D., Heckrath, G., Elsgaard, L., Thomsen, I. K., & Olesen, J. E. (2014). Root and soil carbon distribution at shoulderslope and footslope positions of temperate toposequences cropped to winter wheat. *Catena*, 123, 99–105.

Chorley, R. J., & Kennedy, B. A. (1971). *Physical Geography: A Systems Approach*. London, UK: Prentice Hall.

Cucchiaro, S., Cavalli, M., Vericat, D., Crema, S., Llena, M., Beinat, A., Marchi, L., & Cazorzi, F. (2018). Monitoring topographic changes through 4D-structure-from-motion photogrammetry: Application to a debris-flow channel. *Environmental Earth Sciences*, 77(18), 1–21.

Doyle, M. W., Shields, D., Boyd, K. F., Skidmore, P. B., & Dominick, D. (2007). Channel-forming discharge selection in river restoration design. *Journal of Hydraulic Engineering*, 133(7), 831–837.

Edokpa, D., Milledge, D., Allott, T., Holden, J., Shuttleworth, E., Kay, M., Johnston, A., et al.(2022). Rainfall intensity and catchment size control storm runoff in a gullied blanket peatland. *Journal of Hydrology*, 609, 127688.

Evrard, O., Laceby, J. P., Lepage, H., Onda, Y., Cerdan, O., & Ayrault, S. (2015). Radiocesium transfer from hillslopes to the Pacific Ocean after the Fukushima Nuclear Power Plant accident: A review. *Journal of Environmental Radioactivity*, 148, 92–110.

Ferguson, R. (2008). Gravel-bed rivers at the reach scale. In Habersack, H., Piegay, H. & Rinaldi, M. (eds.), *Gravel-Bed Rivers VI: From Process Understanding to River Restoration*. Amsterdam: Elsevier. 33–53.

Fernández-Raga, M., Palencia, C., Keesstra, S., Jordán, A., Fraile, R., Angulo-Martínez, M. & Cerdà, A. (2017). Splash erosion: A review with unanswered questions. *Earth-Science Reviews*, 171, 463–477. https://doi.org/10.1016/j.earscirev.2017.06.009

Forsmoo, J., Anderson, K., Macleod, C. J., Wilkinson, M. E., & Brazier, R. (2018). Drone-based structure-from-motion photogrammetry captures grassland sward height variability. *Journal of Applied Ecology*, 55(6), 2587–2599.

Fryirs, K. A., Brierley, G. J., Preston, N. J., & Kasai, M. (2007). Buffers, barriers and blankets: The (dis) connectivity of catchment-scale sediment cascades. *Catena*, 70(1), 49–67.

Fryirs, K. A. (2013). (Dis)Connectivity in catchment sediment cascades: A fresh look at the sediment delivery problem. *Earth Surface Processes and Landforms*, 38, 30–46.

García-Ruiz, J. M., Beguería, S., Lana-Renault, N., Nadal-Romero, E. & Cerdà, A. (2017). Ongoing and emerging questions in water erosion studies. *Land Degradation & Development*, 28, 5–21. https://doi.org/10.1002/ldr.2641

Heckmann, T. & Vericat, D. (2018). Computing spatially distributed sediment delivery ratios: Inferring functional sediment connectivity from repeat high-resolution digital elevation models. *Earth Surface Processes and Landforms*, 43, 1547–1554. https://doi.org/10.1002/esp.4334

Heidbüchel, I., Troch, P. A., Lyon, S. W., & Weiler, M. (2012). The master transit time distribution of variable flow systems. *Water Resources Research*, 48(6).

Hikel, H., Yair, A., Schwanghart, W., Hoffmann, U., Straehl, S. & Kuhn, N. J. (2013): Experimental investigation of soil ecohydrology on rocky desert slopes in the Negev Highlands, Israel. *Zeitschrift für Geomormorphologie*, Suplementary Issue, 57, 39–58.

Hooke, J. & Souza, J. (2021). Challenges of mapping, modelling and quantifying sediment connectivity. *Earth-Science Reviews*, 223. https://doi.org/10.1016/j.earscirev.2021.103847

Kalantari, Z., Ferreira, C. S. S., Koutsouris, A. J., Ahmer, A.-K., Cerdà, A. & Destouni, G. (2019). Assessing flood probability for transportation infrastructure based on catchment characteristics, sediment connectivity and remotely sensed soil moisture. *Science of the Total Environment*, 661, 393–406. https://doi.org/10.1016/j.scitotenv.2019.01.009

Keesstra, S., Pereira, P., Novara, A., Brevik, E. C., Azorin-Molina, C., Parras-Alcántara, L., Jordán, A. & Cerdà, A. (2016). Effects of soil management techniques on soil water erosion in apricot orchards. *Science of the Total Environment*, 551–552, 357–366. https://doi.org/10.1016/j.scitotenv.2016.01.182

Keesstra, S., Nunes, J. P., Saco, P., Parsons, T., Pöppl, R., Masselink, R. & Cerdà, A. (2018). The way forward: Can connectivity be useful to design better measuring and modelling schemes for water and sediment dynamics?. *Science of the Total Environment*, 644, 1557–1572. https://doi.org/10.1016/j.scitotenv.2018.06.342

Keesstra, S. D., Davis, J., Masselink, R. H., Casalí, J., Peeters, E. T. H. M. & Dijksma, R. (2019). Coupling hysteresis analysis with sediment and hydrological connectivity in three agricultural catchments in Navarre, Spain. *Journal of Soils Sediments*, 19, 1598–1612. https://doi.org/10.1007/s11368-018-02223-0

Klaus, J., & McDonnell, J. J. (2013). Hydrograph separation using stable isotopes: Review and evaluation. *Journal of Hydrology*, 505, 47–64.

Kleine, L., Tetzlaff, D., Smith, A., Goldhammer, T. & Soulsby, C. (2021). Using isotopes to understand landscape-scale connectivity in a groundwater-dominated, lowland catchment under drought conditions. *Hydrological Processes*, 35. https://doi.org/10.1002/hyp.14197

Keesstra, S. D., Temme, A. J. A. M., Schoorl, J. M. & Visser, S. M. (2014). Evaluating the hydrological component of the new catchment-scale sediment delivery model LAPSUS-D. *Geomorphology*, 212, 97–107. https://doi.org/10.1016/j.geomorph.2013.04.021

Koch, J. C., Dornblaser, M. M., & Striegl, R. G. (2021). Storm-scale and seasonal dynamics of carbon export from a nested subarctic watershed underlain by permafrost. *Journal of Geophysical Research: Biogeosciences*, 126(8), e2021JG006268.

Kröpfl, A. I., Cecchi, G. A., Villasuso, N. M. & Distel, R. A. (2013). Degradation and recovery processes in Semi-Arid patchy rangelands of northern Patagonia, Argentina. *Land Degradation & Development*, 24: 393–399. doi:10.1002/ldr.1145

Lecce, S. A., & Pavlowsky, R. T. (2014). Floodplain storage of sediment contaminated by mercury and copper from historic gold mining at Gold Hill, North Carolina, USA. *Geomorphology*, 206, 122–132.

Li, S., Lu, J., Liang, G., Wu, X., Zhang, M., Plougonven, E., Wang, Y., Gao, L., Abdelrhman, A. A., Song, X., Liu, X. & Degré, A. (2021). Factors governing soil water repellency under tillage management: The role of pore structure and hydrophobic substances. *Land Degradation & Development*, 32, 1046–1059. https://doi.org/10.1002/ldr.3779

Li, X., Fu, S., Hu, Y., & Liu, B. (2022). Effects of rock fragment coverage on soil erosion: Differ among rock fragment sizes?. *CATENA*, 214, 106248.

Lizaga, I., Gaspar, L., Blake, W. H., Latorre, B., & Navas, A. (2019). Fingerprinting changes of source apportionments from mixed land uses in stream sediments before and after an exceptional rainstorm event. *Geomorphology*, 341, 216–229.

Llena, M., Vericat, D., Cavalli, M., Crema, S., & Smith, M. W. (2019). The effects of land use and topographic changes on sediment connectivity in mountain catchments. *Science of the Total Environment*, 660, 899–912.

López-Vicente, M., Kramer, H. & Keesstra, S. (2021a). Effectiveness of soil erosion barriers to reduce sediment connectivity at small basin scale in a fire-affected forest. *Journal of Environmental Management*, 278. https://doi.org/10.1016/j.jenvman.2020.111510

López-Vicente, M., Cerdà, A., Kramer, H. & Keesstra, S. (2021b). Post-fire practices benefits on vegetation recovery and soil conservation in a Mediterranean area. *Land Use Policy*, 111, 105776. https://doi.org/10.1016/j.landusepol.2021.105776

López-Vicente, M., González-Romero, J., & Lucas-Borja, M. E. (2020). Forest fire effects on sediment connectivity in headwater sub-catchments: Evaluation of indices performance. *Science of the Total Environment*, 732, 139206.

Lugato, E., Smith, P., Borrelli, P., Panagos, P., Ballabio, C., Orgiazzi, A., Fernandez-Ugalde, O., Montanarella, L. & Jones, A (2018). Soil erosion is unlikely to drive a future carbon sink in Europe. *Science Advances*, 4(11), eaau3523.

Luk, S. H., Abrahams, A. D., & Parsons, A. J. (1993). Sediment sources and sediment transport by rill flow and interrill flow on a semi-arid piedmont slope, southern Arizona. *Catena*, 20(1–2), 93–111.

Martinez-Agirre, A., Álvarez-Mozos, J., Milenković, M., Pfeifer, N., Giménez, R., Valle, J. M., & Rodríguez, Á. (2020). Evaluation of Terrestrial Laser Scanner and Structure from Motion photogrammetry techniques for quantifying soil surface roughness parameters over agricultural soils. *Earth Surface Processes and Landforms*, 45(3), 605–621.

Masselink, R. J. H., Temme, A. J. A. M., Giménez, R., Casalí, J. & Keesstra, S. D. (2017). Assessing hillslope-channel connectivity in an agricultural catchment using rare-earth oxide tracers and random forests models. *Cuadernos de Investigación Geográfica / Geographical Research*, 43. https://doi.org/10.18172/cig.3169

Mekonnen, M., Keesstra, S. D., Baartman, J. E., Stroosnijder, L., & Maroulis, J. (2017). Reducing sediment connectivity through man-made and natural sediment sinks in the Minizr catchment, Northwest Ethiopia. *Land Degradation & Development*, 28(2), 708–717.

Miller, J. R., Watkins, X., O'Shea, T., & Atterholt, C. (2021). Controls on the spatial distribution of trace metal concentrations along the bedrock-dominated South Fork New River, North Carolina. *Geosciences*, 11(12), 519.

Mouyen, M., Steer, P., Chang, K. J., Le Moigne, N., Hwang, C., Hsieh, W. C., Jeandet, L, et al. (2020). Quantifying sediment mass redistribution from joint time-lapse gravimetry and photogrammetry surveys. *Earth Surface Dynamics*, 8(2), 555–577.

Novara, A., Pulido, M., Rodrigo-Comino, J., Di Prima, S., Smith, P., Gristina, L., Giménez-Morera, A., Terol E., Salesa, D. & Keesstra, S. (2019). Long-term organic farming on a citrus plantation results in soil organic matter recovery. *Cuadernos de Investigación Geográfica*, 45, 271–286. http://doi.org/10.18172/cig.3794

Old, G., Naden, P., Granger, S. J., Bilotta, G. S., Brazier, R. E. & Macleod, C. J. A. (2012). A novel application of natural fluorescence to understand the sources and transport pathways of pollutants from livestock farming in small headwater catchments. *Science of the Total Environment*, 417, 169–182.

Owens, P. N., Blake, W. H., Giles, T. R., & Williams, N. D. (2012). Determining the effects of wildfire on sediment sources using 137Cs and unsupported 210Pb: The role of landscape disturbances and driving forces. *Journal of Soils and Sediments*, 12(6), 982–994.

Oyewumi, O., Cavanaugh, C., Guzzardi, D., & Costa, M. (2022). Geochemical assessment of trace element concentrations in the Farmington River, Connecticut, Northeastern, USA. *Environmental Monitoring and Assessment*, 194(5), 1–15.

Palacio, R. G., Bisigato, A. J. & Bouza, B. J. (2014). Soil erosion in three grazed plant communities in northeastern Patagonia. *Land Degradation and Development*, 25, 594–603. doi:10.1002/ldr.2289

Parsons, A. J., Brazier, R. E., Wainwright, J., & Powell, D. M. (2006). Scale relationships in hillslope runoff and erosion. *Earth Surface Processes and Landforms: The Journal of the British Geomorphological Research Group*, 31(11), 1384–1393.

Parsons, A. J., Bracken, L., Pöppl, R. E., Wainwright, J. & Keesstra, S. D. (2015). Introduction to special issue on connectivity in water and sediment dynamics. *Earth Surface Processes and Landforms*, 40, 1275–1277. https://doi.org/10.1002/esp.3714

Petit, S. & Burel, F. (1998). Effects of landscape dynamics on the metapopulation of a ground beetle (Coleoptera, Carabidae) in a hedgerow network. *Agriculture, Ecosystems and Environment*, 69, 243–252.

Powell, D., Brazier, R., Parsons, A., Wainwright, J. & Nichols, M. (2007). Sediment transfer and storage in dryland headwater streams. *Geomorphology*, 88 (1–2), 152–166.

Prats, S. A., Malvar, M. C., Simões-Vieira, D. C., MacDonald, L. & Keizer, J. J. (2013). Effectiveness of hydro- mulching to reduce runoff and erosion in a recently burnt pine plantation in central Portugal. *Land Degradation & Development*, doi: 10.1002/ldr.2236.

Priddy, C. L., Pringle, J. K., Clarke, S. M., & Pettigrew, R. P. (2019). Application of photogrammetry to generate quantitative geobody data in ephemeral fluvial systems. *The Photogrammetric Record*, 34(168), 428–444.

Puttock, A., Macleod, C., Bol, R., Sessford, P., Dungait, J. & Brazier, R. E. (2013). Changes in ecosystem structure, function and hydrological connectivity control water, soil and carbon losses in semi-arid grass to woody vegetation transitions. *Earth Surface Processes and Landforms*, 38 (13), 1602–1611.

Richards, G., Gilmore, T. E., Mittelstet, A. R., Messer, T. L., & Snow, D. D. (2021). Baseflow nitrate dynamics within nested watersheds of an agricultural stream in Nebraska, USA. *Agriculture, Ecosystems & Environment*, 308, 107223.

Rodrigo-Comino, J., Ponsoda-Carreres, M., Salesa, D., Terol, E., Gyasi-Agyei, Y., & Cerdà, A. (2020). Soil erosion processes in subtropical plantations (Diospyros kaki) managed under flood irrigation in Eastern Spain. *Singapore Journal of Tropical Geography*, 41(1), 120–135.

Rodrigo-Comino, J., Terol, E., Mora, G., Gimenez-Morera, A., & Cerdà, A. (2020). Vicia sativa Roth. can reduce soil and water losses in recently planted vineyards (Vitis vinifera L.). *Earth Systems and Environment*. https://doi.org/10.1007/s41748-020-00191-5

Rodrigo Comino, J., Keesstra, S. D., & Cerdà, A. (2018). Connectivity assessment in Mediterranean vineyards using improved stock unearthing method, LiDAR and soil erosion field surveys. *Earth Surface Processes and Landforms*, 43(10), 2193–2206.

Rodrigo Comino, J. & Cerdà, A. (2018). Improving stock unearthing method to measure soil erosion rates in vineyards. *Ecological Indicators*, 85, 509–517. https://doi.org/10.1016/j.ecolind.2017.10.042

Sepehri, M., Ghahramani, A., Kiani-Harchegani, M., Ildoromi, A. R., Talebi, A. & Rodrigo-Comino, J. (2021). Assessment of drainage network analysis methods to rank sediment yield hotspots. *Hydrological Sciences Journal*, 66, 904–918. https://doi.org/10.1080/02626667.2021.1899183

Saco, P. M., Rodríguez, J. F., Moreno-de las Heras, M., Keesstra, S., Azadi, S., Sandi, S., Baartman, J., Rodrigo-Comino, J. & Rossi, M. J. (2020). Using hydrological connectivity to detect transitions and degradation thresholds: Applications to dryland systems. *Catena* 186. https://doi.org/10.1016/j.catena.2019.104354

Saco, P. M. & Moreno-De Las Heras, M. (2013). Ecogeomorphic coevolution of semiarid hillslopes: Emergence of banded and striped vegetation patterns through interaction of biotic and abiotic processes. *Water Resources Research*, 49, 115–126. https://doi.org/10.1029/2012WR012001

Salesa, D. & Cerdà, A. (2020). Soil erosion on mountain trails as a consequence of recreational activities. A comprehensive review of the scientific literature. *Journal of Environmental Management*, 271. https://doi.org/10.1016/j.jenvman.2020.110990

Sharma, N., Kaushal, A., Yousuf, A., Sood, A., Kaur, S., & Sharda, R. (2022). Geospatial technology for assessment of soil erosion and prioritization of watersheds using RUSLE model for lower Sutlej sub-basin of Punjab, India. *Environmental Science and Pollution Research*, 1–17.

Shoshany, M. (2012). Identifying desert thresholds by mapping inverse and recovery potentials in patch patterns using spectral and morphological algorithms. *Land Degradation & Development*, 23: 331–338. doi:10.1002/ldr.2146

Soulsby, C., Birkel, C., Geris, J., Dick, J., Tunaley, C., & Tetzlaff, D. (2015). Stream water age distributions controlled by storage dynamics and nonlinear hydrologic connectivity: Modeling with high-resolution isotope data. *Water Resources Research*, 51(9), 7759–7776.

Smith, P., Soussana, J. F., Angers, D., Schipper, L., Chenu, C., Rasse, D. P., Batjes, N. H., et al. (2020). How to measure, report and verify soil carbon change to realize the potential of soil carbon sequestration for atmospheric greenhouse gas removal. *Global Change Biology*, 26(1), 219–241.

Smith, M. W., & Vericat, D. (2014). Evaluating shallow-water bathymetry from through-water terrestrial laser scanning under a range of hydraulic and physical water quality conditions. *River Research and Applications*, 30(7), 905–924.

Smith H. G., Hopmans P., Sheridan G. J., Lane P. N. J., Noske P. J. & Bren L. J. (2012). Impacts of wildfire and salvage harvesting on water quality and nutrient exports from radiata pine and eucalypt forest catchments in south-eastern Australia. *Forest Ecology and Management*, 263, 160–169.

Smith, H. G., & Blake, W. H. (2014). Sediment fingerprinting in agricultural catchments: A critical re-examination of source discrimination and data corrections. *Geomorphology*, 204, 177–191.

Smith, P. (2004). Soils as carbon sinks: The global context. *Soil Use and Management*, 20(2), 212–218.

Sombrero, A. & de Benito, A. (2010). Carbon accumulation in soil. Ten-year study of conservation tillage and crop rotation in a semi-arid area of Castile-Leon, Spain. *Soil and Tillage Research*, 107, 64–70. https://doi.org/10.1016/j.still.2010.02.009

Tekwa, I. J., Laflen, J. M., Kundiri, A. M., & Alhassan, A. B. (2021). Evaluation of WEPP versus EGEM and empirical model efficiencies in predicting ephemeral gully erosion around Mubi area, Northeast Nigeria. *International Soil and Water Conservation Research*, 9(1), 11–25.

Tessler, N., Wittenberg, L. & Greenbaum, N. (2013). Soil water repellency persistence after recurrent forest fires on Mount Carmel, Israel. *International Journal of Wildland Fire*, 22, 515–526.

Tischendorf, L. (2001). Can landscape indices predict ecological processes consistently?. *Landscape Ecology*, 16(3), 235–254.

Turnbull, L., Hütt, M.-T., Ioannides, A. A., Kininmonth, S., Pöppl, R., Tockner, K., Bracken, L. J., Keesstra, S., Liu, L., Masselink, R., Masselink, R. & Parsons, A. J. (2018). Connectivity and complex systems: Learning from a multi-disciplinary perspective. *Applied Network Science*, 3. https://doi.org/10.1007/s41109-018-0067-2

Turnbull, L., Wainwright, J. & Brazier, R. E. (2010). Hydrology, erosion and nutrient transfers over a transition from semi-arid grassland to shrubland in the South-Western USA: A modelling assessment. *Journal of Hydrology*, 388 (3–4), 258–272.

Turski, M., Lipiec, J., Chodorowski, J., Sokołowska, Z., & Skic, K. (2022). Vertical distribution of soil water repellency in ortsteinic soils in relation to land use. *Soil and Tillage Research*, 215, 105220.

Vannoppen, W., Vanmaercke, M., De Baets, S. & Poesen, J. (2015). A review of the mechanical effects of plant roots on concentrated flow erosion rates. *Earth-Science Reviews*, 150, 666–678. https://doi.org/10.1016/j.earscirev.2015.08.011

Walling, D. E. (2013). The evolution of sediment source fingerprinting investigations in fluvial systems. *Journal of Soils and Sediments*, 13, 1658–1675.

Walter, T. R., Salzer, J., Varley, N., Navarro, C., Arámbula-Mendoza, R., & Vargas-Bracamontes, D. (2018). Localized and distributed erosion triggered by the 2015 Hurricane Patricia investigated by repeated drone surveys and time lapse cameras at Volcán de Colima, Mexico. *Geomorphology*, 319, 186–198.

With, K. A., & King, A. W. (1997). The use and misuse of neutral landscape models in ecology. *Oikos*, 219–229.

Wohl, E., Brierley, G., Cadol, D., Coulthard, T. J., Covino, T., Fryirs, K. A., Grant, G., Hilton, R. G., Lane, S. N., Magilligan, F. J., Meitzen, K. M., Passalacqua, P., Pöppl, R. E., Rathburn, S. L., & Sklar, L. S. (2019). Connectivity as an emergent property of geomorphic systems. *Earth Surface Processes and Landforms*, 44, 4–26.

Wolman, M. G. & Miller, J. P. (1960). Magnitude and frequency of forces in geomorphic processes. *Journal of Geology*, 68(1), 54–74. https://doi.org/10.1086/626637

Yan, R., & Gao, J. (2021). Key factors affecting discharge, soil erosion, nitrogen and phosphorus exports from agricultural polder. *Ecological Modelling*, 452, 109586.

Zhang, G., Mahale, V. N., Putnam, B. J., Qi, Y., Cao, Q., Byrd, A. D., Bukovcic, P, et al. (2019). Current status and future challenges of weather radar polarimetry: Bridging the gap between radar meteorology/hydrology/engineering and numerical weather prediction. *Advances in Atmospheric Sciences*, 36(6), 571–588.

10

Indices

TOBIAS HECKMANN AND SAEED NAJAFI

10.1 Introduction

Connectivity is an abstract construct, unlikely to be measurable in a single number (cf. Chapter 9 of this book), and governed by a multitude of static (structural) and dynamic (functional) factors. Therefore, like in other fields of research where measures are sought for complex properties (sustainability, vulnerability, resilience, to name a few), we resort to indices. An index is a 'type of composite measure that summarises and rank-orders several specific observations and represents some more-general dimension' (Babbie, 2010: 162; see also Liverman et al., 1988). The 'specific observations' in our case refer to either variables known or thought to influence connectivity or to representations of connectivity that can be mapped or measured. In Section 10.2, we will discuss these variables and observables. In Section 10.3, we present a broad overview of sediment connectivity indices used in recent studies, with a focus on different approaches rather than on a complete listing; Heckmann et al. (2018), Wohl et al. (2019) and Najafi et al. (2021a) provide extensive reviews of connectivity metrics and their implementation in indices. Section 10.4 also contains example applications (Section 10.4.2); following a recent paper by Hooke and Souza (2021), we distinguish diagnostic/descriptive and explanatory/predictive applications. Considering the application of indices also touches on their validation (in fact a research gap; Sections 10.4.2 and 10.5). Finally, in Section 10.5 we summarise and give an outline of future avenues for application and research.

10.2 Connectivity Indices: Ingredients and Issues

A connectivity index consists of multiple variables conceptually known to control the spatial configuration and (potential) magnitude of sediment fluxes in a landscape. Some of these variables are explicit and easily measured on digital elevation models (DEMs), some of them are proxies for variables that are not or cannot be measured at the desired spatial and temporal scale (i.e., extent and resolution). Like in digital geomorphometry, some variables are derived directly from DEMs (primary parameters; Wilson & Bishop, 2013) or remote sensing data, while others represent more complex combinations of primary parameters (secondary parameters). In this section, we list the most important parameters, explain in brief their computation and the rationale behind their implementation in connectivity indices.

10.2.1 Variables for Connectivity Indices

Slope governs the proportion of gravity that is directed in the downslope direction. Steeper gradient is associated with increasing slope instability and an acceleration of processes that detach and transfer sediment, and a higher flow velocity provides more energy for sediment transport. A decrease of slope, for example, where a steep tributary channel enters the floodplain of a higher order river, leads to a decrease in transport capacity and to deposition. Such changes of slope with distance along a flow path on a hillslope or within a channel are measured by (profile or vertical) **curvature**.

Surface roughness affects connectivity in two ways. First, microtopographic surface roughness is associated with ponding storage that potentially delays or impedes the formation of continuous surface runoff as an agent of sediment transport (Antoine et al., 2009; Straffelini et al., 2021). Second, more energy is dissipated when sediment is transported over rough surfaces. At this spatial scale, roughness is generated by the **characteristics of the surface (material, grain sizes)** on the one hand, and by **landcover/vegetation** on the other. Both can be represented in connectivity indices, separately or combined. The weighting factor W represents impedance to water and sediment fluxes in the popular IC index (Borselli et al., 2008; Cavalli et al., 2013). While Borselli et al. (2008) use the landcover-related crop (=C) factor of the universal soil loss equation as W, Cavalli et al. (2013) compute surface roughness from high-resolution DEMs and use it as W. In their modification of IC, Trevisani & Cavalli (2016) suggest anisotropic roughness indices to better account for the flow-directional characteristics of sediment transport and connectivity, and Lizaga et al. (2018) combine vegetation and topographic roughness into one single W parameter. Microtopographic (or soil surface) roughness can be quantified from DEMs with very high-resolution, usually generated from terrestrial laser scans or photogrammetric analysis of drone-based aerial imagery (see Martinez-Agirre et al., 2020, and references cited therein). For larger study areas, however, DEMs usually have a spatial resolution of >= 1 m and represent roughness at a larger spatial scale (form roughness rather than surface roughness); at this scale, **topographic sinks** can be detected that effectively prevent sediment transfer. Lane et al. (2017) distinguish sinks that can be attributed to uncertainties in the DEM ('methodological disconnection') from sinks that exist in reality and cause sediment pathways to end ('process disconnection'); the filling of only those sinks that constitute 'methodological disconnections' leads to an assessment of potentially (de-)coupled parts of a catchment.

Sediment connectivity, at least with respect to water-mediated sediment transfer, depends on the strength (amount and velocity) of **surface runoff**. Surface run-off occurrence and pathways can be mapped at discrete points in time (e.g., Meerkerk et al., 2009; Calsamiglia et al., 2020), but not measured quantitatively in a spatially distributed and temporally continuous manner. The local **size of the contributing area** A, computed as **flow accumulation**, is related to the amount of surface run-off Q, for example, $Q = aA^b$ with a and b being empirical constants, or $Q = A^{0.5}$ (Dalla Fontana & Marchi, 2003; Marchi & Dalla Fontana, 2005; the expression $A^{0.5}$ is also contained in the upslope component of the IC index, Borselli et al., 2008). Thus, flow accumulation is being used as a proxy for surface runoff in many hydro-geomorphological indices. The computation of A depends on the choice of a flow routing algorithm; Cavalli et al. (2013), for example, preferred the DInf algorithm (Tarboton, 1997) over the single-flow D8 algorithm (steepest descent) for its ability to represent diverging flow. The contributing area A can be multiplied with local slope, which yields the stream power index SPI = A tan(beta) and is considered a proxy for sediment transport capacity. Variants of the SPI have been used in early concepts of connectivity indices (Marchi & Dalla Fontana, 2005), and the principle is reflected also in recent ones (e.g., the index of connectivity (IC) after Borselli et al., 2008; Cavalli et al., 2013).

The explanation for using the contributing area (flow accumulation) as a proxy for surface runoff and related processes or properties highlights the fact that purely local surface properties do not suffice for connectivity assessment. Indices assessing either the geomorphic coupling between two landforms or sediment connectivity at the larger spatial scale of a geomorphic system, for example, a (sub-)catchment, must include **non-local parameters**. These relate to the sizes and properties of contributing areas in upslope/upstream direction and to the length and properties of potential **sediment pathways** in the downslope/ downstream direction. This concept has been explicitly implemented in several indices: The DEBAS and DENET indicators (Marchi & Dalla Fontana, 2005) evaluate the number of cells that fall below a specified SPI threshold within the contributing area (DEBAS) or channel network upslope/upstream of each raster cell (DENET), respectively. These indices evaluate a deficit ('DE') with respect to stream power on basin slopes ('BAS') and within the channel network ('NET') that is thought to cause decoupling. The modified topographic wetness index (TWI; Lane et al., 2004, 2009) combines the local (slope) and upslope (flow accumulation) component with a downslope perspective by evaluating the lowest TWI value along the downslope/downstream flow path. This index reflects the notion that water-mediated sediment transfer across the surface requires continuously saturated flow paths. The lowest propensity to saturation (represented by the lowest TWI) along a flow path determines the propensity of the whole flow path to be continuously saturated and hence to provide enough runoff to achieve sediment connectivity. Reid et al. (2007) successfully used this index to model the delivery of sediment from landslides to the channel network. The IC (Borselli et al., 2008; Cavalli et al., 2013; see Section 3.1.1) has an upslope and a downslope component referring to the contributing area and the downstream/downslope flow path, respectively.

A high density of pathways along which sediment can be transported (e.g., a high **drainage density**) supports high connectivity (Brunsden and Thornes, 1979). Gay et al. (2016) compare, for each cell of a raster DEM, the flow path distance to the nearest thalweg (automatically delineated) with the flow path distance to the nearest channel (mapped channel network). This metric is related to the relation of infiltration vs. surface runoff and hence to drainage density, and was used to amend the IC in flat (i.e., slope below 7%) lowland areas. Other approaches to implement the potential surface runoff in connectivity indices include a rainfall erosivity factor (Chartin et al., 2017) or a curve number factor that reflects the run-off generation depending on soil and landuse properties (Hooke et al., 2017; Kalantari et al., 2017).

10.2.2 Spatial Units and Implications for Connectivity Indices

The majority of existing connectivity indices works on the basis of raster data (see reviews by Heckmann et al., 2018; Najafi et al., 2021a). Like other parameters computed from DEMs and remote sensing data, these connectivity indices are influenced by the raster resolution, that is, the size of the raster cells. Cantreul et al. (2018), for example, studied the effect of cell size on the IC (Cavalli et al., 2013) and found a systematic increase of IC values with decreasing resolution (larger cell sizes); moreover, topographic details only contained in fine-scale DEMs (such as field limits) had a strong effect on IC spatial patterns. The same authors concluded that high-resolution DEMs are to be preferred as they are able to represent such features and provide more accurate flow paths; higher resolutions than 1 m increased computational time but did not improve connectivity information.

Raster cells of a DEM, however, represent artificial discretisations of the Earth surface and are not, *per se,* geomorphologically meaningful spatial units, unlike landscape features such as landforms (river reaches, floodplains, alluvial cones, etc.). Moreover, (dis-)connectivity not necessarily arises at the scale of cell-sized areas in a landscape. Nevertheless, the aforementioned variables are acquired, and the corresponding connectivity indices are often computed, on a raster cell basis. In general, the choice of a **fundamental spatial unit** is key to the assessment of connectivity (Turnbull et al., 2018; Pöppl & Parsons, 2018). This has two implications for connectivity indices: First, meaningful spatial units have to be defined and delineated. Second, either connectivity-relevant variables need to be collected at or upscaled to the scale of these spatial units in order to compute indices, or raster-based indices must be aggregated/upscaled in order to represent these larger units. The latter can be done, for example, based on the mean or median index value for the respective unit. However, the mean or median are not the only options for characterising a larger area with respect to connectivity. De Walque et al. (2017), for example, systematically assessed what statistical parameters yielded the best coefficient of determination in a regression model predicting the susceptibility to muddy floods in a Belgian catchment. The best model included the 99th percentile of a revised version of the IC index, which means that the highest, rather than the average, connectivity within a (sub)catchment had the highest predictive capacity with respect to this hydro-geomorphic hazard.

Here, we collect a list of possible spatial units, delineated by field-, DEM- or remote sensing-based mapping, that have been used either in theoretical considerations or in connectivity studies:

- **Landforms** are a basic constituent of geomorphological maps, hence the latter lend themselves for connectivity analysis (Buter et al., 2020, 2022; Figure 10.1). The (dis-)connectivity framework conceived by Fryirs et al. (2007a) focuses on landforms (with slope as an important characteristic) that impede or block sediment fluxes: Low-gradient channel reaches lead to longitudinal disconnectivity (forming **barriers** to sediment transfer), while low-gradient landforms such as floodplains act as **buffers** that decouple adjacent slopes from the main river (lateral disconnectivity).
- Jasiewicz and Stepinski (2013) proposed **geomorphons** that represent landform elements automatically identified by pattern recognition (neighbourhood analysis) on a DEM.
- **Agricultural fields** (Gascuel-Odoux et al., 2011) represent homogeneous (with respect to landuse/landcover) spatial units where relevant processes occur (runoff generation, soil erosion and deposition, transfer of sediment-bound nutrients and pollutants). They are frequently bordered by hedges, ditches or roads that modify runoff and sediment routing. Hence, they are very important in a hydrological, geomorphic, and management context.

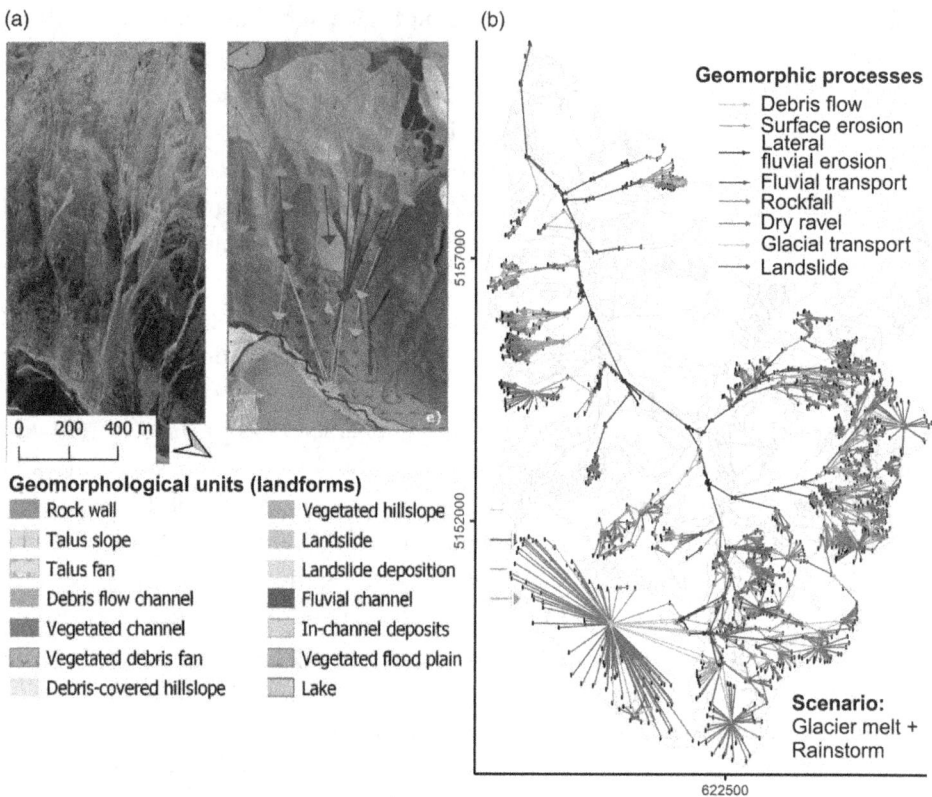

Figure 10.1 Landforms depicted on a geomorphological map (a) as the
fundamental unit of a sediment connectivity assessment. On the map (b), landforms
are represented by nodes and linked by edges that represent geomorphic processes
under a 'glacier melt + rainstorm' scenario. Figure put together from original
figures in Buter et al. (2022). A black and white version of this figure will appear
in some formats. For the colour version, refer to the plate section.

- Singh et al. (2017) extend the concept of **hydrologic response units** (HRU; Flügel,
  1995) with a connectivity perspective, delineating **connectivity response unit**s
  (CRU). HRUs are spatial units for which a similar hydrological response to meteorological forcing can be expected, and they are created by overlaying relevant
  spatial datasets (topography, soil, vegetation, etc.) and delineating homogeneous
  spatial objects characterised by combinations of these variables. CRUs not only
  include variables relevant for run-off generation but also for sediment generation
  and transfer (Singh et al., 2017). A somewhat similar approach is the '**geomorphic
  cell**' for which Pöppl and Parsons (2018) provide a conceptual background as a
  fundamental unit that includes system state and different types of connectivity.
- When analysing longitudinal connectivity, **river reaches** (Wohl et al., 2017) or
  valley segments (Wohl & Beckman, 2014) represent suitable spatial units.

Such spatial units are structurally connected if sediment transfer is possible between them, for example, if a potential sediment pathway connects neighbouring units. In order to address functional connectivity, we need evidence of geomorphic processes being (or having been) active that affect (or have affected) sediment flux between these units. With respect to such processes, we can distinguish between indices that focus on **lateral connectivity** (hillslope-channel coupling) and those that focus on **longitudinal connectivity** (within-channel, between tributary and trunk streams; Brierley et al., 2006). The IC (Borselli et al., 2008; Cavalli et al., 2013) is computed with respect to user-specified targets (see Section 10.3.1.1). If the channel network is selected as the target, the index will address lateral connectivity; if the catchment outlet is the target, the index addresses both lateral and longitudinal connectivity. Wohl et al. (2017) propose an index that specifically focuses on longitudinal connectivity; it is computed at the river reach scale from a weighted overlay of digital elevation data (elevation as a proxy for flood generation mechanism; channel gradient), lithology and vegetation data, and information on human influence (flow diversion, roads and culverts).

While some of the aforementioned ingredients of connectivity indices include the routing of potential flows implicitly (e.g., flow accumulation), a somewhat intuitive approach to explicitly representing a system that consists of interconnected spatial entities is the **network**; the corresponding analytical framework is graph theory (e.g., Newman, 2010; reviews with respect to geoscientific applications have been provided by Heckmann et al., 2015; Phillips et al., 2015). The fundamentals of network-based connectivity indices are explained in Section 10.3.1.3.

10.2.3 Implications of Connectivity Indices with Respect to Time

The application of connectivity indices is constrained also by temporal issues. An index can be a valid indicator of connectivity as long as the properties represented by its ingredients do not change. Most indices are static, because their variables represent properties rather than processes. Changes in connectivity induced by changing vegetation on a seasonal (Foerster et al., 2014) to decadal scale (López-Vicente et al., 2017) can be investigated by computing multitemporal maps of connectivity indices that include vegetation or its properties as variables (Figure 10.2).

Slope and flow accumulation, for example, are only proxies for surface run-off formation and its erosive capacity; they do not represent surface runoff at a particular point in time. However, a high-magnitude low-frequency event can lead to full connectivity in a system that is disconnected under 'average' forcing. Such extreme events, or the cumulative activity of geomorphic processes on the long

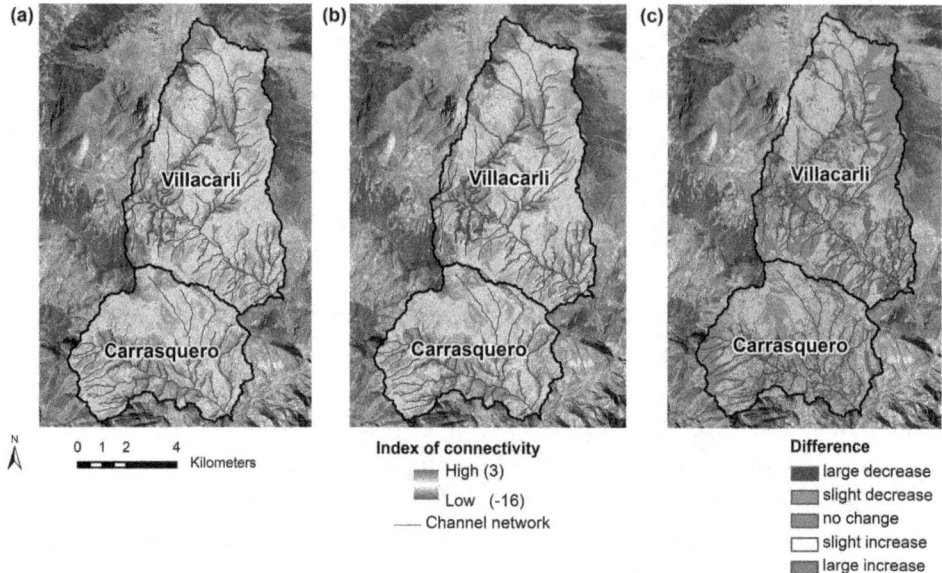

Figure 10.2   Multitemporal map of the IC for a catchment in the Spanish Pyrenees. Panel (a) reflects the situation in April, (b) in August and (c) shows the change in IC between the two points of time. Figure taken from Foerster et al. (2014). A black and white version of this figure will appear in some formats. For the colour version, refer to the plate section.

temporal scale, can imply process-form-feedbacks that effectively change structural connectivity (interaction of functional and structural connectivity, see Chapter 2); an index computed for the pre-event landscape would no longer be valid any more in the post-event landscape. Calsamiglia et al. (2020) observed such feedbacks; in their study area, flood events induced morphological changes, and hence changes in structural connectivity, that affected functional connectivity in consecutive events. Harvey (2002) investigated the frequency of processes that affect geomorphic coupling in his study area and found return periods between tens of days and 30 years; a full coupling of the whole system was found to occur once or twice during the past 5,000 years. Moreover, some events induce decoupling or re-coupling of system components that may last for years to decades. The dependence of coupling relations, and hence system-wide connectivity, on the magnitude of forcing events is depicted in Figure 10.3; through the magnitude-frequency relationship of the forcing events, the degree of connectivity is linked to the temporal scale – the larger the event (triggering high connectivity), the less frequent it occurs. Wohl et al. (2019) consider fluvial sediment transport to be connected at time scales longer than decades. These considerations and an overview of existing indices have led to the conclusion that the relevant temporal scale for connectivity indices spans minutes (minimum duration of an event) to several decades (Heckmann et al., 2018).

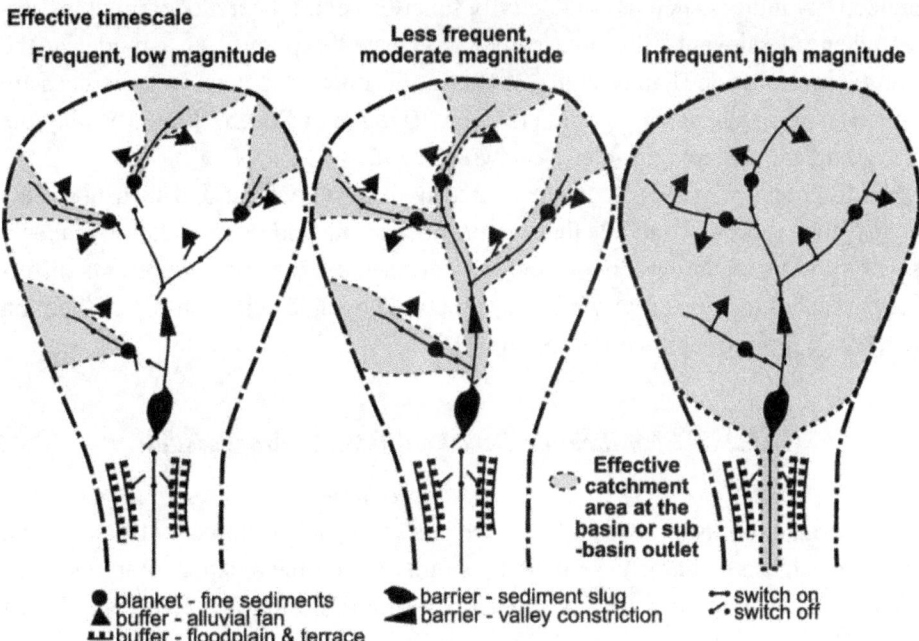

Effective timescale

Frequent, low magnitude   Less frequent,   Infrequent, high magnitude
                            moderate magnitude

Effective
catchment
area at the
basin or sub
-basin outlet

● blanket - fine sediments   ◗ barrier - sediment slug   ⇝ switch on
▲ buffer - alluvial fan   ◀ barrier - valley constriction ⤳ switch off
⊥⊥ buffer - floodplain & terrace

Figure 10.3   Event magnitude (associated with frequency, or return period) governs the size of the effective catchment area (grey) by triggering 'switches' (=geomorphic coupling relations between subcatchments or landforms). Figure taken from Fryirs et al. (2007b). This figure has been re-used by Fryirs (2013) to illustrate the 'effective timescale', that is, the time frame over which (dis-) connectivity occurs. In the latter publication, the three event magnitudes are associated with post-World War II, Holocene, and Last Glacial Maximum time scales, respectively.

10.3 Indices of Connectivity

Connectivity assessment is tackled in three interrelated chapters of this book: Measurements (Chapter 9), Indices (Chapter 10) and Modelling (Chapter 11). Indices are related to measuring connectivity because some are computed from field observations, for example, maps and remote sensing imagery containing information about hydrological or geomorphic coupling at some point or period in time. This is what Hooke and Souza (2021) termed 'descriptive' application of indices (see Section 10.4.1, Figure 10.7): One could also say that such indices treat connectivity as the dependent variable (Goodwin, 2003).

Indices are related to modelling connectivity in a twofold manner: First, they can be the outcome of modelling water and sediment pathways. In this case, connectivity emerges from model results (Nunes et al., 2018), and models are used for connectivity assessment, for example, under different forcing scenarios. Second,

connectivity indices such as connectivity functions could be a model input parameter when all relevant processes cannot be treated explicitly in a model at the
required spatial scale (Nunes et al., 2018). In this context, connectivity indices are
applied in an explanatory or predictive way (Hooke & Souza, 2021), or play the
role of an independent variable (Goodwin, 2003).

Finally, connectivity indices are related both to modelling and measurements
because they represent models that require validation, and this validation makes it
necessary to use field data, maps and measurements. This section presents different approaches to connectivity indices, structured by index characteristics (Section
10.3.1) and types of application (Section 10.4).

10.3.1 A Classification Based on Index Characteristics

10.3.1.1 Raster-based Indices

As an example for raster-based indices, we present the IC proposed by Borselli et al.
(2008) which can be named as the first quantitative sediment connectivity index.

The index is computed for each raster cell of a DEM and has two components
(Figure 10.4):

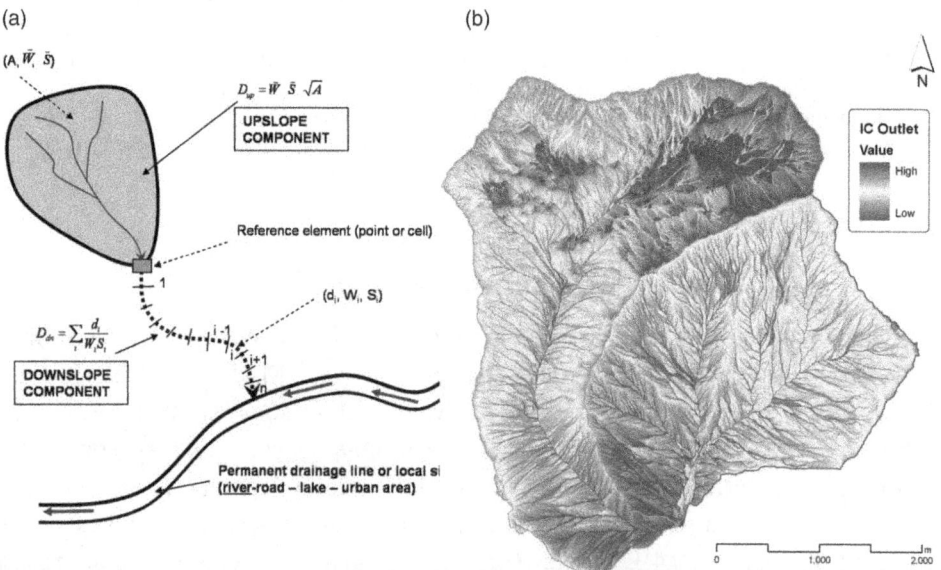

Figure 10.4   (a) Conceptual diagram explaining the computation of IC for each
raster cell; the upslope component refers to the contributing area, the downslope
component refers to the flow pathway until the proximal user-specified target is
reached. (b) IC map (resolution: 1 m, target = catchment outlet) of the Gadria and
Strimm catchments, Val Venosta, South Tyrol, Italy. Figures taken from Cavalli
et al. (2013). A black and white version of this figure will appear in some formats.
For the colour version, refer to the plate section.

The upslope component D_{up} is computed as

$$D_{up} = \bar{W}\,\bar{S}\,\sqrt{A}.$$

It represents the upslope contributing area A (as a proxy for surface runoff) and its properties, that is the average slope $\bar{S}$ [m m^{-1}] and the average impedance factor $\bar{W}$. In the original publication of IC, the C factor of the USLE erosion model was used as W, which implies that low W reflects high impedance. When W is derived from surface roughness (Cavalli et al., 2013), high roughness translates into low W (0.001;1) and high impedance. It can be seen from the equation that cells draining large, steep and low-impedance catchments are associated with high D_{up} values and hence high connectivity.

The downslope component D_{dn} is computed as

$$D_{dn} = \sum_i^n \frac{d_i}{W_i S_i}.$$

It evaluates the length d_i, slope S_i and impedance W_i of the ith of n raster cells along the flow path that connects the respective raster cell to a user-specified target. The target is usually set as the channel network (then, the index refers to hillslope-channel coupling), or the catchment outlet (then, the index encompasses overall connectivity). The equation shows that short, steep and low-impedance (=high W) flow paths lead to low D_{dn} values, which in this case signifies a higher likelihood of coupling of the respective cell and the target.

Finally, D_{up} and D_{dn} are combined to yield the IC index:

$$IC = \log_{10}\left(\frac{D_{up}}{D_{dn}}\right).$$

It has a range of $[-\infty, +\infty]$, larger values being caused by high D_{up} and/or low D_{dn} values and signifying high connectivity.

Over time and with increasing number of applications of IC, different modifications have been proposed by some researchers. For example, Cavalli et al. (2013) adapted the index for mountain and alpine watersheds where high-resolution DEMs lend themselves for deriving roughness measures that can be translated into the weighting factor W. The reason for this modification was based on the fact that mountain areas are frequently unvegetated, so the C factor used in Borselli's original formula could not be representative. Ortíz-Rodriguez et al. (2017) made use of each variant's advantages by using C factor values on vegetated areas, and W derived from surface roughness on unvegetated areas. Kalantari et al. (2017) proposed a modified version of the Cavalli et al. (2013) index: Maps of cumulative surface run-off generation were computed using the SCS Curve Number method, and the ratio Q/Q_{max} was used as W in the computation of IC under the assumption that high run-off facilitates sediment transport and, hence, connectivity. In this way, soil properties and hydrologic/climatic conditions could be included in the IC, which according to the authors made IC more appropriate for lowland areas. With a similar aim, Gay et al. (2016) adapted the IC to include infiltration and saturation properties as reflected in the drainage density.

IC is an index of structural connectivity; that means it highlights the spatial distribution of potential sediment pathways and facilitates structural connectivity to be compared on the raster-cell to catchment scale. López-Vicente and Ben-Salem (2019) attempted to amend IC with additional factors that relate to soil (permeability) and forcing properties (rainfall erosivity). Depending on the temporal scale of application, the resulting aggregated index of flow and sediment connectivity (AIC) has the subscript 'st' when longer-term data (e.g., annual averages) are used, and the subscript 'fn' when it refers to a particular event, week or month. In this case, AIC is thought to better reflect functional connectivity because it takes into account temporally variable factors and especially the forcing magnitude.

10.3.1.2 Effective Catchment Area (ECA) and Sediment Contributing Area (SCA)

'Effective catchment area' (Fryirs et al., 2007a, b; Souza et al., 2016) and 'sediment contributing area' (Haas et al., 2011; Najafi et al., 2021b; Altmann et al., 2021) represent related approaches that are based on the assumption that sediment delivered to a catchment outlet is derived only from parts of the drainage area rather than from its full extent (Parsons et al., 2006), and that these subareas can be delineated on the basis of a DEM and a set of rules.

ECA is based on a single slope threshold (below which sediment is more likely to be deposited rather than being transported further) and the presence of specific landforms (buffers and barriers; Fryirs et al., 2007a, b). While Fryirs et al. (2007b) implement a 2° slope threshold to delineate the ECA, Souza et al. (2016) use a set of thresholds that are related to different degrees of decoupling and hence also to different forcing magnitudes needed to establish coupling: Slope gradients below 0.5° represent major disconnecting landscape elements, and gradients between 0.5° and 2° minor disconnecting elements; in these cases, major or minor events are needed to effect geomorphic coupling, respectively. Slope gradients between 2° and 25° are seen as 'non-limiting', while gradients >25° represent boosters to sediment transmission, as sediment storage in terrain that steep is unlikely. Schopper et al. (2019) used thresholds of 2° and 8°, and compared the spatial pattern of the resulting ECA with field observations and an IC map; they found both ECA at 2° and IC to be consistent with field observations.

In the SCA approach, a channel gradient threshold (below which adjacent channel reaches are decoupled with respect to sediment transfer), the distance to channel network and vegetation cover have been included in addition to the slope threshold (see Haas et al., 2011; Altmann et al., 2021). Thus, the SCA approach takes both lateral (i.e., hillslope-channel) and longitudinal (i.e., within-channel) coupling into account. The areal extent of SCA shows areas prone to sediment transport and also a 'potential' for sediment supply. Some studies (Haas et al., 2011; Najafi et al., 2021b; Altmann et al., 2021) have successfully established regression functions relating log-transformed SCA size to log-transformed sediment yield (SY) measured in sediment traps in channels that drain catchment areas of between 10^1 and $10^7 \, \mathrm{m}^2$. An interesting observation is how the slope thresholds found to yield the best correlation between logSCA and logSY vary between study areas: Haas et al. (2011) used $25°$ in two northern Alpine study areas, Altmann et al. (2021) established $23°$ through optimisation in their western Alpine study area. In contrast, Najafi et al. (2021b) and Figure 10.5 selected a threshold as low as $5°$ in their semi-arid upland study area in NW Iran.

The findings reported here suggest that approaches that implement structural connectivity can be used to a certain extent to predict SY, which is a result of sediment erosion, transport and delivery, that is, functional connectivity. Note, however, that this is only true for transport-limited systems; both ECA and SCA approaches do not take the sediment availability into account, that is, they only refer to sediment accessibility (Najafi et al., 2021b). Where sediment supply is limited, the correlation between ECA or SCA size and SY is not expected to exist. Where it does exist, the comparative simplicity of the delineation of both ECA and SCA, plus the widespread availability of the necessary elevation and landcover data make these approaches attractive both for research and application.

10.3.1.3 Network-based Indices

Geomorphic systems have long been perceived as cascading systems (Chorley & Kennedy, 1971), and corresponding diagrams have been drawn to illustrate the coupling between their components through sediment fluxes (e.g., Schrott et al., 2003). A comparatively new approach is the set up of network models of geomorphic systems and their analysis within a graph theory framework. A network consists of nodes that are connected by edges. Technically, graph theory is a means to mathematically describe the structure of systems. In this section, we give a brief overview of network-based analyses with respect to (sediment) connectivity (see also Heckmann et al., 2015).

In the context of sediment connectivity, the network nodes represent spatial entities (cf. Section 10.2.2). Edges between two nodes then represent either adjacency in the direction of flow or the (potential) occurrence of sediment transfer. In the first case, nodes (=e.g., landforms) being coupled in the downslope direction form toposequences (Otto, 2006); where geomorphic processes transfer sediment

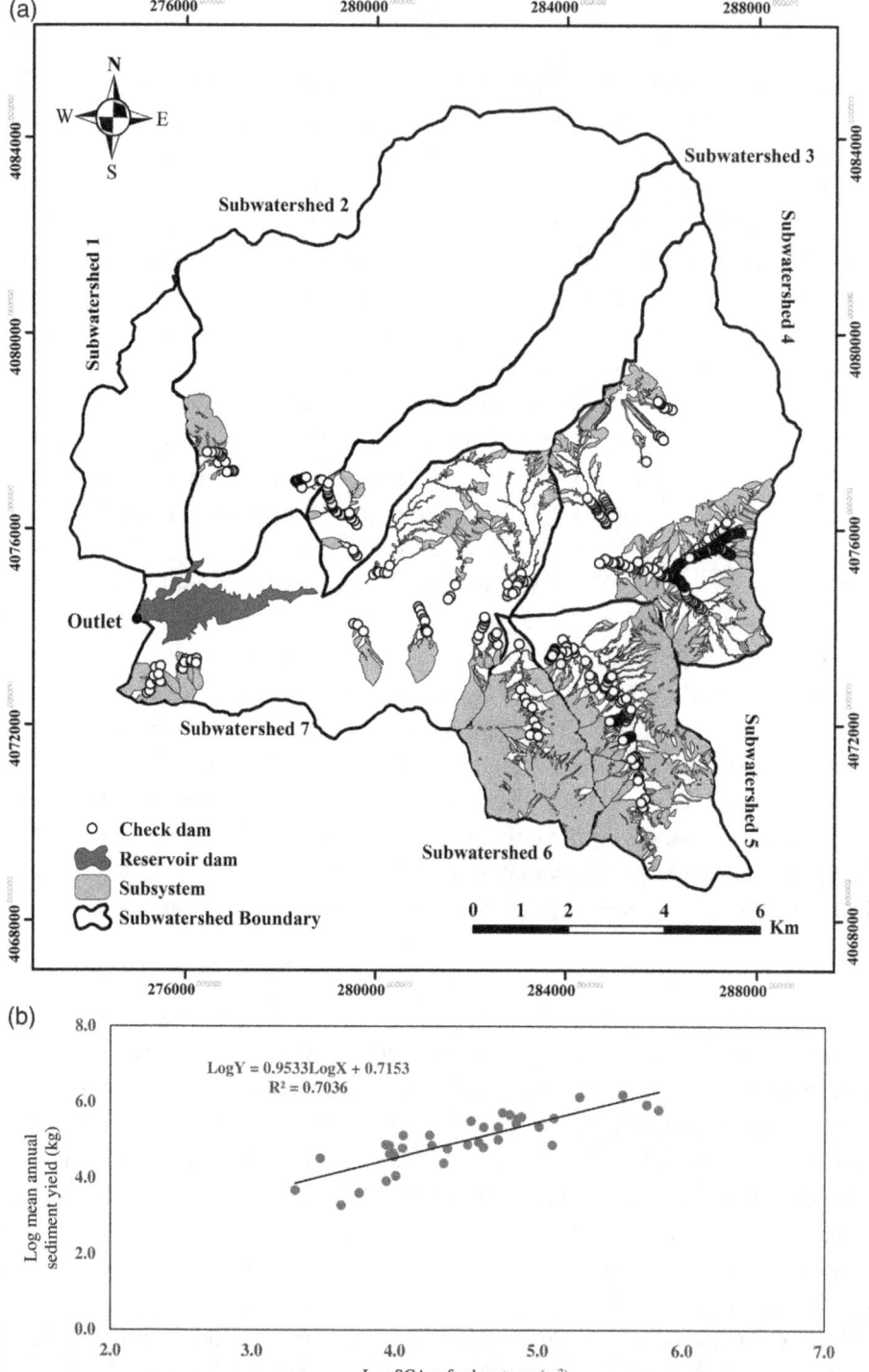

Figure 10.5 (a) Map of the Taham watershed, Zanjan province, NW Iran, with check dams and their corresponding SCA delineated on a DEM. (b) Regression analysis of log-transformed mean annual sediment yield on log-transformed SCA size. Figures taken from Najafi et al. (2021b). A black and white version of this figure will appear in some formats. For the colour version, refer to the plate section.

along a series of such units, we address them as sediment cascades. Edges can be constructed using a number of approaches: Expert-based geomorphological mapping can infer sediment transfer from diagnostic features visible in the field or on orthophotos (e.g., Cossart & Fressard, 2017; Buter et al., 2020, 2022), and draw edges by hand accordingly. Flow routing and morphometric parameters derived from DEMs can help to establish edges automatically. In the most basic form, flow routing algorithms are applied to DEMs to construct (multiple) flow direction matrices, a representation of a flow network that can be used, for example, to model fluxes (Schwanghart & Kuhn, 2010). Aggregating raster cells to larger objects such as landforms or agricultural fields (e.g., Aurousseau et al., 2009) reduces the number of nodes and edges, while the flow routing information of the original DEM is retained. The resulting network could even be amended manually to represent flow enhancement or impedance that cannot be derived from DEMs, such as sub-grid scale channels, hedges, or culverts below a road.

Here, we present as an example the work by Cossart & Fressard (2017) who proposed the 'network structural connectivity index' (NSC). First, a comparatively coarse regular grid of points is set up across the study area; these nodes are assigned landform types derived from a geomorphological map, and are connected by edges in the direction of gravity. The geomorphological map is then used to infer where flows are decoupled by barriers, buffers or blankets (see also Fryirs et al., 2007a, b).

The resulting network model is represented mathematically as adjacency and flow distance matrices, and graphically as network diagrams. Cossart and Fressard (2017) compute two indices from which the index is calculated as a ratio. The first index is the potential flow index that computes, for each node i, the proportion of paths (a path is an edge sequence that connects any node to the catchment outlet) that include i. This metric F_i is related to the node betweenness centrality index from graph theory (see Newman, 2010). According to this concept, the importance of a node (a landform) in the network (geomorphic system) is higher as more paths (sediment pathways, cascades) run through it. The second index addresses the accessibility from sources to sinks. It is computed from the distance matrix that contains the distances from any node to any other node in the network. First, the upslope/-stream distances to the node i and the downstream distances from i to its downslope/-stream neighbours are summed. This sum is divided by the total distance of all paths within the network. This so-called Shimbel index (Newman, 2010) Shi reflects the importance of confluences and the assumption that the influence of nodes closer to the outlet is higher on sediment delivery. A lower Shi implies that node i participates in shorter paths and is hence more accessible and more connected. The final NSC index is computed as the ratio F_i / Shi_i, so that the most important nodes share a high F_i and a low Shi_i value (Figure 10.6).

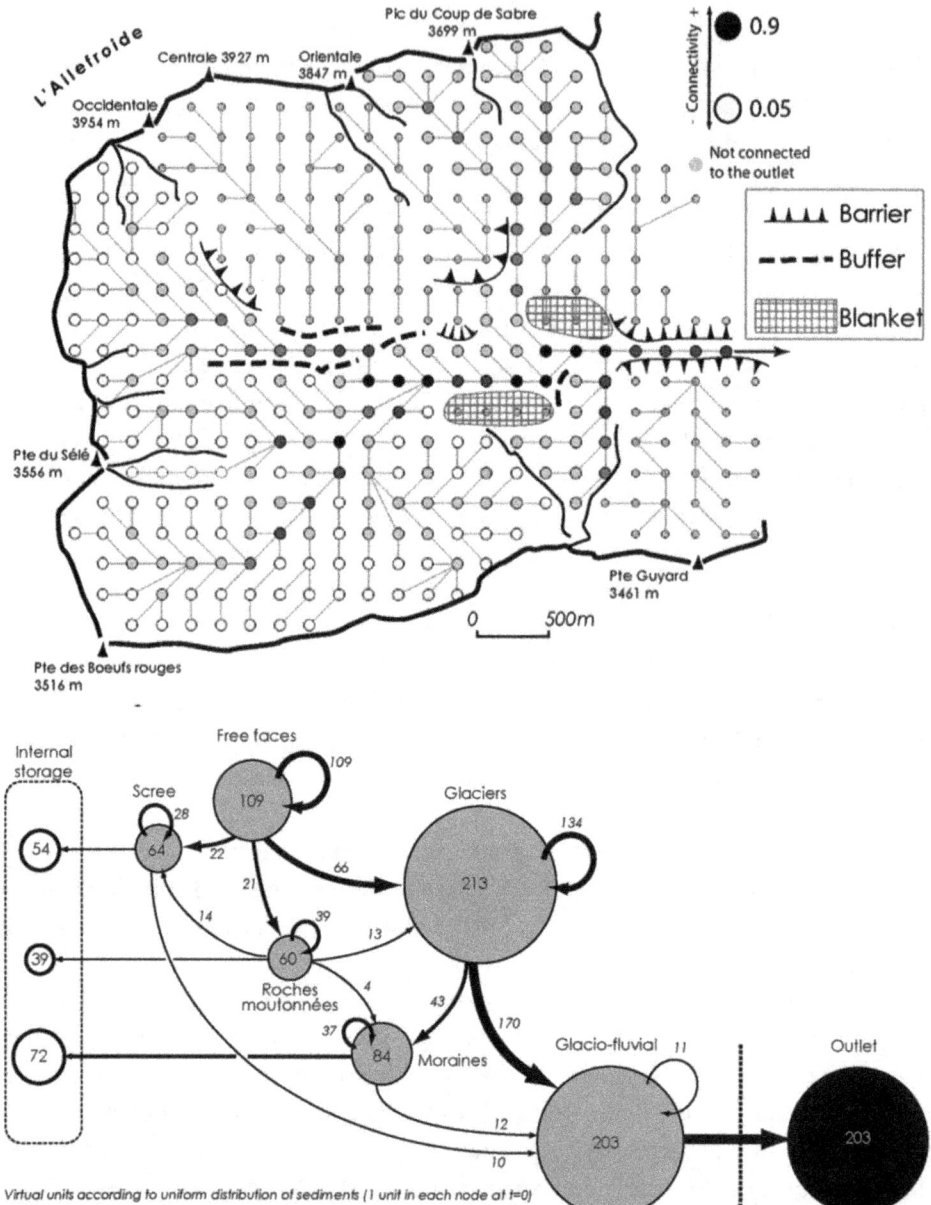

Figure 10.6 Connectivity assessment of the Celse-Nière catchment, French Alps, under current conditions (the original study also shows scenarios with a decoupling and a re-coupling of previously decoupled landscape units). Top panel shows the network of sediment pathways according to flow directions, and disrupted by buffers, barriers and blankets according to a geomorphological map. The nodes are coloured according to their connectivity index value. Bottom panel shows simulated rates of sediment transfer between system compartments, assuming initial (t = 0) uniform distribution of sediment across nodes. Figure taken from Cossart and Fressard (2017; selected parts of their Figure 7). A black and white version of this figure will appear in some formats. For the colour version, refer to the plate section.

The NSC index is a node property that takes into account the role of the respective node in the network. Other node properties include the node *degree* (number of incoming and/or outgoing edges) and the *betweenness centrality* (see earlier). In addition, network representations of cascading geomorphic systems can be analysed (see also Heckmann et al., 2015) with respect to...

- *Edges*: For example, the aforementioned betweenness *centrality* can also be computed for edges. A central, that is, important, edge is one that forms part of many paths (edge sequences) connecting nodes to others (or, for a geomorphic system, to the catchment outlet).
- Sequences of edges (=*paths*): For example, the longest path of a network is called its *diameter*. Longer paths could mean lower connectivity as more processes (with long transport distances) need to be active in order to transfer sediment through the catchment. Heckmann and Schwanghart (2013) extracted and analysed statistically types of sediment pathways that are driven by sequences of different geomorphic processes.
- Components: A graph *component* is a group of nodes that are connected by paths within the group, but not connected to other groups of nodes. Conceptually, a geomorphic system consisting of one or few components has higher connectivity than a system that consists of multiple smaller components. *Motifs* are recurring local patterns of node interactions. In geomorphometry, this concept is used for the construction and analysis of geomorphons (Jasiewicz & Stepinski, 2013); it could be applicable to connectivity research, for example, by identifying 'typical' configurations of system components that are associated with different types or intensities of sediment transfer.
- Graph properties: Important properties that refer to the whole graph are its *degree distribution* (the probability distribution of the 'degree' property of its nodes) or its *path length distribution*. The *degree distribution*, for example, is related to the sensitivity of a network to disturbance (Newman, 2010).

An important application of such metrics is their reaction to changes in system structure. In their application the NSC index, for example, Cossart and Fressard (2017) changed the studied geomorphic system in order to reflect changes such as a disruption of the sediment cascade, or a reconnection of previously decoupled compartments.

To conclude, the data structure of a network and the methodological toolbox of graph theory make it possible to 'keep the whole in mind while studying the parts and vice versa' (Jordán & Scheuring, 2004). The approach creates new opportunities for the analysis of geomorphological maps (e.g., Buter et al., 2020, 2022); networks can also be used for models of sediment routing in the context of connectivity (see Section 10.4).

10.4 Types of Index Application

Hooke and Souza (2021) proposed an application-based classification of connectivity indices (Figure 10.7), distinguishing descriptive and predictive applications. Following this classification, we collect in this section applications where indices serve as descriptors (or the dependent variable) and predictors (or the independent variable) of connectivity, respectively.

10.4.1 Descriptive Use of Indices

In the descriptive or diagnostic context, indices can be used to assess and quantify (i) spatial patterns of sediment pathways and (ii) coupled and decoupled sections of a watershed, for a certain (present or past) point or period of time. Therefore, this way of index applications is related to connectivity measurements (see Chapter 9). It makes use of field observations, measurements and maps. The corresponding indices are important for the investigation of factors of connectivity, in terms of properties and processes; in this way, such applications may inform the development of robust, integrated and validated indices that can be applied in a predictive manner (Section 10.4.2).

Although no indices were derived in the corresponding studies, we need to mention here the first approaches to investigating geomorphic coupling and sediment connectivity. These were based on geomorphological mapping and produced maps

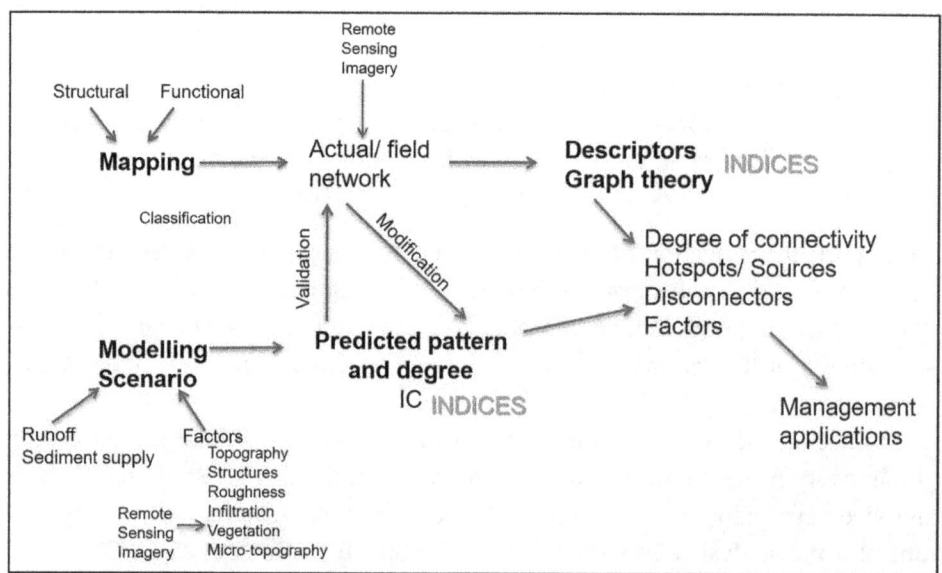

Figure 10.7 Classification of connectivity index approaches and applications proposed by Hooke and Souza (2021). A black and white version of this figure will appear in some formats. For the colour version, refer to the plate section.

of landforms and diagnostic features that indicate sediment transfer or deposition at different spatial scales, for example, for river reaches (Hooke, 2003) or catchments (Harvey, 2001). Repeated mapping allows for the assessment of effects of extreme events and changes in vegetation cover, river reaches and manmade structures on sediment connectivity patterns. Schrott et al. (2003) summarised the details of a geomorphological map of an alpine catchment to a more abstract graphical representation of a cascading system, containing subsystems and sediment transfer between storage landforms through the activity of geomorphic processes typically occurring in those (sub-)systems. Recently, Buter et al. (2022) constructed network models based on the landforms in a geomorphological map of the Solda river basin in the Italian Alps. Linkages representing the coupling of adjacent landforms with respect to sediment transfer were drawn for different scenarios of hydrometeorological forcing, including for example, enhanced discharge due to glacial meltwater production during heatwaves or due to rainstorms (see Figure 10.1 and Sections 10.2.2 and 10.3.1.3). Network indices such as the one proposed by Cossart and Fressard (2017; see Section 10.3.1.3) can be used for quantifying structural connectivity for different real or hypothetical scenarios.

The 'effective' or 'sediment contributing' portion of a catchment area can be mapped through the mapping of field evidence and the interpretation of aerial photographs (see Section 10.3.1.2), for example, vegetation cover, presence of ephemeral gullies, local sinks, small breaches, sediment flow paths, and so on. Knowledge gained from descriptive investigations, for example, slope thresholds below which sediments are deposited, can strongly support DEM-based approaches to automatically delineating these areas and thus to assessing connectivity on larger spatial scales.

Functional connectivity patterns, that is the extent and configuration of areas of erosion and deposition, and the flow paths that link these areas, can be observed and mapped in the field during or after forcing events, such as heavy rain (Figure 10.8). Such observations can be linked to measurements of discharge and/or sediment flux during the study period. Repeated after-event identification of functional connectivity leads to a frequency-based analysis and the investigation of influencing factors (Marchamalo et al., 2016; Calsamiglia et al., 2020; Najafi et al., 2021a). This approach has some limitations; it is costly and time-consuming, which restricts its applicability to (very) small watersheds. However, modern remote sensing from uncrewed aerial vehicles facilitates mapping, but a certain limitation in the size of the study area still applies. Moreover, mapping that involves the interpretation of field evidence is somewhat subjective.

The sediment delivery ratio (SDR; Walling, 1983; de Vente et al., 2007), that is, the proportion of gross erosion within a catchment that is being exported beyond its outlet, can be interpreted as a metric of functional connectivity. A large proportion (SDR close to one) of eroded material will be delivered to and beyond the

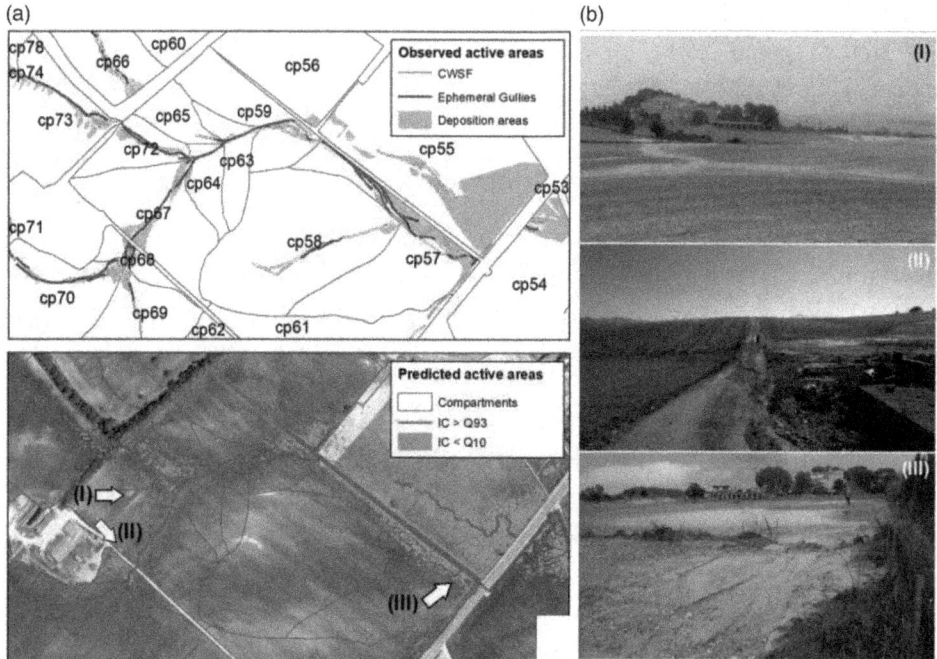

Figure 10.8 (a) Ephemeral gullies, deposition areas and concentrated water and sediment flows mapped from aerial photos after a rainstorm event at Can Revull catchment, Mallorca, Spain. (b) Aerial image with high- and low-IC areas highlighted and the locations of photos I, II and III (right). Figure taken and rearranged from Calsamiglia et al. (2020). A black and white version of this figure will appear in some formats. For the colour version, refer to the plate section.

outlet only if the catchment is highly connected; a poorly connected catchment will yield only small amounts of sediment at the outlet as even lots of eroded sediments would be re-deposited within the catchment (cf. Heckmann et al., 2018). The role of SDR has been contested in the literature (Kinnell, 2004; Parsons et al., 2006; Hoffmann, 2015), for example, as considerable uncertainty is caused by gross erosion, the denominator of SDR, being modelled rather than measured (e.g., Najafi et al., 2021b). Heckmann and Vericat (2018) proposed a spatially distributed SDR estimated from DEMs of difference (i.e., the difference of two DEMs acquired for the same area at different points of time). This is feasible where high erosion rates lead to surface changes that can be measured using repeat DEMs acquired by LiDAR or photogrammetric analysis of aerial photos. The study by Turley et al. (2021) is a recent example where this approach was applied.

10.4.2 Predictive Use of Indices

Under the assumption that the geofactors and/or models used to compute a connectivity index are in fact related to the activity of processes in response to external forcing, the corresponding index should exhibit predictive capacity with respect to observable properties and behaviour of a catchment, for example, sediment fluxes following a rainfall event. As many indices refer to structural connectivity, and sediment fluxes constitute functional connectivity, the question could also be how functional connectivity (in response to a specific forcing scenario) can be predicted by an assessment of structural connectivity. One problem with this prediction is that functional connectivity varies considerably with the intensity of external forcing, while structural connectivity indices consist of mostly static factors. To this end, Section 2.1 already mentioned attempts to include factors that are more dynamic and more related to functional connectivity.

An important implication of predictive uses of connectivity indices is the need for validation using field data. Up to now, most studies either lack validation at all or inspect the correspondence of index maps with the field situation depicted on photos or maps (Martini et al., 2022). Such qualitative, visual validation, and the use of connectivity indices in predictive models have considerable merits, but there is a lack of a more quantitative validation. Recent studies have made some progress, and their results suggest that IC could, in fact, be a valid and useful connectivity predictor: De Walque et al. (2017) evaluated a modified version of IC as a predictor of the occurrence of muddy floods in the Belgian loess belt and found that models implementing the connectivity index performed better than models that included only an assessment of sediment production by the USLE erosion model; this was attributed to the IC representing a 'measure of the ease of transfer of the sediment towards the outlet'. Calsamiglia et al. (2020) found that observed erosional or depositional features that formed during flood events were substantially correlated with areas identified as prone to erosion or deposition, respectively, by the IC index (see Figure 10.8). Moreover, they identified IC thresholds above which areas became effective in terms of sediment delivery. These thresholds were found to vary with event magnitude, such that bigger events required lower thresholds and led to larger effective areas and longer connected pathways, while only areas above a very high IC threshold became active during events of lower magnitude. Besides the spatial correlation of IC and geomorphic activity, this consistency suggests that the index could be valid.

Martini et al. (2022) used data from the well-researched Rio Cordon catchment in the Italian Alps. Sediment sources identified by geomorphological mapping were categorised into 'connected' and 'disconnected' with respect to sediment delivery to the channel network, and a DEM of differences revealed whether there was evidence of sediment being transported to the stream (functionally connected). Logistic regression analysis was then performed, using the median IC of each sediment source as independent variable and its coupling status as the dependent variable. This model was able to distinguish coupled from decoupled sediment sources, with an IC threshold of −2.32 (sources with mean IC greater than this threshold were coupled); however, it was not possible to predict those areas where the DEM of difference indicated active morphodynamics and functional connectivity. Similarly, Koci et al. (2019) concluded that areas of high IC correctly indicated likely flow paths and provide a relative measure of how likely major gullies would experience headcut retreat. However, the amount of measured headcut retreat did not show a clear relationship with mean IC values. Barneveld et al. (2019) found only very weak ($r^2 < 0.15$) predictive capacity in regression models of annual sheet erosion sediment delivery ratio on mean IC. Moreover, these authors concluded that IC was not useful for upscaling soil loss from the field scale to the catchment scale. These findings corroborate the capacity of IC of predicting spatial patterns of geomorphic activity and sediment delivery; as an index of structural connectivity, it appears not to be related quantitatively (at least not everywhere and not strongly) to the amount of this activity, which constitutes functional connectivity.

While this issue is quite obvious on the short time scale, there are indications that structural connectivity, as quantified using an index, could be correlated with 'average' functional connectivity at longer time scales: Arabkhedri et al. (2021) compared the specific sediment yield (SSY = SY divided by contributing area) of eleven adjacent catchments in the Loess region of Iran (Figure 10.9); these catchments have identical topographic, soil and landcover properties which led the authors to assume equal SSY. A linear regression revealed that 71% of the observed variance in SSY (with a factor of 11 between the lowest and the highest SSY) could be explained by the average IC computed on a 30 m DEM, while single geofactors contained in the IC only had coefficients of determination up to 50%. SDR as a measure of functional connectivity has been linked to the IC in the study by Vigiak et al. (2016) who assumed an s-shaped (sigmoid) relationship between IC (Borselli et al., 2008) and a hillslope's SDR. However, this relationship is based on theoretical considerations rather than on empirical evidence.

The size of the sediment contributing area (SCA, see Section 10.3.1.2) is not a connectivity index itself; however, it implements the concepts of lateral and longitudinal structural connectivity in the delineation of SCA. In that respect, Altmann et al. (2021, and references therein) show that connectivity can be related quantitatively to SY from mountain catchments, with the limitation that the correlation works for mean annual SY (see also Haas et al., 2011; Najafi et al., 2021b; Figure 10.5) and is unlikely to be applicable on the event time scale.

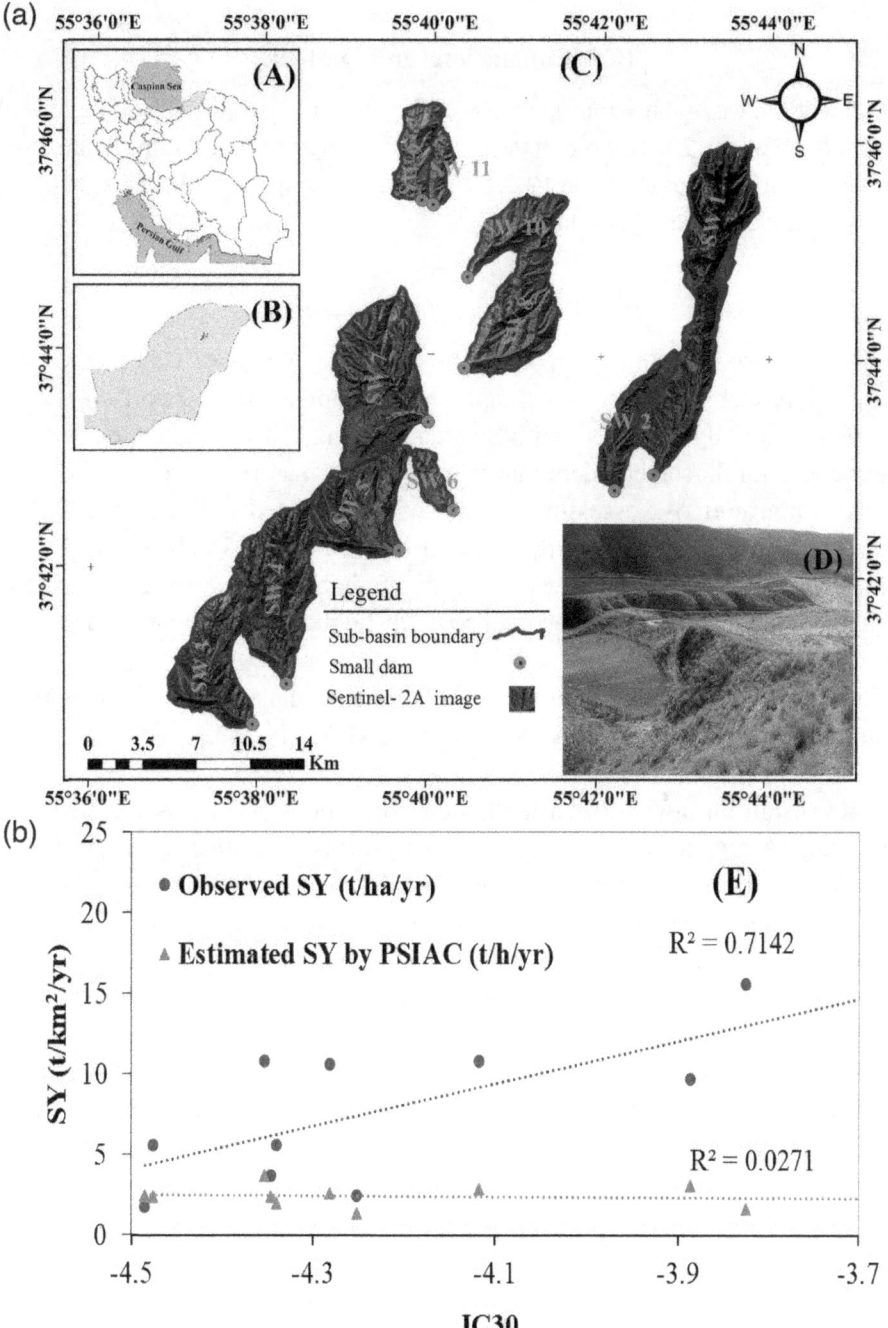

Figure 10.9 (a) Map of eleven check dams and their contributing areas in the Shour-Dareh basin, Golestan province, N. Iran. (b) Regression analysis of mean annual specific sediment yield (SSY) from these catchments on their mean IC (blue circles). The estimated SSY by PSIAC refers to an empirical, semi-quantitative method to assess sediment production. Figures taken from Arabkhedri et al. (2021). A black and white version of this figure will appear in some formats. For the colour version, refer to the plate section.

10.5 Conclusions and Outlook

In this section, we first use our overview of the literature (see also Heckmann et al., 2018; Najafi et al., 2021a) to collate present-day and possible future applications of connectivity indices (Section 10.5.1). Then, we attempt to collect a list of opportunities and challenges (Section 10.5.2).

10.5.1 Prospects for Application

We identify two major fields in which the use of connectivity indices should be evaluated, further developed and disseminated for application. The different applications are partially consistent with Smetanova et al. (2018) who concluded from interviews with 85 stakeholders that there was a demand for 'erosion, flood and sediment transport risk assessment maps with visualised sediment pathways in relation to existing infrastructures ..., and limits/thresholds expressing the conditions under which these hazards are most probable. Furthermore ... maps for diffuse pollution of ground water along with predictive functions to assess the impact of new projects'.

The first major field of application is sediment management in catchments, including (soil) erosion, sediment delivery to channels and within the channel network.

Soil erosion on agricultural fields depends, among others, on the slope gradient and the propensity to experience surface runoff. Both are also factors of (hydrological and) sediment connectivity, so that some correlation of soil erosion model estimations and connectivity indices such as IC can be expected. However, connectivity is also a function of offsite, downslope topography and landcover (not taken into account in the estimation of local erosion). Consequently, two sites with the same local erosion potential can differ considerably with respect to connectivity. Given the different consequences of soil erosion in poorly vs. strongly connected systems, it makes sense to investigate spatial patterns of combinations of high/low erosion hazard and high/low connectivity for a more holistic erosion risk assessment and management (Liu et al., 2020). Pöppl et al. (2019) aimed at detecting hotspots of soil erosion and sediment delivery to channels in a 138 km^2 agricultural study area; they used the GeoWEPP erosion model in combination with field-based connectivity mapping and IC-based connectivity assessment. López-Vicente et al. (2015) combined the Morgan–Morgan–Finney erosion model with the IC; they found a good spatial correlation of the map of potential soil redistribution and geomorphic evidence mapped in the field. Najafi et al. (2021b) combined IC-based structural connectivity assessment with an assessment of functional connectivity based on SDR; the latter, in turn, was estimated by relating observed SY to erosion estimated using the RUSLE erosion model.

Surface runoff and sediment delivery are often associated with **nutrients and pollutants,** hence connectivity indices are likely to support the assessment of eutrophication/pollution of rivers and lakes through nutrient/pollutant input. Researchers in nutrient and pollutant transfer to channels have delineated 'critical source areas', areas from which transfer of such substances (together with water and sediment) to the channel network is likely (e.g., Thompson et al., 2012; Shore et al., 2013). Shore et al. (2013) used the network TWI (Lane et al., 2004) to identify, via a threshold, highly connected subareas; critical source areas were delineated where such highly connected areas featured poorly drained soils with excess phosphorus. Alder et al. (2015) used the flow distance (as opposed to Euclidean distance) to the channel network as a simple connectivity index. According to this metric, they found that 55% of the agricultural areas in Switzerland are directly (21%) or indirectly (34%) coupled to surface waters (including drained roads, farm tracks and slope depressions). Dupas et al. (2015) correlated catchment properties with transfer rates of phosphorus and nitrogen. They found that the IDPR connectivity index (Mardhel et al., 2004; also used in an IC modification by Gay et al., 2016) was more correlated with phosphorus transfer rates than with nitrogen transfer rates; this was attributed to phosphorous being transferred mainly by surface and subsurface runoff and erosion.

Managing **coarse sediment transport** in Alpine rivers, and its (dis-)connectivity, creates a field of tension between often conflicting goals. Skolaut et al. (2015) state that 'sediment continuity has a notable impact on several management issues in alpine river basins and poses multiple use conflicts related to e.g. small hydropower, ecology, fishing, flood control, morphology, or the good status according to the EU Water Framework Directive'. In this context, indices represent building blocks of holistic environmental management. Depending on the assessment of the respective environmental system conditions and on perspectives and needs of different stakeholders, management goals may include the enhancement/restoration (e.g., Skolaut et al., 2015; Magilligan et al., 2016) or decrease (e.g., Hooke et al., 2017) of sediment connectivity.

The second major field of application refers to the sensitivity of catchments to change. In the Anthropocene, there's hardly any place on Earth that has not been affected by past and ongoing landcover and/or climatic change. Such changes can have localised effects, but localised changes may propagate through a geomorphic system. This **propagation of changes** is one aspect of **catchment sensitivity,** such that rapid and/or complete propagation characterises sensitive systems, while insensitive systems are capable of absorbing or buffering changes, and hence the transmission of changes is hampered ('transmission sensitivity', see Fryirs, 2016). This also affects the recording of signals of change in sedimentary archives (Bernhardt et al., 2017). Connectivity is therefore a modifier of how systems transmit changes, and hence of catchment sensitivity. Lisenby et al. (2018: 8) state that, among others, connectivity is likely to alter the relationship between the magnitude of a disturbance event and the geomorphic response. They conclude that 'landscape sensitivity, connectivity, recovery, event sequencing and extrinsic/intrinsic thresholds are fundamentally linked … and together can exert significant control on the geomorphic effectiveness of a natural disturbance event'. This is another reason for including connectivity assessment in management and decision-making workflows. For example, González-Romero et al. (2022, see also references therein) used a modified and extended IC (see López-Vicente and Ben-Salem, 2019) for the evaluation of post-fire soil stabilisation measures in Spain. Connectivity as a driver of offsite changes has been investigated using the landscape evolution model CAESAR (Coulthard & van de Wiel, 2017); the authors showed that landuse changes imposed on one half of their study area caused up- and, above all, downstream geomorphic changes in the other half, respectively.

10.5.2 Opportunities and Challenges

Liverman et al. (1988) name eight criteria for indices of global sustainability; here, we briefly address six of them that in our opinion apply best to our topic, in light of the current state of research in connectivity indices.

(1) **Sensitivity to change in time.** The current state of research shows the need to take time into account (Heckmann et al., 2018). The consideration of the temporal scale at which a connectivity index is valid leads to two avenues for research and application. First, in order to be applicable in a predictive manner (and to be better related to functional connectivity), indices need to include factors that reflect external forcing (López-Vicente & Ben-Salem, 2019; Pöppl et al., 2019). Second, indices can be applied, in a more descriptive manner, to investigate the change of (structural) connectivity with time. These changes may refer to the past (evolution of a system through time along with land-cover changes), the present (reaction to structural changes through disturbance) or even the future (scenarios, as a decision support for catchment sediment management).

(2) **Sensitivity to change across space or within groups**: Most connectivity indices are computed on a raster cell basis, with notable exceptions regarding network approaches applied to larger spatial units such as landforms that are geomorphologically more meaningful (e.g., Cossard & Fressard, 2017; Buter et al., 2020, 2022). Raster-based indices can be aggregated to arbitrary scales, from landform to (sub-)catchment, and some studies have shown that spatial units, for example, with different mean IC, are characterised by different

geomorphic behaviour. An important advantage of IC and related indices is the user-specified choice of a target that makes it possible to address lateral (hillslope-channel) and total (i.e., lateral and longitudinal combined) coupling. Network-based indices excel in considering both local and larger-scale factors; they reflect the notion of catchments as cascading systems where sediment is being transferred between system elements, and in which connectivity is a system property that emerges from those coupling relationships.

(3) **Predictive ability**: Here we identify a research gap that is only beginning to be filled and needs more attention. Up to now, index maps have usually been validated qualitatively by comparing them visually to field observations or photos; therefore, semiquantitative or quantitative validation is required most. There are works in which connectivity indices (the well known IC, specifically) were successfully used as independent variables for the predictive assessment of geomorphic processes (muddy floods, De Walque et al., 2017), sediment transfer and sediment delivery (Arabkhedri et al., 2021; Martini et al., 2022). Present-day research indicates that the predictive capacity of existing indices for functional connectivity is limited (e.g., Martini et al., 2022). Moreover, the character of the geomorphic system under study in terms of supply or transport limitation has to be taken into account. In supply limited systems, sediment mobilisation and delivery can be very low even under very high structural connectivity – this is why Najafi et al. (2021b) distinguish sediment availability and accessibility in their study. In transport-limited systems, where geomorphic activity produces surface changes that can be measured by morphological budgeting approaches, the spatially distributed sediment delivery ratio (Heckmann & Vericat, 2018; Turley et al., 2021) may lend itself for validation of the predictive capacity of connectivity indices.

(4) **Availability of reference or threshold values**: Most of present-day indices yield a relative assessment of sediment connectivity. That is, index values are frequently rank-ordered (from poor to high connectivity), using classification schemes based on the frequency distribution, e.g., quantiles or Jenk's natural breaks algorithm implemented in GIS software packages. Such a classification cannot be compared between disjunct areas, especially not when the latter differ with respect to environmental conditions or data quality. Consequently, threshold values above which one would speak of 'high connectivity' have not been published, even for the widely used indices. Recent work implementing IC suggests the existence of thresholds above which functional connectivity occurs under a given forcing (e.g., Calsamiglia et al., 2018a, b, 2020; Martini et al., 2022). While such findings corroborate the usability of connectivity indices, we assume that such thresholds reflect the behaviour of individual study areas and are unlikely to be transferable to different areas (see also Keesstra et al., 2018).

(5) **Relative ease of collection and use**: Thanks to the widespread, almost global availability of DEMs and landuse/landcover data at suitable resolution and of high quality, data collection for use in connectivity indices has never been easier than today. Additionally, affordable uncrewed aerial vehicles allow for the rapid and cost-effective generation of very high-resolution data. From pre-processing to index computation, user-friendly software solutions exist, many of them free and open-source (e.g., SAGA GIS: Conrad et al., 2015). Bespoke software tools have been developed and published for the well known IC index by Crema and Cavalli (2018). Hence, the data and software tools criterion is probably the one that is met best. However, connectivity is not controlled by topography and landcover alone, so it might be necessary to collect and imple-ment different data, in addition to DEMs and remote sensing-based landcover maps; this refers, for example, to soil and lithological data representing infil-tration capacity and other properties related to the formation and propagation of surface runoff (see review by Heckmann et al., 2018).

Ease of data collection and the availability of tools, however, have not led (yet) to a more widespread use of connectivity indices in applied sediment management: Smetanova et al. (2018) conducted a questionnaire survey among 85 stakeholders and found that, while half of the respondents considered connectivity management important, there is a lack of implementation of the connectivity concept and con-nectivity indices. This is in spite of the fact that connectivity indices should be more easily accessible than models; moreover, most existing models lack the nec-essary amount of detail needed to see connectivity emerge from their application (Nunes et al., 2018). The suitability of a model, however, not only depends on the degree of abstraction and the factors chosen to be part of the model, but also on the purpose of the model. The same is likely to be true for indices of sediment connec-tivity. Ali and Roy (2009), who compared hydrological connectivity indices, con-clude that no single technique has emerged as being useful in all studies. Turley et al. (2021) applied several connectivity indices to a proglacial catchment on the flanks of Mt. Rainier, USA. Among these indices were expert-based approaches to assess hillslope-channel coupling (e.g., Whiting & Bradley, 1993) from geo-morphological maps, ECA, IC and derivatives, network structure (Cossard & Fressard, 2017) and the spatially distributed sediment delivery ratio (Heckmann & Vericat, 2018). From the comparison of index results and field observations, the authors stress (i) the importance of implementing vegetation in connectivity indices, (ii) the role of system structure that can be assessed using network-based indices and (iii) the dependence of sediment delivery ratios on event magnitude. They conclude that all indices provide useful, though not individually complete, connectivity information; which index to choose depends on research objectives, geomorphic setting and data availability. Finally, Turley et al. (2021) ask for the development of a holistic index combining lateral, longitudinal, structural and functional connectivity. While we have highlighted, above all, the dearth of studies to validate connectivity indices, and the need to relate them to observed or modelled system behaviour, we join Turley et al. (2021) in this conclusion. Moreover, we encourage the use of connectivity indices in applied science, for example, in studies assessing offsite effects of soil erosion.

References

Alder, S., Prasuhn, V., Liniger, H., Herweg, K., Hurni, H., Candinas, A. & Gujer, H.U. (2015). A high-resolution map of direct and indirect connectivity of erosion risk areas to surface waters in Switzerland – A risk assessment tool for planning and policy-making. *Land Use Policy*, 48, 236–49. doi:10.1016/j.landusepol.2015.06.001.

Altmann, M., Haas, F., Heckmann, T., Liébault, F., & Becht, M. (2021). Modelling of sediment supply from torrent catchments in the Western Alps using the sediment contributing area (SCA) approach. *Earth Surface Processes and Landforms*, 46(5), 889–906.

Antoine, M., Javaux, M., & Bielders, C. (2009). What indicators can capture runoff-relevant connectivity properties of the micro-topography at the plot scale? *Advances in Water Resources*, 32(8), 1297–1310. https://doi.org/10.1016/j.advwatres.2009.05.006

Arabkhedri, M., Heidary, K., & Parsamehr, M. R. (2021). Relationship of sediment yield to connectivity index in small watersheds with similar erosion potentials. *Journal of Soils and Sediments*, 21(7), 2699–2708.

Aurousseau, P., Gascuel-Odoux, C., Squividant, H., Trepos, R., Tortrat, F., & Cordier, M. O. (2009). A plot drainage network as a conceptual tool for the spatial representation of surface flow pathways in agricultural catchments. *Computers & Geosciences*, 35(2), 276–288. https://doi.org/10.1016/j.cageo.2008.09.003

Babbie, E. R. (2010). *The Practice of Social Research* (12th ed.). Belmont, CA: Wadsworth Cengage Learning.

Bernhardt, A., Schwanghart, W., Hebbeln, D., Stuut, J-B. W. & Strecker, M. R. (2017). Immediate propagation of deglacial environmental change to deep-marine turbidite systems along the Chile convergent margin. *Earth and Planetary Science Letters*, 473, 190–204. doi:10.1016/j.epsl.2017.05.017

Borselli, L., Cassi, P., & Torri, D. (2008). Prolegomena to sediment and flow connectivity in the landscape: A GIS and field numerical assessment. *Catena*, 75(3), 268–277. https://doi.org/10.1016/j.catena.2008.07.006

Brierley, G., Fryirs, K., & Jain, V. (2006). Landscape connectivity. The geographic basis of geomorphic applications. *Area*, 38(2), 165–174.

Brunsden, D., & Thornes, J. B. (1979). Landscape sensitivity and change. *Transactions of the Institute of British Geographers*, 4(4), 463–484. Retrieved from www.jstor.org/stable/622210

Burt, T. P., & Allison, R. J. (eds.) (2010). *Sediment Cascades*. Chichester, UK: John Wiley & Sons.

Buter, A., Heckmann, T., Fillisetti, L., Savi, S., Mao, L., Gems, B., & Comiti, F. (2022). Effects of catchment characteristics and hydro-meteorological scenarios on sediment connectivity in glacierised catchments. *Geomorphology*, 108128. https://doi.org/10.1016/j.geomorph.2022.108128

Buter, A., Spitzer, A., Comiti, F., & Heckmann, T. (2020). Geomorphology of the Sulden River basin (Italian Alps) with a focus on sediment connectivity. *Journal of Maps*, 16(2), 890–901. https://doi.org/10.1080/17445647.2020.1841036

Calsamiglia, A., Fortesa, J., García-Comendador, J., Lucas-Borja, M. E., Calvo-Cases, A., & Estrany, J. (2018a). Spatial patterns of sediment connectivity in terraced lands: Anthropogenic controls of catchment sensitivity. *Land Degradation & Development*, 29(4), 1198–1210.

Calsamiglia, A., García-Comendador, J., Fortesa, J., López-Tarazón, J. A., Crema, S., Cavalli, M., ... & Estrany, J. (2018b). Effects of agricultural drainage systems on sediment connectivity in a small Mediterranean lowland catchment. *Geomorphology*, 318, 162–171.

Calsamiglia, A., Gago, J., Garcia-Comendador, J., Bernat, J. F., Calvo-Cases, A., & Estrany, J. (2020). Evaluating functional connectivity in a small agricultural catchment under contrasting flood events by using UAV. *Earth Surface Processes and Landforms*, 56, 1427. https://doi.org/10.1002/esp.4769

Cantreul, V., Bielders, C., Calsamiglia, A., & Degré, A. (2018). How pixel size affects a sediment connectivity index in central Belgium. *Earth Surface Processes and Landforms*, 43(4), 884–893.

Cavalli, M., Trevisani, S., Comiti, F., & Marchi, L. (2013). Geomorphometric assessment of spatial sediment connectivity in small Alpine catchments. *Geomorphology*, 188, 31–41. https://doi.org/10.1016/j.geomorph.2012.05.007

Chartin, C., Evrard, O., Laceby, J. P., Onda, Y., Ottlé, C., Lefèvre, I., & Cerdan, O. (2017). The impact of typhoons on sediment connectivity: Lessons learnt from contaminated coastal catchments of the Fukushima Prefecture (Japan). *Earth Surface Processes and Landforms*, 42(2), 306–317. https://doi.org/10.1002/esp.4056

Chorley, R., & Kennedy, B. (1971). *Physical Geography: A Systems Approach*. London: Prentice-Hall.

Conrad, O., Bechtel, B., Bock, M., Dietrich, H., Fischer, E., Gerlitz, L., ... Böhner, J. (2015). System for Automated Geoscientific Analyses (SAGA) v. 2.1.4. *Geoscientific Model Development*, 8(7), 1991–2007. https://doi.org/10.5194/gmd-8-1991-2015

Cossart, É., & Fressard, M. (2017). Assessment of structural sediment connectivity within catchments: insights from graph theory. *Earth Surface Dynamics*, 5(2), 253–268.

Coulthard, T. J., & van de Wiel, M. J. (2017). Modelling long term basin scale sediment connectivity, driven by spatial land use changes. *Geomorphology*, 277, 265–281. https://doi.org/10.1016/j.geomorph.2016.05.027

Crema, S., & Cavalli, M. (2018). SedInConnect: A stand-alone, free and open source tool for the assessment of sediment connectivity. *Computers and Geosciences*, 111, 39–45. https://doi.org/10.1016/j.cageo.2017.10.009

Dalla Fontana, G. D., & Marchi, L. (2003). Slope-area relationships and sediment dynamics in two alpine streams. *Hydrological Processes*, 17(1), 73–87. https://doi .org/10.1002/hyp.1115

De Vente, J., Poesen, J., Arabkhedri, M., & Verstraeten, G. (2007). The sediment delivery problem revisited. *Progress in Physical Geography*, 31(2), 155–178. https://doi .org/10.1177/0309133307076485

De Walque, B., Degré, A., Maugnard, A., & Bielders, C. L. (2017). Artificial surfaces characteristics and sediment connectivity explain muddy flood hazard in Wallonia. *Catena*, 158, 89–101 https://doi.org/10.1016/j.catena.2017.06.016.

Dupas, R., Delmas, M., Dorioz, J-M., Garnier, J., Moatar, F. & Gascuel-Odoux, C. (2015). Assessing the impact of agricultural pressures on N and P loads and eutrophication risk. *Ecological Indicators*, 48, 396–407. doi:10.1016/j.ecolind.2014.08.007.

Flügel, W.-A. (1995). Delineating hydrological response units by geographical information system analyses for regional hydrological modelling using PRMS/MMS in the drainage basin of the River Bröl, Germany. *Hydrological Processes*, 9, 423–436.

Foerster, S., Wilczok, C., Brosinsky, A., & Segl, K. (2014). Assessment of sediment connectivity from vegetation cover and topography using remotely sensed data in a dryland catchment in the Spanish Pyrenees. *Journal of Soils and Sediments*, 14(12), 1982–2000.

Fryirs, K. (2013). (Dis)Connectivity in catchment sediment cascades: A fresh look at the sediment delivery problem. *Earth Surface Processes and Landforms*, 38(1), 30–46.

Fryirs, K. A., Brierley, G. J., Preston, N. J., & Kasai, M. (2007a). Buffers, barriers and blankets: The (dis)connectivity of catchment-scale sediment cascades. *Catena*, 70(1), 49–67. https://doi.org/10.1016/j.catena.2006.07.007

Fryirs, K. A., Brierley, G. J., Preston, N. J., & Spencer, J. (2007b). Catchment-scale (dis)connectivity in sediment flux in the upper Hunter catchment, New South Wales, Australia. *Geomorphology*, 84(3–4), 297–316. https://doi.org/10.1016/j.geomorph.2006.01.044

Fryirs, K. A., Wheaton, J. M. & Brierley, G. J. (2016). An approach for measuring confinement and assessing the influence of valley setting on river forms and processes. *Earth Surface Processes and Landforms*, 41(5): 701–10. doi:10.1002/esp.3893.

Fryirs, K. A. (2017). River sensitivity: a lost foundation concept in fluvial geomorphology: A lost foundation concept in fluvial geomorphology. *Earth Surface Processes and Landforms*, 42(1), 55–70. https://doi.org/10.1002/esp.3940

Gascuel-Odoux, C., Aurousseau, P., Doray, T., Squividant, H., Macary, F., Uny, D., & Grimaldi, C. (2011). Incorporating landscape features to obtain an object-oriented landscape drainage network representing the connectivity of surface flow pathways over rural catchments. *Hydrological Processes*, 25(23), 3625–3636. https://doi.org/10.1002/hyp.8089

Gay, A., Cerdan, O., Mardhel, V., & Desmet, M. (2016). Application of an index of sediment connectivity in a lowland area. *Journal of Soils and Sediments*, 16(1), 280–293. https://doi.org/10.1007/s11368-015-1235-y

González-Romero, J., López-Vicente, M., Gómez-Sánchez, E., Peña-Molina, E., Galletero, P., Plaza-Álvarez, P., Fajardo-Cantos, A., Moya, D., de las Heras, J. & Lucas-Borja, M. E. (2022). Post-fire management effects on hillslope-stream sediment connectivity in a Mediterranean forest ecosystem. *Journal of Environmental Management*, 316, 115212. doi:10.1016/j.jenvman.2022.115212.

Goodwin, B. (2003). Is landscape connectivity a dependent or independent variable? *Landscape Ecology*, 18, 687–699.

Haas, F., Heckmann, T., Wichmann, V., & Becht, M. (2011). Quantification and modeling of fluvial bedload discharge from hillslope channels in two alpine catchments (Bavarian Alps, Germany). *Zeitschrift für Geomorphologie, Supplementary Issues*, 147–168.

Harvey, A. M. (2002). Effective timescales of coupling within fluvial systems. *Geomorphology*, 44(3–4), 175–201.

Harvey, A. M. (2001). Coupling between hillslopes and channels in upland fluvial systems: Implications for landscape sensitivity, illustrated from the Howgill Fells, northwest England. *Catena*, 42(2–4), 225–250. https://doi.org/10.1016/S0341-8162(00)00139-9

Heckmann, T., & Vericat, D. (2018). Computing spatially distributed sediment delivery ratios: Inferring functional sediment connectivity from repeat high-resolution digital elevation models. *Earth surface processes and landforms*, 43(7), 1547–1554.

Heckmann, T., Cavalli, M., Cerdan, O., Foerster, S., Javaux, M., Lode, E., … & Brardinoni, F. (2018). Indices of sediment connectivity: Opportunities, challenges and limitations. *Earth-Science Reviews*, 187, 77–108.

Heckmann, T., & Schwanghart, W. (2013). Geomorphic coupling and sediment connectivity in an alpine catchment – Exploring sediment cascades using graph theory. *Geomorphology*, 182, 89–103. https://doi.org/10.1016/j.geomorph.2012.10.033

Heckmann, T., Schwanghart, W., & Phillips, J. D. (2015). Graph theory – recent developments of its application in geomorphology. *Geomorphology*, 243, 130–146. https://doi.org/10.1016/j.geomorph.2014.12.024

Hoffmann, T. (2015). Sediment residence time and connectivity in non-equilibrium and transient geomorphic systems. *Earth-Science Reviews*, 150, 609–627. https://doi.org/10.1016/j.earscirev.2015.07.008

Hooke, J. (2003). Coarse sediment connectivity in river channel systems: a conceptual framework and methodology. *Geomorphology*, 56(1–2), 79–94. https://doi.org/10.1016/S0169-555X(03)00047-3

Hooke, J. M., Sandercock, P., Cammeraat, L. H., Lesschen, J. P., Borselli, L., Torri, D., … Navarro-Cano, J. A. (2017). Mechanisms of degradation and identification of connectivity and erosion hotspots. In J. Hooke & P. Sandercock (eds.), *Combating Desertification and Land Degradation* (pp. 13–37). Cham: Springer International Publishing.

Hooke, J., & Souza, J. (2021). Challenges of mapping, modelling and quantifying sediment connectivity. *Earth-Science Reviews*, 223, 103847. https://doi.org/10.1016/j.earscirev.2021.103847

Jasiewicz, J., & Stepinski, T. F. (2013). Geomorphons – A pattern recognition approach to classification and mapping of landforms. *Geomorphology*, 182, 147–156. https://doi.org/10.1016/j.geomorph.2012.11.005

Jordán, F., & Scheuring, I. (2004). Network ecology: Topological constraints on ecosystem dynamics. *Physics of Life Reviews*, 1(3), 139–172. https://doi.org/10.1016/j.plrev.2004.08.001

Kalantari, Z., Cavalli, M., Cantone, C., Crema, S., & Destouni, G. (2017). Flood probability quantification for road infrastructure: Data-driven spatial-statistical approach and case study applications. *Science of the Total Environment*, 581, 386–398.

Keesstra, S., Nunes, J. P., Saco, P., Parsons, T., Pöppl, R., Masselink, R., & Cerdà, A. (2018). The way forward: Can connectivity be useful to design better measuring and modelling schemes for water and sediment dynamics? *Science of the Total Environment*, 644, 1557–1572. https://doi.org/10.1016/j.scitotenv.2018.06.342

Kinnell, P. I. A. (2004). Sediment delivery ratios: A misaligned approach to determining sediment delivery from hillslopes. *Hydrological Processes*, 18(16), 3191–3194. https://doi.org/10.1002/hyp.5738

Koci, J., Sidle, R. C., Jarihani, B., & Cashman, M. J. (2019). Linking hydrological connectivity to gully erosion in savanna rangelands tributary to the great barrier reef using structure-from-motion photogrammetry. *Land Degradation & Development*, 64(11), 223. https://doi.org/10.1002/ldr.3421

Lane, S. N., Bakker, M., Gabbud, C., Micheletti, N., & Saugy, J. N. (2017). Sediment export, transient landscape response and catchment-scale connectivity following rapid climate warming and Alpine glacier recession. *Geomorphology*, 277, 210–227.

Lane, S. N., Reaney, S. M., & Heathwaite, A. L. (2009). Representation of landscape hydrological connectivity using a topographically driven surface flow index. *Water Resources Research*, 45(8), 1–10.

Lane, S. N., Brookes, C. J., Kirkby, M. J., & Holden, J. (2004). A network-index-based version of top model for use with high-resolution digital topographic data. *Hydrological Processes*, 18(1), 191–201. https://doi.org/10.1002/hyp.5208

Lisenby, P. E., Croke, J. & Fryirs, K. A. (2018). Geomorphic effectiveness: A linear concept in a non-linear world. *Earth Surface Processes and Landforms*, 43(1): 4–20. doi:10.1002/esp.4096.

Liverman, D., Hanson, M. E., Brown, B. J., & Merideth Jr., R. W. (1988). Global sustainability: Toward measurement. *Environmental Management*, 12(2), 133–143.

Liu, Y., Zhao, L. & Yu, X. (2020). A sedimentological connectivity approach for assessing on-site and off-site soil erosion control services. *Ecological Indicators*, 115, 106434. doi:10.1016/j.ecolind.2020.106434.

Lizaga, I., Quijano, L., Palazón, L., Gaspar, L., & Navas, A. (2018). Enhancing connectivity index to assess the effects of land use changes in a mediterranean catchment. *Land Degradation & Development*, 29(3), 663–675. https://doi.org/10.1002/ldr.2676

López-Vicente, M., Quijano, L., Palazón, L., Gaspar, L., & Navas, A. (2015). Assessment of soil redistribution at catchment scale by coupling a soil erosion model and a sediment connectivity index (central Spanish pre-pyrenees). *Cuadernos De Investigación Geográfica*, 41(1), 127. https://doi.org/10.18172/cig.2649

López-Vicente, M., Nadal-Romero, E., & Cammeraat, E. L. H. (2017). Hydrological connectivity does change over 70 years of abandonment and afforestation in the Spanish Pyrenees. *Land Degradation & Development*, 28(4), 1298–1310. https://doi.org/10.1002/ldr.2531

López-Vicente, M., & Ben-Salem, N. (2019). Computing structural and functional flow and sediment connectivity with a new aggregated index: A case study in a large Mediterranean catchment. *Science of the Total Environment*, 651, 179–191.

Magilligan, F. J., Graber, B. E., Nislow, K. H., Chipman, J. W., Sneddon, C. S. & Fox, C. A. (2016). River restoration by dam removal: Enhancing connectivity at watershed scales. *Elementa: Science of the Anthropocene*, 4, 108.

Marchamalo, M., Hooke, J. M., & Sandercock, P. J. (2016). Flow and sediment connectivity in semi-arid landscapes in SE Spain: Patterns and controls. *Land Degradation & Development*, 27(4), 1032–1044. https://doi.org/10.1002/ldr.2352

Marchi, L., & Dalla Fontana, G. (2005). GIS morphometric indicators for the analysis of sediment dynamics in mountain basins. *Environmental Geology*, 48(2), 218–228. https://doi.org/10.1007/s00254-005-1292-4

Maŕdhel, V., Frantar, P., Uhan, J., & Mio, A. (2004). *Index of development and persistence of the river networks as a component of regional groundwater vulnerability assessment in Slovenia*. International Conference on groundwater vulnerability assessment and mapping

Martinez-Agirre, A., Álvarez-Mozos, J., Milenković, M., Pfeifer, N., Giménez, R., Valle, J. M., & Rodríguez, Á. (2020). Evaluation of Terrestrial Laser Scanner and Structure from Motion photogrammetry techniques for quantifying soil surface roughness parameters over agricultural soils. *Earth Surface Processes and Landforms*, 45(3), 605–621. https://doi.org/10.1002/esp.4758

Martini, L., Cavalli, M., & Picco, L. (2022). Predicting sediment connectivity in a mountain basin: A quantitative analysis of the index of connectivity. *Earth Surface Processes and Landforms*, 47(6), 1500–1513. https://doi.org/10.1002/esp.5331

Meerkerk, A. L., van Wesemael, B., & Bellin, N. (2009). Application of connectivity theory to model the impact of terrace failure on runoff in semi-arid catchments. *Hydrological Processes*, 23(19), 2792–2803. https://doi.org/10.1002/hyp.7376

Najafi, S., Dragovich, D., Heckmann, T., & Sadeghi, S. H. (2021a). Sediment connectivity concepts and approaches. *Catena*, 196, 104880.

Najafi, S., Sadeghi, S. H., & Heckmann, T. (2021b). Analysis of sediment accessibility and availability concepts based on sediment connectivity throughout a watershed. *Land Degradation & Development*, 32(10), 3023–3044.

Newman, M. (2010). *Networks: An Introduction*. Oxford: Oxford University Press.

Nunes, J. P., Wainwright, J., Bielders, C. L., Darboux, F., Fiener, P., Finger, D., & Turnbull, L. (2018). Better models are more effectively connected models. *Earth Surface Processes and Landforms*, 32(4), 1297. https://doi.org/10.1002/esp.4323

Ortíz-Rodríguez, A. J., Borselli, L., & Sarocchi, D. (2017). Flow connectivity in active volcanic areas: Use of index of connectivity in the assessment of lateral flow contribution to main streams. *Catena*, 157, 90–111. https://doi.org/10.1016/j.catena.2017.05.009

Otto, J.-C. (2006). *Paraglacial sediment storage quantification in the Turtmann Valley, Swiss Alps* (Doctoral Dissertation), Bonn. Retrieved from http://hss.ulb.uni-bonn.de/diss_online/

Parsons, A. J., Wainwright, J., Brazier, R. E., & Powell, D. M. (2006). Is sediment delivery a fallacy? *Earth Surface Processes and Landforms*, 31(10), 1325–1328. https://doi.org/10.1002/esp.1395

Phillips, J. D., Schwanghart, W., & Heckmann, T. (2015). Graph theory in the geosciences. *Earth-Science Reviews*, 143, 147–160. https://doi.org/10.1016/j.earscirev.2015.02.002

Pöppl, R. E., & Parsons, A. J. (2018). The geomorphic cell: A basis for studying connectivity. *Earth Surface Processes and Landforms*, 43(5), 1155–1159. https://doi.org/10.1002/esp.4300

Pöppl, R. E., Dilly, L. A., Haselberger, S., Renschler, C. S. & Baartman, J. E. M. (2019). Combining soil erosion modeling with connectivity analyses to assess lateral fine sediment input into agricultural streams. *Water*, 11(9), 1793. https://doi.org/10.3390/w11091793

Reid, S. C., Lane, S. N., Montgomery, D. R., & Brookes, C. J. (2007). Does hydrological connectivity improve modelling of coarse sediment delivery in upland environments? *Geomorphology*, 90(3–4), 263–282. https://doi.org/10.1016/j.geomorph.2006.10.023

Schopper, N., Mergili, M., Frigerio, S., Cavalli, M., & Pöppl, R. (2019). Analysis of lateral sediment connectivity and its connection to debris flow intensity patterns at different return periods in the Fella River system in northeastern Italy. *Science of the Total Environment*, 658, 1586–1600. https://doi.org/10.1016/j.scitotenv.2018.12.288

Schrott, L., Hufschmidt, G., Hankammer, M., Hoffmann, T., & Dikau, R. (2003). Spatial distribution of sediment storage types and quantification of valley fill deposits in an alpine basin, Reintal, Bavarian Alps, Germany. *Geomorphology*, 55, 45–63.

Schwanghart, W., & Kuhn, N. J. (2010). TopoToolbox: A set of Matlab functions for topographic analysis. *Environmental Modelling & Software*, 25(6), 770–781.

Shore, M., Murphy, P. N. C., Jordan, P., Mellander, P.-E., Kelly-Quinn, M., Cushen, M., Mechan, S., Shine, O. & Melland, A.R. (2013). Evaluation of a surface hydrological connectivity index in agricultural catchments. *Environmental Modelling & Software*, 47, 7–15. doi:10.1016/j.envsoft.2013.04.003.

Singh, M., Tandon, S. K., & Sinha, R. (2017). Assessment of connectivity in a water-stressed wetland (Kaabar Tal) of Kosi-Gandak interfan, north Bihar Plains, India. *Earth Surface Processes and Landforms*, 42(13), 1982–1996. https://doi.org/10.1002/esp.4156

Skolaut, C., Liébault, F., Habersack, H., Lenzi, M. A., Rusjan, S., Sodnik, J. & Pichler, A. (2015). *Synthesis Report: Sediment Management in Alpine Basins (SedAlp): Integrating sediment continuum, risk mitigation and hydropower.* Accessed September 15, 2017. www.sedalp.eu/download/reports.shtml

Smetanova, A., Paton, E. N., Maynard, C., Tindale, S., Fernández-Getino, A. P., Perez, M. J. M., Bracken, L., Le Bissonnaid, L. & Keesstra, S. (2018). Stakeholders' perception of the relevance of water and sediment connectivity in water and land management. *Land Degradation and Development*, 29, 1541–2036, doi:10.1002/ldr.2934.

Souza, J. O., Correa, A. C. & Brierley, G. J., (2016). An approach to assess the impact of landscape connectivity and effective catchment area upon bedload sediment flux in Saco Creek Watershed, Semiarid Brazil. *Catena*. 138, 13–29. https://doi.org/10.1016/j.catena.2015.11.006

Straffelini, E., Cucchiaro, S., & Tarolli, P. (2021). Mapping potential surface ponding in agriculture using UAV-SfM. *Earth Surface Processes and Landforms.* Advance online publication. https://doi.org/10.1002/esp.5135

Tarboton, D. (1997). A new method for the determination of flow directions and upslope areas in grid digital elevation models. *Water Resources Research,* 33(2), 309–319.

Thompson, J. J. D., Doody, D. G., Flynn, R. & Watson, C. J. (2012). Dynamics of critical source areas: does connectivity explain chemistry? *The Science of the Total Environment,* 435-436: 499–508. doi:10.1016/j.scitotenv.2012.06.104.

Trevisani, S., & Cavalli, M. (2016). Topography-based flow-directional roughness: Potential and challenges. *Earth Surface Dynamics,* 4(2), 343–358.

Turley, M., Hassan, M.A. & Slaymaker, O. (2021). Quantifying sediment connectivity: Moving toward a holistic assessment through a mixed methods approach. *Earth Surface Processes and Landforms,* 46(12): 2501–2519. doi:10.1002/esp.5191.

Turnbull, L., Hütt, M.-T., Ioannides, A. A., Kininmonth, S., Pöppl, R., Tockner, K., ... Parsons, A. J. (2018). Connectivity and complex systems: Learning from a multi-disciplinary perspective. *Applied Network Science,* 3(1), 47. https://doi.org/10.1007/s41109-018-0067-2

Vigiak, O., Beverly, C., Roberts, A., Thayalakumaran, T., Dickson, M., McInnes, J. & Borselli, L., (2016). Detecting changes in sediment sources in drought periods: The Latrobe River case study. *Environmental Modelling & Software,* 85, 42–55. https://doi.org/10.1016/j.envsoft.2016.08.011

Walling, D. E. (1983). The sediment delivery problem: Scale Problems in Hydrology. *Journal of Hydrology,* 65(1–3), 209–237. https://doi.org/10.1016/0022-1694(83)90217-2

Wilson, J. P., & Bishop, M. P. (2013). 3.7 Geomorphometry. In J. Shroder, A. D. Switzer, & D. M. Kennedy (eds.), *Treatise on Geomorphology* (pp. 162–186). Elsevier. https://doi.org/10.1016/B978-0-12-374739-6.00049-X

Wohl, E., Brierley, G., Cadol, D., Coulthard, T. J., Covino, T., Fryirs, K. A., ... & Sklar, L. S. (2019). Connectivity as an emergent property of geomorphic systems. *Earth Surface Processes and Landforms,* 44(1), 4–26.

Wohl, E., & Beckman, N. D. (2014). Leaky rivers: Implications of the loss of longitudinal fluvial disconnectivity in headwater streams. *Geomorphology,* 205, 27–35. https://doi.org/10.1016/j.geomorph.2011.10.022

Wohl, E., Rathburn, S., Chignell, S., Garrett, K., Laurel, D., Livers, B., ... Wegener, P. (2017). Mapping longitudinal stream connectivity in the North St. Vrain Creek watershed of Colorado. *Geomorphology,* 277, 171–181. https://doi.org/10.1016/j.geomorph.2016.05.004

11

Modelling

JOHN WAINWRIGHT

11.1 Introduction

The initial step in any modelling exercise is to ask the question why we want to model the system in question. Models may be used to understand systems and their behaviour, to make predictions about the future, to provide support for decision-making and policy development, and to communicate science amongst numerous other rationales (Kirkby, 1987; Epstein, 2008; Mulligan & Wainwright, 2013; Edmonds et al., 2019). The answer to this question can then be used to guide how connectivity should be included in the modelling process.

The question of what connectivity is has been addressed in the previous 10 chapters but it might be useful to characterize two end-members forming a definition. The first considers connectivity to be an explicit property of the system in question, or more correctly an emergent property of that system (in the 'weak emergence' sense of Bedau (1997). In other words, the behaviour of the system components produces a set of interactions that define the system in a way that cannot necessarily be predicted from knowledge of the components themselves. (Nunes et al., 2018) define this end-member as 'emerging connectivity'. Although not explicitly in a connectivity framework, Newman et al. (2006) frame this style of modelling as 'Darwinian' in the sense that it can better represent the evolution of the system in question with fewer subjective decisions being taken to build the model. This approach to modelling can also be considered to be based more on an inductive framing of research questions, aiming for the modelling process to tell us about the system. It is useful for applications where feedbacks between functional and structural connectivity (e.g. Wainwright et al., 2011) occur over a shorter timescale than the duration of the model application, and for developing understanding based on model development and application.

The second end-member imposes on the system a specific definition of what connectivity is. This definition may come directly from a conceptual model of the system,

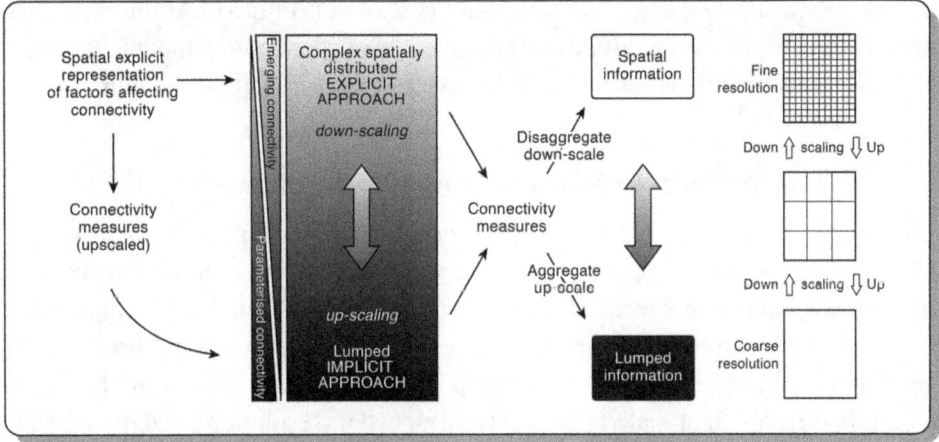

Figure 11.1 Schematic overview of the ways in which landscape connectivity is represented in different modelling approaches, reflecting the two end-members of approach discussed (based on Nunes et al., 2018 and Zhang et al., 2013).

or indirectly through the application of a conceptual model to indices used to characterize the system. Alternatively, it may also come from previous applications of the first type of model. This end-member is called 'parameterized connectivity' by Nunes et al. (2018), with parallels to the 'Newtonian' approach of Newman et al. (2006). This style of model has a more rigid structure with the components and their interconnexions predefined, requiring more subjective decisions on the part of the modeller. Although it can represent how structural connectivity might affect functional connectivity, it is less well able to represent the evolution of the system, and thus feedbacks between functional and structural connectivity. It represents a more deductive approach to modelling, suitable for timescales over which there are not major evolutions or changes in structural connectivity as a result of functional connectivity. While it might be appropriate for testing understanding of the system in particular contexts, the approach might be better used where prediction or application of the model results is required.

As Nunes et al. (2018) demonstrate, these end-members also reflect the nature of spatial information in a model (Figure 11.1). Emerging connectivity approaches require spatially distributed models in which model elements and related parameters vary across the environment represented. Lumped models on the other hand contain no spatial information and thus any connectivity is represented by parameterization. In reality, most environmental models contain elements of lumping, so the modelling process requires the definition of scales of application, and how to move between scales. These scales are both spatial, and as noted earlier, temporal in the way they represent dynamics of change. Once the basic units of spatial scale can be defined, development of the model can move on to considering what data are most appropriate at that scale, and how to collect them. At this stage, the model structure can be identified and applications of connectivity models assessed. However, this process is not straightforward and requires careful thought.

11.2 Defining the Model Scale and Moving Between Scales

Blöschl and Sivapalan (1995) demonstrated in (hydrological) modelling that the idea of scale is used differently according to process, observation and working scales. Developing these ideas, Zhang et al. (2013: 71) suggested that '[a]pplication of environmental modelling always involves four different scales. These are the geographic scale of a research area, temporal scale related to the time period of research, measurement scale of parameters (input data resolution), and model scale referring to both temporal and spatial scales when a model was established'. In other words, as well as returning to the question of why we want to model in the first place, it is necessary to answer questions about space and time scales of the application and data available to support it, and to reflect on what to do if applying a model that was defined for a different location or scale. To assert that a model is 'valid' because it has been tested on data from elsewhere or at different scales is logically untenable (Oreskes et al., 1994) – there is never such a thing as 'just' applying a model.

Returning to the idea of basic units of a model, it is useful to reflect on the inter-disciplinary overview of Turnbull et al. (2018) of fundamental units in connectivity studies. They state that in geomorphology there is '[n]o clear or consistent definition of the fundamental unit (it is dependent on the research question)' (p. 4) and that it thus presents a major challenge in the discipline. Because it depends on the research question, it could represent either end of the spectrum defined earlier, for example, at the lumped end, the idea of land elements of consistent sets of properties. Pöppl & Parsons (2018) take this idea forwards and define the 'geomorphic cell' defined as having specific states (source, sink, or steady-state) with 'connecteins' of different types, noting that it is important to consider what the correct resolution is for the research question rather than just look at what data are available. However, they suggest that implementation is more likely in a Geomorphic Information Systems (GIS) framework, without recognizing that this approach will constrain to certain ways of thinking (until we develop appropriate *Geomorphic* Information Systems). Beyond the laboured terminology, there are also inconsistencies with definitions, for example, in terms of pathways and fluxes being constant within a fundamental unit. It is hard to see how such constraints could be reasonable except under rather limiting steady-state assumptions, and thus how the definition could allow the evaluation and interpretation of functional connectivity.

These discussions reflect ongoing debates in hydrological modelling about representative elementary volumes (REV) and representative elementary areas (REA). The REV comes from assumptions of continuity in groundwater flow theory and developed to mean the scale beyond which any function of porosity changed smoothly with scale (with an implied upper limit for example, across lithological boundaries) (Hubbert, 1956 and Bear, 1972 cited in Blöschl (1999). Wood et al. (1988) generalized to the idea of the REA, with the definition of a representative length scale greater than rapid fluctuations in the system properties but smaller than slowly varying changes or significant heterogeneities in the system. The REA must account for all relevant field properties and be static, or only slowly evolving. Limitations with these assumptions were found by Blöschl et al. (1995) and highlighted more clearly by Fan & Bras (1995) who suggested that the concept could not be used simply for moving from small to large scales in hydrological models. Clark et al. (2017) reiterated these points and suggested they relate more to ideas about parameterization of specific models rather than providing founding principles for the discipline. Beven (2000, 2001) took these critiques further based on the ideas of the uniqueness of place, heterogeneity and its characterization, and scale-dependency of parameters without an underlying theory of how to deal with such dependency. The idea of a link to parameterization is often also circular e.g. Defina (2000) uses REA in sense of being the same size as a model cell, which is similar to the comment Pöppl and Parsons make about geomorphologists using available scales of data rather than appropriate ones. Refsgaard et al. (2016) generalized further to the idea of representative elementary scale, although their approach does not add much fundamentally beyond adding uncertainty analysis into the mix. At the lumped end of hydrological modelling, England & Onstad (1968) introduced the idea of the hydrologic response unit (HRU), based on the areal distribution of soils on the hypsometric curve. Later applications added further limitations including slope, aspect and vegetation type (Leavesley et al., 1983) and land-use and pedotransfer properties (Flügel, 1995) thus assuming 'a homogeneous hydrological response under equivalent meteorological forcing' (Poblete et al., 2020: 1). Flügel (1995) recommended the use of GIS, with again the constraints discussed earlier as a result. Furthermore, the HRU can only be a static representation, and thus limited in being able to assess functional connectivity or non-evolving systems.

These various debates suggest that (even closely related) disciplines need to ensure there is ongoing dialogue, and that there is no simple answer to the definition of a fundamental unit in geomorphology. Care is required to consider the scale of the connectivity question to be answered and how it relates to the broader research programme being undertaken. In relation to the two end-members outlined earlier, to define fundamental units, it is useful first to return to the definition of (weak) emergence according to Bedau (1977). In this definition, it is necessary to distinguish between the macrostate of the overall system and the microstates of each system component. Those microstates evolve according external forcing and to local conditions and processes. A system is defined as weakly emergent if and only if the macrostate can be derived from the microstates and external forcing but only by simulation. In other words, there must be an understanding of structures and interactions, but outcomes are only representable if those structures and interactions can be simulated (successfully). A logical extension is that the conditions and processes are a function of the connectivity of the components. Note that there is no limitation in the definition that the small-scale components must be static, and thus such a model can be evolutionary both in terms of components and overall structure. This approach integrates ideas of complex systems and overcomes the limitation of the 'palimpsest [sic] of traditional system thinking based on general systems theory' (Turnbull et al., 2018: 19). There is no requirement either that the components need to be as small as possible, and while it is increasingly possible to simulate individual raindrops or sediment particles, the decision of the appropriate scale requires a return to the question to be answered and is often also a compromise with available data and computing power. Furthermore, if there is no limitation for components to be small, there is no reason the same approach could not be used in defining fundamental units at the lumped scale. The decision then is whether to allow evolution of the lumped elements, or to have a static representation and use scaling methods to represent and/or parameterize the connectivity.

Basic definitions of upscaling – moving from fine to coarse resolution – and downscaling, from coarse to fine resolution are given by Blöschl & Sivapalan (1995) (Figure 11.1). Zhang et al. (2013) provide a detailed overview of approaches taken. In upscaling, kriging and other methods of interpolation can be used, as well as using model results at one scale to provide information in more detail either by nesting the models or using metamodeling to produce statistical models, which may involve complicated distribution-based transforms. Statistical models based on point measurements are also often used in climate model upscaling. Often models applied at different scales will have a calibration applied and it may not be clear from the discussion of results that this step has taken place. Calibration is an implicit scaling method and its use requires caution when interpreting results (Mulligan & Wainwright, 2013). Downscaling properties may involve averaging, use of dominant values, thinning or linear convolution. Some methods such as fractal approaches are suitable for both upscaling and downscaling. As will be seen later, some of the simpler methods are inappropriate because they fail to account for connectivity.

11.3 Scales of Data and Connectivity

As noted by Winsberg (2018), it has long been appreciated that data are theory-laden (see further discussions in Frigg et al. (2015; Kitcher (1995)). Whether making a measurement of distance, or a compound property such as hydraulic conductivity, we are making assumptions about how the measurement devices work, are calibrated against standards and have a theoretical meaning in-built (and not forgetting the possibility of operator error and uncertainty). For these reasons, it is useful not to consider data as absolutes, but to reflect that what we work with in testing theories is a model of the data, following the arguments of Suppes (1962). The key point is that we do not make basic errors in interpreting theories when confronting them with (models of) data – what Bogen & Woodward (1988) call 'saving the phenomena'. For present purposes, our models of data need to be consistent with representations of connectivity, either as ways in which properties of fundamental units interact in the case of emergent connectivity models, or as ways in which connectivity is represented within fundamental units in parameterized connectivity models. The emphasis here is on modelling of water flows and consequent erosion, though some of the examples will also be relevant in other geomorphic settings.

11.3.1 Topographic Data and Flow Routing

Topography is the basic building block of any geomorphic model using a Digital Terrain Model (DTM), and a lot of effort in recent years has been put into developing techniques for collecting high resolution data (e.g. Passalacqua et al., 2015). These data allow us to go beyond simple assumptions of flow across the surface using calculations of the local slope angle. Antoine et al. (2009) introduced the idea of the Relative Surface Connection (RSC: Figure 11.2) function and developed it in Antoine et al. (2011) as a method for hydrograph scaling. Peñuela et al. (2013, 2015) elaborated these ideas further and developed variogram-based techniques for estimating the shape of the curve on different microtopographies, in particular for noting when the connectivity threshold is reached and rapid runoff develops. Thus, it should be feasible to use the RSC when moving between scales of flow estimates, although further work is required to address how the patterns translate into sediment fluxes. Lane et al. (2017) used the approach to interpret whether pits in a real landscape DTM are natural or not, however, more work is required to evaluate what expected RSCs should look like for different terrains with different resolutions of topographic data. Müller (2007) also demonstrated that scaling topography affects estimated path lengths and therefore flow travel times. Her approach was to scale the roughness coefficient in the model.

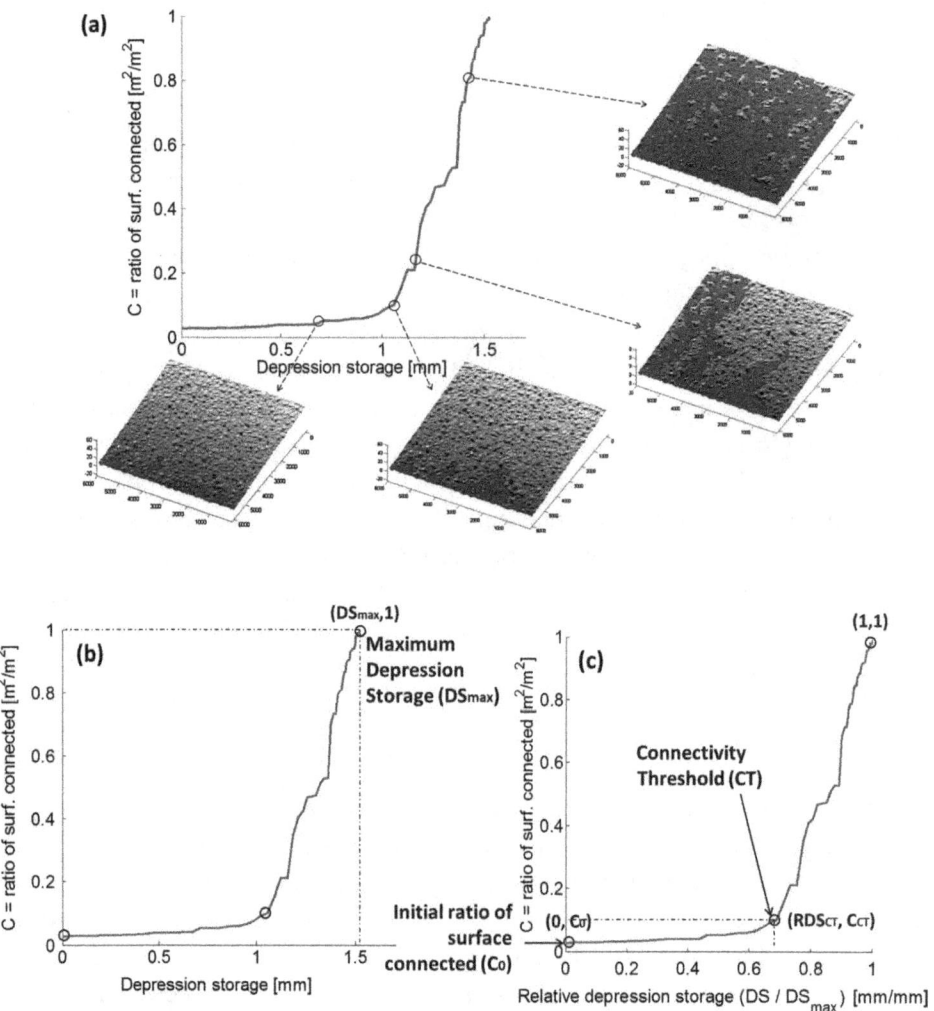

Figure 11.2   Definition of the RSC function (from Peñuela et al. (2015)).
(a) As the depression storage fills, the relative proportion of the surface becomes
connected to the outflow edge. (b) shows the RSC function; and (c) the normalized
RSC function. The connectivity threshold is the inflexion point in the RSC
function, and together with the initially connected surface a maximum depression
storage, these three parameters can be used to estimate sub-grid patterns of flow
for models of connectivity. A black and white version of this figure will appear in
some formats. For the colour version, refer to the plate section.

In comparing DTMs based on contour, LiDAR and GPS data, Casas et al. (2006)
found improved estimates of flood extents, but also that the model was more sensitive to Manning's n with the higher resolution data, so topography and other
parameters are again not independent. Thomsen et al. (2015) compared different
measurement techniques to estimate depression storage from microtopography

and demonstrated they had significant impacts on results from the LImburg Soil Erosion Model (LISEM). The link between topography, microtopography and roughness will be discussed in more detail later.

Numerous methods have been applied for scaling DTMs. For example, Her et al. (2015) used a range of interpolation methods for generating 30 m DTMs from 90 m resolution SRTM data, and comparing them with observed 30 m SRTM data in New Mexico. They found splines, kriging and cubic convolution gave the best results with 1.29–9.83% of cells misclassified as pits or not pits. Crema et al. (2020) used heat-diffusion based model for inpainting of DTMs, albeit without a clear justification of the appropriateness of this method, which seems to produce more smoothing than seen with higher-resolution data. For extracting subtle channels on a low-angle floodplain in South Carolina, Xu et al. (2020) found a bottom hat transform with adaptive threshold the most useful method. The issue with these scaling methods is that they make *a priori* assumptions about the connectivity of the surface (or usually the lack of it) and do not build information such as that from the RSC into the procedure. Callaghan & Wickert (2019) attempted to produce a depression-filling algorithm 'that fills only those depressions on a landscape which would become filled under reasonable runoff conditions. This allows for the existence of real depressions and hydrologic disconnects in the landscape'. However, the definition of 'real' here is dependent on a user-defined parameter that is static across the landscape to be evaluated, with the potential for circularity in arguing whether resulting patterns of flow connectivity are realistic. Using high-resolution data over large extents is a non-trivial task just to define patterns of flow routing, even when intensive computational methods are employed (McGough et al., 2012). Conversely, Lisenby & Fryirs (2017) argue for the use of coarse-scale DTMs, which they suggest include fewer artefacts and thus create less disconnectivity. However, they admit their approach is only really useful for initial analyses and still does not address the question of which pits are real or not.

From a lumped perspective, approaches based on HRU might be a good starting point for catchments dominated by surface flows (see review in Poblete et al., 2020) with the caveats noted earlier or Connectivity Response Units, with connectivity in-built as a threshold (Singh et al., 2017)). Spence & Hosler (2007) have considered how catchments might be discretized as a sequence of cascading reservoirs in a watershed of lakes with interconnecting channels in Canada. Their study is relevant to the broader category of 'fill-and-spill' models of subsurface-stormflow-dominated catchments (Tromp-van Meerveld & McDonnell, 2006). Both drainage density and sampling frequency affected the number and patterns of reservoirs defined. However, there is a paradox in that drainage density is a function of water content so that catchment saturation is needed when defining reservoirs, or in other words connectivity in these systems must be modelled as a dynamic property.

11.3.2 Soil Moisture and Infiltration

Definition of moisture patterns and scaling relations was carried out by Western & Blöschl (1999) who tested whether standard geostatistical procedures were robust when confronted by connected patterns of soil moisture that violate standard assumptions. While they found that the methods worked in their application to a site in Australia, further work is required to generalize these results. Using a model-based approach, Ivanov et al. (2010) have argued for the nonuniqueness of soil-moisture distributions in that moisture variability is a complex function of mean moisture at the slope scale, with hysteresis shown in wetting and drying conditions. Fatichi et al. (2015) elaborated these results again at a 1 × 1 m cell size for a range of different environmental régimes, comparing the effects of abiotic parameters such as climate and soil properties with biotic effects. They found that abiotic controls were always more significant than biotic ones, with biotic controls having most effect in Mediterranean climates compared to either wetter or drier climates. Kim et al. (2016) elaborated the results further comparing 1 × 1 m and adaptive grid models with 0.3 m horizontal spacing and 0.02–0.1 m vertical spacing. They found significant effects from the watershed structure and, again, effects of hysteresis. Thus, when scaling moisture patterns, either to provide initial conditions for detailed models or for upscaling lumped models, connectivity and temporal patterns of rainfall need to be taken into account.

Once initial conditions are defined infiltration is modelled to estimate runoff. It has long been recognized that infiltration rates are scale-dependent (Hawkins and Cundy (1987) and e.g. Wainwright (1996a)) and that patterns of runoff would depend on the amount and pattern of infiltration variability. To address this question from a connectivity perspective, Harel & Mouche (2013) applied a one-dimensional queuing-theory model. They showed that different patterns of connectivity develop as a function of different probability distributions of infiltrability. Furthermore, they considered the importance of a nugget effect (due to uncorrelated errors or variability at scales smaller than the model cell) and small-scale correlation and found both were important in defining the connectivity function of infiltration. They also generalized the result of Renard & Allard (2013) from subsurface flows that connectivity isexponential for long distances. In Harel & Mouche (2014), the approach was expanded to a two-dimensional queuing model, accounting for lateral effects and soil heterogeneity. Again, different distributions and correlated fields of infiltrability show significant differences in resultant connectivity (Figure 11.3). They provided some generalizations in relation to percolation theory for scaling infiltrability, but noted their results were not straightforward and that increased connectivity did not necessarily lead to increased flow even where functional connectivities are correlated with flow rates.

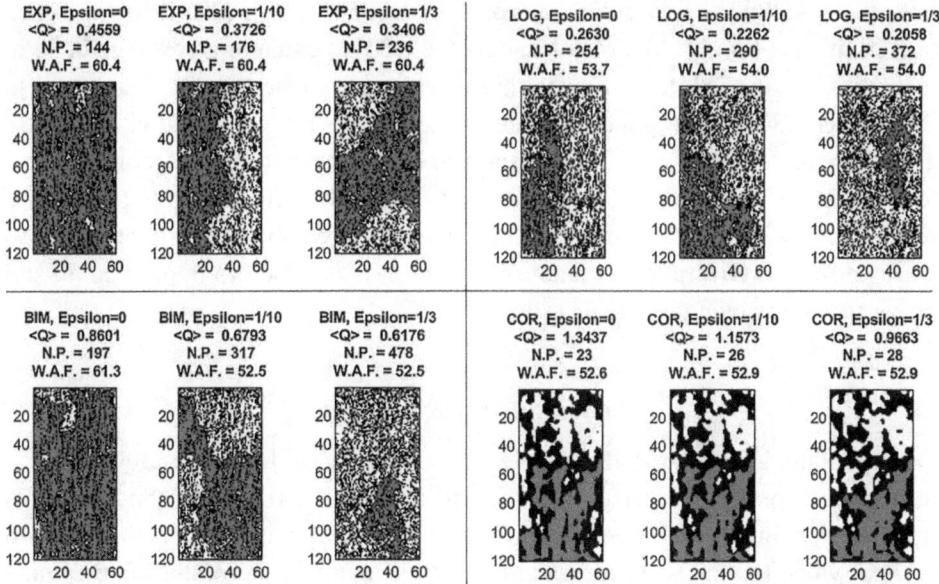

Figure 11.3 Simulations by Harel & Mouche (2014) comparing how uncorrelated and correlated infiltrability fields affect the runoff from a rectangular surface (120 cells in the downslope direction from the top to the bottom of the plots by 60 cells across the slope) The three uncorrelated surfaces are defined by exponential (EXP), log-normal (LOG) and bimodal (BIM) distribution functions. The correlated (COR) surfaces are generated using an isotropic multi-Gaussian field. The Epsilon values specify the amount of cross-slope flow in the downslope direction: Epsilon = 0 means that all flow goes into the immediately adjacent downslope cell; Epsilon = 1/3 means that flow is equally distributed between the immediately adjacent downslope cell and the two lateral cells in the downslope direction. <Q> is the mean runoff flow rate, N.P. is the number of runoff patterns and W.A.F. is the wet area fraction, defined as the number of wet pixels divided by the total number of pixels. A black and white version of this figure will appear in some formats. For the colour version, refer to the plate section.

As a result, further investigations are needed to understand how infiltration rates might be scaled as a function of connectivity.

While measurements of infiltration rate can be used at finer resolutions, it is often predicted using soil properties. Availability of widespread, high-resolution soils data is limited despite the work since McBratney et al. (2003) in developing methods for digital mapping of soils. Minasny & McBratney (2016) reviewed progress and noted most work was being carried out at a grid spacing of 20–200 m for studies at catchment and regional scales although promising techniques in proximal soil sensing are being developed at field scale. Global coverage at 30 m resolution is anticipated by 2030, which has clear implications for definitions of fundamental units in the meantime. High resolution data vary between locations, with the so-called 'hyper-scale' approach of Behrens et al. (2014), being based on 100 m and 90 m DTMs, with Mulder et al. (2016) extending coverage at 90 m resolution for the whole of France, but an example of heterogeneity mapped in Brazil used 280 sample points in study area of 1,378 km^2 (Bonfatti et al., 2020). The study of Atkinson et al. (2020) showed the promising use of 'geomorphons' to estimate soil properties at different scales in South Africa. Despite good progress, Arrouays et al. (2020) note the importance of producing uncertainty maps so that widespread coverage at whatever resolution is not used to imply that all soil properties are known globally.

11.3.3 Surface Roughness

Once ponding is generated at the surface, the velocity of flow is usually estimated using some form of roughness coefficient. Smith (2014) defines the roughness window as being that between measurement scale and what he calls the partition scale of when roughness becomes topography (Figure 11.4). While some authors consider the distinction to be synonymous with microtopography as discussed earlier, Smith notes that it is important to use process-based distinctions. Furthermore, the choice of the upper limit can be arbitrary, which he notes Lane (2005) has previously called problematic, although there seems not to be a significant body of work attempting to address this issue. Moving between scales is most commonly carried out using fractal techniques (reviewed in (Smith, 2014; Zhang et al., 1999) at lower resolutions). There is a much more developed literature in relative resistance to wind flow than to water flow, and this is another limitation that needs to be addressed in the scaling and connectivity fields. Parsons et al. (1994) demonstrated that scale-dependence of flow resistance on hillslopes was also a function of measurement technique, further illustrating the 'model of data' concept discussed

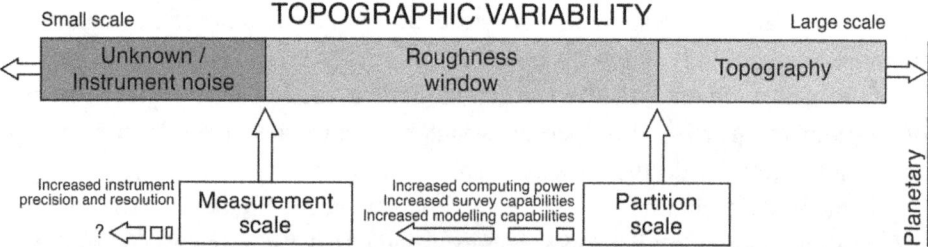

Figure 11.4 Definition of how information about topographic variability is dependent on the scale at which measurement is possible (and may thus change as technology improves) and the scale between which macroscale topography is represented, as opposed to surface roughness (Smith, 2014). Different model applications or changes in technology may also cause this distinction of the 'partition scale' to vary. A black and white version of this figure will appear in some formats. For the colour version, refer to the plate section.

earlier. Smith also cites cases in the aeolian (Skidmore, 1997) and fluvial (Darby et al., 2010) literatures on scale dependency of resistance estimates.

From a modelling perspective Hughes et al. (2011) used sub-grid formulations of Manning's *n* to demonstrate that sub-grid roughness was more important than sub-grid topography in simplified canal simulations (but again note that (Müller, 2007) demonstrated these factors are correlated). Özgen et al. (2015) explicitly built a roughness-based formulation of shallow-water flows using the coefficient of variation of 9 the microtopography and parameters related to the Chézy coefficient. Although they demonstrated that improved predictions could be provided from coarse grids, they cautioned wider use of their parameters because calibration was required. It would be interesting to investigate the impacts of connectivity on the parameters and whether the need for calibration could be reduced in this way. Hutton et al. (2014) used explicit scaling approaches to correct a hillslope model running at a coarse resolution based on simulations using high-resolution inputs. They developed rating curves to improve flow estimates and modified the friction parameter to improve erosion estimates. Again, it would be interesting to see if the results of the scaling factors could be generalized by taking account of connectivity characteristics.

It is not just spatial connectivity that needs to be considered in scaling processes. Wainwright & Parsons (2002) showed that temporal connectivity or continuity of rainfall translates to spatial connectivity effect. The effect is exacerbated by surface roughness, interpreted as providing more opportunity time for infiltration. Caviedes-Voullième et al. (2021) also suggested feedback of microtopography on infiltration using a modelling approach, demonstrating the effect was most pronounced at intermediate scales and amplitudes of microtopography.

11.4 Model Structures

As all modellers know, it is the preparation of the model and its data that generally takes the most time. Thus, it is not unreasonable that we are nearly 2/3 of the way through the chapter before discussing actual models. In this section, we will consider different model structures that might be considered to reflect the emerging and parameterized connectivity end-members outlined earlier.

11.4.1 Emerging Connectivity Approaches

While the explicit representation of specific 'drops' or 'packets' or 'parcels' of water or sediment has been used in a range of models to look at emergent surface properties since the work of Smith (1991) (see also Forrest & Haff, 1992; Murray & Paola, 1994; Favis-Mortlock, 1998), these approaches did not consider the evolution of connectivity directly. Haff (2001) also discusses the evolution of the 'waterbots' or 'precipiton' approaches in the cellular automata literature. Favis-Mortlock (1998; Favis-Mortlock et al., 2000) come closest in this sense in discussing how the emergent behaviour of water parcels produce connected networks of flow and ultimately rill networks on hillslopes in the RillGrow model, and Crave and Davy (2001) similarly discuss network evolution at landscape scale in their 'precipiton'-based model. Braun & Sambridge (1997) passed parcels of water in their CASCADE model of landscape evolution, again with a focus on the evolution of network systems, and their approach influenced the initial form of the CAESAR model, in which Coulthard et al. (2002) do (briefly) discuss connectivity explicitly in relation to tributary and mainstream connectivity. It should be noted, though, that later developments of CAESAR employ the LISFLOOD approach to flow routing (Coulthard et al., 2013), and later applications are thus based on a hydrodynamic rather than a cellular flow model. A similar, cellular approach to flow routing is also used in the CYBEROSION models of Wainwright (2008, 2015; Wainwright & Millington, 2010), in which connectivity is explicitly considered in the way the direct and indirect actions of people affect landscape evolution.

Reulier et al. (2016, 2017) explicitly designed an agent-based model (ABM) to look at connectivity at the catchment scale. Their Landscape StruCture And Runoff (LASCAR) model uses 'agentgouttes' ('dropagents') which are placed on agricultural landscapes, where the patches of the model are neutral, grass or ditches. Movement of the dropagents is then carried out as a function of the interaction between themselves and the different types of patch. They later applied the model to a small agricultural watershed in Normandy using aerial photographs to look at the evolution of connectivity as a result of land-use change over a 67-year period (Reulier et al., 2019). In particular, they demonstrated that the number of locations where runoff entered the ditch and channel network increased over time, and especially between 1967 and 2014, with important implications for understanding impacts on water quality and flooding (Figure 11.5).

A detailed subsurface-flow ABM has been developed by (Mewes & Schumann, 2018) and, although inspired by connectivity-based models, the initial application does not discuss connectivity in detail. One issue with the ABM approach is the heavy computational load introduced by simulating individual parcels of water, and the resulting balance between choice of parcel size and detail of flows that can be reconstructed. As an intermediate approach, Reaney (2008) used a combination of the distributed Connectivity of Runoff Model (CRUM: (Reaney et al., 2007, 2014)) with 'hydroagents', which are used to trace the patterns of sources and sinks of flow while still representing the overall fluxes. A similar approach for tracing sediment particles was developed by Cooper et al. (2012) using a Marker-in-Cell extension of the MAHLERAN model, and earlier by Wainwright (1994) in

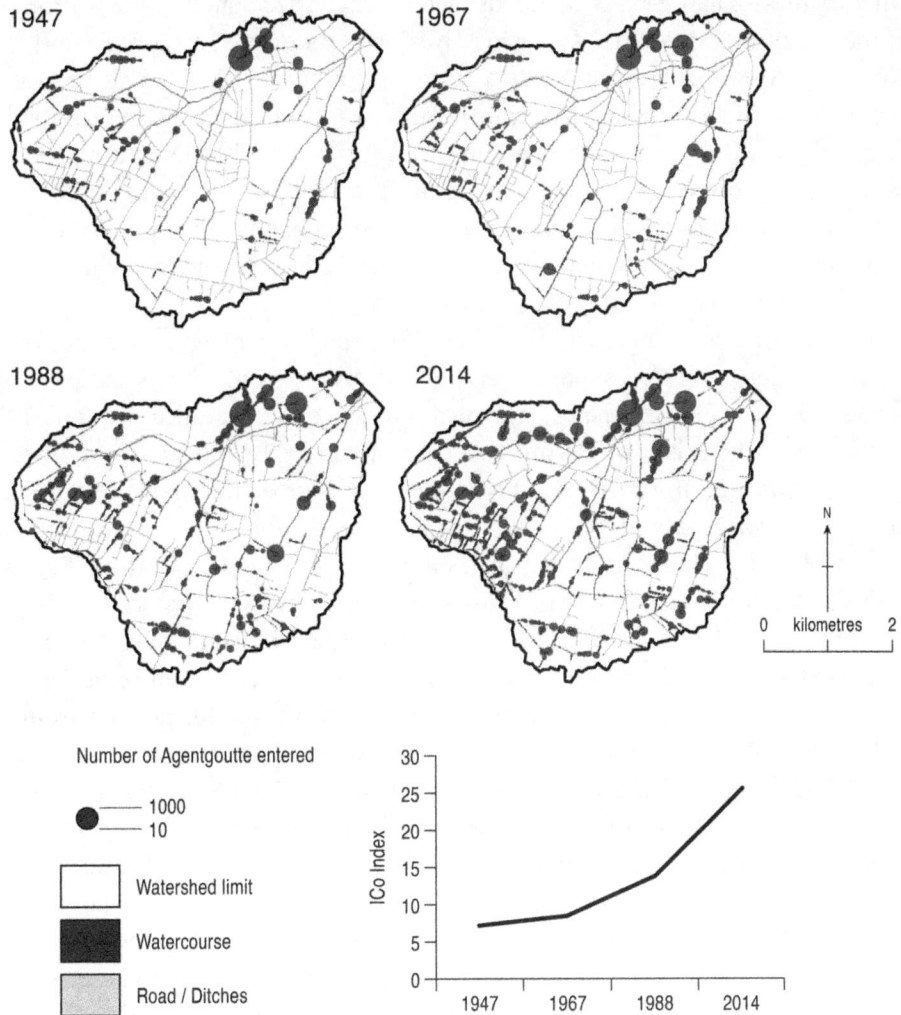

Figure 11.5   Use of the LASCAR model to demonstrate how the connectivity of the 17.6 km^2 Lingèvres catchment in Normandy, France, evolved over a period of 67 years (Reulier et al., 2019). The ICo index is the percentage of cultivated areas in the catchment that is connected hydrologically to the ditches and channels in the catchment, thus potentially allowing outflow.

tracing the movement of archaeological materials on slopes. Neill et al. (2020a, b) have also followed this approach to track faecal indicator organisms in runoff to understand patterns of water contamination in relation to the connectivity between infield sources and the stream network. This sort of hybrid modelling approach emphasizes the different styles of process representation, wherein the cellular or agent-based methods tend to emphasize the discontinuous nature of processes, whereas mass-balance-based methods tend to emphasize continuity. For example, in the use of the Model for Assessing Hillslope-Landscape Erosion, Runoff And Nutrients (MAHLERAN) model at a high resolution of 0.5 × 0.5 m to look at structural and functional connectivity feedbacks in land degradation, Turnbull & Wainwright (2019) had to specify thresholds of flow and sediment transport in estimating connectivity metrics to avoid the apparent persistence of connectivity at the start or end of events. In relation to understanding the evolution of connectivity, this process-representation distinction can be critical and in deciding on an approach, it is necessary to avoid potential bias.

The mass-balance approach can be used to represent emergent connectivity if the cell or element size is small enough. The decision about what is small enough is based on local characteristics, and the discussion about parameterization earlier suggests that there is always some element of sub-grid parameterization necessary. Baartman et al. (2020) compared the behaviour of five models to a range of scenarios of connectivity in what they called a semi-virtual catchment. The topography of a real catchment was modified to remove all anthropic features, with these baseline conditions then used to compare with the model results with different sets of features affecting the connectivity of the landscape. The models employed were process-based water and soil-erosion models, applied at the scale of a rainfall event. Different flow-routing methods were used, ranging from the full shallow-water equations (FullSWOF_2D), to kinematic wave approximations (EROSION3D, OpenLISEM) and D8 approximations (LandSoil, Watershed), and in each case a cell size of 1 × 1 m was used. Despite the similarities of model approach and process representations, Baartman et al. (2020) found a surprisingly low level of agreement in the runoff and sediment fluxes produced, and the total areas connected to the catchment outflow (Figure 11.6). They interpreted these results as suggesting that at even at this resolution, there needs to be a greater attention to detail of the model parameterization and thus it cannot be assumed *a priori* that simulation at this resolution using this style of model is entirely emergent.

11.4.2 Parameterized Connectivity Approaches

As the size of application increases and resolution of implementation decreases, the parameterization of connectivity becomes increasingly important, even if it is only carried out implicitly. For example, Millares-Valenzuela et al. (2022) use a process-based runoff and erosion model to look at events in a 45.9 km^2 catchment in southern Spain, with a 5 × 5 m cell size, but surface information at a resolution of 250 m. For a series of events, they then calibrate the model against catchment outflow data and plot erosion data. Issues of model equifinality with this approach are significant (Parsons et al., 1997; Mulligan & Wainwright, 2013) and may explain

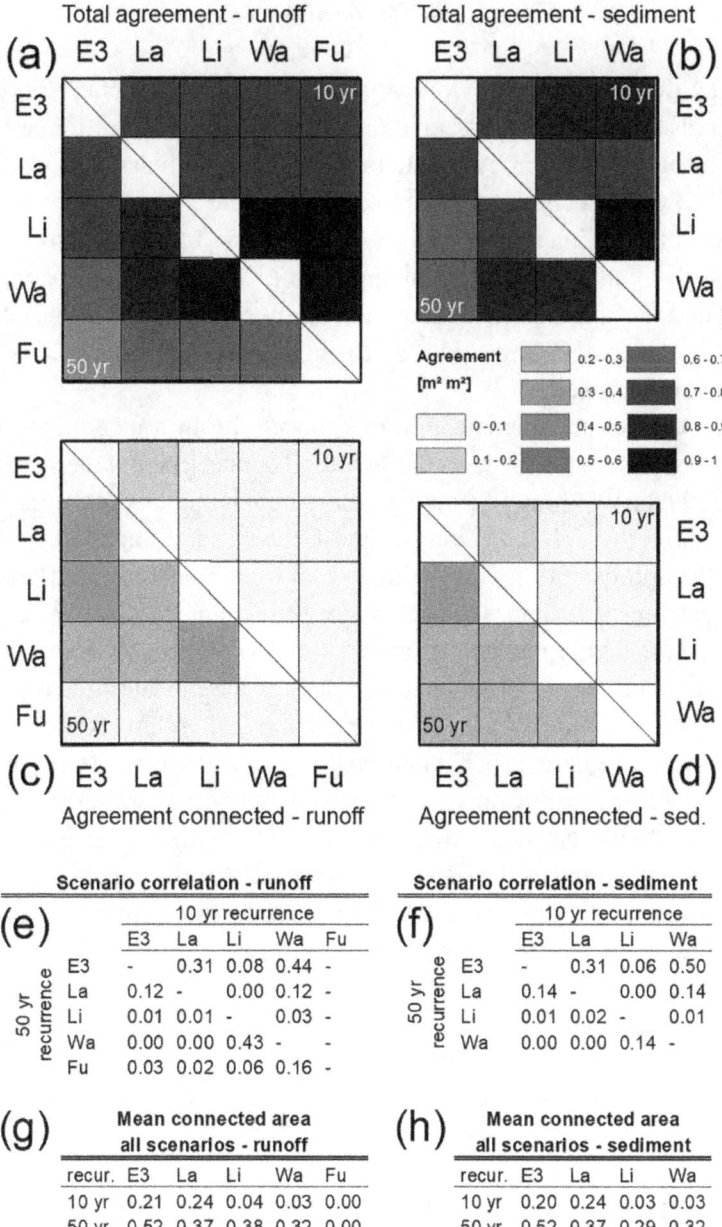

Figure 11.6 Agreement between the five different runoff and erosion models evaluated in terms of their connectivity characteristics by Baartman et al. (2020). The extent to which the models agree regarding the total area (connected and unconnected) are shown for (a) runoff and (b) sediment flux. For the connected areas only, model agreement is shown in (c) for runoff (c) and (d) for sediment flux. Colours indicate the mean area of agreement averaged over all scenarios. Tables (e) and (f) show the correlation in total connected area between model pairs in different scenarios for runoff (e) and sediment (f). (g) and (h) show the connected area, averaged over all scenarios, for runoff (g) and sediment (h). E3: Erosion3D, LA: LandSoil, Li: OpenLISEM, Wa: Watershed, Fu: FullSWOF_2D. A black and white version of this figure will appear in some formats. For the colour version, refer to the plate section.

some of the divergence in the comparisons with different connectivity indices, although it also has to be remembered that those indices have different modelling assumptions built into them. Any calibration carried out must be evaluated in the context of how it affects the sub-grid-scale parameterization of connectivity. In the context of modelling surface and subsurface flows on floodplains, Poole et al. 2004; 2008) have used a hierarchical approach to structuring and parameterizing the model and this approach may prove useful by expansion into other catchment settings. For example, Zhang et al. (2020) simulate solute flows in karst in China using a similar approach.

A further explicit parameterization of connectivity in a model is to define an underlying network that represents the structural connectivity of the system. Benda & Dunne (1997a, 1997b) used a network-based model to look at the movement of sediment pulses through a catchment, and Czuba & Foufoula-Georgiou (2014) expanded the approach to produce a model for how mud, sand and gravel travel through a catchment system. One advantage of this approach is that large catchments can be efficiently modelled – in this case by application to the 44,000 km^2 of the Minnesota river. Subsequent applications of the model (Czuba & Foufoula-Georgiou, 2015; Czuba et al., 2017; Gran & Czuba, 2017) show how complex sediment dynamics and emergent behaviour can be represented over large spatial scales. The CAtchment Sediment Connectivity And Delivery (CASCADE) – though not to be confused with the landform-evolution model with the same acronym discussed earlier of Schmitt et al. (2016; Tangi et al., 2019; Bizzi et al., 2021) uses the network approach too (Figure 11.7), and allows sediment from multiple sources to be traced through large catchment systems.

While network-based approaches are most commonly applied in considering channel networks, there is no reason it cannot be extended to other catchment domains. For example, Cossart (2016) and Heckmann & Schwanghart (2013) provide conceptual representations of how other slope processes might be considered from a network perspective, and Passalacqua (2017) provides a similar overview for deltaic environments. From an urban perspective, Jovanovic et al. (2019) have used network-based approaches to understand how urban stormwater flows have evolved as the city of Scotsdale, Arizona, expanded, and Paton & Haacke (2021) use the approach for understanding sources and pathways of diffuse pollution in cities. Further reviews of network-based methods in geomorphology are presented in Voutsa et al. (2021).

Some models integrate channel network and catchment behaviour by using a hybrid approach. The KINEROS2 model defines the network as a series of trapezoidal channels with kinematic planes representing individual hillslopes (Goodrich et al., 2012). Ziegler et al. (2007) used this model to demonstrate the impact of connectivity of field and channel systems in Vietnam, and Wainwright (1996b) used a similar approach to

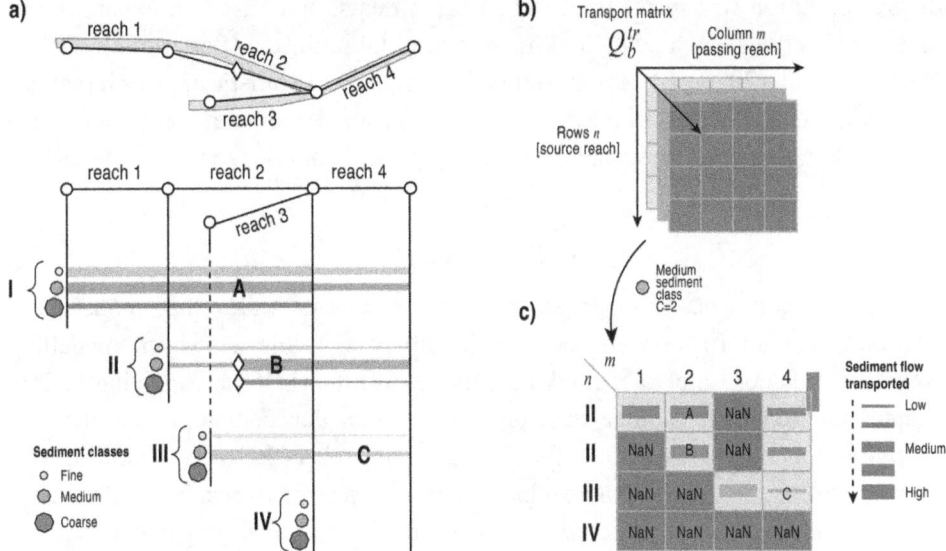

Figure 11.7   Visual representation of the basis of the CASCADE model (Tangi et al., 2019). (a) shows how a simplified network of channel reaches might produce multiple cascades of sediment and how each cascade is made up of different particle sizes (simplified here as fine, medium and coarse grained). In (b), the model data structure is shown as a three-dimensional matrix, showing each reach of the system as the matrix columns, the source reach of each cascade as the matrix rows, repeated for each size class of sediment. An example of the outcome of sediment-flux calculations is shown in (c). The values of m are the reach numbers in Arabic numerals, and n are the source reaches in Roman numerals. The cells A, B and C show how the calculations relate back to the catchment structure in (a). A black and white version of this figure will appear in some formats. For the colour version, refer to the plate section.

look at the impacts of connected badland and agricultural areas in generating an extreme flood in southern France. The COUP2D approach of Michaelides & Wainwright (2002, 2008) uses a cell-based representation of hillslopes and floodplains in a catchment, and links them to the dynamics of flow using a 1D channel-network model. Testing the model using sensitivity analysis and laboratory experiments demonstrates the importance of different surface parameters in different parts of a catchment system, further implying that simple outlet-based calibrations of cell-based models will produce a poor representation of the connectivity processes.

Risk- or probability-based representations of connectivity at catchment scale have also been used. Reaney et al. (2011) developed the Sensitive Catchment Integrated Modelling and Analysis Platform (SCIMAP) model using a variant of TOPMODEL and an erodibility estimate to estimate the risk of sediment production at a point, together with an estimate of probability of delivery to a particular point in the channel network. The SCIMAP tool is commonly used for estimating diffuse pollution risk in relatively large catchments, and has been expanded for other applications, such as the risk from microbial pollution (Porter et al., 2017). Mahoney et al. (2020a, b) have developed a similar probabilistic approach that has an explicit representation of the channel network, and demonstrates the importance of understanding both slope and channel components of connectivity in detail.

11.5 Conclusions

General developments in open science and open data mean that models and environmental information are increasingly freely available. However, modelling connectivity is not simply about taking a model 'off the shelf' and applying it. This chapter has emphasized a three-stage process, in which conceptualization of the connectivity problem takes the fore. Two broad end-members have been outlined, following Nunes et al. (2018) that can broadly be categorized as relating to the use of modelling for assessing connectivity and the use of connectivity to guide modelling. The former can be used to address questions of both structural and functional connectivity, and the ways in which there are feedbacks between them. Thus, it is useful when considering the evolution of system behaviour. The latter tends to impose a static, structural connectivity on the system and can be used to evaluate functional connectivity, but is less suited for looking at system dynamics and structural-functional feedbacks. Alternatively, information from field observations can be used to downscale to provide information about the connectivity of coarser scale results. Care is needed in ensuring that there is not a circular argument in the interpretation of connectivity that has been imposed on the conceptual structure of the approach taken.

Secondly, there should be a strong emphasis on data acquisition and interpretation in the modelling process. Although purely hypothetical models may be useful in the development of connectivity theory, that theory will only be geomorphologically useful if confronted by evidence from the real world. Equally, developments in high-resolution data collection mean that information underpinning connectivity modelling can be better developed at least from a topographic and roughness perspective. Similar developments in relation to other parameters are needed, though, as there is commonly a mismatch between data support for different parameters. Given issues of equifinality with making the links between the different data sources, this is a critical issue. The general dictum in relation to computation of Garbage in-Garbage out holds here.

The third element is to consider the model structure and how it fits into the conceptual and data contexts of the research questions to be answered. The structure affects ways in which the connectivity dynamics and system evolution can be addressed, and in so doing consider the links between structural and functional connectivity. It should be recognized that there is not a one-size-fits-all strategy to be taken, and in some instances very simple models may be appropriate for answering specific questions. They may even be better in some circumstances in propagating less error through the model system. Model complexity and related data complexity need to be closely interrelated and developed in the process of deciding how connectivity science is informed by but also informs our understanding of geomorphic systems.

References

Antoine, M., Javaux, M., & Bielders, C., 2009. What indicators can capture runoff-relevant connectivity properties of the micro-topography at the plot scale? *Advances in Water Resources*. 32, 1297–1310. https://doi.org/10.1016/j.advwatres.2009.05.006

Antoine, M., Javaux, M., & Bielders, C. L., 2011. Integrating subgrid connectivity properties of the micro-topography in distributed runoff models, at the interrill scale. *Journal of Hydrology*. 403, 213–223. https://doi.org/10.1016/j.jhydrol.2011.03.027

Arrouays, D., Poggio, L., Salazar Guerrero, O. A., & Mulder, V. L., 2020. Digital soil mapping and GlobalSoilMap. Main advances and ways forward. *Geoderma Regional*. 21, e00265. https://doi.org/10.1016/j.geodrs.2020.e00265

Atkinson, J., de Clercq, W., & Rozanov, A., 2020. Multi-resolution soil-landscape characterisation in KwaZulu Natal: Using geomorphons to classify local soilscapes for improved digital geomorphological modelling. *Geoderma Regional*. 22, e00291. https://doi.org/10.1016/j.geodrs.2020.e00291

Baartman, J. E. M., Nunes, J. P., Masselink, R., Darboux, F., Bielders, C., Degré, A., Cantreul, V., Cerdan, O., Grangeon, T., Fiener, P., et al. 2020. What do models tell us about water and sediment connectivity? *Geomorphology*. 367, 107300. https://doi.org/10.1016/j.geomorph.2020.107300

Bedau, M. A., 1997. Weak emergence. *Noûs*. 31, 375–399. https://doi.org/10.1111/0029-4624.31.s11.17

Behrens, T., Schmidt, K., Ramirez-Lopez, L., Gallant, J., Zhu, A.-X., & Scholten, T., 2014. Hyper-scale digital soil mapping and soil formation analysis. *Geoderma*. 213, 578–588. https://doi.org/10.1016/j.geoderma.2013.07.031

Benda, L., & Dunne, T., 1997a. Stochastic forcing of sediment routing and storage in channel networks. *Water Resources Research*. 33, 2865–2880. https://doi.org/10.1029/97WR02387

Benda, L., & Dunne, T., 1997b. Stochastic forcing of sediment supply to channel networks from landsliding and debris flow. *Water Resources Research*. 33, 2849–2863. https://doi.org/10.1029/97WR02388

Beven, K., 2001. How far can we go in distributed hydrological modelling? *Hydrology and Earth System Sciences*. 5, 1–12. https://doi.org/10.5194/hess-5-1-2001

Beven, K. J., 2000. Uniqueness of place and process representations in hydrological modelling. *Hydrology and Earth System Sciences*. 4, 203–213. https://doi.org/10.5194/hess-4-203-2000

Bizzi, S., Tangi, M., Schmitt, R. J. P., Pitlick, J., Piégay, H., & Castelletti, A. F., 2021. Sediment transport at the network scale and its link to channel morphology in the braided Vjosa River system. *Earth Surface Processes and Landforms*. 46, 2946–2962. https://doi.org/10.1002/esp.5225

Blöschl, G., 1999. Scaling issues in snow hydrology. *Hydrological Processes*. 13, 2149–2175. https://doi.org/10.1002/(SICI)1099-1085(199910)13:14/15<2149::AID-HYP847>3.0.CO;2-8

Blöschl, G., Grayson, R. B., & Sivapalan, M., 1995. On the representative elementary area (REA) concept and its utility for distributed rainfall-runoff modelling. *Hydrological Processes.* 9, 313–330. https://doi.org/10.1002/hyp.3360090307

Blöschl, G., & Sivapalan, M., 1995. Scale issues in hydrological modelling: A review. *Hydrological Processes.* 9, 251–290. https://doi.org/10.1002/hyp.3360090305

Bogen, J., & Woodward, J., 1988. Saving the phenomena. *The Philosophical Review.* 97, 303. https://doi.org/10.2307/2185445

Bonfatti, B. R., Demattê, J. A. M., Marques, K. P. P., Poppiel, R. R., Rizzo, R., Mendes, W. de S., Silvero, N. E. Q., & Safanelli, J. L., 2020. Digital mapping of soil parent material in a heterogeneous tropical area. *Geomorphology.* 367, 107305. https://doi.org/10.1016/j.geomorph.2020.107305

Braun, J., & Sambridge, M., 1997. Modelling landscape evolution on geological time scales: A new method based on irregular spatial discretization. *Basin Research.* 9, 27–52. https://doi.org/10.1046/j.1365-2117.1997.00030.x

Callaghan, K. L., & Wickert, A. D., 2019. Computing water flow through complex landscapes – Part 1: Incorporating depressions in flow routing using FlowFill. *Earth Surface Dynamics.* 7, 737–753. https://doi.org/10.5194/esurf-7-737-2019

Casas, A., Benito, G., Thorndycraft, V. R., & Rico, M., 2006. The topographic data source of digital terrain models as a key element in the accuracy of hydraulic flood modelling. *Earth Surface Processes and Landforms.* 31, 444–456. https://doi.org/10.1002/esp.1278

Caviedes-Voullième, D., Ahmadinia, E., & Hinz, C., 2021. Interactions of microtopography, slope and infiltration cause complex rainfall-runoff behavior at the hillslope scale for single rainfall events. *Water Resources Research.* 57. https://doi.org/10.1029/2020WR028127

Clark, M. P., Bierkens, M. F. P., Samaniego, L., Woods, R. A., Uijlenhoet, R., Bennett, K. E., Pauwels, V. R. N., Cai, X., Wood, A. W., & Peters-Lidard, C. D., 2017. The evolution of process-based hydrologic models: historical challenges and the collective quest for physical realism. *Hydrology and Earth System Sciences.* 21, 3427–3440. https://doi.org/10.5194/hess-21-3427-2017

Cooper, J. R., Wainwright, J., Parsons, A. J., Onda, Y., Fukuwara, T., Obana, E., Kitchener, B., Long, E. J., & Hargrave, G. H., 2012. A new approach for simulating the redistribution of soil particles by water erosion: A marker-in-cell model: Soil erosion, marker-in-cell model. *Journal of Geophysical Research.* 117. https://doi.org/10.1029/2012JF002499

Cossart, E., 2016. L'(in)efficacité géomorphologique des cascades sédimentaires en question: les apports d'une analyse réseau. *Cybergeo: European Journal of Geography.* https://doi.org/10.4000/cybergeo.27625

Coulthard, T. J., Macklin, M. G., & Kirkby, M. J., 2002. A cellular model of Holocene upland river basin and alluvial fan evolution. *Earth Surface Processes and Landforms.* 27, 269–288. https://doi.org/10.1002/esp.318

Coulthard, T. J., Neal, J. C., Bates, P. D., Ramirez, J., de Almeida, G. A. M., & Hancock, G. R., 2013. Integrating the LISFLOOD-FP 2D hydrodynamic model with the CAESAR model: implications for modelling landscape evolution: Integrating hydrodynamics in landscape evolution models. *Earth Surface Processes and Landforms.* 38, 1897–1906. https://doi.org/10.1002/esp.3478

Crave, A., & Davy, P., 2001. A stochastic "precipiton" model for simulating erosion/sedimentation dynamics. *Computers & Geosciences.* 27, 815–827. https://doi.org/10.1016/S0098-3004(00)00167-9

Crema, S., Llena, M., Calsamiglia, A., Estrany, J., Marchi, L., Vericat, D., & Cavalli, M., 2020. Can inpainting improve digital terrain analysis? Comparing techniques for void filling, surface reconstruction and geomorphometric analyses. *Earth Surface Processes and Landforms*. 45, 736–755. https://doi.org/10.1002/esp.4739

Czuba, J. A., & Foufoula-Georgiou, E., 2015. Dynamic connectivity in a fluvial network for identifying hotspots of geomorphic change. *Water Resources Research*. 51, 1401–1421. https://doi.org/10.1002/2014WR016139

Czuba, J. A., & Foufoula-Georgiou, E., 2014. A network-based framework for identifying potential synchronizations and amplifications of sediment delivery in river basins. *Water Resources Research*. 50, 3826–3851. https://doi.org/10.1002/2013WR014227

Czuba, J. A., Foufoula-Georgiou, E., Gran, K. B., Belmont, P., & Wilcock, P. R., 2017. Interplay between spatially explicit sediment sourcing, hierarchical river-network structure, and in-channel bed material sediment transport and storage dynamics. *Journal of Geophysical Research*. 122, 1090–1120. https://doi.org/10.1002/2016JF003965

Darby, S. E., Trieu, H. Q., Carling, P. A., Sarkkula, J., Koponen, J., Kummu, M., Conlan, I., & Leyland, J., 2010. A physically based model to predict hydraulic erosion of fine-grained riverbanks: The role of form roughness in limiting erosion. *Journal of Geophysical Research*. 115, F04003. https://doi.org/10.1029/2010JF001708

Defina, A., 2000. Two-dimensional shallow flow equations for partially dry areas. *Water Resources Research*. 36, 3251–3264. https://doi.org/10.1029/2000WR900167

Edmonds, B., Le Page, C., Bithell, M., Chattoe-Brown, E., Grimm, V., Meyer, R., Montañola-Sales, C., Ormerod, P., Root, H., & Squazzoni, F., 2019. Different modelling purposes. *Journal of Artificial Societies and Social Simulation*. 22, 6. https://doi.org/10.18564/jasss.3993

England, C. B., Onstad, C. A., 1968. Isolation and characterization of hydrologic response units within agricultural watersheds. *Water Resources Research*. 4, 73–77. https://doi.org/10.1029/WR004i001p00073

Epstein, J. M., 2008. Why model? *Journal of Artificial Societies and Social Simulation*. 11, 12.

Fan, Y., & Bras, R. L., 1995. On the concept of a representative elementary area in catchment runoff. *Hydrological Processes*. 9, 821–832. https://doi.org/10.1002/hyp.3360090708

Fatichi, S., Katul, G. G., Ivanov, V. Y., Pappas, C., Paschalis, A., Consolo, A., Kim, J., & Burlando, P., 2015. Abiotic and biotic controls of soil moisture spatiotemporal variability and the occurrence of hysteresis. *Water Resources Research*. 51, 3505–3524. https://doi.org/10.1002/2014WR016102

Favis-Mortlock, D., 1998. A self-organizing dynamic systems approach to the simulation of rill initiation and development on hillslopes. *Computers & Geosciences*. 24, 353–372. https://doi.org/10.1016/S0098-3004(97)00116-7

Favis-Mortlock, D. T., Boardman, J., Parsons, A. J., & Lascelles, B., 2000. Emergence and erosion: A model for rill initiation and development. *Hydrological Processes*. 14, 2173–2205. https://doi.org/10.1002/1099-1085(20000815/30)14:11/12<2173::AID-HYP61>3.0.CO;2-6

Flügel, W.-A., 1995. Delineating hydrological response units by geographical information system analyses for regional hydrological modelling using PRMS/MMS in the drainage basin of the River Bröl, Germany. *Hydrological Processes*. 9, 423–436. https://doi.org/10.1002/hyp.3360090313

Forrest, S. B., & Haff, P. K., 1992. Mechanics of wind ripple stratigraphy. *Science*. 255, 1240–1243. https://doi.org/10.1126/science.255.5049.1240

Frigg, R., Thompson, E., & Werndl, C., 2015. Philosophy of climate science part I: Observing climate change: Observing climate change. *Philosophy Compass*. 10, 953–964. https://doi.org/10.1111/phc3.12294

Goodrich, D. C., Burns, I. S., Unkrich, C. L., Semmens, D. J., Guertin, D. P., Hernandez, M., Yatheendradas, S., Kennedy, J. R., & Levick, L. R., 2012. KINEROS2/AGWA: Model use, calibration, and validation. *Transactions of the ASABE.* 55, 1561–1574. https://doi.org/10.13031/2013.42264

Gran, K. B., & Czuba, J. A., 2017. Sediment pulse evolution and the role of network structure. *Geomorphology.* 277, 17–30. https://doi.org/10.1016/j.geomorph.2015.12.015

Haff, P. K., 2001. Waterbots, In Harmon, R. S., & Doe, W. W. (eds.), *Landscape Erosion and Evolution Modeling.* Springer US: Boston, MA, 239–275. https://doi.org/10.1007/978-1-4615-0575-4_9

Harel, M.-A., & Mouche, E., 2014. Is the connectivity function a good indicator of soil infiltrability distribution and runoff flow dimension?: Connectivity function. *Earth Surface Processes and Landforms.* https://doi.org/10.1002/esp.3604

Harel, M.-A., & Mouche, E., 2013. 1-D steady state runoff production in light of queuing theory: Heterogeneity, connectivity, and scale: 1-D Runoff production and queuing theory. *Water Resources Research.* 49, 7973–7991. https://doi.org/10.1002/2013WR013596

Hawkins, R. H., & Cundy, T. W., 1987. Steady-state analysis of infiltration and overland flow for spatially-varied hillslopes. *Journal of the American Water Resources Association.* 23, 251–256. https://doi.org/10.1111/j.1752-1688.1987.tb00804.x

Heckmann, T., & Schwanghart, W., 2013. Geomorphic coupling and sediment connectivity in an alpine catchment – Exploring sediment cascades using graph theory. *Geomorphology.* 182, 89–103. https://doi.org/10.1016/j.geomorph.2012.10.033

Her, Y., Heatwole, C. D., & Kang, M. S., 2015. Interpolating SRTM elevation data to higher resolution to improve hydrologic analysis. *Journal of the American Water Resources Association.* 51, 1072–1087. https://doi.org/10.1111/jawr.12287

Hubbert, M. K. 1956. Darcy's law and the field equations of the flow of underground fluids. *Transactions of the AIME.* 207, 222–239.

Hughes, J. D., Decker, J. D., & Langevin, C. D., 2011. Use of upscaled elevation and surface roughness data in two-dimensional surface water models. *Advances in Water Resources.* 34, 1151–1164. https://doi.org/10.1016/j.advwatres.2011.02.004

Hutton, C., Nicholas, A., & Brazier, R., 2014. Sub-grid scale parameterization of hillslope runoff and erosion processes for catchment-scale models of semi-arid landscapes: Sub-grid scale runoff and erosion modelling. *Hydrological Processes.* 28, 1713–1721. https://doi.org/10.1002/hyp.9712

Ivanov, V. Y., Fatichi, S., Jenerette, G. D., Espeleta, J. F., Troch, P. A., & Huxman, T. E., 2010. Hysteresis of soil moisture spatial heterogeneity and the "homogenizing" effect of vegetation: Soil moisture spatial heterogeneity. *Water Resources Research.* 46. https://doi.org/10.1029/2009WR008611

Jovanovic, T., Hale, R. L., Gironás, J., & Mejia, A., 2019. Hydrological functioning of an evolving urban stormwater network. *Water Resources Research.* 55, 6517–6533. https://doi.org/10.1029/2019WR025236

Kim, J., Dwelle, M. C., Kampf, S. K., Fatichi, S., & Ivanov, V. Y., 2016. On the non-uniqueness of the hydro-geomorphic responses in a zero-order catchment with respect to soil moisture. *Advances in Water Resources.* 92, 73–89. https://doi.org/10.1016/j.advwatres.2016.03.019

Kirkby, M. J., 1987. Models in physical geography, In Clark, M. J., Gregory, K. J., Gurnell, A. M. (eds.), *Horizons in Physical Geography.* MacMillan, Basingstoke, 47–61.

Kitcher, P., 1995. *The Advancement of Science.* Oxford University Press, Oxford. https://doi.org/10.1093/0195096533.001.0001

Lane, S. N., 2005. Roughness – time for a re-evaluation? *Earth Surface Processes and Landforms.* 30, 251–253. https://doi.org/10.1002/esp.1208

Lane, S. N., Bakker, M., Gabbud, C., Micheletti, N., & Saugy, J.-N., 2017. Sediment export, transient landscape response and catchment-scale connectivity following rapid climate warming and Alpine glacier recession. *Geomorphology.* 277, 210–227. https://doi.org/10.1016/j.geomorph.2016.02.015

Leavesley, G. H., Lichty, R. W., Troutman, B. M., & Saindon, L. G., 1983. *Precipitation-Runoff Modeling System: User's Manual (No. 83–4238).* Water-Resources Investigations Report. Denver, CO.

Lisenby, P. E., & Fryirs, K. A., 2017. 'Out with the Old?' Why coarse spatial datasets are still useful for catchment-scale investigations of sediment (dis)connectivity. *Earth Surface Processes and Landforms.* 42, 1588–1596. https://doi.org/10.1002/esp.4131

Mahoney, D. T., Fox, J., Al-Aamery, N., & Clare, E., 2020a. Integrating connectivity theory within watershed modelling part I: Model formulation and investigating the timing of sediment connectivity. *Science of the Total Environment.* 740, 140385. https://doi.org/10.1016/j.scitotenv.2020.140385

Mahoney, D. T., Fox, J., Al-Aamery, N., & Clare, E., 2020b. Integrating connectivity theory within watershed modelling part II: Application and evaluating structural and functional connectivity. *Science of the Total Environment.* 740, 140386. https://doi.org/10.1016/j.scitotenv.2020.140386

McBratney, A. B., Mendonça Santos, M. L., & Minasny, B., 2003. On digital soil mapping. *Geoderma.* 117, 3–52. https://doi.org/10.1016/S0016-7061(03)00223-4

McGough, A. S., Liang, S., Rapoportas, M., Grey, R., Vinod, G. K., Maddy, D., Trueman, A., & Wainwright, J., 2012. Massively parallel landscape-evolution modelling using general purpose graphical processing units. In *2012 19th International Conference on High Performance Computing.* Presented at the 2012 19th International Conference on High Performance Computing (HiPC), IEEE, Pune, India, pp. 1–10. https://doi.org/10.1109/HiPC.2012.6507488

Mewes, B., & Schumann, A. H., 2018. IPA (v1): a framework for agent-based modelling of soil water movement. *Geoscientific Model Development.* 11, 2175–2187. https://doi.org/10.5194/gmd-11-2175-2018

Michaelides, K., & Wainwright, J., 2008. Internal testing of a numerical model of hillslope–channel coupling using laboratory flume experiments. *Hydrological Processes.* 22, 2274–2291. https://doi.org/10.1002/hyp.6823

Michaelides, K., & Wainwright, J., 2002. Modelling the effects of hillslope-channel coupling on catchment hydrological response. *Earth Surface Processes and Landforms.* 27, 1441–1457. https://doi.org/10.1002/esp.440

Millares-Valenzuela, A., Eekhout, J. P. C., Martínez-Salvador, A., García-Lorenzo, R., Pérez-Cutillas, P., & Conesa-García, C., 2022. Evaluation of sediment connectivity through physically-based erosion modeling of landscape factor at the event scale. *CATENA.* 213, 106165. https://doi.org/10.1016/j.catena.2022.106165

Minasny, B., & McBratney, Alex.B., 2016. Digital soil mapping: A brief history and some lessons. *Geoderma.* 264, 301–311. https://doi.org/10.1016/j.geoderma.2015.07.017

Mulder, V. L., Lacoste, M., Richer-de-Forges, A. C., & Arrouays, D., 2016. GlobalSoilMap France: High-resolution spatial modelling the soils of France up to two meter depth. *Science of the Total Environment.* 573, 1352–1369. https://doi.org/10.1016/j.scitotenv.2016.07.066

Müller, E. N., 2007. *Scaling Approaches to the Modelling of Water, Sediment and Nutrient Fluxes within Semi-arid Landscapes, Jornada Basin, New Mexico.* Logos-Verl, Berlin.

Mulligan, M., & Wainwright, J., 2013. Modelling and model building, In Wainwright, J., Mulligan, M. (eds.), *Environmental Modelling.* John Wiley & Sons, Ltd: Chichester, UK, 7–26. https://doi.org/10.1002/9781118351475.ch2

Murray, A. B., & Paola, C., 1994. A cellular model of braided rivers. *Nature.* 371, 54–57. https://doi.org/10.1038/371054a0

Neill, A. J., Tetzlaff, D., Strachan, N. J. C., Hough, R. L., Avery, L. M., Kuppel, S., Maneta, M. P., & Soulsby, C., 2020a. An agent-based model that simulates the spatio-temporal dynamics of sources and transfer mechanisms contributing faecal indicator organisms to streams. Part 1: Background and model description. *Journal of Environmental Management.* 270, 110903. https://doi.org/10.1016/j.jenvman.2020.110903

Neill, A. J., Tetzlaff, D., Strachan, N. J. C., Hough, R. L., Avery, L. M., Maneta, M. P., & Soulsby, C., 2020b. An agent-based model that simulates the spatio-temporal dynamics of sources and transfer mechanisms contributing faecal indicator organisms to streams. Part 2: Application to a small agricultural catchment. *Journal of Environmental Management.* 270, 110905. https://doi.org/10.1016/j.jenvman.2020.110905

Newman, B. D., Wilcox, B. P., Archer, S. R., Breshears, D. D., Dahm, C. N., Duffy, C. J., McDowell, N. G., Phillips, F. M., Scanlon, B. R., & Vivoni, E. R., 2006. Ecohydrology of water-limited environments: A scientific vision: Opinion. *Water Resources Research.* 42. https://doi.org/10.1029/2005WR004141

Nunes, J. P., Wainwright, J., Bielders, C. L., Darboux, F., Fiener, P., Finger, D., Turnbull, L., & Team, C. W. T.-T., 2018. Better models are more effectively connected models. *Earth Surface Processes and Landforms.* 43, 1355–1360. https://doi.org/10.1002/esp.4323

Oreskes, N., Shrader-Frechette, K., & Belitz, K., 1994. Verification, validation, and confirmation of numerical models in the earth sciences. *Science.* 263, 641–646. https://doi.org/10.1126/science.263.5147.641

Özgen, I., Teuber, K., Simons, F., Liang, D., & Hinkelmann, R., 2015. Upscaling the shallow water model with a novel roughness formulation. *Environmental Earth Sciences.* 74, 7371–7386. https://doi.org/10.1007/s12665-015-4726-7

Parsons, A. J., Abrahams, A. D., & Wainwright, J., 1994. On determining resistance to interrill overland flow. *Water Resources Research.* 30, 3515–3521. https://doi.org/10.1029/94WR02176

Parsons, A. J., Wainwright, J., Abrahams, A. D., & Simanton, J. R., 1997. Distributed dynamic modelling of interrill overland flow. *Hydrological Processes.* 11, 1833–1859. https://doi.org/10.1002/(SICI)1099-1085(199711)11:14<1833::AID-HYP499>3.0.CO;2-7

Passalacqua, P., 2017. The Delta Connectome: A network-based framework for studying connectivity in river deltas. *Geomorphology.* 277, 50–62. https://doi.org/10.1016/j.geomorph.2016.04.001

Passalacqua, P., Belmont, P., Staley, D. M., Simley, J. D., Arrowsmith, J. R., Bode, C. A., Crosby, C., et al. 2015. Analyzing high resolution topography for advancing the understanding of mass and energy transfer through landscapes: A review. *Earth-Science Reviews.* 148, 174–193. https://doi.org/10.1016/j.earscirev.2015.05.012

Paton, E., & Haacke, N., 2021. Merging patterns and processes of diffuse pollution in urban watersheds: A connectivity assessment. *WIREs Water.* 8. https://doi.org/10.1002/wat2.1525

Peñuela, A., Javaux, M., & Bielders, C. L., 2015. How do slope and surface roughness affect plot-scale overland flow connectivity? *Journal of Hydrology.* 528, 192–205. https://doi.org/10.1016/j.jhydrol.2015.06.031

Peñuela, A., Javaux, M., & Bielders, C. L., 2013. Scale effect on overland flow connectivity at the plot scale. *Hydrology and Earth System Sciences.* 17, 87–101. https://doi.org/10.5194/hess-17-87-2013

Poblete, D., Arevalo, J., Nicolis, O., & Figueroa, F., 2020. Optimization of hydrologic response units (HRUs) using gridded meteorological data and spatially varying parameters. *Water.* 12, 3558. https://doi.org/10.3390/w12123558

Pöppl, R. E., & Parsons, A. J., 2018. The geomorphic cell: A basis for studying connectivity: The geomorphic cell. *Earth Surface Processes and Landforms.* 43, 1155–1159. https://doi.org/10.1002/esp.4300

Poole, G., Stanford, J., Running, S., Frissell, C., Woessner, W., & Ellis, B., 2004. A patch hierarchy approach to modeling surface and subsurface hydrology in complex floodplain environments. *Earth Surface Processes and Landforms.* 29, 1259–1274. https://doi.org/10.1002/esp.1091

Poole, G. C., O'Daniel, S. J., Jones, K. L., Woessner, W. W., Bernhardt, E. S., Helton, A. M., Stanford, J. A., Boer, B. R., & Beechie, T. J., 2008. Hydrologic spiralling: the role of multiple interactive flow paths in stream ecosystems. *River Research and Applications.* 24, 1018–1031. https://doi.org/10.1002/rra.1099

Porter, K. D. H., Reaney, S. M., Quilliam, R. S., Burgess, C., & Oliver, D. M., 2017. Predicting diffuse microbial pollution risk across catchments: The performance of SCIMAP and recommendations for future development. *Science of the Total Environment.* 609, 456–465. https://doi.org/10.1016/j.scitotenv.2017.07.186

Reaney, S. M., 2008. The use of agent based modelling techniques in hydrology: Determining the spatial and temporal origin of channel flow in semi-arid catchments. *Earth Surface Processes and Landforms.* 33, 317–327. https://doi.org/10.1002/esp.1540

Reaney, S. M., Bracken, L. J., & Kirkby, M. J., 2014. The importance of surface controls on overland flow connectivity in semi-arid environments: results from a numerical experimental approach: Surface controls on overland flow connectivity. *Hydrological Processes.* 28, 2116–2128. https://doi.org/10.1002/hyp.9769

Reaney, S. M., Bracken, L. J., & Kirkby, M. J., 2007. Use of the connectivity of runoff model (CRUM) to investigate the influence of storm characteristics on runoff generation and connectivity in semi-arid areas. *Hydrological Processes.* 21, 894–906. https://doi.org/10.1002/hyp.6281

Reaney, S. M., Lane, S. N., Heathwaite, A. L., & Dugdale, L. J., 2011. Risk-based modelling of diffuse land use impacts from rural landscapes upon salmonid fry abundance. *Ecological Modelling.* 222, 1016–1029. https://doi.org/10.1016/j.ecolmodel.2010.08.022

Refsgaard, J. C., Højberg, A. L., He, X., Hansen, A. L., Rasmussen, S. H., & Stisen, S., 2016. Where are the limits of model predictive capabilities? Representative elementary scale – RES. *Hydrological Processes.* 30, 4956–4965. https://doi.org/10.1002/hyp.11029

Renard, P., & Allard, D., 2013. Connectivity metrics for subsurface flow and transport. *Advances in Water Resources.* 51, 168–196. https://doi.org/10.1016/j.advwatres.2011.12.001

Reulier, R., Delahaye, D., Caillault, S., Viel, V., Douvinet, J., & Bensaid, A., 2016. Mesurer l'impact des entités linéaires paysagères sur les dynamiques spatiales du ruissellement : une approche par simulation multi-agents. *Cybergeo: European Journal of Geography.* https://doi.org/10.4000/cybergeo.27768

Reulier, R., Delahaye, D., & Viel, V., 2019. Agricultural landscape evolution and structural connectivity to the river for matter flux, a multi-agents simulation approach. *CATENA.* 174, 524–535. https://doi.org/10.1016/j.catena.2018.11.036

Reulier, R., Delahaye, D., Viel, V., & Davidson, R., 2017. Connectivité hydro-sédimentaire dans un petit bassin versant agricole du nord-ouest de la France : de l'expertise de terrain à la modélisation par Système Multi-Agent. *Géomorphologie : Relief, Processus, Environnement.* 23, 327–340. https://doi.org/10.4000/geomorphologie.11857

Schmitt, R. J. P., Bizzi, S., & Castelletti, A., 2016. Tracking multiple sediment cascades at the river network scale identifies controls and emerging patterns of sediment connectivity: Tracking multiple sediment cascades at river network scale. *Water Resources Research.* 52, 3941–3965. https://doi.org/10.1002/2015WR018097

Singh, M., Tandon, S. K., & Sinha, R., 2017. Assessment of connectivity in a water-stressed wetland (Kaabar Tal) of Kosi-Gandak interfan, north Bihar Plains, India: Connectivity response units in a wetland. *Earth Surface Processes and Landforms.* 42, 1982–1996. https://doi.org/10.1002/esp.4156

Skidmore, E. L., 1997. Comment on chain method for measuring soil roughness. *Soil Science Society of America Journal.* 61, 1532–1533. https://doi.org/10.2136/sssaj1997.03615995006100050034x

Smith, M. W., 2014. Roughness in the earth sciences. *Earth-Science Reviews.* 136, 202–225. https://doi.org/10.1016/j.earscirev.2014.05.016

Smith, R., 1991. The application of cellular automata to the erosion of landforms. *Earth Surface Processes and Landforms.* 16, 273–281. https://doi.org/10.1002/esp.3290160307

Spence, C., & Hosler, J., 2007. Representation of stores along drainage networks in heterogenous landscapes for runoff modelling. *Journal of Hydrology.* 347, 474–486. https://doi.org/10.1016/j.jhydrol.2007.09.035

Suppes, P. 1962. Models of data. In Nagel, E., Suppes, P., Tarski, A. (eds) *Logic, Methodology and Philosophy of Science: Proceedings of the 1960 International Congress,* 252–261. Stanford University Press, Stanford, CA.

Tangi, M., Schmitt, R., Bizzi, S., & Castelletti, A., 2019. The CASCADE toolbox for analyzing river sediment connectivity and management. *Environmental Modelling & Software.* 119, 400–406. https://doi.org/10.1016/j.envsoft.2019.07.008

Thomsen, L. M., Baartman, J. E. M., Barneveld, R. J., Starkloff, T., & Stolte, J., 2015. Soil surface roughness: Comparing old and new measuring methods and application in a soil erosion model. *SOIL.* 1, 399–410. https://doi.org/10.5194/soil-1-399-2015

Tromp-van Meerveld, H. J., & McDonnell, J. J., 2006. Threshold relations in subsurface stormflow: 2. The fill and spill hypothesis: Threshold flow relations, 2. *Water Resources Research.* 42. https://doi.org/10.1029/2004WR003800

Turnbull, L., Hütt, M.-T., Ioannides, A. A., Kininmonth, S., Pöppl, R., Tockner, K., Bracken, L. J., et al. 2018. Connectivity and complex systems: Learning from a multi-disciplinary perspective. *Applied Network Science.* 3, 11. https://doi.org/10.1007/s41109-018-0067-2

Turnbull, L., & Wainwright, J., 2019. From structure to function: Understanding shrub encroachment in drylands using hydrological and sediment connectivity. *Ecological Indicators.* 98, 608–618. https://doi.org/10.1016/j.ecolind.2018.11.039

Voutsa, V., Battaglia, D., Bracken, L. J., Brovelli, A., Costescu, J., Díaz Muñoz, M., Fath, B. et al., 2021. Two classes of functional connectivity in dynamical processes in networks. *The Journal of the Royal Society Interface.* 18, 20210486. https://doi.org/10.1098/rsif.2021.0486

Wainwright, J., 2015. Stability and instability in Mediterranean landscapes: A geoarchaeological perspective, In Dykes, A. P., Mulligan, M., Wainwright, J. (eds.), *Monitoring and Modelling Dynamic Environments.* Wiley, Chichester, UK, 22.

Wainwright, J., 2008. Can modelling enable us to understand the rôle of humans in landscape evolution? *Geoforum.* 39, 659–674. https://doi.org/10.1016/j.geoforum.2006.09.011

Wainwright, J., 1996a. Infiltration, runoff and erosion characteristics of agricultural land in extreme storm events, SE France. *CATENA.* 26, 27–47. https://doi.org/10.1016/0341-8162(95)00033-X

Wainwright, J., 1996b. Hillslope response to extreme storm events: The example of the Vaison-la-Romaine event, In Anderson, M. G., Brooks, S. M. (eds.), *Advances in Hillslope Processes.* John Wiley and Sons, Chichester, 997–1026.

Wainwright, J., 1994. Erosion of archaeological sites: Results and implications of a site simulation model. *Geoarchaeology.* 9, 173–201. https://doi.org/10.1002/gea.3340090302

Wainwright, J., & Millington, J. D. A., 2010. Mind, the gap in landscape-evolution modelling. *Earth Surface Processes and Landforms.* 35, 842–855. https://doi.org/10.1002/esp.2008

Wainwright, J., & Parsons, A. J., 2002. The effect of temporal variations in rainfall on scale dependency in runoff coefficients: Temporal variations in rainfall. *Water Resources Research.* 38, 7-1–7-10. https://doi.org/10.1029/2000WR000188

Wainwright, J., Turnbull, L., Ibrahim, T. G., Lexartza-Artza, I., Thornton, S. F., & Brazier, R. E., 2011. Linking environmental régimes, space and time: Interpretations of structural and functional connectivity. *Geomorphology.* 126, 387–404. https://doi.org/10.1016/j.geomorph.2010.07.027

Western, A. W., & Blöschl, G., 1999. On the spatial scaling of soil moisture. *Journal of Hydrology.* 217, 203–224. https://doi.org/10.1016/S0022-1694(98)00232-7

Winsberg, E., 2018. *Philosophy and Climate Science*, 1st ed. Cambridge University Press, Cambridge. https://doi.org/10.1017/9781108164290

Wood, E. F., Sivapalan, M., Beven, K., & Band, L., 1988. Effects of spatial variability and scale with implications to hydrologic modeling. *Journal of Hydrology.* 102, 29–47. https://doi.org/10.1016/0022-1694(88)90090-X

Xu, H., van der Steeg, S., Sullivan, J., Shelley, D., Cely, J. E., Viparelli, E., Lakshmi, V., & Torres, R., 2020. Intermittent channel systems of a low-relief, low-gradient floodplain: Comparison of automatic extraction methods. *Water Resources Research.* 56. https://doi.org/10.1029/2020WR027603

Zhang, X., Drake, N. A., & Wainwright, J., 2013. Spatial modelling and scaling issues, In Wainwright, J., Mulligan, M. (eds.), *Environmental Modelling.* John Wiley & Sons, Ltd, Chichester, UK, 69–90. https://doi.org/10.1002/9781118351475.ch5

Zhang, X., Drake, N. A., Wainwright, J., & Mulligan, M., 1999. Comparison of slope estimates from low resolution DEMs: Scaling issues and a fractal method for their solution. *Earth Surface Processes and Landforms.* 24, 763–779. https://doi.org/10.1002/(SICI)1096-9837(199908)24:9<763::AID-ESP9>3.0.CO;2-J

Zhang, Z., Chen, X., Cheng, Q., Li, S., Yue, F., Peng, T., Waldron, S., Oliver, D. M., & Soulsby, C., 2020. Coupled hydrological and biogeochemical modelling of nitrogen transport in the karst critical zone. *Science of the Total Environment.* 732, 138902. https://doi.org/10.1016/j.scitotenv.2020.138902

Ziegler, A. D., Giambelluca, T. W., Plondke, D., Leisz, S., Tran, L. T., Fox, J., Nullet, M. A., Vogler, J. B., Minh Troung, D., & Tran Duc V., 2007. Hydrological consequences of landscape fragmentation in mountainous northern Vietnam: Buffering of Hortonian overland flow. *Journal of Hydrology.* 337, 52–67. https://doi.org/10.1016/j.jhydrol.2007.01.031

Part IV

Managing Connectivity

12

Agricultural Land

MANUEL LÓPEZ-VICENTE, JESÚS RODRIGO-COMINO,
GAO-LIN WU AND YI-FAN LIU

12.1 Global Land Cover, and Farmland Expansion and Connectivity

Human activities have been modifying the natural dynamics in the earth's systems, and among these modifications, conversion of forests, rangeland, grassland, and wetland into agricultural land and pastures are the most relevant (Poesen, 2018). By combining different methods (e.g., multi-proxy study of lake sediment cores, radiocarbon and radiometric techniques, charcoal, biological proxies (e.g., diatoms, chironomids, and pollen)) it has been shown that the expansion of agriculture and grassland has fostered the occurrence of mudflow, flooding, soil erosion, and sediment connectivity. For instance, the expansion of agriculture in South America in several stages associated with human settlements and technological evolution (Armesto et al., 2010), the maximum expansion of agriculture in Europe during the nineteenth century (Morellón et al., 2011), and the intensification in Southeast Asia (from the sixteenth to the mid nineteenth century) of rice terrace agriculture and demographic changes acted as major catalysts for deforestation, soil loss, and biodiversity change (Findley et al., 2022). Modernization of agriculture during the twentieth-century included an intensive use of fertilizers, herbicides and pesticides, causing nonpoint source pollution of soils and water bodies linked to the movement of sediments and runoff (Martín, 2017). Besides, industrialization of agriculture has led to a change in the C footprint, of N fertilization, including all the life cycle greenhouse gas emissions related to N fertilization (Aguilera et al., 2021). Conversely, in many countries studies of soil and water conservation practices in croplands appeared mostly in the second half of the past century: USA (irrigation-practices; Dreibelbis, 1944), Australia (softwood plantations on gentle slopes; Shoobridge, 1951), Japan (sweet potato plants cultured after tobacco crop; Kawai et al., 1957) and Kenya (cultivation of *Cynodon dactylon* to protect soil; Pereira et al., 1967). All in all, the consequences of these land use and management changes include the expansion and creation of new soil erosion-prone areas, the acceleration of overland

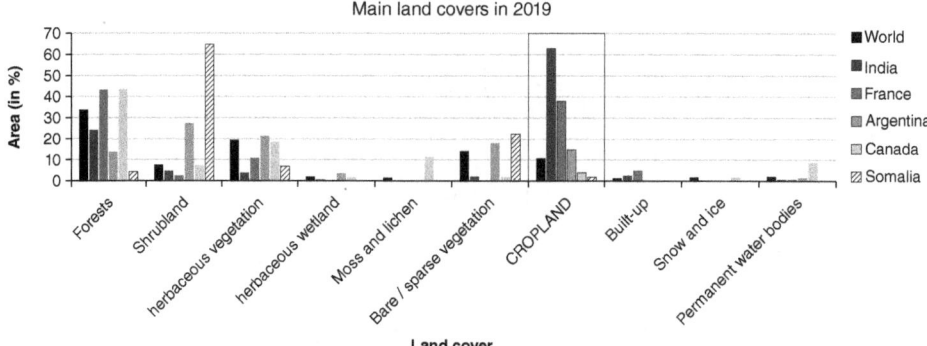

Figure 12.1   Main land covers in the world and in selected countries (Data source: Copernicus, Land monitoring service, Global Land Cover). A black and white version of this figure will appear in some formats. For the colour version, refer to the plate section.

flow during rainfall-runoff events, changes in the natural patterns of sediment connectivity, and overall, an increase in the total amount of sediment that reaches the main water bodies via ephemeral and permanent streams, with serious consequences on ecosystems, human infrastructures and environmental sustainability. Emphasis on the environmental effects of agriculture have focused largely on increased rates of soil erosion. Such increases, and how such increases are measured, are strongly mediated by changes in water and sediment connectivity.

Currently and according to Copernicus Global Land Service (a product of the European Commission that provides bio-geophysical products of global land surface), in 2019, 10.12% of the Earth's surface was cropland (Land Cover Viewer), a significant amount when considering the vast areas covered by deserts, ice caps, high mountain ranges, forests and wetlands (including marshes and swamps). However, differences appear when the percentages of cropland are compared between countries (Figure 12.1).

Depending on plant type, in particular its morphology and phenology, crops can be classified between herbaceous crops (arable land), woody or permanent crops, and grasslands. Based on crop yield and management practices, farmland can be classified as such with extensive and intensive agriculture:

• *Woody crops* can be defined as a class of plants composed of permanent species such as trees and/or shrubs, including all types of orchards and plantations with hard stems that occupy the soil for several years. These crops are not included in the rotation regime, and this fact is relevant in soil erosion dynamics. Woody crops include citrus trees, fruit trees from temperate climate and subtropical climate, berries, nut trees, olive groves, vineyards, nurseries of non-forest woody crops, greenhouse woody crops, and other permanent crops (INE, 2009). The regular planting of trees presents an ordered or geometric distribution, with a determined density that depends on water availability, following parallel rows. Monfreda et al. (2008) estimated that 8.6% of the total global area is harvested by woody crops, for example, Africa (14.4%), Latin America (13.1%), Middle East (12%) and Asia (10.3%).

- *Herbaceous crops* comprise those plants whose air part has herbaceous consistency and are harvested yearly. They include cereals, dried legumes, potatoes, industrial crops (e.g., cotton, oily crops, tobacco, spices), vegetables, decorative flowers and plants, as well as seeds and seedlings destined for sale.
- *Grasslands* are areas in which the vegetation is dominated by a nearly continuous cover of grasses, where the presence of trees, shrubs and other crops is prevented. This type of vegetation covers specific areas in the world, due to climate and topography, and are mainly devoted to forage/fodder crops. There are natural grasslands and artificial grasslands.
- *Extensive agriculture* is characterized by lower crop yields, includes traditional farming and tillage practices, covers large surface areas, and is more environmentally friendly than intensive agriculture. Most rainfed crops can be included in this category.
- *Intensive agriculture* aims to obtain as much crop yield as possible in limited areas by means of using technology, fertilizers, herbicides, pesticides, and irrigation. The most extreme version is agriculture without soil in greenhouses with hydroponic cultivation.

12.2 Sediment Connectivity and Delivery in Agricultural Landscapes

Soil erosion and transport are key geomorphic processes in agricultural areas. The amount of detached soil that is transported, the distance traveled, runoff duration and particle-size selectivity are aspects all included in the process of sediment transport. Over time, different approaches have been presented to conceptually and mathematically assess soil erosion that seriously threatens farmland sustainability due to the irreplaceable loss of fertile soil. One of these concepts is the sediment-delivery ratio which is defined as "the percentage relationship between annual sediment yield and annual gross erosion in the watershed, the percentage being derived with both sediment yield and erosion expressed in tons" (Glymph, 1954). This concept has been widely used to calculate sediment yield and (temporal) sediment storage in catchment systems with variable land uses, including arable lands, abandoned fields, woody crops, scrubland and forests, in combination with soil erosion models, such as with the Revised Universal Soil Loss Equation (RUSLE) (López-Vicente et al., 2011) or Revised Morgon, Morgon and Finney (RMMF) model (López-Vicente and Navas, 2010). However, this concept has some limitations that emerge when both sediment yield and storage are evaluated at medium- and long-term scales (Parsons et al., 2006).

More recently, the concept of "sediment connectivity" is used to assess the water-mediated transfer of soil and sediment particles in landscape systems (see also Chapter 2). The introduction of arable crops, the establishment of land drainage and the widespread intensification and mechanization of agriculture as well as the establishment of different water pathways such as roads induce significant increases in sediment connectivity (Lexartza-Artza & Wainwright, 2011). Wohl et al. (2019) provided a definition of the term "dis-connectivity" that concerns features or processes that are too remote from each other in space or time, so that a change in one component or process does not influence another. In the next sections, we elaborate on sediment transport in the three main cropping systems: Herbaceous, woody, and grasslands.

12.2.1 Sediment Connectivity in Herbaceous Cropping Systems

Regarding the processes of soil detachment and sediment transport in fields with herbaceous crops, the main concern is the occurrence of bare soil for a certain period throughout the year. This situation is especially relevant because it affects the entire field, and thus, for several weeks or months, the unique protection of topsoil against soil particle detachment and transport by raindrop impact and runoff is the presence of harvest residues (e.g., straw). Homogeneous ground cover conditions and the lack of barriers make sediment connectivity totally dependent on different topographic (slope length and gradient, concavity, and convexity), edaphic (soil depth, permeability, and cohesion) and rainfall (magnitude and intensity) parameters, further being influenced by different types of land management (e.g., irrigation, fallowing, number of crops per year/crop rotation) (Table 12.1). Crop rotation does not significantly influence duration of bare soil conditions because most crops included in a crop rotation system share similar phenology.

In French fields cultivated with corn and beet, Nouaim et al. (2022) found that the majority of erosion occurs during fall on fields left bare after harvest, but those fields that maintained a fall and winter cover reduced the soil losses significantly. In Spanish fields with winter barley, López-Vicente et al. (2008) described a more complex scenario, where the interactions between the RUSLE rainfall erosivity, soil erodibility, and cover-management factors explain similar predicted soil losses for the first (from July to October, it includes the harvesting and ploughing tillage practices; 0.25 Mg/ha month) and the third (from November to April; it has the maximum rainfall intensity with the highest number of freeze-thaw cycles, and soil moisture content; 0.24 Mg/ha month) erosive periods in spite of the strong temporal differences in the values of the three RUSLE factors. Intensive agriculture strongly

Table 12.1 *Levels (written in italics) of bare soil (BS) and ground cover (GC) conditions in herbaceous crops as a function of agronomic (number of crops per year, rotation, fallow, rainfed or irrigated) practices, and climatic (semi-arid, humid) conditions.*

	One crop per year			Two crops per year (double-cropping)		
Rainfed: Semi-arid (low biomass)	*Lasting BS: Low GC several months*	*Lasting BS: Low GC several months*	*Low BS: High GC several months*	*not the case*	*not the case*	*not the case*
Rainfed: Humid (high biomass)	*Medium BS: Medium GC a few months*	*Medium BS: Medium GC a few months*	*No BS: Very high GC all months*	*Low BS: High GC several months*	*Low BS: High GC several months*	*No BS: Very high GC all months*
Irrigated (high biomass)	*Medium BS: Medium GC a few months*	*Medium BS: Medium GC a few months*	*not the case*	*Low BS: High GC several months*	*Low BS: High GC several months*	*not the case*
	Rotation: No	Rotation: Yes → Minor temporal changes	Fallow	Rotation: No	Rotation: Yes → Minor temporal changes	Fallow

impacts on soil and groundwater quality, especially in the deep vadose zone, such as Liu et al. (2022) recently observed in a wheat-maize double-cropping field. The study of these processes is complex due to the different transport velocities observed in the alluvial-proluvial zones and the occurrence of temporal nitrogen storage. In semi-arid areas, soil erosion by wind can become an important process of topsoil loss in fields with annual crops. For instance, Buschiazzo et al. (2007) observed important rates of wind erosion and concentrations of nitrogen and phosphorus in the sediment originated from fields located in Pampas of Argentina after seeding sunflower and rye. These authors also found that wind speed influence on the concentration and enrichment of nutrients in the sediments.

Rice cultivation is carried out in paddy fields that are characterized by very low slope gradients, in almost flat areas. Under these conditions, soil erosion is low (<1 Mg/ha yr) and is commonly evaluated by means of assessing suspended solid in runoff samples (Chen et al., 2012). Estimating soil erosion and nutrient losses from surface runoff in paddy fields is essential for the assessment of sustainable rice production and water quality protection. For subtropical China, Zhou et al. (2019) demonstrated that the conversion of traditional manual transplanting to direct seeding increased soil erosion and P losses. The economic aspect of soil erosion and nutrient loss receives priority attention from farmers as higher or lower incomes can be obtained by means of following good or bad management practices. For instance, Addis et al. (2020) carried out a cost-benefit analysis of soil and water conservation (SWC) measures for teff, sorghum and chickpeas fields in the highlands of Ethiopia and found that crop yield increased after adoption of SWC measures and the cost of production inputs (mainly fertilizers) clearly reduced compared with the values obtained in fields without SWC measures. Therefore, agronomic good management practices reduce soil nutrient depletion by means of minimizing soil erosion and sediment transport. In some cases, soil and nutrient do not travel long distances and the predominant spatial scale of soil redistribution is at sub-catchment scale. Neto et al. (2008) studied soil redistribution using ^{137}Cs in soy and corn fields located on slopes, and found great deposition of sediments in the valley, coming from both the soil under soy (23 Mg/ha yr) and corn (21–38 Mg/ha yr) crop. Another soil loss type on arable land is soil removal by crop harvest which occurs when harvesting root and tuber crops. Field observations and measurements have shown that considerable amounts of soil can be removed from the field due to soil sticking to the harvested roots and the export of soil clods during the crop harvest. At European Union (EU) scale, Panagos et al. (2019) estimated that the total soil loss by crop harvesting in sugar beets and potato fields was at ca. 14.7 million tons/yr in an EU-28 arable land estimated at 110 million ha.

12.2.2 Sediment Connectivity in Woody Cropping Systems

When a woody crop plantation is installed, it must be considered that it will permanently stand there, in the same soil; therefore, if land degradation occurs, it will be difficult to recover, for example, soil functions associated with quality and depth. Seeds and fruits are harvested annually by humans or machines and biomass is pruned. Therefore, the soil surface is always receiving impacts by wheel tracks and footsteps, significantly altering connectivity. Agricultural areas with woody crops are most affected by the loss of sediments and water, which generates degradation processes as a consequence of unsustainable land uses and favorable environmental conditions to accelerate erosion rates (García-Ruiz et al., 2015). As a consequence, high erosion rates are recorded that exceed the permitted sustainable levels, which were estimated at around 1.1 Mg/ha yr by Verheijen et al. (2009). Some of the woody crops with the highest rates of erosion are located in temperate climates with steep slopes and nonconsolidated soils on erodible parent materials such as olive groves, vineyards or citrus orchards with values greater than

30 Mg/ha yr (Rodrigo-Comino, 2018; Taguas et al., 2013). Specifically, this vulnerability to water erosion is still unknown for areas where there has been a drastic change in land use to other types of woody crops due to their greater profitability in the market such as the subtropical crops such as mango, cherimoya or avocado (Atucha et al., 2013; Durán Zuazo et al., 2005). However, all of them coincide in the same soil management system: Bare soils, the intensive use of tillage with machinery that compacts the surface and the abuse of chemicals (Liu et al., 2016).

Subsequently, inappropriate management highly affects soil water retention capacity and thus sediment connectivity because macro- and micropores are destroyed after collapsing and water, and even nutrients cannot be stored or absorbed by the roots (Li et al., 2018; Menzies Pluer et al., 2020). This is also an issue to be solved since other soil properties are also directly affected such as bulk density, making the soil heavier and more difficult to be tilled, or aggregate stability (Hakansson & Lipiec, 2000). When the soil is compacted, runoff coefficients tend to rise, and, then, chemicals used to control weeds and vegetation cover can also be transported to water bodies (Ferreira et al., 2020; García-Díaz et al., 2017). Vicente-Vicente et al. (2016) pointed out that woody crops are usually cultivated in soils with low organic matter due to the absence of vegetation cover, which transforms these soils into net sources of CO_2, especially, throughout soil erosion and organic carbon mineralization.

Considering the earlier-mentioned points, it is clear that the transmission of water, sediments nutrients and water pollutants from different locations along the rows and interrows of woody crop plantations will condition soil quality as well as the productivity and quality of woody cropping systems (Figure 12.2). Therefore, applying accurate methods to consistently assess sediment connectivity and dis-connectivity is key to better understand the related mechanisms. In-situ measurements or monitoring techniques, for example, show that connectivity processes are governed by a range of different factors such as the spatial distribution of plants, either following a linear or an irregular pattern. Tillage on bare surfaces redistributes the soil generating sinks of sediments that appear underneath the woody crops and serve the sources in the interrow parts, although they are the highest and lowest local topographical terrain, respectively (Rodrigo-Comino et al., 2018). At the hillslope scale in linear patterns, the lower part can collect sediments from the upper positions, which can be transported along the backslope.

The decision to cultivate on "undisturbed" parts of hillslope morphology or the use of terraces can either decrease (Calsamiglia et al., 2018) or increase sediment connectivity (Pijl et al., 2019a). Another factor is the application of ground covers. The use of spontaneous vegetation, cover crops, mulching, pruning residues, or geotextiles will condition the generation of water ponds, acting as barriers and canalizing surface runoff across the soil profile since they are able to improve soil

Figure 12.2   Spatial and human-decision factors affecting connectivity processes in woody crops. (A) Linear framework plantation in steep slopes at the Mosel Valley (Germany); (B) Irregular framework plantation in steep slopes at the Montes de Málaga (Spain); (C) "Undisturbed" hillslopes cultivated with conventional soil managed olive orchards in the Jaén province (Spain); (D) Agricultural terraces stopping flow connectivity processes on steep slopes in volcanic soils at Canary Islands (Spain); (E) Spontaneous vegetation cover growing along the interrow areas in steep slope vineyards in the Saar Valley (Germany); and, (F) Application of straw mulch in organic farming vineyards plantations in Valencia province (Spain). All pictures were taken by J. Rodrigo-Comino. A black and white version of this figure will appear in some formats. For the colour version, refer to the plate section.

infiltration, soil water retention capacity, organic matter, and aggregate stability (Cerdà et al., 2022; Knapen et al., 2009).

Vineyards are considered one of the most vulnerable woody crops to soil erosion (Rodrigo-Comino, 2018) (Figure 12.3). Structural connectivity at the catchment scale can vary highly due to the regular framework patterns and necessary peripheral infrastructures necessary to drive the tractors. Authors such as Fressard and Cossart (2019) claimed that unexpected behaviors of water and sediment dynamics can result from dysfunctions in the management structures. To assess connectivity processes in vineyards, one of the most commonly used methods is (a) the improved stock unearthing method (ISUM), which uses the graft union as a passive bioindicator of land surface changes (Rodrigo-Comino et al., 2019) combined with connectivity indices (Rodrigo-Comino et al., 2020); (b) structure from motion techniques (López-Vicente and Álvarez, 2018); (c) sediment collection/sampling using devices placed in different slope positions (Rodrigo-Comino et al., 2017); and, (d) the analysis of high-resolution Digital Elevation Models (Pijl et al., 2019b).

In olive orchards, where soil erosion is also commonly detected, especially throughout the Mediterranean strip, this problem is becoming a debate among researchers, farmers and land managers (Fleskens & Stroosnijder, 2007; Gómez et al., 2008). In

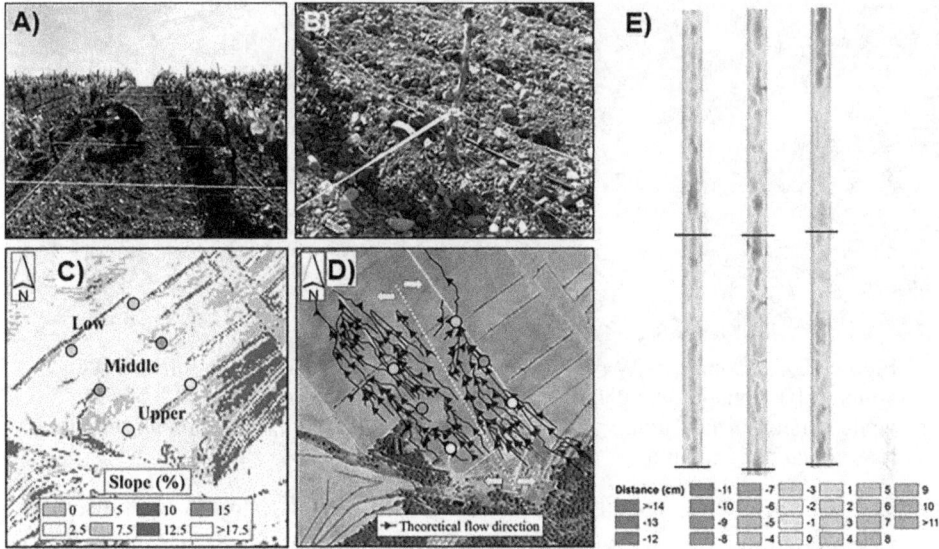

Figure 12.3 Example of ISUM applications and maps to assess connectivity processes. (A) Field campaign measuring cross sectional transects; (B) Detection of the graft union on the vine; (C) Localization and inclination map of three experimental plots at different hillslope positions; (D) Using the tool flow direction of ArcGIS 10.5 (ESRI, USA) to contextualize the resulting ISUM map; and, (E) Examples of ISUM maps. All images were taken/generated by J. Rodrigo-Comino. A black and white version of this figure will appear in some formats. For the colour version, refer to the plate section.

Italy, specifically in the north of the Calabria region, Conforti et al. (2013) detected erosion rates higher than 6 Mg/ha yr for a basin with olive groves and reported on problems with the generation of streams and gullies increasing sediment connectivity. In Jordan, using passive bioindicators, Kraushaar et al. (2014) recorded up to 95 Mg/ha yr. In the case of Spanish olive groves, there are numerous studies from Andalusia (Southern Spain), which have estimated a total of 1.52 million hectares cultivated, with an average production between the 2006/2007 and 2013/2014 campaigns of 5.48 million tons (Andalusian Master Plan for Olive Groves). Specifically, in the Córdoba province, the conventional olive groves located in areas with a certain slope, which maintain a good part of the surface of their bare soils and in which the farmers use machinery to eliminate the vegetal cover, are registering serious problems of hydrophobicity, compaction and reduction of the fertile horizon increasing soil erosion and connectivity (Burguet et al., 2016; Di Stefano et al., 2016). Using the curve number method has been shown to be a suitable way to assess connectivity in olive orchards, for example, for the estimation of surface runoff according to soil type, use, management and hydrological condition. For example, in Puente Genil (Spain), Taguas et al. (2015) demonstrated potential runoff rates that will generate rills and gullies increasing the connectivity in catchments cultivated with olive

Figure 12.4   Connectivity processes activated on bare surfaces in olive (A) and almond (B) orchards. All pictures were taken by J. Rodrigo-Comino. A black and white version of this figure will appear in some formats. For the colour version, refer to the plate section.

orchards, further being conditioned by the recurrence of extreme rainfall events. Also, Taguas et al. (2013), pioneering research in monitoring streams and gullies in the same area, detected runoff rates for events greater than 7.2 and 17.0 Mg with a return period of three years. In this way, the use of plant cover systems or planned tillage to stop the connection among ephemeral gullies and rills should be mandatory measures beyond the obligations derived from the conditionality of the Common Agricultural Policy, or even subsidized by many farmers (Figure 12.4).

In areas with citrus and subtropical trees, connectivity assessments are limited to some studies that applied indices of connectivity for large catchments with some plantations of woody crops, however not being the main scope of research (Ortíz-Rodríguez et al., 2017). Recently, the application of ISUM again revealed that connectivity should be considered by land managers, policymakers and stakeholders. Results have shown that terraced and flood-irrigated persimmons and citrus orchards register constant mobilization of soil and water despite the fact that these areas are closed (Bayat et al., 2019). The findings revealed that at the pedon scale, the area between rows of woody crops is not flat due to microtopographic changes generated by tillage practices, wildlife burrows, plant growth or the effect of surface flow and splash. The roughness generated between the width interrows shows linear erosion features, sediment deposits or specific topographic changes (Figure 12.4) due to soil consolidation (Moradi et al., 2020). Something similar is inferred for almonds (Becerra et al., 2010), chestnuts (Barrena-González et al., 2020), coffee (Blanco Sepúlveda & Aguilar Carrillo, 2015) or tea plantations, where soil erosion is confirmed but connectivity investigations are scarce or nonexistent.

12.2.3 Sediment Connectivity in Grasslands

Grasslands are one of the world's most widespread terrestrial ecosystems (White et al., 2000; Dixon et al., 2014). There are many definitions of "grassland" and the boundaries between them and other ecosystems are often fuzzy. Some defined grasslands have relatively high tree cover and are not classified as forests until they reach 30% (Mermoz et al., 2018). In many regions, grasslands, forests and their distinct transition zones are not clearly delineated, whereas these transition zones are defined forest steppe in Eurasia (Erdős et al., 2018), considering grassland as an area with a few trees or shrubs but dominated by grasses or other herbaceous plants, that is, open grasslands, grassy shrubland, and savannahs (Parr et al., 2014; Wilsey, 2018). This broad definition of "grassland" includes areas occupying 40% of the world's land cover (Guo et al., 2012; Bardgett et al., 2021). In terms of grassland types, it involves the North American prairies, the African savannas, the Eurasian steppes and the South American pampas, as well as various artificial grasslands and rangelands around the world (White et al., 2000). Grasslands play key roles in global biodiversity and provide kinds of material and nonmaterial benefits, such as water supply, carbon storage, climate mitigation, and erosion control (Suttie et al., 2005; O'Mara, 2012).

It has been reported that as much as 49% of global grasslands are facing extensive degradation due to climatic factors and human interferences (Abberton, 2010; Gang et al., 2014). For example, in "intensively managed grasslands" used for farming and livestock generally inorganic fertilizers are utilized in order to achieve high stocking densities of livestock breeds and frequent resowing of plants. These practices have dominated in some parts of the world over the past century, now facing degradation problems (Wilsey, 2018; Bardgett et al., 2021). On the Qinghai-Tibet Plateau in China, in recent decades climate warming, glacier and permafrost degradation, overgrazing and rat damage are pushing up to 90% of the grasslands toward degradation and possible collapse (Liu et al., 2018; Cao et al., 2019). Wind and water erosion, iterative freeze and thaw also play an activation role, further speeding up the degradation process, sometimes even resulting in "black-soil land" (Dong et al., 2013) (Figure 12.5). Brown et al. (2011) reported that nearly 90% of seminatural species-rich grasslands in the UK have been degraded since the 1940s due to intensive agriculture and land conversion. In the southern Brazil, inappropriate management and land use change have led to a loss of more than 60% of the original grassland area (Andrade et al., 2015).

A land-degradation-related decline in soil organic matter leads to increased soil erosion. Although grasslands are generally considered to be less prone to soil erosion than farmlands due to the protective vegetation cover, high erosion rates are recorded in grasslands around the world (Borrelli et al., 2017)., increasing sediment load of streams by increasing water and sediment connectivity (Poesen et al., 2003). Grassland sediment was found to have a considerable contribution to suspended sediment in the River Aire, UK (Vercruysse and Grabowski, 2019). These high contributions are a consequence of the mobilization of sediment particles caused by livestock trampling within the catchment (Meyles et al., 2006; Bilotta et al., 2007).

Figure 12.5   Climate and human-decision factors affecting connectivity processes
in alpine grasslands. (A) Fragmented meadow patches; (B) Sediment transport in
erosion gullies; (C) Artificial planting practices in extremely degraded grasslands.
All pictures were taken by G.L. Wu and Y.-F. Liu. A black and white version of this
figure will appear in some formats. For the colour version, refer to the plate section.

Grasslands play a key role in water retention as the existence of grass root systems improves soil structure, increases water infiltration, and reduces surface runoff.
Honigova et al. (2012) reported that the ability of grasslands to reduce surface runoff
was 20% and 50% higher than that of farmlands and urban areas, respectively. Saha
and Kukal (2015) found that grasslands had the maximum water holding capacity
(38.8%) in 0–15 cm surface soil depth compared to farmlands and forests in the lower
Himalayas of the northwestern India. In the upper reaches of the Yangtze River in
China, the mean flood mitigation value of grasslands was about 29.66 mm/a, which
was about 1.88 times and 0.38 times of that in paddy field and forests, respectively
(Fu et al., 2013). As a consequence, the water retention capacity of the soil decreases
once these grasslands are degraded, because of soil structure deterioration directly
affecting soil infiltration performance, soil porosity, and soil compactness (Liu et al.,
2022a). In addition, sediment delivery in grasslands are often accompanied by the
transport and delivery of nutrients and pollutants, especially in managed grasslands.
In Europe, USA, Australia and New Zealand, grassland covers extensive areas that
are mostly managed for dairy and meat production (Brazier et al., 2007; Bilotta et al.,
2008). Some studies have indicated that permanent grasslands contribute significantly to diffuse pollution, even exceeding EU water quality thresholds. Ploughing
and reseeding of grasslands exacerbated the already significant diffuse pollution
rates (Bilotta et al., 2008; Granger et al., 2010). However, planting artificial grasses
in waterways can also be a measure to reduce sediment connectivity by means of
increasing sediment-trapping efficiencies (Mekonnen et al., 2017).

   Therefore, it is necessary to apply specific management practices to mitigate
earlier mentioned processes in grasslands. Specific actions include biotic and
abiotic improvement measures to restore grasslands, such as the introduction of
native species, elimination of invasive species, to control fertilization, irrigation,
reseeding, mowing, topsoil transplantation, fencing, implement better grazing systems, and build green ecological barriers (Khalil et al., 2021) (Figure 12.6). In arid
and semi-arid areas, management practices including revegetation, grazing fallow

Figure 12.6   Example of managed practices in grazing grasslands. (A) fencing management; (B) reseeding in extremely degraded grasslands. All pictures were taken by G.L. Wu and Y.-F. Liu. A black and white version of this figure will appear in some formats. For the colour version, refer to the plate section.

and grassland fencing are applicable for lightly to severely degraded grasslands. Several studies suggested moderate-intensity rotational grazing could maintain or improve soil fertility, carbon and nitrogen storage, promote plant growth, and thus reduce soil erosion (Pilon et al., 2017; Hancock et al., 2020). In dry Mediterranean mountain pasturelands, management practices seek to optimize resources on random paths for economic benefit, and spatial optimization of the vegetation canopy to minimize erosion rates (Thornes, 2007). In the Qinghai-Tibet Plateau of China, where the harsh climatic conditions with alternating freeze-thaw cycles and short growing seasons pose challenges to the restoration of degraded grasslands, artificial planting is an effective practice for restoring severely degraded grasslands (Ma et al., 2002; Wu et al., 2010). Since most grass species cannot grow well in such environments (Li et al., 2019), species selection in this practice needs to follow criteria, such as sufficient adaptability to harsh climatic conditions, high ecological and feeding values, native species preferred, seeds available, perennial grass species preferred, and considering the matching of grass morphological traits and living habits to improve resource utilization efficiency (Liu et al., 2022b). In most of grasslands of Europe, USA, Australia, and New Zealand, intense management, for example, over-fertilization, ploughing and reseeding, results in high rates of sediment and nutrient loss. Caution should therefore be taken in any sustainable intensification practices, including ploughing and reseeding.

12.3 Sediment Dis-connectivity in Agricultural Land

A break in the continuity of overland flow can happen due to natural and/or man-made features and activities (Figure 12.7). In many cases, runoff and sediment dis-connectivity is associated with local topographic thresholds, created by either barriers (higher elevation) or depressions (lower elevation), which impede runoff connectivity. Barriers, such as traditional stonewalls form an important feature in

Figure 12.7   Traditional elevated drystone wall in abandoned fields in Huesca province, Spanish Pyrenees (A). Steep viticulture in terraces in Orense province, DO Ribeira Sacra, NW Spain (B). Ponding in a cereal field in the central Ebro river valley in Zaragoza province, NE Spain (C). Dis-connectivity of a mud-flow generated in a cereal field (right part of the picture) and the adjacent vineyard (left part of the picture) due to the presence of elevated narrow buffer strips of grass and an unpaved trail; picture taken near Barbastro, Huesca province (D). Large check dam built in an abandoned agro-ecosystem to reduce the sediment yield; picture taken in Albacete province, SE Spain (E). Buffer strips of forest between the cultivated hillslopes and the Duero river; picture taken near Pinhão, northern Portugal (F). All pictures were taken by M. López-Vicente. A black and white version of this figure will appear in some formats. For the colour version, refer to the plate section.

traditional agricultural landscapes in many regions around the world and have an important and clear effect on soil loss rates, runoff coefficients and connectivity (López-Vicente et al., 2017). Building agricultural terraces is a cultivation technique, used since ancient times, leaving a clear imprint on the landscape. Terraces are generally built to retain soil and water, to mitigate soil erosion, and to support agriculture in-steep landscapes. Agricultural terraces are artificial landforms or areas where the slope gradient decreases significantly with regard to the general trend, and act as effective areas of sediment deposition (Cucchiaro et al., 2021). Nowadays, land abandonment, new farming practices are affecting many terraced landscapes. As a consequence, there is a progressive increase in land degradation in such areas (Tarolli, 2018). Flat and concave areas (depressions) can also create disconnected zones without a direct link with the overland flow pathway and the outlet. This is especially relevant in farmlands where ponding may create environmental and agronomic problems. In the last few years, high-resolution digital elevation models (DEM) generated from data obtained from unmanned aerial vehicles (UAV) have allowed to accurately quantify this phenomenon (Straffelini et al., 2021). Trails and unpaved roads are among the most important man-made features that modify overland flow pathways in farmland, forestland, and range-land. These features can, on the one hand, favor the occurrence of concentrated flow and lateral connections between different hillslope sectors, while, on the other hand, trails and roads may disconnect the upslope and downslope sectors (Lana-Renault et al., 2018).

Particularly noteworthy is the use of buffer strips, which are made of planted rows and dramatically reduce the speed of overland flow and sediment connectivity (e.g., Poeppl et al., 2012). In some places, multispecies riparian buffer strips, consisting of rows of fast-growing trees, shrubs and grasses, have demonstrated their ability to increase streambank stability and act as effectively nutrient, pesticide and sedi-ment sink for their associated pollutants coming from the upslope agricultural fields (Schultz et al., 1995). Buffer strips favor soil stabilization, absorption of infiltrated water, and soil-root-microbe-nutrient-pesticide-sediment interaction characteristics than those observed in the cropped fields. Buffer strips can also reduce soil erosion in pastures affected by high grazing, especially fenced riparian buffer strips that diminish runoff yield and sediment transport from pasture soils (Pilon et al., 2017). In fruit orchards, buffer strips can be located in the boundary of the fields in order to separate the orchard areas from tributary streams. In this case, the buffer strip width appears as a key aspect to reduce the entrance of sediments and pollutants into the tributaries during rainfall events (Stehle et al., 2016). However, erosion rills may substantially reduce pesticide retention efficacies of buffer strips during intense run-off events and suggest that the capability of buffer strips as a risk mitigation tool for runoff may be overestimated in some regulatory risk assessment procedures conducted for pesticide authorization. In almond-tree plantations located in Sierra Nevada (south-eastern Spain), Martínez-Raya et al. (2006) evaluated the effect of different 3 m width plant-cover strips (thyme, barley and lentils) to regulate soil loss and surface-runoff patterns, finding reductions of soil loss by 97%, 87% and 58% for thyme, barley and lentils, compared to bare soil.

12.4 Soil Erosion and Connectivity Control by Means of Governmental Programs

Soil and water conservation programs by government agencies and nongovern-mental initiatives focus on preventing soil erosion and manage sediment transport. Essentially this means they try to impact the connectivity of the landscape, imped-ing fast water and sediment fluxes through the landscape. In the following sec-tion, a summary is given on how actions developed by the European Commission, Chinese government, USA, FAO and the African Great Green Wall movement use (dis)connectivity to prevent erosion, sediment accumulation in reservoirs and ensure healthy soils on site.

12.4.1 European Union

The policies of the European Commission have been sensitive to the different soil issues in farmland, and include a list of initiatives, regulations and programs. The EU's Common Agricultural Policy (CAP), launched in 1962, is a partnership between agriculture and society, and between Europe and its farmers. This concern about soil protection inspired the celebration of the European Community Workshop on Soil Erosion Protection in Freising, Germany, in May 1988 (Schwertmann et al., 1989). Regarding soils, CAP declares that "While being cost-effective, farmers should work in a sustainable and environmentally friendly manner, and maintain our soils and biodiversity" (European Commission, 2022). In 2015, the mean European soil loss rate by sheet and rill erosion was estimated at 2.46 Mg/ha yr, exceeding by 1.6 the average soil formation rates (Panagos et al., 2015a). Although this rate is too high, the application of the Good Agricultural Environmental Conditions (reduced tillage, winter cover crops, plant residues, stone walls, grass margins and contouring) reduced soil loss from European arable lands by ca. 20% between 2005 and 2015 (Panagos et al., 2015b). More recently, Borrelli and Panagos (2020) estimated that the soil conservation measures (i.e., reduced tillage, cover crops and plant residues; reported in the EU Farm Structure Survey – 2016, all aiming for reducing the connectivity of the landscape) in reducing soil erosion in Europe, especially key CAP policies such as Good Agricultural and Environmental Conditions and Greening allowed an overall decrease of C-factor (included in the RUSLE soil erosion model) of ca. −0.84% compared to the 2010 survey. The new CAP was formally adopted on December 2, 2021; and it is due to be implemented from January 1, 2023. To respond to changing needs and challenges, the 2023–2027 EU agricultural policy will assign 40% of the new CAP budget to environmentally respectful farming. Greener farming practices will be promoted through eco-schemes, which are specific payments that will be provided to farmers that adopt climate-sensitive and nature-friendly practices, in line with the European Green Deal objectives. Examples of these actions are: organic farming, crop rotation, carbon farming and agroecology. These actions aim to have a better soil cover and a more diverse landscape that provokes the soil, hillslopes and finally the whole landscape to be less connected in terms of water and sediment fluxes. The reformed CAP will benefit from this increased investment, incorporating stronger agricultural knowledge and innovation systems to boost the development of innovation projects, disseminate their results, and encourage their use as widely as possible.

In November 2021, the European Commission presented the EU Soil Strategy for 2030 – Reaping the benefits of healthy soils for people, food, nature and climate. The vision for soil: By 2050, all EU soil ecosystems are in healthy condition and are thus more resilient, which will require very decisive changes in this decade. A healthy soil contributes to a less connected system by increasing the infiltration capacity and rougher surface due to better aggregates. By then, protection, sustainable use and restoration of soil has become the norm. As a key solution, healthy soils contribute to address our big challenges of achieving climate neutrality and becoming resilient to climate change, developing a clean and circular (bio)economy, reversing biodiversity loss, safeguarding human health, halting desertification and reversing land degradation (European Commission, 2021). Especially for the latter two elements the connectivity knowledge is essential to prevent erosion and desertification and restore soils and landscapes. Other initiatives to protect soils in agricultural areas are LIFE projects. LIFE ("The Financial Instrument for the Environment") is a program launched by the European Commission and coordinated by the Environment Directorate-General. Although soil has not been a core theme of LIFE, the program has funded many soil-related projects since its launch in 1992, and there has been an increasing focus on soil protection since the publication of the Thematic Strategy in 2006. LIFE has cofinanced actions targeting erosion, landslides, contamination, loss of organic matter, sealing, compaction, and other soil management issues (European Commission, 2014), clearly using the connectivity concepts as a key element to target these soil threats. Recently, the new European Green Deal was presented with the ambition to make the European Union the first climate-neutral continent by 2050. It has an ambitious package of measures within the Biodiversity Strategy 2030, the Farm to Fork and the European Climate Law including actions to protect soils. The Farm to Fork strategy addresses soil pollution with 50% reduction in use of chemical pesticides by 2030 and aims 20% reduction in fertilizer use plus a decrease of nutrient losses by at least 50%. The Biodiversity Strategy has the ambition to set a minimum of 30% of the EU's land area as protected areas, limit urban sprawl, reduce the pesticides risk, bring back at least 10% of agricultural area under high-diversity landscape features, put forward the 25% of the EU's agricultural land as organically farmed, and progress in the remediation of contaminated sites (Montanarella & Panagos, 2021). This is the first time the landscape diversity is specifically targeted in European policy. By acknowledging the importance of the landscape level and the usefulness to make the landscape diverse shows that connectivity has a role to play in designing future landuse/scape planning.

12.4.2 China

In 1998, after the devastating floods along the Yangtze River, considerable attention was focused on the possible causes, and the crisis clearly prompted action to advance nature conservation and environmental protection in China. The central government announced the establishment of the Nature Forest Protection Project in an effort to halt the destruction of natural forests, whereas the "Grain-for-Green" Project (GFGP) was initiated to return cultivated land on slopes of 25° or more to forestland in the upper Yangtze River Basin, the middle and upper Yellow River Basin and the Loess Plateau (Ye et al., 2003). The high connectivity of the steep slopes was acknowledged as the key problem to target in the GFDP. After two decades of GFGP implementation and despite its success (e.g., average soil erosion decreased from 4,884.49 to 4,087.57 t/km^2 yr in Yan'an county), Ning et al. (2021) found that soil erosion attenuation differed among sites, even observing lower soil conservation effect in counties with higher implementation. These authors proposed the concept of soil conservation potential, and put forward a dynamic implementation mechanism for the formulation and optimization of ecological programs and projects in future: First, using the soil conservation potential to determine the implementation intensity in each region; second, adjusting the intensity to the changes of the soil conservation potential in the following implementation; and third, repeating earlier steps to ensure high efficiency of soil erosion control, and achieving the sustainability and effectiveness of the ecological projects. A better understanding of the connectivity of water and sediment fluxes in these different areas would have helped to optimize the interventions taken. In semi-arid areas on the Loess Plateau, in particular in the Jinghe River basin, Xu et al. (2022) found that the GFGP caused a decrease in the farmland area of 466 km^2 and of the degraded grassland of 394 km^2, whereas the area of forestland and scrubland increased by 339 km^2. Even more interesting was the decrease of annual water yield (by 36.2%) and sediment yield (by 60.7%) between the average values from 2000 to 2015 and those of the previous 50 years, which corresponds to findings in other studies such as Keesstra et al. (2009, 2018a, 2018b) where the decrease in connectivity reduced both water fluxes and sediment fluxes by similar amounts. The former authors also found a clear increase of carbon storage, about 2.4%. The water yield and sediment control of grassland restoration area and forestland restoration area were better than those of farmland, and grasses, it appeared, were a better vegetation type of GFGP compared to trees for the Loess Plateau.

The "Grain for Green" ecological project has led to the widespread distribution and succession of biocrusts in the open spaces between vascular plants in the abandoned farmlands. Yang et al. (2022) found that (a) the mean weight diameter in the subsurface soil increased with the biocrust succession; (b) the later successional stage biocrusts (mixed crust and moss crust) improved the deeper soil aggregate stability compared with the early successional stage biocrusts; and (c) in the 0–5 cm layer, biocrust succession indirectly improved the stability of aggregates by increasing soil organic matter. Since 1999, China's GFGP has greatly reduced the runoff and sediment load, causing a change in the tendency of soil erosion and sediment transport in farmland. Prior to the Reform and Opening-up in China in 1978, climatic change played a leading role in check dams sedimentation and hydrological changes, which are known to impact the connectivity of the catchment system. Subsequently, despite accelerated soil erosion associated with the family contract responsibility system, the check dams, afforestation and a warmer climate mitigated the soil erosion to some extent, runoff and sediment load continued to decrease, in part as a result of the disconnecting effect of the check-dams. However, the policy of filling gullies and creating land since 2010 has created non-solved questions in terms of eco-environment, hydrological cycle, and economics aspects (Dong et al., 2022). In a study in Ethiopia, the effect of dis-connectivity in an agricultural area showed the potential of these types of measures to mitigate gully erosion as well (Mekonnen et al., 2017; Amare et al., 2021)

References

Abberton, M., Conant, R., & Batello, C., 2010. *Grassland Carbon Sequestration: Management, Policy and Economics*. Food and Agriculture of the United Nations, Rome.

Addis, H.K., Abera, A., & Abebaw, L., 2020. Economic benefits of soil and water conservation measures at the sub-catchment scale in the northern Highlands of Ethiopia. *Progress in Physical Geography* 44(2), 251–266.

Aguilera, E., Sanz-Cobena, A., Infante-Amate, J., García-Ruiz, R., Vila-Traver, J., Guzmán, G.I., González de Molina, M., Rodríguez, A., Piñero, P., & Lassaletta, L., 2021. Long-term trajectories of the C footprint of N fertilization in Mediterranean agriculture (Spain, 1860–2018). *Environmental Research Letters* 16, 085010.

Amare, S., Langendoen, E., Keesstra, S., Ploeg, M.V.D., Gelagay, H., Lemma, H., & Zee, S.E.V.D., 2021. Susceptibility to gully erosion: applying random forest (RF) and frequency ratio (FR) approaches to a small catchment in Ethiopia. *Water*, 13(2), 216.

Andrade, B.O., Koch, C., Boldrini, I.I., Vélez-Martin, E., Hasenack, H., Hermann, J.M., Kollmann, J., Pillar, V.D., & Overbeck, G.E., 2015. Grassland degradation and restoration: a conceptual framework of stages and thresholds illustrated by southern Brazilian grasslands. *Natureza & Conservação* 13, 95–104.

Armesto, J.J., Manuschevich, D., Mora, A., Smith-Ramirez, C., Rozzi, R., Abarzúa, A.M., & Marquet, P.A., 2010. From the Holocene to the Anthropocene: A historical framework for land cover change in southwestern South America in the past 15,000 years. *Land Use Policy* 27(2), 148–160.

Atucha, A., Merwin, I.A., Brown, M.G., Gardiazabal, F., Mena, F., Adriazola, C., & Lehmann, J., 2013. Soil erosion, runoff and nutrient losses in an avocado (*Persea americana* Mill) hillside orchard under different groundcover management systems. *Plant Soil* 368, 393–406.

Bardgett, R.D., Bullock, J.M., Lavorel, S., Manning, P., Schaffner, U., Ostle, N., Chomel, M., et al. 2021. Combatting global grassland degradation. *Nature Reviews Earth & Environment* 2, 720–735.

Barrena-González, J., Lozano-Parra, J., Alfonso-Torreño, A., Lozano-Fondón, C., Abdennour, M.A., Cerdà, A., & Pulido-Fernández, M., 2020. Soil erosion in Mediterranean chestnut tree plantations at risk due to climate change and land abandonment. *Central European Forestry Journal* 66, 85–96.

Bayat, F., Monfared, A.B., Jahansooz, M.R., Esparza, E.T., Keshavarzi, A., Morera, A.G., Fernández, M.P., & Cerdà, A., 2019. Analyzing long-term soil erosion in a ridge-shaped persimmon plantation in eastern Spain by means of ISUM measurements. *Catena* 183, 104176.

Becerra, A.T., Botta, G.F., Bravo, X.L., Tourn, M., Melcon, F.B., Vazquez, J., Rivero, D., Linares, P., & Nardon, G., 2010. Soil compaction distribution under tractor traffic in almond (*Prunus amigdalus* L.) orchard in Almería España. *Soil and Tillage Research* 107, 49–56.

Bilotta, G.S., Brazier, R.E., & Haygarth, P.M., 2007. The impacts of grazing animals on the quality of soils, vegetation, and surface waters in intensively managed grasslands. *Advances in Agronomy* 94, 237–280.

Bilotta, G.S., Brazier, R.E., Haygarth, P.M., Macleod, C.J.A., Butler, P., Granger, S., Krueger, T., Freer, J., & Quinton, J., 2008. Rethinking the contribution of drained and undrained grasslands to sediment-related water quality problems. *Journal of Environmental Quality* 37, 906–914.

Blanco Sepúlveda, R., & Aguilar Carrillo, A., 2015. Soil erosion and erosion thresholds in an agroforestry system of coffee (*Coffea arabica*) and mixed shade trees (*Inga spp* and *Musa spp*) in Northern Nicaragua. *Agriculture, Ecosystems & Environment* 210, 25–35.

Borrelli, P., & Panagos, P., 2020. An indicator to reflect the mitigating effect of Common Agricultural Policy on soil erosion. *Land Use Policy* 92, 104467.

Borrelli, P., Robinson, D.A., Fleischer, L.R., Lugato, E., Ballabio, C., Alewell, C., Meusburger, K., et al. 2017. An assessment of the global impact of 21st century land use change on soil erosion. *Nature Communications* 8, 2013.

Brazier, R.E., Bilotta, G.S., & Haygarth, P.M., 2007. A perspective on the role of lowland, agricultural grasslands in contributing to erosion and water quality problems in the UK. *Earth Surface Processes and Landforms* 32, 964–967.

Brown, C., Walpole, M., Simpson, L., & Tierney, M., 2011. Introduction to the UK National Ecosystem Assessment. In: *The UK National Ecosystem Assessment Technical Report*. UK National Ecosystem Assessment, UNEP-WCMC, Cambridge.

Burguet, M., Taguas, E.V., Cerdà, A., &Gómez, J.A., 2016. Soil water repellency assessment in olive groves in Southern and Eastern Spain. *Catena* 147, 187–195.

Buschiazzo, D.E., Zobeck, T.M., & Abascal, S.A., 2007. Wind erosion quantity and quality of an Entic Haplustoll of the semi-arid pampas of Argentina. *Journal of Arid Environments* 69(1), 29–39.

Calsamiglia, A., Fortesa, J., García-Comendador, J., Lucas-Borja, M.E., Calvo-Cases, A., & Estrany, J., 2018. Spatial patterns of sediment connectivity in terraced lands: Anthropogenic controls of catchment sensitivity. *Land Degradation and Development* 29, 1198–1210.

Cao, J.J., Adamowski, J.F., Deo, R.C., Xu, X.Y., Gong, Y.F., & Feng, Q., 2019. Grassland degradation on the Qinghai-Tibetan Plateau: Reevaluation of causative factors. *Rangeland Ecology & Management* 72, 988–995.

Cerdà, A., Franch-Pardo, I., Novara, A., Sannigrahi, S., & Rodrigo-Comino, J., 2022. Examining the effectiveness of catch crops as a nature-based solution to mitigate surface soil and water losses as an environmental regional concern. *Earth Systems and Environment* 6, 29–44.

Chen, S.-K., Liu, C.-W., & Chen, Y.-R., 2012. Assessing soil erosion in a terraced paddy field using experimental measurements and universal soil loss equation. *Catena* 95, 131–141.

Conforti, M., Buttafuoco, G., Leone, A.P., Aucelli, P.P.C., Robustelli, G., & Scarciglia, F., 2013. Studying the relationship between water-induced soil erosion and soil organic matter using Vis–NIR spectroscopy and geomorphological analysis: A case study in southern Italy. *Catena* 110, 44–58.

Cucchiaro, S., Paliaga, G., Fallu, D.J., Pears, B.R., Walsh, K., Zhao, P., Van Oost, K., et al. 2021. Volume estimation of soil stored in agricultural terrace systems: A geomorphometric approach. *Catena* 207, 105687.

Di Stefano, C., Ferro, V., Burguet, M., & Taguas, E.V., 2016. Testing the long term applicability of USLE-M equation at an olive orchard microcatchment in Spain. *Catena* 147, 71–79.

Dixon, A.P., Faber-Langendoen, D., Josse, C., Morrison, J., & Loucks, C.J., 2014. Distribution mapping of world grassland types. *Journal of Biogeography* 41, 2003–2019.

Dong, H., Song, Y., Chen, L., Liu, H., Fu, X., & Xie, M. 2022. Soil erosion and human activities over the last 60 years revealed by magnetism, particle size and minerals of check dams sediments on the Chinese Loess Plateau. *Environmental Earth Sciences* 81(5), article 162.

Dong, Q.M., Zhao, X.Q., Wu, G.L., Shi, J.J., & Ren, G.H., 2013. A review of formation mechanism and restoration measures of "black-soil-type" degraded grassland in the Qinghai-Tibetan Plateau. *Environmental Earth Sciences* 70, 2359–2370.

Dreibelbis, F.R., 1944. A summary of soil-moisture data useful in soil- and water-conservation investigations. *Eos, Transactions American Geophysical Union* 25(6), 1041–1047.

Durán Zuazo, V.H., Aguilar Ruiz, J., Martínez Raya, A., & Franco Tarifa, D., 2005. Impact of erosion in the taluses of subtropical orchard terraces. *Agriculture Ecosystems & Environment* 107, 199–210.

Erdős, L., Ambarlı, D., Anenkhonov, O.A., Bátori, Z., Cserhalmi, D., Kiss, M., Kröel-Dulay, G., et al. 2018. The edge of two worlds: A new review and synthesis on Eurasian forest-steppes. *Applied Vegetation Science* 21, 345–362.

European Commission, 2014. *LIFE and Soil protection*. Publications Office of the European Union (2014). Luxembourg. DOI:10.2779/64447.

European Commission, 2021. EU Soil Strategy for 2030 – Reaping the benefits of healthy soils for people, food, nature and climate. COM (2021) 699 final. Brussels, 17.11.2021.

European Commission, 2022. *The common agricultural policy at a glance*. Website: https://ec.europa.eu/info/food-farming-fisheries/key-policies/common-agricultural-policy/cap-glance_en (visited on 25 May 2022).

Ferreira, C.S., Veiga, A., Caetano, A., Gonzalez-Pelayo, O., Karine-Boulet, A., Abrantes, N., Keizer, J., & Ferreira, A.J., 2020. Assessment of the impact of distinct vineyard management practices on soil physico-chemical properties. *Air, Soil and Water Research* 13, https://doi.org/10.1177/1178622120944847.

Findley, D.M., Acabado, S., Amano, N., Kay, A.U., Hamilton, R., Barretto-Tesoro, G., Bankoff, G., Kaplan, J.O., & Roberts, P., 2022. Land use change in a pericolonial society: Intensification and diversification in Ifugao, Philippines between 1570 and 1800 CE. *Frontiers in Earth Science* 10, 680926.

Fleskens, L., & Stroosnijder, L., 2007. Is soil erosion in olive groves as bad as often claimed? *Geoderma* 141, 260–271.

Fressard, M., & Cossart, E., 2019. A graph theory tool for assessing structural sediment connectivity: Development and application in the Mercurey vineyards (France). *Science of The Total Environment* 651, 2566–2584.

Fryirs, K.A., 2013. (Dis) Connectivity in catchment sediment cascades: a fresh look at the sediment delivery problem. *Earth Surface Processes and Landforms* 38, 30–46.

Fu, B., Wang, Y., Xu, P., & Yan, K., 2013. Mapping the flood mitigation services of ecosystems-a case study in the Upper Yangtze River Basin. *Ecological Engineering* 52, 238–246.

Gang, C.C., Zhou, W., Chen, Y.Z., Wang, Z.Q., Sun, Z.G., Li, J.L., Qi, J.G., & Odeh, I., 2014. Quantitative assessment of the contributions of climate change and human activities on global grassland degradation. *Environmental Earth Sciences* 72, 4273–4282.

García-Díaz, A., Bienes, R., Sastre, B., Novara, A., Gristina, L., & Cerdà, A., 2017. Nitrogen losses in vineyards under different types of soil groundcover. A field runoff simulator approach in central Spain. *Agriculture, Ecosystems & Environment* 236, 256–267.

García-Ruiz, J.M., Beguería, S., Nadal-Romero, E., González-Hidalgo, J.C., Lana-Renault, N., & Sanjuán, Y., 2015. A meta-analysis of soil erosion rates across the world. *Geomorphology* 239, 160–173.

Glymph, L.M., 1954. Studies of sediment yields from watersheds. *International Association for Hydrological Sciences Publication* 36, 173–191.

Gómez, J.A., Giráldez, J.V., & Vanwalleghem, T., 2008. Comments on "Is soil erosion in olive groves as bad as often claimed?" by L. Fleskens and L. Stroosnijder. *Geoderma* 147, 93–95.

Granger, S.J., Hawkins, J.M.B., Bol, R., Whitem, S.M., Naden, P., Old, G., Bilotta, G.S., Brazier, R.E., Macleod, C.J.A., & Haygarth, P.M., 2010. High temporal resolution monitoring of multiple pollutant responses in drainage from an intensively managed grassland catchment caused by a summer storm. *Water, Air and Soil Pollution* 205, 377–393.

Guo, Q., Hu, Z., Li, S., Li, X., Sun, X., & Yu, G., 2012. Spatial variations in aboveground net primary productivity along a climate gradient in Eurasian temperate grassland: effects of mean annual precipitation and its seasonal distribution. *Global Change Biology* 18, 3624–3631.

Hakansson, I., & Lipiec, J., 2000. A review of the usefulness of relative bulk density values in studies of soil structure and compaction. *Soil & Tillage Research* 53, 71–85.

Hancock, G.R., Ovenden, M., Sharma, K., Rowlands, W., & Wells, T., 2020. Soil erosion-the impact of grazing and regrowth trees. *Geoderma* 361, 114102.

Honigova, I., Vackar, D., Lorencova, E., Melichar, J., Gotzl, M., Sonderegger, G., Ouskova, V., Hosek, M., & Chobot, K., 2012. Survey on grassland ecosystem services. *Report to the EEA-European Topic Centre on Biological Diversity.* Nature Conservation Agency of the Czech Republic, Prague, p78.

Hooke, J., 2003. Coarse sediment connectivity in river channel systems: a conceptual framework and methodology. *Geomorphology* 56(1–2), 79–94.

INE (Instituto Nacional de Estadística), 2009. *Agrarian Census 2009 Project.* Madrid, Spain, 124 pages.

Kawai, K., Okada, M., & Ikemune, K., 1957. Studies of Cropping Systems for Soil Conservation against Erosion HI. Influences of tobacco and potato cultivation on soil erosion and conservative measures. *Japanese Journal of Crop Science* 26(1), 63–64.

Keesstra, S. D., Bruijnzeel, L. A., & Van Huissteden, J., 2009. Meso-scale catchment sediment budgets: combining field surveys and modeling in the Dragonja catchment, southwest Slovenia. *Earth Surface Processes and Landforms*, 34(11), 1547–1561.

Keesstra, S., Nunes, J. P., Saco, P., Parsons, T., Poeppl, R., Masselink, R., & Cerdà, A., 2018a. The way forward: Can connectivity be useful to design better measuring and modelling schemes for water and sediment dynamics? *Science of the Total Environment*, 644, 1557–1572.

Agricultural Land 309

Keesstra, S., Nunes, J., Novara, A., Finger, D., Avelar, D., Kalantari, Z., & Cerdà, A., 2018b. The superior effect of nature based solutions in land management for enhancing ecosystem services. *Science of the Total Environment*, 610, 997–1009.

Khalil, M.I., Cordovil, C.M.D.S., Francaviglia, R., Beverley, H., Klumpp, K.; Koncz, P., Llorente, M., Madari, B.E., Muñoz-Rojas, M., & Nerger, R., 2021. Grasslands. In *Recarbonizing Global Soils: A Technical Manual of Recommended Sustainable Soil Management*; FAO, Italy, Rome, Volume 3, ISBN 978-92-5-134893-2.

Knapen, A., Smets, T., & Poesen, J., 2009. Flow-retarding effects of vegetation and geotextiles on soil detachment during concentrated flow. *Hydrological Processes* 23, 2427–2437.

Kraushaar, S., Herrmann, N., Ollesch, G., Vogel, H.-J., & Siebert, C., 2014. Mound measurements – Quantifying medium-term soil erosion under olive trees in Northern Jordan. *Geomorphology* 213, 1–12.

Lana-Renault, N., López-Vicente, M., Nadal-Romero, E., Ojanguren, R., Llorente, J.A., Errea, P., Regüés, D., et al., 2018. Catchment based hydrology under post farmland abandonment scenarios. *Geographical Research Letters* 44(2), 503–534.

Lexartza-Artza, I., & Wainwright, J., 2011. Making connections: Changing sediment sources and sinks in an upland catchment. *Earth Surface Processes and Landforms* 36 (8), 1090–1104.

Li, X., Gao, J., Zhang, J., Wang, R., Jin, L., & Zhou, H., 2019. Adaptive strategies to overcome challenges in vegetation restoration to coalmine wasteland in a frigid alpine setting. *Catena* 182, 104142.

Li, Z., Schneider, R.L., Morreale, S.J., Xie, Y., Li, C., & Li, J., 2018. Woody organic amendments for retaining soil water, improving soil properties and enhancing plant growth in desertified soils of Ningxia, China. *Geoderma* 310, 143–152.

Liu, H., Blagodatsky, S., Giese, M., Liu, F., Xu, J., & Cadisch, G., 2016. Impact of herbicide application on soil erosion and induced carbon loss in a rubber plantation of Southwest China. *Catena* 145, 180–192.

Liu, J., Milne, R.I., Cadotte, M.W., Wu, Z.Y., Provan, J., Zhu, G.F., Gao, L.M., & Li, D.Z., 2018. Protect third pole's fragile ecosystem. *Science* 362, 1368.

Liu, M., Min, L., Wu, L., Pei, H., & Shen, Y., 2022. Evaluating nitrate transport and accumulation in the deep vadose zone of the intensive agricultural region, North China Plain. *Science of the Total Environment* 825, 153894.

Liu, Y., Li, S.Y., Shi, J.J., Niu, Y.L., Cui, Z., Zhang, Z.C., Wang, Y.L., Ma, Y.S., Lopez-Vicente, M., & Wu, G.L., 2022b. Effectiveness of mixed cultivated grasslands to reduce sediment concentration in runoff on hillslopes in the Qinghai-Tibetan Plateau. *Geoderma* 422, 115933.

Liu, Y.F., Zhang, Z.C., Liu, Y., Cui, Z., Leite, P.A.M., Shi, J.J., Wang, Y.L., & Wu, G.L., 2022a. Shrub encroachment enhances the infiltration capacity of alpine meadows by changing the community composition and soil conditions. *Catena* 213,106222.

López-Vicente, M., & Álvarez, S. 2018. Influence of DEM resolution on modelling hydrological connectivity in a complex agricultural catchment with woody crops. *Earth Surface Processes and Landforms* 43(7), 1403–1415.

López-Vicente, M., Lana-Renault, N., García-Ruiz, J.M., & Navas, A., 2011. Assessing the potential effect of different land cover management practices on sediment yield from an abandoned farmland catchment in the Spanish Pyrenees. *Journal of Soils and Sediments* 11, 1440–1455.

López-Vicente, M., Nadal-Romero, E., & Cammeraat, E.L.H. 2017. Hydrological connectivity does change over 70 years of abandonment and afforestation in the Spanish Pyrenees. *Land Degradation & Development* 28, 1298–1310.

López-Vicente, M., & Navas, A., 2010. Relating soil erosion and sediment yield to geomorphic features and erosion processes at the catchment scale in the Spanish Pre-Pyrenees. *Environmental Earth Sciences* 61, 143–158.

López-Vicente, M., Navas, A., & Machín, J., 2008. Identifying erosive periods by using RUSLE factors in mountain fields of the Central Spanish Pyrenees. *Hydrology and Earth System Sciences* 12, 523–535.

Ma, Y.S., Lang, B.N., Li, Q.Y., Shi, J.J., & Dong, Q.M., 2002. Study on rehabilitating and rebuilding technologies for degenerated alpine meadow in the Changjiang and Yellow river source region. *Acta Prataculturae Sinica* 19, 1–4.

Martín, W.S., 2017. Nitrogen, science, and environmental change: The politics of the Green Revolution in Chile and the global nitrogen challenge. *Journal of Political Ecology* 24(1), 777–796.

Martínez-Raya, A., Durán-Zuazo, V.H., & Francia-Martínez, J.R., 2006. Soil erosion and runoff response to plant-cover strips on semiarid slopes (SE Spain). *Land Degradation & Development* 17(1), 1–11.

Mekonnen, M., Keesstra, S. D., Stroosnijder, L., Baartman, J. E., & Maroulis, J., 2015. Soil conservation through sediment trapping: a review. *Land Degradation & Development*, 26(6), 544–556.

Mekonnen, M., Keesstra, S.D., Baartman, J.E.M., Stroosnijder, L., & Maroulis, J., 2017. Reducing Sediment Connectivity Through man-Made and Natural Sediment Sinks in the Minizr Catchment, Northwest Ethiopia. *Land Degradation & Development* 28(2), 708–717.

Menzies Pluer, E.G., Schneider, R.L., Morreale, S. J., Liebig, M.A., Li, J., Li, C.X., & Walter, M.T., 2020. Returning degraded soils to productivity: An examination of the potential of coarse woody amendments for improved water retention and nutrient holding capacity. *Water Air Soil Pollut* 231, 15.

Mermoz, S., Bouvet, A., Toan, T. L., & Herold, M., 2018. Impacts of the forest definitions adopted by African countries on carbon conservation. *Environmental Research Letters* 13, 104014.

Meyles, E.W., Williams, A.G., Ternan, J.L., Anderson, J.M., & Dowd, J.F., 2006. The influence of grazing on vegetation, soil properties and stream discharge in a small Dartmoor catchment, southwest England, UK. *Earth Surface Processes and Landforms* 31, 622–631.

Monfreda, C., Ramankutty, N., & Foley, J.A., 2008. Farming the planet: 2. Geographic distribution of crop areas, yields, physiological types, and net primary production in the year 2000. *Global Biogeochemical Cycles* 22, https://doi.org/10.1029/2007GB002947.

Montanarella, L., & Panagos, P., 2021. The relevance of sustainable soil management within the European Green Deal. *Land Use Policy* 100, 104950.

Moradi, E., Rodrigo-Comino, J., Terol, E., Mora-Navarro, G., Marco da Silva, A., Daliakopoulos, I.N., Khosravi, H., Pulido Fernández, M., & Cerdà, A., 2020. Quantifying soil compaction in persimmon orchards using ISUM (improved stock unearthing method) and core sampling methods. *Agriculture* 10, 266.

Morellón, M., Valero-Garcés, B., González-Sampériz, P., Vegas-Vilarrúbia, T., Rubio, E., Rieradevall, M., Delgado-Huertas, A., et al. 2011. Climate changes and human activities recorded in the sediments of Lake Estanya (NE Spain) during the Medieval Warm Period and Little Ice Age. *Journal of Paleolimnology* 46, 423–452.

Neto, J.P.S., De Souza, N.M., Andrello, A.C., & Appoloni, C.R., 2008. Loss and soil deposition estimate by means of the cesium 137 concentration in the Rio das Ondas Basin, BA. *Soils and Rocks* 31(3), 137–142.

Ning, J., Zhang, D., & Yu, Q. 2021. Quantifying the efficiency of soil conservation and optimized strategies: A case-study in a hotspot of afforestation in the Loess Plateau. *Land Degradation and Development* 32(3), 1114–1126.

Nouaim, W., Rambourg, D., Merzouki, M., El Harti, A., & Karaoui, I., 2022. Assessing the intra-annual variability of agricultural soil losses: a RUSLE application in Nord-Pas-de-Calais, France. *Journal of Water and Land Development* 52, 210–220.

O'Mara, F. P., 2012. The role of grasslands in food security and climate change. *Annals of Botany* 110, 1263–1270.

Ortíz-Rodríguez, A.J., Borselli, L., & Sarocchi, D., 2017. Flow connectivity in active volcanic areas: Use of index of connectivity in the assessment of lateral flow contribution to main streams. *Catena* 157, 90–111.

Panagos, P., Borrelli, P., & Poesen, J., 2019. Soil loss due to crop harvesting in the European Union: A first estimation of an underrated geomorphic process. *Science of the Total Environment* 664, 487–498.

Panagos, P., Borrelli, P., Poesen, J., Ballabio, C., Lugato, E., Meusburger, K., Montanarella, L., & Alewell, C., 2015a. The new assessment of soil loss by water erosion in Europe. *Environmental Science & Policy* 54, 438–447.

Panagos, P., Borrelli, P., & Robinson, D.A., 2015b. Common Agricultural Policy: tackling soil loss across Europe. *Nature* 526, 195.

Parr, C.L., Lehmann, C.E.R., Bond, W.J., Hoffmann, W.A., & Andersen, A.N., 2014. Tropical grassy biomes: misunderstood, neglected, and under threat. *Trends in Ecology & Evolution* 29, 205–213.

Parsons, A.J., Wainwright, J., Brazier, R.E., & Powell, D.M., 2006. Is sediment delivery a fallacy? *Earth Surface Processes and Landforms* 31(10), 1325–1328.

Pereira, H.C., Hosegood, P.H., & Dagg, M., 1967. Effects of tied ridges, terraces and grass leys on a lateritic soil in Kenya. *Experimental Agriculture* 3(2), 89–98.

Pijl, A., Tosoni, M., Roder, G., Sofia, G., & Tarolli, P., 2019a. Design of terrace drainage networks using UAV-based high-resolution topographic data. *Water* 11, 814.

Pijl, A., Barneveld, P., Mauri, L., Borsato, E., Grigolato, S., & Tarolli, P., 2019b. Impact of mechanisation on soil loss in terraced vineyard landscapes. *Cuadernos de Investigación Geográfica* 45, 287–308.

Pilon, C., Moore Jr., P.A., Pote, D.H., Pennington, J.H., Martin, J.W., Brauer, D.K., Raper, R.L., Dabney, S.M., & Lee, J. 2017. Long-term effects of grazing management and buffer strips on soil erosion from pastures. *Journal of Environmental Quality* 46(2), 364–372.

Poesen, J., 2018. Soil erosion in the Anthropocene: Research needs. *Earth Surface Processes and Landforms* 43(1), 64–84.

Poesen, J., Nachtergaele, J., Verstraeten, G., & Valentin, C., 2003. Gully erosion and environmental change: importance and research needs. *Catena* 50, 91–133.

Poeppl, R.E., Keiler, M., von Elverfeldt, K., Zweimueller, I., & Glade, T., 2012. The influence of riparian vegetation cover on diffuse lateral sediment connectivity and biogeomorphic processes in a medium-sized agricultural catchment, Austria. *Geografiska Annaler: Series A, Physical Geography* 94(4), 511–529.

Rodrigo-Comino, J., 2018. Five decades of soil erosion research in "terroir." The State-of-the-Art. *Earth-Science Reviews* 179, 436–447.

Rodrigo-Comino, J., Keesstra, S.D., & Cerdà, A., 2018. Connectivity assessment in Mediterranean vineyards using improved stock unearthing method, LiDAR and soil erosion field surveys. *Earth Surface Processes and Landforms* 43, 2193–2206.

Rodrigo-Comino, J., Keshavarzi, A., Zeraatpisheh, M., Gyasi-Agyei, Y., & Cerdà, A., 2019. Determining the best ISUM (Improved stock unearthing Method) sampling point number to model long-term soil transport and micro-topographical changes in vineyards. *Computers and Electronics in Agriculture* 159, 147–156.

Rodrigo-Comino, J., Lucas Borja, M., Bertalan, L., & Cerdà, A., 2020. Integrating *in situ* measurements of an index of connectivity to assess soil erosion processes in vineyards. *Hydrological Sciences Journal* 65(4), 671–679.

Rodrigo-Comino, J., Senciales, J.M., Ramos, M.C., Martínez-Casasnovas, J.A., Lasanta, T., Brevik, E.C., Ries, J.B., & Ruiz Sinoga, J.D., 2017. Understanding soil erosion processes in Mediterranean sloping vineyards (Montes de Málaga, Spain). *Geoderma* 296, 47–59.

Saha, D., & Kukal, S.S., 2015. Soil structural stability and water retention characteristics under different land uses of degraded lower Himalayas of North-west India. *Land Degradation and Development* 26, 263–271.

Schultz, R.C., Collettil, J.P., Isenhart, T.M., Simpkins, W.W., Mize, C.W., & Thompson, M.L. 1995. Design and placement of a multi-species riparian buffer strip system. *Agroforestry Systems* 29, 201–226.

Schwertmann, U., Rickson, R.J., & Auerswald, K., 1989. Soil erosion protection measures in Europe. In: *Proceedings of the European Community Workshop on Soil Erosion Protection*. Freising, Germany, May 24–26, 1988. ISBN: 978-3-510-65384-3. 216 pages.

Shoobridge, D.W., 1951. Mechanical clearing and ground preparation for softwood plantations in the Australian capital territory. *Australian Forestry* 15(2), 105–109.

Stehle, S., Dabrowski, J.M., Bangert, U., & Schulz, R. 2016. Erosion rills offset the efficacy of vegetated buffer strips to mitigate pesticide exposure in surface waters. *Science of The Total Environment* 545–546, 171–183.

Straffelini, E., Cucchiaro, S., & Tarolli, P. 2021. Mapping potential surface ponding in agriculture using UAV-SfM. *Earth Surface Processes and Landforms* 46, 1926–1940.

Sun, L., Liu, Y.F., Wang, X.T., Liu, Y., & Wu, G.L., 2022. Soil nutrient loss by gully erosion on sloping alpine steppe in the northern Qinghai-Tibetan Plateau. *Catena* 208, 105763.

Suttie, J.M. Reynolds, S.G., & Batello, C., 2005. *Grasslands of the World*. FAO, Italy, Rome.

Taguas, E.V., Ayuso, J.L., Pérez, R., Giráldez, J.V., & Gómez, J.A., 2013. Intra and inter-annual variability of runoff and sediment yield of an olive micro-catchment with soil protection by natural ground cover in Southern Spain. *Geoderma* 206, 49–62.

Taguas, E.V., Yuan, Y., Licciardello, F., & Gómez, J.A., 2015. Curve numbers for olive orchard catchments: case study in Southern Spain. *Journal of Irrigation and Drainage Engineering* 141, 05015003.

Tarolli, P. 2018. Agricultural terraces special issue preface. *Land Degradation & Development* 29(10), 3544–3548.

Thornes, J.B., 2007. Modelling soil erosion by grazing: recent developments and new approaches. *Geographical Research* 45, 13–26.

Vercruysse, K., & Grabowski, R.C., 2019. Temporal variation in suspended sediment transport: linking sediment sources and hydro-meteorological drivers. *Earth Surface Processes and Landforms* 44, 2587–2599.

Verheijen, F.G.A., Jones, R.J.A., Rickson, R.J., & Smith, C.J., 2009. Tolerable versus actual soil erosion rates in Europe. *Earth-Science Reviews* 94, 23–38.

Vicente-Vicente, J.L., García-Ruiz, R., Francaviglia, R., Aguilera, E., & Smith, P., 2016. Soil carbon sequestration rates under Mediterranean woody crops using recommended management practices: A meta-analysis. *Agriculture, Ecosystems & Environment* 235, 204–214.

Wang, G.X., Hu, H.C., Li, G.S., & Li, N., 2009. Impacts of changes in vegetation cover on soil water heat coupling in an alpine meadow of the Qinghai-Tibet Plateau, China. *Hydrology and Earth System Sciences* 13, 327–341.

White, R., Murray, S., & Rohweder, M., 2000. *Pilot analysis of global ecosystems: grassland ecosystems*. World Resources Institute, Washington, DC.

Wilsey, B.J., 2018. *The Biology of Grasslands*. Oxford University Press, Oxford.

Wohl, E., Brierley, G., Cadol, D., Coulthard, T.J., Covino, T., Fryirs, K.A., Grant, G., et al. 2019. Connectivity as an emergent property of geomorphic systems. *Earth Surface Processes and Landforms* 44(1), 4–26.

Wu, G.L., Liu, Z.H., Zhang, L., Hu, T.M., & Chen, J.M., 2010. Effects of artificial grassland establishment on soil nutrients and carbon properties in a black-soil-type degraded grassland. *Plant Soil* 333, 469–479.

Xu, C., Jiang, Y., Su, Z., Liu, Y., & Lyu, J. 2022. Assessing the impacts of Grain-for-Green Programme on ecosystem services in Jinghe River basin, China. *Ecological Indicators* 137, article 108757.

Yang, K., Zhao, Y., & Gao, L. 2022. Biocrust succession improves soil aggregate stability of subsurface after "Grain for Green" Project in the Hilly Loess Plateau, China. *Soil and Tillage Research* 217, article 105290.

Ye, Y.-G., Chen, G.-J., & Fan, F., 2003. Impacts of the "Grain for Green" project on rural communities in the Upper Min River Basin, Sichuan, China. *Mountain Research and Development* 23(4), 345–352.

Zhou, W., Guo, Z., Chen, J., Jiang, J., Hui, D., Wang, X., Sheng, J., et al. 2019. Direct seeding for rice production increased soil erosion and phosphorus runoff losses in subtropical China. *Science of The Total Environment* 695, 133845.

13

Rivers and Wetland Systems

KIRSTIE FRYIRS, GARY BRIERLEY, TIM RALPH AND IAN FULLER

This chapter provides a brief summary of the concept of switches that regulate sediment (dis)connectivity in rivers and catchments (Section 13.1) and wetlands as disconnectors (Section 13.2). We use these summaries to set up how these concepts can be applied and used in management practice. We then present three case studies that apply these concepts in practice as part of on-ground sediment management (Sections 13.3, 13.4 and 13.5). We conclude with a brief commentary on considerations for implementation of switch management in practice (Section 13.6).

13.1 Managing Sediment (Dis)Connectivity: The Concept of Switches that Regulate (Dis)Connectivity in Rivers and Catchments

Water, sediment (and vegetation) interactions are fundamental drivers of riverscape ecosystems as they determine the physical template. While flow and vegetation can sometimes be readily managed, managing the sediment regime of rivers is fundamentally challenging given the highly varied nature of sediment source, transfer and storage patterns in catchments, the need to understand the 'expected' function of highly complex sediment interactions over various spatio-temporal scales, and to determine what can realistically be managed in systems that have often been highly disturbed by humans (Sear et al., 1995; Florsheim et al., 2006; Wohl et al., 2015).

The sediment budgets and sediment storage, transfer and accumulation dynamics of rivers and catchments are often understood in the context of (dis)connectivity (Fryirs et al., 2007; Fryirs, 2013; Bracken et al., 2015; Singh et al., 2021). When managing (dis)connectivity it is important to understand which of these levers (sediment storage, transfer and accumulation) can be realistically adjusted to achieve a desired outcome (Pöppl et al., 2020). The desired outcome may be, for example, to limit sediment supply coming off hillslopes or enhance sediment trapping and storage in an overwidened channel. In this context it is important to consider structural compartments of landscapes, at a range of scales, and the strength of functional linkages between these compartments, and whether it is possible to adjust these levers at-scale (Wainwright et al., 2011; Pöppl et al., 2017; Wohl et al., 2019).

Structural compartments can be hillslopes, floodplains, channels, and wetlands, whereas functional components consider the processes that occur within-compartment, between-compartment and catchment – or landscape-scale to supply, store or transport sediment in a landscape (Brierley et al., 2006, 2021; Fryirs et al., 2007; Singh et al., 2021). When considering catchments, the combination, sequence and pattern of compartments construct the sediment conveyor belt or cascade of a system (Walling, 1983; Burt and Allison, 2010). The interfaces between each of these compartments can be considered as a series of switches (analogous to taps) that are on, off, or dripping depending on the type and strength of processes that are occurring (Fryirs et al., 2007; Fryirs, 2013). When turned on a switch allows sediment supply and transport within- and between compartments. When the switch is turned off storage occurs and no supply or transport is happening. When the switch is dripping (i.e., a leaky system; Wohl and Beckman, 2014), partial supply and transport is occurring.

At any position along the cascade, blockages (or disconnectors) can regulate the functioning of the sediment cascade. Fryirs et al. (2007) describe buffers, barriers and blankets as landforms that disconnect lateral, longitudinal and vertical linkages in catchments (Fryirs et al., 2007). These disconnectors can be either natural (e.g., floodplains and/or wetlands) or anthropogenic (dams and reservoirs, stopbanks including artificial/constructed levees) (Pöppl et al., 2020). The position and residence time of blockages/disconnectors (e.g., years, decades, centuries, thousands of years) regulates the operation of the sediment cascade in each catchment and/or landscape (Brierley et al., 2006; Fryirs, 2013). Depending on the temporal attributes of each blockage or disconnector, and the magnitude-frequency relationships that drive the sediment cascade, switches are turned on, off or drip over various timescales (Hooke, 2003; Jain and Tandon, 2010). The contributing area that is active at any given time with switches on has been called the effective catchment (or contributing) area of (dis)connectivity (Harvey, 2001, 2002; Fryirs et al., 2007). The frequency-magnitude relationships that dictate the timeframe over which blockages are breached and switches turned on or dripping determines the effective timescales of (dis)connectivity in a catchment or landscape (Harvey, 2002; Fryirs et al., 2007; Fryirs, 2013; Lisenby et al., 2020).

Structural and functional attributes of a catchment or landscape sediment cascade provide a critical physical template that is a key control on habitat and ecological change in river systems, and socio-cultural connections to, and associations with, rivers (Boardman et al., 2019; Brierley and Fryirs, 2016; Fausch et al., 2002; Fuller and Death, 2018; Pöppl et al., 2017; Hillman et al., 2008). Understanding this physical template provides the basis for understanding what may or may not need to be managed, and forecasting how catchments and landscapes may/will change under pressures such as climate and landuse alterations. Therefore, there is a growing need to incorporate conceptual and quantitative understandings of (dis) connectivity to support informed decision-making in river and catchment management (Keesstra et al., 2018; Turley et al., 2021; Wohl et al., 2019). However, to date, most management plans overlook the role of sediment (dis)connectivity as drivers of disturbance and treatment responses in rivers and catchments (cf., Brierley and Fryirs, 2009; Fryirs et al., 2009), and only recently have models and workflows been developed that can quantify (dis)connectivity (Brierley et al., 2021). Although such approaches support forecasting exercises that provide managers and practitioners with an evidence base with which to treat sediment problems, or improve the health of catchments and landscapes, uptake of such practices remains in its infancy.

In managing sediment (dis)connectivity, concern lies with both the amount and the composition of sediment in different component parts of the river system (Pöppl et al., 2020). Therefore, in reality and on-the-ground, the only way that management of sediment (dis)connectivity can occur is to regulate the switches – whether they are within-compartment, between-compartment or catchment/landscape-scale. However, before any treatment response is considered (Schmidt et al., 1998), consideration needs to be given to whether the sediment problem being identified is 'real' – that is, is it really a problem, or just a perceived problem, and whether the compartment or system is operating in a way that is not 'expected' (Wohl et al., 2019; Pöppl et al., 2020). This decision-making process highlights the critical need to know your catchment – knowing what is being managed and why (Brierley & Fryirs, 2005). To achieve this requires understanding of the expected or background sediment flux dynamics to determine, first, whether management is needed at all, and second, if management is needed, what can realistically be managed (Fryirs & Brierley, 2016; Lisenby et al., 2020). With this, decisions can be made about where is it best to intervene and what is the required timeline of interventions that will produce a desired response. Managers must understand the spatio-temporal dynamics of the sediment cascade for each catchment or landscape they are working in (not a modelled version). Even the basic understanding of whether the system being managed is naturally disconnected or more strongly connected is vital (Fryirs, 2013).

Once the system being managed is understood, different management plans can be developed relative to management goals (Lisenby et al., 2020). When using (dis)connectivity concepts in rivers and catchments, river recovery potential and enhancement principles are often used to develop these plans (Brierley & Fryirs, 2005; Fryirs & Brierley, 2016). Careful consideration is given to the role of pressures and limiting factors that may enhance or inhibit the potential for recovery and how sediment (dis)connectivity of the system regulates the expression of these processes on-the-ground (Brierley & Fryirs, 2009; Fryirs & Brierley, 2005, 2016; Scorpio et al., 2015; Ziliani & Surian, 2012, 2016). Depending on the compartment being worked with, and the scale of impact required, various treatment options are available to manage the timing and frequency of sediment flux dynamics – from the highly engineered with heavy maintenance, to the do-nothing option that allows systems to self-heal via their own internal regulation (Warrick et al., 2015; Pöppl et al., 2019; Wheaton et al., 2019). For example, if the system is self-healing, precautionary, 'leave it alone' and passive approaches may be appropriate (Fryirs et al., 2018). Elsewhere, the efforts (or time needed) for river recovery may outweigh the benefits of management actions (Kondolf, 2011). In other places, inappropriate implementation of treatments (either on-site or off-site) may result in the stalling or reversal of recovery. For example, legacy sediment release and associated downstream geomorphic effects following dam removal or impacts of mining, water mills/dams or landuse change (e.g., Evrard et al., 2011; James, 2010, 2013; Pöppl et al., 2015; Walter & Merritts, 2008; Wohl, 2015). Situating each and every action within its system context is critical (Wohl, 2018; Brierley & Fryirs, 2009).

In this chapter, we use three sediment (dis)connectivity case studies to highlight the sediment 'problems' being managed, and possible management responses at switches that could be used in practice. Where available we show how some of these activities have worked (or failed) to improve the recovery potential and condition of these systems. We draw on examples from a spectrum of landscapes with highly varied sediment flux dynamics, with two very distinct end members. We use an example from a tectonically active-high relief setting (Waipaoa catchment in Aotearoa New Zealand), a tectonically passive–moderate relief setting (Bega catchment, Australia), and a tectonically passive-low relief setting in which managing wetlands as landscape disconnectors is vital (Macquarie catchment, Australia). Prior to presenting the three case studies, we provide a conceptual overview of wetlands as landscape disconnectors.

13.2 Managing Sediment (Dis)Connectivity: The Concept of Wetlands as Landscape Disconnectors

Wetlands are areas inundated by shallow water and where soils and biological activities are adapted to wet conditions (Mitsch & Gosselink, 2015). Wetlands exist in almost all landscapes and climate zones around the world and often bridge the gap between fully aquatic and terrestrial systems (Brinson, 1993). A combination of extrinsic environmental controls (e.g., geology, tectonics, climate, hydrology, sea- and base-level fluctuations) and intrinsic geomorphic processes (e.g., sedimentation, erosion) and human activities influence their formation, maintenance and modification over time (Larkin et al., 2017; Ralph & Hesse, 2010). These environmental factors also influence hydrological and sediment (dis)connectivity, which are critical for the operation of biophysical processes and feedbacks in wetlands. Both hydrological and sediment (dis)connectivity can influence longitudinal, lateral and vertical linkages at larger scales between wetlands and their catchments, and also at smaller scales within individual wetlands.

Wetlands can act as stores and sources of water and sediment in catchments, and therefore as both disconnectors and connectors in a landscape at different scales of consideration, where they inhibit the transmission of sediment from one part of a system to another (see Fryirs et al., 2007). Sediment buffering and storage occur in wetlands through depositional landforms, such as splays, mounds, and levees on floodplains (Florsheim et al., 2006; Lisenby et al., 2019). Erosional processes such as channel incision and bank erosion facilitate sediment excavation (Florsheim et al., 2008), and therefore increase longitudinal sediment connectivity while reducing lateral connectivity. Wetlands can be important connectors in a landscape where they release water and sediment (e.g., when the surrounding catchment is dry), or when they cease to trap and store sediment (e.g., due to channelisation; Lisenby et al., 2019).

In catchment upland zones, typically confined valleys with high slope-channel connectivity that produce water and sediment, wetlands may take the form of perched or hanging swamps on a hillslope, bogs in depressions, impounded peat swamps behind landslides, valley-fill swamps, or groundwater fed springs (e.g., Brinson, 1993; Fryirs et al., 2019; Figure 13.1). In piedmont or hilly zones, often partly confined valleys with moderate slope-channel connectivity and high longitudinal channel connectivity allowing water and sediment transfer downstream, wetlands may occur as in-stream pools, waterholes, backswamps on floodplain pockets, or chain-of-ponds (e.g., Williams & Fryirs, 2020; Figure 13.1). In lowland zones, usually unconfined valleys with minimal slope-channel connectivity and high longitudinal channel and lateral floodplain connectivity, and where water and sediment tend to deposit and accumulate, wetlands may take the form of meander cutoffs, abandoned channels, backswamps, lakes, and other floodplain wetlands including lagoons and marshes (Semeniuk & Semeniuk, 1995; Figure 13.1). In some dry, lowland zones, where tributary inputs are exceeded by losses (e.g., infiltration, evaporation, outflows), wetlands form on floodplains with distributary or anastomosing channels, and where channel breakdown occurs at floodouts (e.g., Gore et al., 2000; Ralph & Hesse, 2010) Wetlands may also act as long-term

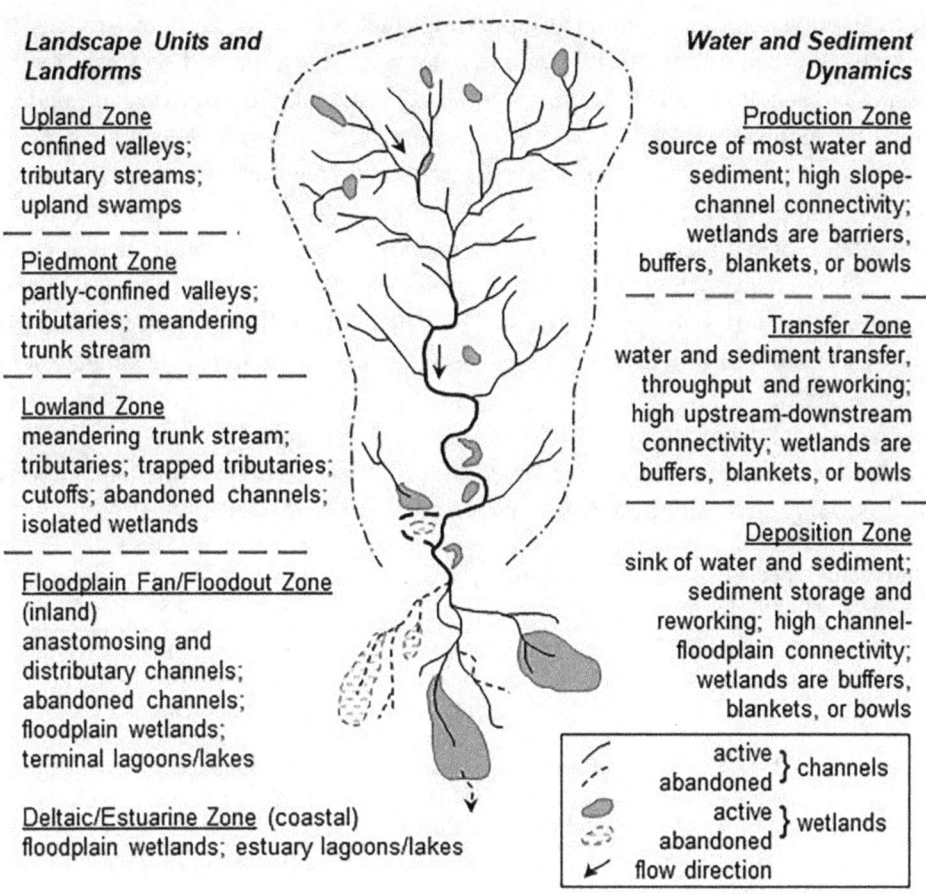

Landscape Units and Landforms

Upland Zone
confined valleys;
tributary streams;
upland swamps
— — — — — — — — .

Piedmont Zone
partly-confined valleys;
tributaries; meandering
trunk stream
— — — — — — — — —

Lowland Zone
meandering trunk stream;
tributaries; trapped tributaries;
cutoffs; abandoned channels;
isolated wetlands
— — — — — — — — —

Floodplain Fan/Floodout Zone
(inland)
anastomosing and
distributary channels;
abandoned channels;
floodplain wetlands;
terminal lagoons/lakes

Deltaic/Estuarine Zone (coastal)
floodplain wetlands; estuary lagoons/lakes

Water and Sediment Dynamics

Production Zone
source of most water and
sediment; high slope-
channel connectivity;
wetlands are barriers,
buffers, blankets, or bowls
— — — — — — — — —

Transfer Zone
water and sediment transfer,
throughput and reworking;
high upstream-downstream
connectivity; wetlands are
buffers, blankets, or bowls
— — — — — — — — —

Deposition Zone
sink of water and sediment;
sediment storage and
reworking; high channel-
floodplain connectivity;
wetlands are buffers,
blankets, or bowls

active ⎱
abandoned ⎰ channels

active ⎱
abandoned ⎰ wetlands

flow direction

Figure 13.1   The role of wetlands as disconnectors in landscape units and sediment process zones in catchments. A black and white version of this figure will appear in some formats. For the colour version, refer to the plate section.

sinks for sediment, or 'bowls' in terms of sediment (dis)connectivity (Figure 13.1). These wetlands include depressions or craters isolated from other water sources (e.g., meteorite impact craters, volcanic craters) or terminal lakes occurring at base-level in a catchment (e.g., playa lakes, impounded wetlands) (e.g., Rayburg & Thoms, 2009). Where wetlands occur at the coast, wetlands take the form of estuaries, mangrove flats, impounded swamps in dunefields, and freshwater seeps.

Wetlands are usually considered and managed in terms of their local water regime (e.g., flow volume, timing, frequency, rate of rise and fall; Ralph & Rogers, 2011), despite providing larger-scale hydrological and ecological connectivity between terrestrial and aquatic ecosystems (Rittenhouse & Peterman, 2018). Wetland commence-to-flow levels and inundation patterns often used to predict conditions for biota, and to support management of other essential habitat and sources of energy and nutrients (Rogers and Ralph, 2011). Wetlands also provide critical provisioning, regulating, supporting and cultural ecosystem services to humans (Costanza et al., 1998), for example, water filtration and sediment retention (Clarkson et al., 2013; Larkin et al., 2020; Xu et al., 2020). However, understanding sediment (dis)connectivity in wetlands also informs their management and conservation (Table 13.1). Wetlands in upland zones are susceptible to headwater changes in rainfall/runoff and fires, potentially leading to destabilisation and release of sediment and nutrients. Wetlands in piedmont zones may be more affected by changes to upstream flow regimes, which affect sediment supply and transfer. Wetlands in lowland zones will be much more reliant on upstream inputs, and lateral connection from overbank flooding may be reduced leading to less

Table 13.1   *Important aspects of the geomorphology of wetlands and implications for managing sediment (dis)connectivity (modified from Tooth et al., 2015).*

Important aspect of wetland geomorphology	Landform and process explanation	Key messages for managing sediment (dis)connectivity
Wetlands are shaped by movements of mass.	The physical structures of wetlands are shaped by geomorphic processes, involving the movement of rock, sediment, water, and organic matter across the Earth's surface.	Wetlands influence forms and rates of hydrological and sediment (dis)connectivity in catchments.
Wetlands form in a range of landscape and climatic settings, including drylands.	Topographic setting and sources of water availability influence morphodynamics.	Wetlands, whether predominantly wet or dry, are important agents of (dis)connectivity that may be switched on or off, with residence types and ease/rates of sediment release vary markedly dependent upon wetland type.
Earth System interactions influence wetland type and formative processes.	Topographic relationships influence wetland processes associated with interactions between the atmosphere, hydrosphere, geosphere and biosphere.	Processes that create and change wetlands are controlled by broader-scale environmental conditions and (dis)connectivity in catchments.
Wetlands are naturally dynamic.	Disturbance events such as tectonic, geological, climatic, sea level changes impact upon wetland morphodynamics.	Wetlands evolve over time, driven partly by changes in broad-scale (dis)connectivity and hydroclimatic teleconnections.

Table 13.1   *(cont.)*

Important aspect of wetland geomorphology	Landform and process explanation	Key messages for managing sediment (dis)connectivity
Wetland dynamics may be complex.	In addition to changing external conditions, wetland development is driven by changing ecological conditions or internal geomorphological adjustments.	Dynamic interplay of biophysical processes relies inherently on (dis)connectivity within a wetland system.
Wetlands are archives of past environmental change.	The history of wetlands development can be reconstructed from study of their landforms, sediments and biological remains.	Wetlands can yield information on past and present regimes of (dis)connectivity.
Global environmental change is influencing wetlands.	Climate change drives changes to wetland structure and function.	Wetlands respond to both subtle and more acute environmental changes, some related directly to (dis)connectivity.
Human activities alter wetlands.	Direct and indirect impacts drive changes to wetland structure and function.	Disturbance and regulation of wetlands can upset the balance of (dis)connectivity required for their survival.
Wetlands may be vulnerable to geohazards but may also buffer the wider landscape from their impacts.	Global environmental change and human activities exacerbate impacts of extreme floods, droughts, fires, and storm surges. Some wetlands are able to absorb these impacts when other parts of the landscape cannot.	Wetland resilience to some geohazards may be enhanced because they are able to switch between being disconnectors and connectors.
Geomorphological knowledge is vital to sustainable use of wetlands.	Geomorphic insights support conservation goals, protecting and enhancing the delivery of wetland ecosystem services. Wetland management is a key part of flood management programmes.	Knowledge and understanding of the role of (dis)connectivity in wetland systems allows improved applications and options for management.

inundation and depauperation of flood-reliant wetlands (Florsheim et al., 2006). Channel erosion in floodplain wetlands may cause reduced channel-floodplain connectivity, exacerbating fire risk (Graves et al., 2019). Some lowland wetlands may be more resilient in the face of upstream changes due to their long history of intermittent allogenic inputs, that is, they are better adapted to variable inundation regimes. Isolated wetlands relying on groundwater or direct rainfall may be affected by larger-scale geological or climatic factors, since they are not supplied from larger catchments with the potential for new or shifting water sources.

The supply of sediment to wetlands and phases of sediment accumulation and release by wetlands are important considerations in wetland management strategies. This builds upon geomorphological understandings of how wetland landforms have developed in the past, how they function under current conditions, and how they are likely to operate in the future. Management options may vary markedly dependent upon the extent to which wetlands rely on water and sediment inputs from other parts of a landscape or an upstream catchment and/or when flow regulation and other human activities alter process interactions in or around wetlands.

13.3 Managing Sediment (Dis)Connectivity in a Tectonically Active-High Relief Setting (Waipaoa Catchment)

Tectonically active, steepland relief of Aotearoa's East Coast Region, coupled with weak, erodible lithologies, widespread indigenous forest clearance in the early twentieth century and a humid climate mean this region delivers some of the highest sediment yields to the global ocean (Marden et al., 2018). The region contributes 33% of New Zealand's total suspended sediment yield from 2.5% of the land area (Hicks et al., 2011). The 2,208 km^2 Waipaoa delivers 15 Mt yr^{-1}, equating to 13% of yield from the North Island and a specific sediment yield of 6,797 t km^{-2} yr^{-1} (Hicks et al., 2011). High erosion rates in the Waipaoa have been the defining catchment management issue since the mid twentieth-century (Marden, 2012). Hillslopes are primary sources of sediment, especially gullies, which include classic deep, unstable, incising linear channels along incipient drainage lines, as well as much larger gully mass movement complexes ('badass' gullies, Marden et al., 2018). Short, steep slopes in the Waipaoa mean shallow landslides, common in large storm events, couple effectively with the channel network. Sediment is also fed continuously to channels that intersect with earthflows. Slope-channel coupling is high, with very efficient sediment delivery to channels.

The upper Waipaoa trunk river and its tributaries have been transformed from single-thread, clear-flowing cobble-bedded rivers, to multi-thread, sediment-laden, rapidly aggrading channels with matrix-rich gravelly beds (Gomez & Livingston, 2012). Channel metamorphosis is a response to increased sediment loads following forest clearance (Marden et al., 2014). The upper river has aggraded up to 20 m (Marden, 2011). The lower Waipaoa River, once gravel-bedded and laterally active, has become fixed between cohesive, silty banks; floodplain sedimentation is dominated by rapid vertical accretion (Gomez et al., 1998). Channel dimensions in the lower Waipaoa have accommodated increased fine sediment supply by reducing bankfull width and cross-section area, leading to channel contraction (Gomez et al., 2007). High suspended load is derived partly from rapid breakdown of gravel-sized clasts of highly erodible shales and mudstones supplied from hillslope erosion in the upper catchment (Gomez et al., 2001), as well as the direct feed of fine sediment from gullies (Hicks et al., 2000). Reduced channel capacity and floodplain accretion in the lower Waipaoa compromises the Waipaoa Flood Control Scheme, which was designed in 1949 and completed in 1969 to provide '100-year' (1% AEP) flood protection for over 10,000 ha of intensive horticulture on the Poverty Bay Flats. A ~$35 million upgrade of the Scheme is underway to maintain '100 year' protection accounting for climate change impact to 2090 (GDC, 2020).

The sediment budget and flux dynamics of the Waipaoa catchment and various compartments have been widely published and are not repeated here. The most significant compartments contributing sediment are 'badass' gullies. These have developed in terrain underlain by highly weathered and crushed Cretaceous shales associated with the East Coast Allochthon, and often form alluvial fans, overwhelming streams. Gullies represent a continuous sediment supply and the single largest source of sediment to the system (Marden et al., 2018) … Earthflows are prominent through much of the upper catchment (Marden et al., 2014). Most of the remainder of the hillcountry is prone to shallow landsliding (Jones & Preston, 2012), as well as deep-seated landslides (Bilderback et al., 2015). Sediment derived from shallow landslides is significant during storm events, but gully, sheet and riverbank erosion have the potential to generate sediment whenever it rains, so the long-term Waipaoa sediment yield is dominated by cumulative effects of low-magnitude events (Hicks et al., 2000,). In the upper catchment, channel widening contributes further sediments, particularly through bank and cliff erosion (DeRose & Basher, 2011). The fine-grained nature of sediment supplied from the upper catchment, especially from gully erosion, accentuates downstream delivery of materials and throughput to the coast (Gomez et al., 2007; Kuehl et al., 2016). In the lower Waipaoa in-channel accretion of this fine-grained sediment has resulted in narrowing of channel cross-sections and reduction in-channel capacity (Gomez et al., 1998, 2007).

Four key sediment management aims in this catchment are directly linked to management of switches (Table 13.2; Figure 13.2). Aim 1: reducing incidence and impact of mass wasting as origination of sediment. Aim 2: modulating release of sediment stores associated with gully complexes in the form of short, steep, fans. Aim 3: retaining sediment on hillslopes. Aim 4: allowing sediment transfer and throughput along tributary and trunk stream channels to offshore. The switch

Table 13.2   *Approaches to managing sediment flux and connectivity relationships in the East Cape of Aotearoa. Managing sediment (dis)connectivity and switches in the Waipaoa catchment (modified from Pöppl et al., 2020).*

What are the sediment 'problems' in this catchment? What is trying to be achieved?		
(1) Retain sediment on hillslopes. (2) Reduce shallow landslide and gully complex formation. (3) Allow sediment transport and throughput along tributary and trunk stream channels to offshore.		
Switch and reaches to be managed	**Response and switch status**	**Techniques that could be used**
Managing sediment supply from hillslopes.	Ensuring sediment is retained on-hillslope or trapped on floodplains and terraces at base of hillslopes. Turn red ● (4) switches **off** to enhance disconnectivity.	Reforestation (exotic forestry). Direct vegetation planting. Fast-growing shrubs and willows Poplar upslope and willow on lower slope, indigenous (*nb.* manuka). Fencing off and stock exclusion on susceptible terrain.
* Managing **gully complexes**	Reduce slope-channel connectivity, ensuring sediment remain on-slope. Turn red ● (1) switch **off** to enhance dysconnectivity – feasible in smaller (<10 ha) gullies.	Reforestation (exotic forestry). Direct vegetation planting. Fast-growing shrubs and willows Poplar upslope and willow on lower slope, indigenous (*nb.* manuka). Fencing off and stock exclusion on susceptible terrain. Construction of check dams, debris dams, detention bunds to store sediment, reduce incision and stabilise gully walls: effective in small gullies only.
* Managing fans	Reduce fan-channel connectivity, enhance buffering capacity of fan. Turn red ● (1) switch **off** to enhance dysconnectivity	Reforestation (exotic forestry) Direct vegetation planting, *nb* manuka Fencing off and stock exclusion. Avoid over-narrowing of active fan. Protect distal fan from lateral channel erosion (bank stabilization, planting)

Table 13.2   (*cont.*)

* Managing **shallow landslides**.	Reduce slope-channel connectivity, ensuring sediment remain on-slope. Turn red ● (1) switch **off** to enhance disconnectivity	Reforestation. Radiata, indigenous species (manuka). Direct vegetation planting. Fast-growing shrubs and willows Poplar upslope and willow on lower slope, indigenous (*nb.* manuka). Fencing off and stock exclusion on susceptible terrain. Riparian planting (natives, willows) to buffer streams
Managing **channel infilling and widening**.	Allow sediment transfer downstream at sustainable rates. Leave green ● (2), (3), (A) switch **on** or allow **dripping** and connectivity upstream–downstream and tributary-trunk.	Managing the problem at source, that is, treat gully complexes and landslides first. Maintain bank defences as a last resort, set stopbanks back to allow room for flooding and dispersal of sediment across floodplain, enhancing sequestration out-of-channel. Monitor aggradation and incision trends, particular as upstream reaches begin to incise.
Managing **channel contraction**.	Allow sediment transfer downstream at sustainable rates. Allow floodplain to act as a sediment sink. Remove all critical activities from the floodplain so far as possible. Leave green ● (2), (3), (A) switch **on** or allow **dripping** and connectivity channel-floodplain.	Create riparian buffer zone. Provide more room for flooding and sediment sequestration. Direct planting of willows, deep-rooting natives, robust grasses, flaxes. Build riprap, dolos, groynes, spurs at strategically important points and reaches.

Symbols and colour codes are matched to Figure 13.2 for cross-referencing.

arrangement and potential regulation for the highly disturbed Waipaoa catchment is illustrated in Figure 13.2. Table 13.2 outlines in detail the desired operation of each switch and rehabilitation techniques that are being used (or could be used) to regulate these switches.

Aim 1 manages mass wasting processes generating sediment from hillslopes (Table 13.2). Best managed via reforestation practices using *Pinus radiata*,

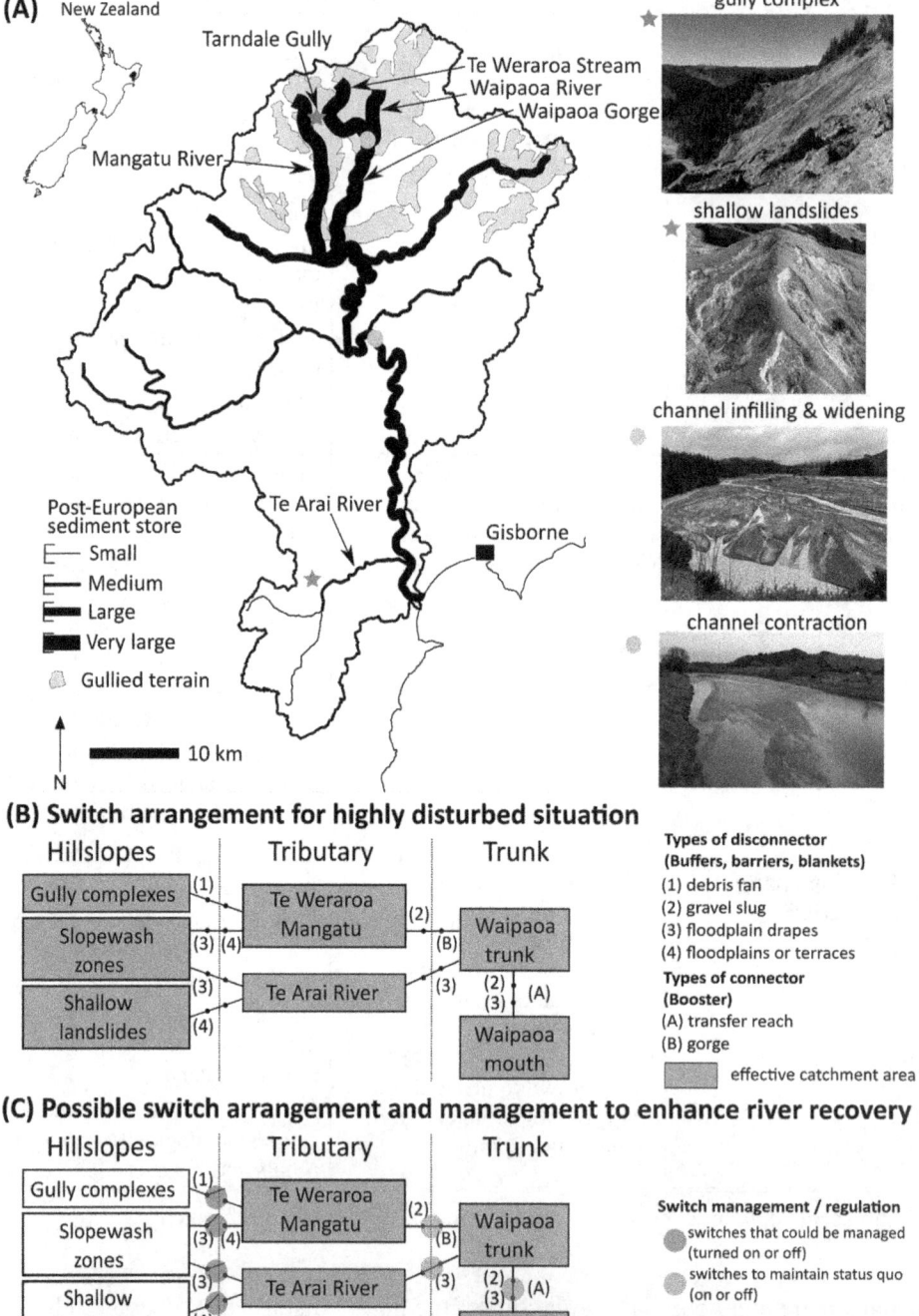

Figure 13.2   Regulating switches to manage sediment (dis)connectivity in the Waipaoa catchment. (B, C) modified from Fryirs et al. (2007). Landslides photo: Mike Crozier. A black and white version of this figure will appear in some formats. For the colour version, refer to the plate section.

re-vegetating hillslopes with a sufficient density of stems per hectare (1,250) very effectively reduces sediment delivery from unvegetated or susceptible terrain (Marden et al., 2014). The switches in these locations need to be turned off to disconnect slope-channel coupling (Figure 13.2C). Treatment of gullies in this approach requires early intervention, since large, 'badass' gully complexes are beyond remediation (Marden et al., 2018). The challenge in the Waipaoa is that over 50 badass gullies remain, however, sediment supply does appear to be diminishing from these systems as structural resistance increases with system complexity and growth of buffering fans (Fuller et al., 2020).

Aim 2 targets management of fans buffering badass gully systems from channels. Fans buffer sediment via two independent switches. The first switch between the gully complex and upper fan proximal to the gully is activated by gully processes (Aim 1), which reflect an array of intrinsic and extrinsic variables (Fuller & Marden, 2011). The second switch sited at the junction between the fan and channel is critical to sediment supply to the channel network. Loading of the fan by sediment supplied from the gully complex steepens the fan profile, enhancing connectivity and sediment delivery (Fuller & Marden, 2011,). During these phases, the distal fan grows because sediment delivery exceeds transport capacity of the stream channel and may block it. Sediment from the distal fan is actively eroded by flood events in the stream channel – the fan in this case is passive since the process conveying sediment to the channel is driven by stream processes. Distal fan trimming by lateral stream erosion steepens the lower fan, generating a knickpoint, which recedes up-fan, evacuating sediment to the channel. Operation of this critical fan-stream junction switch is complex and event-dependent. Considerable year-to-year variability in volumes of sediment supplied from gully complexes reflects intrinsic landform adjustment, as well as extrinsic drivers, so gullies and their associated fans are not in sync across the catchment. The adjacent Tarndale and Mangatu complexes are asynchronous in their behaviour (Fuller et al., 2020). Mangatu fan also has less buffering capacity than Tarndale fan, emphasizing connectivity variability as a consequence of landform configuration and structural resistance. In turn, switches between gully fan-stream are not in-phase and the residence time of material delivered from gully to fan may range from decades to months (Taylor et al., 2018). Nevertheless, these fans offer the possibility for targeted management, the key being retaining sediment in storage in these buffers and turning this switch off. Stabilisation of fans can be achieved by reforestation, which restricts the width of the active fan (Gomez et al., 2003). However, Taylor et al. (2018) suggest that residence times for sediment in the now much smaller active part of the Tarndale fan have likely reduced, enhancing sediment connectivity. Perversely, reforestation of the Tarndale fan may have enhanced sediment volume delivered to the stream channel by some 17,000 t yr^{-1} (Marden et al., 2005).

These distal fan switches need to be turned off, but cannot be managed in isolation, from either processes in the contributing gully complex, or stream channel processes. Leenman and Tunnicliffe (2020) observed that the potential for fans in the adjacent Waiapu catchment to act as buffers depends on the relative speed of fan trenching (connectivity enhanced, switch on) and mainstem incision (disconnecting fan, switch off). As gully contribution decreases over time, it will be imperative to stabilise the lower fans to turn this switch off to ensure this store of sediment is not reworked.

Aim 3 involves regulating sediment generated by sheet erosion and shallow landslides from less erodible terrain that is retained in pastoral production. Slope-channel coupling from shallow landslides in small catchments is dependent on catchment morphology, landslide and triggering event characteristics (Jones & Preston, 2012). Retaining soil on the hillslopes (preventing failure in the first place) and buffering streams from landslide tails is critical: space-planting of poplars and willows and/or a sufficient stem density of native manuka or kanuka can be effective to reduce probability of hillslope failure. Particular care is needed to ensure stored toe-slope colluvium is not remobilised into streams, which can be mitigated by riparian planting. Switches in these locations need to be turned off to disconnect slope-channel coupling, which with targeted and strategic planting is achievable (Figure 13.2C).

Aim 4 regulates sediment transfer and throughput dynamics of river channels in the catchment. This manages tributary and trunk stream channels, with the focus addressing channel infilling and widening, as well as reduction in-channel capacity through fine sediment accretion (Table 13.2). Once in the channel, sediment accumulation becomes more difficult to manage. Managing the problem at source, rather than in the river, is the preferred option, especially since the overwhelming majority of sediment in the Waipaoa is sourced from gullies, their associated fans, and other hillslope processes. As more sediment is supplied to stream channels from multiple, out-of-sync gully complexes and associated fans, further active lateral adjustment and channel widening is likely in the upper Waipaoa, as is continued fine-grained sediment accretion of the lower Waipaoa channel and floodplain (Gomez et al., 2009). Adopting erodible corridor concepts and managing adjacent land tenure and landuse is desirable because the river requires room to adjust to ongoing sediment influx in the coming decades. Hard engineering such as gravel extraction can be effective in aggrading systems if carried out carefully and managed adaptively, but poor quality of weak lithologies comprising the bulk of the Waipaoa bedload means this is unviable. Bedload yields are ~1% of the suspended load (Gomez et al., 2001). Switches in these channel locations need to remain on or allowed to drip to maintain upstream–downstream, tributary-trunk stream and channel-floodplain connectivity, allowing export and/or floodplain sequestration of sediment in the system (Figure 13.2C).

Shutting down hillslope-erosion sources in the Waipaoa is having an effect on trunk-river behaviour. Some tributaries in the upper Waipaoa are now incising. The corollary is that this process continues to generate sediment, which is fed downstream. Management of connectivity in this disturbed, tectonically active steepland catchment is multi-centennial in scope.

13.4 Managing Sediment (Dis)Connectivity in a Tectonically Passive–Moderate Relief Setting (Bega Catchment, Australia)

The Bega catchment lies in an escarpment dominated landscape setting with a granitic geology. Wide, alluvial valleys containing swamps, sometimes occur at the base of the escarpment forming sediment accumulation zones. Middle reaches of the catchment are dominated by bedrock confined and partly confined valleys where sediment is throughput (flushed) or transferred (sediment input = sediment output) through the system (see Fryirs & Brierley, 2001, 2005). There is a relatively short lowland plan that acts as a sediment accumulation zone and sink. Like many rivers found in granitic terrain, sand is the primary calibre of material. Hillslopes are not significant sediment sources in these landscapes as slope-channel decoupling is strong (Fryirs & Brierley, 1999). Most sediment is sourced from valley bottoms via channel adjustment and erosion processes. Sediment yield is low by global standards. Post colonisation human disturbance associated with riparian vegetation and wood removal, and drainage of swamps has had a major impact on the sediment regime and the function of the sediment cascade in this system (Brooks & Brierley, 1997, 2000; Fryirs & Brierley, 1998, 2001, 2005).

The Bega sediment budget and sediment (dis)connectivity patterns and changes since European colonisation, along with the history of river evolution, changes in ecological condition and river recovery are well documented in the literature and are not repeated in detail here (e.g., Brooks & Brierley, 1997, 2000; Brierley & Fryirs, 2000, 2009; Brierley et al., 1999; Chessman et al., 2006; Fryirs & Brierley, 1998, 1999, 2001, 2005, 2016; Fryirs et al., 2007, 2018; Fryirs, 2013). The legacy of past landuse changes continues to assert a profound impact on the riverscapes of the Bega Catchment. Incision of channels into swamps and tributary fills released nearly 10 million m^3 of sediment into the tributary stream network (Fryirs & Brierley, 2001). The overwidening of channels (bank erosion) in both the tributary systems and along the lower truck stream contributed a further 4 million m^3 (Fryirs & Brierley, 2001). Along the lower Bega River, the channel expanded by over 150% as a result of riparian vegetation and wood removal (Brooks & Brierley, 1997). The majority of the sediment released from the catchment created a sediment slug with an approximate in-channel volume of 8 million m^3, with over 4 million m^3 of material stored as floodplain sand sheets (Fryirs & Brierley, 2001).

As a result of this history, sediment (dis)connectivity patterns in the catchment have switched over time between disconnected (swamps and small capacity channels), to hyper-connected (channel incision and expansion, sediment slug delivery downstream), to weakly connected (recovery via within-catchment sediment trapping within a supply limited, sediment exhausted catchment) (see Pöppl et al., 2020). Forecasting to 2100 suggests that this trend is likely to continue (Tangi et al., subm).

River management applications in the Bega catchment do not explicitly manage for sediment (dis)connectivity. Rather this concept directly informs the recovery-enhancement approach (Fryirs and Brierley, 2001, 2005). Managing the sediment budget, and therefore (dis)connectivity, is implicit in all passive approaches to management in this catchment, and is often framed (and expressed) in terms of 'refilling the incised channels', 'locking up sediment in the right places', 'treating the sediment slug', 'allowing sediment movement along transfer reaches' and 'getting the vegetation back on the bench' (i.e., stepped, bank-attached features that act to contract over-widened channels). Extensions into practice use expressions such as 'locking up sediment on paddocks', 'whole of farm management', 'revegetating the banks', 'improved water quality and land productivity', and so on (Fryirs et al., 2018; Thompson et al., 2016).

Figure 13.3A presents a summary overview of the current-day sediment budget. The thickness of the black lines depicts where readily available sediment sources can enhance river recovery via 'sediment management'. Here we identify five key sediment management aims that can be directly linked to management of switches (Table 13.3; Figure 13.3). First, preventing incision into valley fills (swamps). Second, reducing sediment supply to the lowland plain sediment slug. Third, retaining sediment and recreating discontinuous watercourses within channelised fills (former swamps). Fourth, retaining and storing sediment in overwidened channels. Fifth, allowing sustainable sediment transport along bedrock reaches and offshore. In a catchment that has experienced sediment exhaustion, 'every grain counts' and this process will be a decadal or century long process.

Most river management strategies in the Bega Catchment have applied passive approaches, using vegetation as the 'engineer' (sensu Gurnell, 2014). This has been accompanied by changes in landuse and farming practices (i.e., fencing-off, stock access and grazing management, weed management). In many cases, the system has been 'allowed' to adjust and self-heal with minimal active intervention. In the last few decades noticeable results have been achieved in enhancing geomorphic and vegetative recovery and by extension, making progress towards achieving the sediment management aims (e.g., see Fryirs et al., 2018).

Figure 13.3B shows the expected switch arrangement for baseline sediment flux in this catchment, and Figure 13.3C shows the possible switch arrangement if the

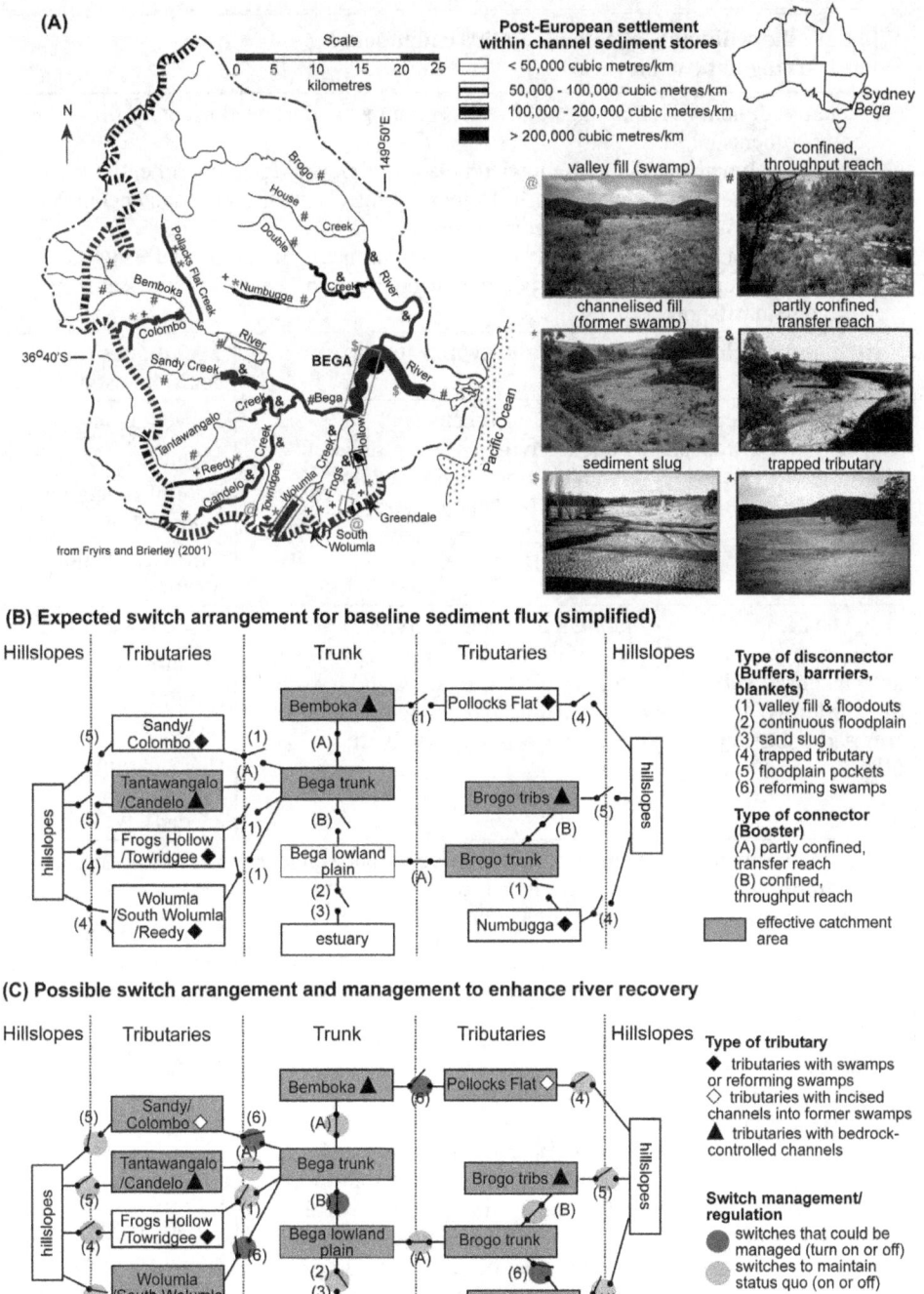

Figure 13.3 Regulating switches to manage sediment (dis)connectivity in the Bega catchment. (A) modified from Fryirs and Brierley (2001) and Pöppl et al. (2020). (B, C) modified from Fryirs et al. (2007). A black and white version of this figure will appear in some formats. For the colour version, refer to the plate section.

Table 13.3 *Managing sediment (dis)connectivity and switches in the Bega catchment (modified from Pöppl et al., 2020).*

What are the sediment 'problems' in this catchment?
What is trying to be achieved?

(1) Retain sediment in remaining valley fills (swamps) and trapped tributary fill compartments.
(2) Reduce sediment supply to the lowland plain sediment slug compartment.
(3) Retain sediment in channelised fill (former swamp) compartments and encourage reformation of discontinuous watercourses.
(4) Retain sediment in overwidened channel compartments in the form of benches.
(5) Allow sediment transport along bedrock reach compartments and some sediment throughput offshore.

Switch and reaches to be managed	Response and switch status	Techniques that could be used
Managing sediment supply from hillslopes.	Maintain slope-channel disconnectivity, ensuring sediment remain on-slope or trapped on floodplain pockets or in trapped tributary fills. Leave green ● (4), (5) switches **off**.	Leave it alone. Monitor. Natural vegetation regeneration. Direct vegetation planting.
@ + Managing sediment storage in unchannelised **valley fills (swamps)** and **trapped tributary fills**.	Maintain disconnectivity in these parts of the system, ensuring sediment remain stored in swamps and trapped tributary fills, and preventing incision. Leave green ● (1) switch **off**.	Leave it alone. Monitor. Natural vegetation regeneration. Headcut prevention and treatment (if needed).
# Managing sediment **throughput** in **confined reaches**.	Maintain sediment supply and transfer (connectivity) in and through these reaches. Leave green ● (A) switch **on**.	Leave it alone. Monitor. Natural vegetation regeneration. Weed management if needed.
& Managing sediment **transfer** in **partly confined reaches**.	Dependent on condition and position in catchment. Managing this switch is a sub-catchment by sub-catchment exercise depending on where sediment trapping, redistribution and transfer is needed and in what volume. Where overwidened channels and sediment exhaustion has occurred, encourage reformation of benches that act to narrow channel capacity. Turn green ● (B) switch **off** to enhance disconnectivity. Where sand sheets smother the channel bed, allow a sustainable sediment supply and transfer to downstream reaches, to scour pools and unbury geomorphic diversity. Turn red ● (B) switch **on** or allow **dripping** and connectivity.	Fencing off (stock access and landuse management). Monitor. Natural vegetation regeneration. Direct vegetation planting. Weed management.

Table 13.3   (*cont.*)

* Managing sediment storage in upstream parts of the catchment where **channelised fills (former swamps)** have formed.	Trap available sediments on the beds of incised channels. Encourage reformation of swamps or discontinuous watercourses within the incised channels. Turn red ● (6) switch **off** to enhance disconnectivity.	Fencing off (stock access and landuse management). Monitor. Natural vegetation regeneration. Direct vegetation planting. Weed management.
$ Managing **sediment slug**.	Encourage locking up and storage of sediment in floodplains and landforms with longer residence time (e.g., benches and in-channel islands). Allow reworking of some sediment on channel bed around vegetated islands. Allow sustainable sediment supply and transfer to estuary and offshore. Leave green ● (2), (3) switch **off** or allow **dripping** and disconnectivity.	Fencing off (stock access and landuse management). Monitor. Natural vegetation regeneration. Direct vegetation planting. Weed management.

Symbols and colour codes are matched to Figure 13.3 for cross-referencing.

5 key sediment management aims in this catchment are to be achieved. Figure 13.3C also shows the switches that can be regulated and Table 13.3 outlines in detail the desired operation of each switch (leave on or off, turn on or off, allow to drip) and rehabilitation techniques that are being used (or could be used) to regulate these switches.

Aim 1 involves working with the remaining valley fills (swamps) and trapped tributary fills in the system (Table 13.3). Many of these compartments have been lost via incision since European colonisation (Fryirs and Brierley, 1998). Those that remain are store significant volumes of sediment. Protecting these compartments and preventing incision into the swamps is a high priority. The switches in these locations (e.g., along Frogs Hollow and Towridgee) (Figure 13.3C) need to remain turned off to maintain slope-channel and tributary-trunk stream disconnectivity. Vegetation-based rehabilitation techniques can be used alongside treatment of headcuts that threaten to incise these compartments.

Aim 2 involves working with a low sinuosity sand bed river that contains the sediment slug (Fryirs and Brierley, 2001) (Table 13.3). This lowland plain compartment is sediment transport capacity limited. The current channel contains extensive, homogenous sand sheets, some benches and elongate vegetated islands that create an anastomosing-like within-channel planform in places (Brooks and Brierley, 1997). Given that the catchment is now largely sediment supply exhausted, the addition of more sediment to the slug has slowed considerably in the last few decades, and is forecast to slow further in coming decades (Tangi et al., subm).

Management now focusses on trapping and enhancing the storage of sediment in well-vegetated islands and benches while allowing for the transfer of small volumes of sediment along the smaller-capacity, anastomosing-like channels. As channel-floodplain connectivity is strong in this part of the catchment, long-term sediment trapping on floodplains (as a sediment sink) is also occurring. The switches in these locations need to be left off or dripping (Figure 13.3C) to maintain upstream–downstream disconnectivity and minimise the potential for negative off-site impacts in the estuary and at the river mouth. Again, vegetation-based rehabilitation techniques can be used, including the ongoing management of willows to, on one hand, prevent channel choking, but on the other maintain sediment storage/binding and stabilisation of the sediment slug.

Aim 3 involves retaining sediment in channelised fill (former swamp) compartments and encourage reformation of discontinuous watercourses (Table 13.3). Sediment release from these compartments has been the primary source of sediment to downstream reaches and the sediment slug since European colonisation. Today, these systems are largely sediment exhausted and under-supplied. Therefore, any sediment that is supplied to these compartments should be locked in place to enhance the re-creation of swamps and discontinuous watercourses within the incised channels (Fryirs et al., 2018). The switches in these locations need to be switched off (Figure 13.3C) to enhance upstream–downstream disconnectivity. Recovery is supported by fencing-off, stock management and re-establishment of natural swamp vegetation.

Aim 4 involves retaining sediment in overwidened channel compartments in the form of benches (Table 13.3). These sediment transfer compartments are located along which channels in partly confined valley settings that have become over-widened since European colonisation. The sediment slug has moved through these reaches on its way to the lowland plain. Dependent upon condition and position in catchment, managing this compartment is a sub-catchment by sub-catchment exercise depending on where sediment trapping, redistribution and transfer is needed and in what volume, as well as the amount of sediment that is currently being supplied to, and available in, each of these reaches (with reference to Figure 13.3A and Fryirs and Brierley, 2001, 2005). Where overwidened channels and sediment exhaustion has occurred, managing these compartments should trap sediment on in-channel benches that narrow channel capacity, turning off the switch; (Figure 13.3C). However, where sand sheets smother the channel bed, managing this compartment involves allowing a sustainable supply and transfer of sediment to downstream reaches, turning on switches or allowing them to drip to enhance upstream–downstream connectivity (Figure 13.3C). Again, vegetation-based rehabilitation techniques are key.

Finally, aim 5 involves allowing sediment transport along bedrock reach compartments and some sediment throughput offshore (Table 13.3). These compartments are most often located in the mid-catchment and act as efficient sediment through-put zones. Depending on condition these reaches can be characterised by large bedrock-based pools, islands and a well-vegetated riparian zone (good condition), or are smothered by sand sheets (poor condition) (Fryirs, 2015; Fryirs et al., 2021). Managing these compartments for geomorphic river recovery depends on their posi-tion in the catchment (with reference to Figure 13.3A) and the amount of sediment that is currently being supplied to, and available in, each of these reaches (Fryirs and Brierley, 2001; Fryirs et al., 2021). In general, however, sediment management aims to maintain a sustainable supply and transfer of sediment downstream in and through these reaches. In these cases switches need to remain turned on to maintain upstream–downstream connectivity (Figure 13.3C). In the most downstream exam-ple of this compartment, sediment released through the dripping switch of the low-land plain can be transported offshore. As most of these compartments are already operating as expected/desired (Figure 13.3A), leaving them alone and focussing on the health of these compartments is a viable rehabilitation strategy.

13.5 Managing Sediment (DIS)Connectivity in a Tectonically Passive-Low Relief Setting (Macquarie Catchment, Australia)

The Macquarie catchment drains an area of ~26,000 km^2 (above Narromine) on the western side of the Great Dividing Range in NSW, Australia (Figure 13.4A). Metamorphic and sedimentary bedrock valleys in the subhumid uplands provide perennial discharge and sediment supply to the river, before it descends through a partly confined alluvial valley in its middle reaches and then debouches onto a semiarid alluvial floodplain-fan in its lower reaches (Ralph and Hesse, 2010). Sandy bedload and suspended load sediment are transferred through the middle reaches, whereas in the lower reaches, the river reworks older alluvium through bank erosion and avulsion processes and fine sediment accumulates on the exten-sive floodplain and in wetlands (Ralph et al., 2016). The meandering single chan-nel of the Macquarie River breaks down on the floodplain-fan into a series of anastomosing and distributary channels within the mud-dominated floodplain wet-lands of the Macquarie Marshes before channels reconverge, and the river joins the Barwon–Darling system.

In low relief, inland settings, large catchments and long rivers with downstream declining trends in discharge and stream power usually have inefficient sediment conveyance (Ralph & Hesse, 2010). Yet extensive land clearance, soil tillage and grazing have created significant legacy effects that have altered sediment and hydrological regimes, disrupting hillslope-channel sediment connectivity and downstream sediment connectivity (Prosser et al., 2001; Wethered et al., 2015). The contemporary sediment regimes of major inland rivers are driven largely by

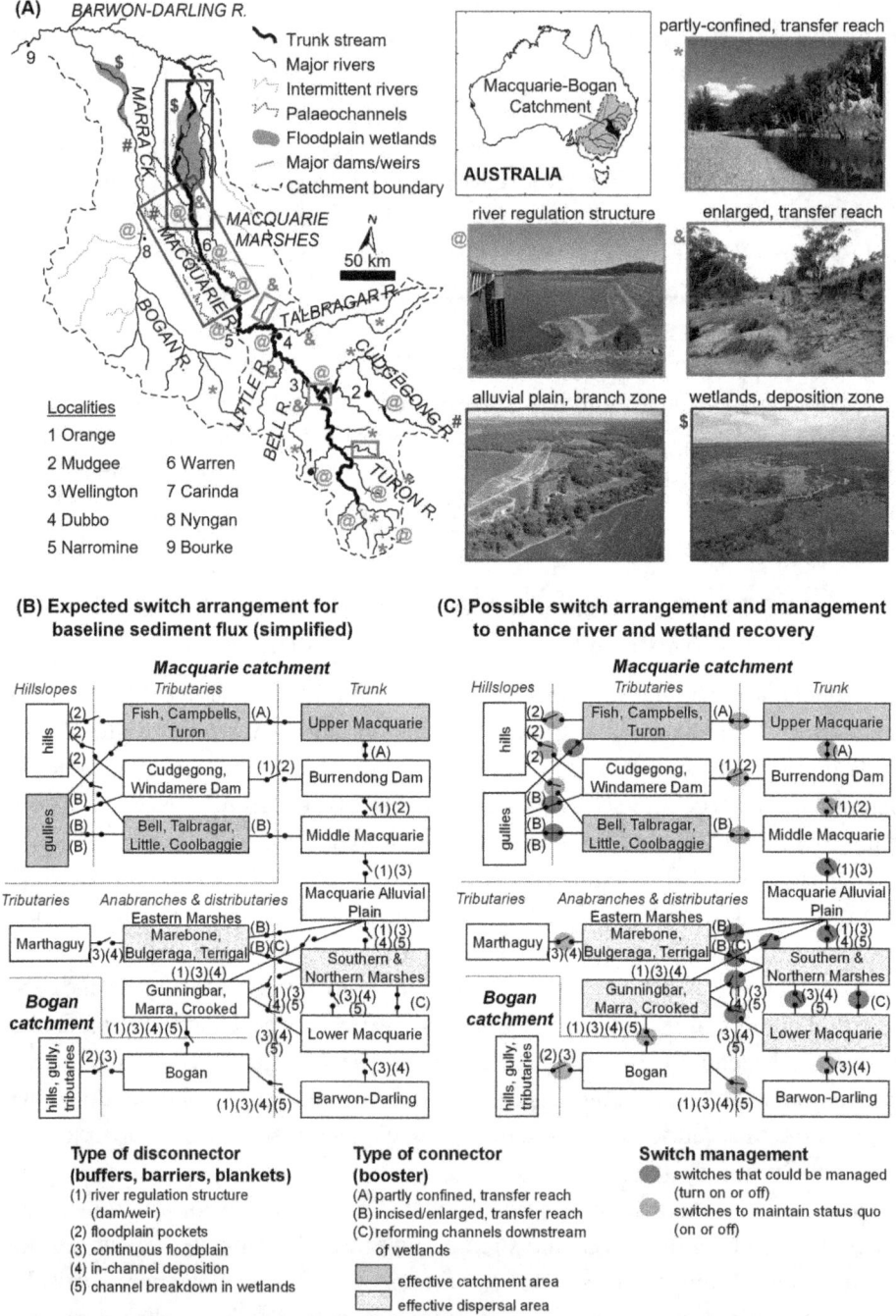

Figure 13.4   Regulating switches to manage sediment (dis)connectivity in the Macquarie catchment. (A) catchment map modified from Ralph and Hesse (2010), sediment budget data from DeRose et al. (2003). A black and white version of this figure will appear in some formats. For the colour version, refer to the plate section.

reworking of materials derived gullies and channel banks (i.e., in-channel and valley margin stores; Olley & Wasson, 2003). Further downstream, large volumes of sediment are locked up in sediment sinks, including reservoirs, river beds, floodplains and wetlands (DeRose et al., 2003). As a result of these process relationships and catchment disconnectivity, sediment export and catchment yields are low. This is shown by the sediment budget for the Macquarie River, where only ~5% of the initial estimated contemporary sediment contributions from hills, gullies and banks are transported through various branches on the alluvial plain and the floodplain wetlands of the Macquarie Marshes to enter the Barwon–Darling River (DeRose et al., 2003; Figure 13.4A).

In the Macquarie catchment, the (dis)connectivity of sediment flux from the headwaters to the middle and lower reaches is governed by both the system configuration (i.e., limited source area, but long river corridor with various buffers, barriers and blankets) and the presence of major river regulation structures. Coarse sediment from the headwaters does not usually make it through dams and weirs in the upper and middle catchment (Kingsford, 2000). In contrast, unregulated tributaries in the middle parts of the valleys typically yield high bedload and suspended sediment loads due to their history of channel incision and expansion in agricultural landscapes that contain considerable stores of Quaternary alluvium. Coolbaggie Creek is the last major unregulated tributary of the Macquarie River, contributing fine-grained sediment to the lower Macquarie River (Wethered et al., 2015). Like several other incised and enlarged tributaries (see Figure 13.4A), subsoil sources (i.e., gullies and banks) dominate Coolbaggie, and by a third of the way down this channel, subsoils account for 100% of fine-grained sediment (Wethered et al., 2015). Limited topsoil supply to Coolbaggie suggests low catchment erosion rates and limited sediment connectivity from the catchment. Entrenchment and enlargement of the lower reaches of Coolbaggie indicate lateral channel-floodplain disconnection, while a small sand slug impedes some longitudinal coarse sediment transfer (Wethered et al., 2015). As with the headwaters, regulated tributaries and the trunk stream, human activities and disturbance events have disrupted the patterns and processes of sediment (dis)connectivity in unregulated tributaries of the Macquarie catchment so that now the proportion of hillslope and in-channel sediment sources and sediment types being transferred downstream are unbalanced.

Downstream reductions in-channel size, discharge and sediment calibre along the trunk stream of the Macquarie River assist channel-floodplain coupling, as floods spread water and sediment onto adjacent floodplain and wetland surfaces in the lower reaches (Ralph & Hesse, 2010). Several large distributary channels (ephemeral creeks, palaeochannels of the Macquarie River; Hesse et al., 2018) occur on the Macquarie Alluvial Plain, some being maintained by significant water diversions at river regulation structures. Dispersal and disintegration of channel-ised flow creates the extensive semi-permanent and ephemeral wetlands of the Macquarie Marshes, within which sedimentation and erosion drive constant geo-ecological adjustments and new channel formation through avulsion several times per hundred years (Ralph et al., 2011). Some channels suffer from erosion due to knickpoint retreat (i.e., head-cutting gullies), while others experience bed incision and/or bank erosion and slumping (Oyston et al., 2014). Stream power and ero-sion risk are variable in the system, but hotspots occur where flows reconverge in anabranches, at the downstream perimeter of wetlands, and where some his-torical erosion control structures have been placed (Ralph et al., 2021). Although erosion is a natural process, understanding and monitoring factors that contrib-ute to erosion at locations that are highly susceptible to change, or that have less resilience to change, will assist river, wetland and water management (Florsheim et al., 2008). Accelerated erosion can have negative impacts including loss of land and resources, and damage to property and infrastructure (Piegay et al., 2005). In the Macquarie Marshes, which have historically had a reduction in the extent and duration of wetland inundation due to river regulation and abstractions (Kingsford, 2000), excessive channel erosion and enlargement poses a significant threat to floodplain wetland inundation and therefore to aquatic processes and ecological values (e.g., Kobayashi et al., 2011, 2015).

River and wetland management in the Macquarie catchment does not explicitly take into account sediment (dis)connectivity, and changes in the sediment bud-get and associated geomorphic and ecological conditions (i.e., degradation and recovery) are not well documented. In such a large catchment, there is usually a disconnect between activities and applications of river and wetland management in the headwaters and those being implemented in the middle and lower reaches. Nevertheless, managing sediment loads and water quality, and therefore some aspects of sediment (dis)connectivity, is implicit to management approaches in this catchment. Five key sediment management aims can be linked to connectivity switches in the Macquarie catchment (Table 13.4), including a simplified baseline switch arrangement (Figure 13.4B) and a possible switch arrangement if the aims are to be achieved (Figure 13.4C).

Aim 1 involves reducing sediment supply from hillslopes and gullies, which are areas of historical land disturbance and erosion (Table 13.4). Protecting hillslopes and valley fills that have not suffered erosion is a high priority, since they are signif-icant stores (potential sources) of sediment. The switches in these locations need to remain turned off to maintain slope-channel disconnectivity, ensuring soils/sediments remain on-slope or trapped on floodplain pockets or in other fills (Figure 13.4C). Sediment deposition and retention in gullies should also be supported by turning some switches off to enhance disconnectivity. A combination of managing stock

Table 13.4 *Managing sediment (dis)connectivity and switches in the Macquarie River catchment, including the Macquarie Marshes.*

What are the sediment 'problems' in this catchment?
What is trying to be achieved?

(1) Reduce sediment supply from hillslopes and gullies.
(2) Maintain sediment transfer in partly confined reaches.
(3) Retain sediment in incised/enlarged channel reaches in the form of benches.
(4) Adjust dispersal of sediment through branches/distributaries to reduce pressure on floodplain wetlands.
(5) Reduce erosion and sediment transport out of floodplain wetlands.

Switch and reaches to be managed	Response and switch status	Techniques that could be used
Managing sediment supply from hillslopes and gullies.	Maintain slope-channel disconnectivity, ensuring soils/sediments remain on-slope or trapped on floodplain pockets or in other fills. Leave green ● (2) switches **off**. Support sediment deposition and retention in gullies. Turn red ● (B) switches **off** for disconnectivity.	Fencing off (stock access and landuse management). Monitor. Natural vegetation regeneration. Direct vegetation planting. Headcut and bank erosion prevention and treatment (if needed).
* Managing sediment transfer in **partly confined reaches**.	Dependent on condition and position in catchment to achieve a balance of sediment connectivity appropriate for the sub-catchment or reach. Maintain sediment supply and transfer in and through the reaches. Leave green ● (A) switch **on** and leave green ● (B) switch **dripping** for connectivity. Large green ● (1) switches cannot easily be managed for sediment connectivity.	Leave it alone. Monitor. Natural vegetation regeneration. Weed management (if needed).
& Managing sediment in **incised/enlarged channel reaches** in the form of benches.	Trap available sediments on the beds of incised/enlarged channels. Encourage reformation of bars and benches or discontinuous watercourses in the channels. Where overwidened channels and sediment exhaustion has occurred, encourage stabilisation of benches that act to narrow channel capacity. Leave green ● (B) switch **dripping** for connectivity.	Fencing off (stock access and landuse management). Monitor. Natural vegetation regeneration. Direct vegetation planting. Weed management. Bed and bank erosion prevention and treatment (if needed).

Table 13.4   (*cont.*)

# Managing sediment in **branches/ distributaries** to reduce pressure on floodplain wetlands.	Managing this switch is a branch by branch exercise depending on where sediment trapping, redistribution and transfer is needed and in what volume. **Adjust** red ● (1) (4) switches for improved balance of connectivity, especially where in-channel deposition is concerned. Encourage stabilisation of incised/enlarged channels and formation of benches that act to narrow channel capacity. **Adjust** red ● (B) (C) switches for improved balance of connectivity. Leave green ● (3) (4) switches **off** for some disconnectivity.	Assessment and modification of river regulation structures (if needed). Fencing off (stock access and landuse management). Monitor. Natural vegetation regeneration. Direct vegetation planting. Weed management. Bed and bank erosion prevention and treatment (if needed).
$ Managing erosion and sediment transport out of **floodplain wetlands**.	Managing this switch is a branch by branch (or wetland by wetland) exercise depending on where channels break down in wetlands and where erosion hotspots occur downstream of wetlands. Manage red ● (1) (4) (5) switches for **improved balance** of connectivity and to take account of sediment deposition and erosion hotspots. Encourage stabilisation of incised/ enlarged channels and sediment retention. Adjust red ● (B) (C) switches for **improved balance** of connectivity. Maintain disconnectivity in these parts of the system, ensuring sediment remain stored in wetlands, and preventing incision. Leave green ● (3) (4) (5) switches **off**.	Assessment and modification of river regulation and erosion control structures (if needed). Fencing off (stock access and landuse management). Monitor. Natural vegetation regeneration. Direct vegetation planting. Weed management. Bed and bank erosion prevention and treatment (if needed).

Symbols and colour codes are matched to Figure 13.4 for cross-referencing.

access and landuse (e.g., fencing off), vegetation-based rehabilitation techniques, and headcut and bank erosion prevention and treatment can be used.

Aim 2 involves maintaining sediment transfer in partly confined reaches (Table 13.4). This is dependent on condition and position in catchment to achieve a balance of sediment connectivity appropriate for the sub-catchment or reach. For example, it is important to maintain the functionality of pool and riffle sequences, exemplified along the Turon River (Figure 13.4A). The switches in these reaches need to be left on or dripping to maintain upstream–downstream connectivity (Figure 13.4C). Large river regulation structures in these reaches cannot easily be managed for sediment connectivity. A 'hands-off' approach to management should suffice, with appropriate riparian zone vegetation maintenance.

Aim 3 involves retaining sediment in incised/enlarged channels in the form of benches (Table 13.4). Sediment released from these reaches, including unregulated tributaries like Coolbaggie Creek (Figure 13.4A), has been the main historical source of sediment to the middle and lower reaches of the system. Today, these channels are either actively eroding their banks or are sediment exhausted, so the goal should be to trap available sediments and to encourage reformation of bars and benches or discontinuous watercourses in the channels. Switches in these settings need to be adjusted, but left dripping to maintain some upstream–downstream connectivity (Figure 13.4C), since the lower trunk stream is reliant on these for sediment, and sediment from the headwaters is largely disconnected by dams and weirs. Recovery can be supported by managing stock access and landuse, vegetation-based rehabilitation techniques, and headcut and bank erosion prevention and treatment.

Aim 4 involves adjusting the dispersal of sediment – sharing the load – through channel branches to reduce pressure on floodplain wetlands, which currently bear the brunt of sediment deposition in hotspots coincident with zones of channel breakdown (Table 13.4). Numerous small channels disperse water and sediment to the floodplain, supporting high channel-floodplain connectivity in these major sediment storage compartments (Figure 13.4A). Managing for this aim is a branch-by-branch exercise depending on where sediment trapping, redistribution and transfer is needed and in what volume. Adjustment of some switches is required to improve the balance of connectivity, especially where excessive in-channel deposition occurs. Upstream–downstream sediment connectivity may be particularly unbalanced where weirs divert water and suspended sediment, but not bedload sediment into secondary channels, leaving the primary channel vulnerable to blockages and potentially avulsion. Otherwise, where channels are enlarged due to regulated flows, options include encouraging channel stabilisation and formation of bars and/or benches to reduce channel capacity. Some switches should remain off to maintain longitudinal disconnectivity, for example, in places where high channel-floodplain connectivity occurs or where instream deposition is appropriate, so as not to overload any branch with sediment (Figure 13.4C). Assessment and modification of river regulation structures may be required, along with managing stock access, vegetation-based rehabilitation, and erosion control techniques. Avulsion can increase lateral and longitudinal connectivity by creating new dispersal pathways for water, sediment, nutrients, and carbon, and can be a beneficial process for ecological succession in low-gradient floodplain wetlands if accepted by management strategies (Ralph et al., 2011, 2016).

Aim 5 involves reducing erosion and sediment transport out of floodplain wetlands, as these processes threaten wetland inundation and associated eco-logical conditions (Table 13.4). Managing for this aim is a branch-by-branch (or wetland-by-wetland) exercise depending on where channels break down and where erosion occurs (Figure 13.4A). Management of switches for improved balance of connectivity and to take account of sediment deposition and erosion hotspots should occur, the latter usually occurring as incisional headcuts at out-lets of wetlands, or bed lowering and bank undercutting in existing channels (Figure 13.4C). Since erosion increases upstream–downstream sediment con-nectivity by reworking previously stored sediment from the wetlands, manage-ment should encourage stabilisation of incised/enlarged channels and sediment retention. Adjustment of some switches for improved balance of downstream connectivity should occur together with maintenance of disconnectivity by leav-ing other switches turned off, preventing excessive incision and ensuring most of the sediment remains stored in the wetlands. Sediment export from the system should, where possible, be appropriate for downstream users (e.g., water quality and channel maintenance) and to minimise the risk of deleterious impacts in the Barwon–Darling River.

13.6 Considerations for Implementation of Switch Management in Practice

Although managing for sediment (dis)connectivity is rarely an explicit river, wet-land or catchment management goal, understandings of sediment (dis)connectivity are sometimes used implicitly in proactive, process-based, recovery enhancement approaches (Beechie et al., 2010; Brierley & Fryirs, 2009). This can only occur where catchment-specific understanding of sediment (dis)connectivity patterns and relationships is available. Context is important, so it pays to understand con-trols upon the sediment regime – where a river gets its sediment from, its volume/calibre, how (dis)connectivity relationships play out in a given catchment, and how they have been disrupted by human disturbance. This determines what is manage-able in a geomorphologically informed way (Fryirs et al., 2021). Such understand-ings are required to situate on-ground interventions (or conservation) to improve the (eco-)geomorphic condition of rivers and wetlands. They can be used to guide efforts to attain the best achievable state under prevailing (and future) conditions. Further work is required to relate sediment (dis)connectivity relations to analy-ses of nutrients, contaminants, biotic and other biophysical fluxes in rivers, wet-lands and catchments. Collectively, these are all critical components of integrated approaches to land and water management at the catchment-scale. To us, there are at least five key factors that practitioners need to consider when embarking on managing sediment (dis)connectivity of rivers and wetlands in practice.

Determining the 'expected' sediment flux dynamics of a catchment is the first consideration in managing sediment (dis)connectivity in practice. Sediment problems are only problems when a system is operating in an unexpected way (Kondolf et al., 2001; Simon et al., 2007) or severely impacts on the ecological function of the river (Boulton, 2007; Kondolf et al., 2006). These problems could be a function of the oversupply of sediment (e.g., the Waipaoa catchment), or an undersupply and exhaustion of sediment (e.g., the Bega catchment). Other problems may be related to the unbalanced distribution of sediment in branches or distributary channels, or the occurrence of hotspots of sediment deposition and/or erosion, the latter including excavation of sediment due to channel incision in unexpected places, in sensitive floodplain wetlands (e.g., the Macquarie catchment). Findings from the case studies in this chapter demonstrate the importance of understanding what is being managed and why, and whether the problem is manageable or not.

A second factor for consideration in managing sediment (dis)connectivity is understanding the effective timescales over which the sediment cascade is operating, and how long it will take to see a response to achieve a management aim or solve a sediment 'problem', and whether this is acceptable in a societal or management context. Are there interventions that will show relatively immediate effects, or will the desired impacts take longer to manifest? Are there proactive or catalytic actions that could be taken now to trigger or accelerate 'natural' recovery processes in the river or wetland system, that may reduce recovery time? For example, in the Bega it is likely to take centuries for the system to supply sufficient sediment to 'fill the holes in the landscape', whereas for the Waipaoa impacts of management actions may be relatively quick: the impacts of forestry planted for erosion control in the upper catchment have shut off and reduced several sediment sources (Marden et al., 2014) and the upper Waipaoa (upstream of the Te Weraroa Stream) is now incising. However, aggradation of the lower Waipaoa is forecast to continue to increase for at least the remainder of the twenty-first century (Gomez et al., 2009) as sediment is reworked from upper reaches and continues to be supplied from badass gully complexes, which will remain for decades. In the Macquarie catchment, it has taken many decades for incised and enlarged tributaries to start to stabilise following the rapid onset of human impacts related to European settlement, yet many still suffer from severe gully and bank erosion (Wethered et al., 2015). In the Macquarie Marshes, channel erosion occurs relatively quickly in some wetlands (e.g., Oyston et al., 2014), but the obvious effects of new channel formation and avulsion may take decades or centuries to play out (Ralph et al., 2016). This is also the case for the lagging ecological successions that occur due to changes in flow dispersal and inundation patterns related to avulsion (e.g., Ralph et al., 2011; Everard, 2016).

A third, and arguably most important, consideration in managing sediment (dis) connectivity is deciding which switches should be left alone, which switches can actually be managed on-the-ground, how many switches need to be managed, and whether managing that switch, or switches, will have the intended at-scale impact on the sediment cascade over the desired management timeframe (Fausch et al., 2002). Some switches can easily be managed, for example, conserving and protecting intact wetlands and discontinuous watercourses and monitoring the risk of any threatening processes to ensure the switch remains off (Bega and Macquarie examples). Other switches may be more difficult to manage, for example, preventing or treating many large-scale gully complexes and shallow landslides to turn a switch off (Waipaoa). Targeted space-planting can mitigate the effects of shallow landslides, although even under indigenous vegetation, sufficiently large storms will continue to generate landslides. The key is strategic buffer planting at source to prevent or minimise off-slope sediment delivery and retain storage in the hillslope. The large gully system is beyond remediation, until structural resistance reduces rates of sediment supply to a point where planting may become effective, but many of these features are associated with deep-seated failures against which reforestation is ineffective. The key message here is treat gullies early to prevent the runaway effect. Effective sediment management practices apply carefully tailored, process-based solutions that tackle the underlying causes of problems at the appropriate spatio-temporal scale, using carefully targeted interventions as required (Wheaton et al., 2019). Blanket applications of the same management practices have proven to be ineffective and prone to failure (Kondolf and Micheli, 1995; Spink et al., 2009). Understanding the process regime also brings a recognition that some processes are simply beyond realistic treatment. For example, options are limited in managing major river regulation structures that are significant barriers to downstream sediment connectivity or where other socio-economic barriers prevent process-based or nature based treatment.

A fourth consideration in all management applications is cost. Is the return on investment in managing switches going to be high? Considering what level of intervention and investment is required to treat a problem, and maintain the intervention over 1–2 years, 5–10 years or multi-decadal horizons is critical. Is there an option to leave it alone to allow the internal dynamics of the cascade to self-adjust and self-heal? Are there low-cost, minimalist interventions that can be used to trigger and enhance recovery (Fryirs et al., 2018; Wheaton et al., 2019)? Could other more radical options be considered such as changing landuse, moving infrastructure or settlements/people out of harm's way, or changing people's perceptions and attitudes to sediment 'problems'? (Brierley, et al., 2021; Mach & Siders,

2021; Tubridy et al., 2021). Framing such perspectives in an adaptive management context, it is important to consider 'moving targets' and guidelines to incorporate manipulation of switches in such practices (e.g., Brierley & Fryirs, 2009, 2016; Downs & Piegay, 2019; Sear et al., 1995; Wohl et al., 2015).

Finally, proactive sediment (dis)connectivity management considers and identifies possible future sediment-related problems and hazards before they occur, strategically targeting the management of some switches before they become a problem (Brierley & Fryirs, 2005). For example, this could entail tackling switches at source that may be prone to, or close to a threshold condition that could create an erosion or sedimentation problem. For example, in the Bega Catchment it is critical to prevent incision into intact trapped tributary fills and (Fryirs & Brierley, 1999), as is forecasting hotspots of erosion potential in the Macquarie Marshes (Ralph et al., 2021). In the Waipaoa, this entails proactive identification and treatment of gullies to prevent development of badass complexes (Marden et al., 2018, Leenman & Tunnicliffe, 2018, Tunnicliffe et al., 2018, Leenman & Tunnicliffe, 2020). Proactive management addresses the underlying causes of possible future problems rather than waiting for the symptoms to emerge.

The functioning of river ecosystems is dependent on catchment structure and connectivity (Fuller & Death, 2018). In highly connected, overloaded systems like the Waipaoa, stream habitats are too unstable for invertebrates or periphyton to survive in such a highly mobile and suspended sediment-dominated environment. This situation has limited the diversity, types, and abundance of plants and animals here (Fuller & Death, 2018). In the Bega, the sediment slug has smothered habitat, until it passes and habitat recovery can occur and ecosystem condition can improve (Chessman et al., 2006). Truncating or reducing sediment supply for example, downstream of dams, can similarly reduce habitat quality and negatively impact stream biota. The flows and transfers of sediment in a catchment must be assessed to implement recovery in river health and transfer pathways *between* habitats are critical for survival of fish and invertebrates: healthy river ecology requires good habitat with appropriate refuge connections (Fuller & Death, 2018).

13.7 Conclusion

River and wetland case studies from contrasting landscape settings with differing sediment cascades and (dis)connectivity relationships in Australia and New Zealand present contrasting sediment 'problems'. A generic approach to managing sediment (dis)connectivity through turning switches on or off provides a consistent platform in these differing settings. This approach can be readily amended to other situations. We show how understanding sediment (dis)connectivity can inform catchment-based sediment management plans and efforts, and provide a set of guiding considerations that all practitioners who embark on managing sediment (dis)connectivity should appraise for their own rivers, wetlands and catchments, framed here as questions:

- Do you understand what the 'expected' sediment flux dynamics of a catchment are? Sediment problems are only problems when a system is operating in an unexpected way.
- Do you understand the effective timescales over which the sediment cascade is operating, and how long it will take to see a response to management? Is this acceptable in a societal or management context?
- Do you know which switches should be left alone, which switches can be managed on-the-ground, how many switches need to be managed? Do you know whether managing that switch, or switches, will have the intended at-scale impact on the sediment cascade over the desired management timeframe?
- How much will it cost and what level of intervention is required to treat the 'problem'? How will interventions be maintained over 1–2 years, 5–10 years or multi-decadal timeframes? Is there an option to leave it alone to allow the internal dynamics of the cascade to self-adjust and self-heal?
- Are you using a proactive management approach that considers and identifies possible future sediment-related problems and hazards before they occur? Are you able to strategically target the management of some switches before they become a problem?

References

Beechie, T.J., Sear, D.A., Olden, J.D., Pess, G.R., Buffington, J.M., Moir, H., Roni, P. & Pollock, M.M. (2010). Process-based principles for restoring river ecosystems. *BioScience*, 60(3), 209–222.

Bilderback, E. L., Pettinga, J. R., Litchfield, N. J., Quigley, M., Marden, M., Roering, J. J., & Palmer, A. S. (2015). Hillslope response to climate-modulated river incision in the Waipaoa catchment, East Coast North Island, New Zealand. *Bulletin*, 127(1–2), 131–148.

Boardman, J., Vandaele, K., Evans, R. & Foster, I. D. (2019). Off-site impacts of soil erosion and runoff: Why connectivity is more important than erosion rates. *Soil Use and Management*, 35(2), 245–256.

Boulton, A. J. (2007). Hyporheic rehabilitation in rivers: restoring vertical connectivity. *Freshwater Biology*, 52(4), 632–650.

Bracken, L.J., Turnbull, L., Wainwright, J. & Bogaart, P. (2015). Sediment connectivity: a framework for understanding sediment transfer at multiple scales. *Earth Surface Processes and Landforms*, 40, 177–188

Brooks, A.P. & Brierley, G.J. (1997). Geomorphic responses of lower Bega River to catchment disturbance, 1851–1926. *Geomorphology*, 18(3–4), 291–304.

Brooks, A. & Brierley, G. (2000). The role of European disturbance in the metamorphosis of the lower Bega River. In Brizga, S. & Finlayson, B. (eds.), *River management: The Australasian experience*, Chichester: John Wiley & Sons, pp. 221–246.

Brierley G.J. & Fryirs, K. (2000). River Styles in Bega Catchment, NSW, Australia: Implications for river rehabilitation. *Environmental Management*, 25 (6), 661–679.

Brierley, G.J. & Fryirs, K.A. (2005). *Geomorphology and River Management: Applications of the River Styles Framework*, Oxford, UK: Blackwell Publications. 398 pp.

Brierley, G. & Fryirs, K. (2009). Don't fight the site: Three geomorphic considerations in catchment-scale river rehabilitation planning. *Environmental Management*, 43 (6), 1201–1218.

Brierley, G.J. & Fryirs, K.A. (2016). The use of evolutionary trajectories to guide 'moving targets' in the management of river futures. *River Research and Applications*, 32 (5), 823–835.

Brierley, G., Fryirs, K. & Jain, V. (2006). Landscape connectivity: The geographic basis of geomorphic applications. *Area*, 38 (2), 165–174.

Brierley, G.J., Cohen, T., Fryirs, K. & Brooks, A. (1999). Post-European changes to the fluvial geomorphology of Bega catchment, Australia: Implications for river ecology. *Freshwater Biology*, 41 (4), 839–848.

Brierley, G., Tunnicliffe, J., Bizzi, S., Lee, F., Perry, G., Pöppl, R. & Fryirs, K. (2021). Quantifying sediment (dis)connectivity in the modelling of river systems. In *Treatise on Geomorphology*. Elsevier.

Brierley, G.J., Hikuroa, D.C.H., Friedrich, H., Fuller, I.C., Brasington, B., Hoyle, J., Tunnicliffe, J., Allen, K. & Measures, R. (2021). Why we should release New Zealand's strangled rivers to lessen the impact of future floods. *The Conversation*. March 1st, 2021. https://theconversation.com/ – 153077.

Brinson, M.M. (1993). *A hydrogeomorphic classification for Wetlands. U.S. Army Corps of Wetlands*. Wetlands Research Program Technical Report WRP-DE-4. Vicksburg, MS: U.S. Army Corps of Engineers Waterways Experiment Station.

Chessman, B.C., Fryirs, K. A. & Brierley, G.J. (2006). Linking geomorphic character, behaviour and condition to fluvial biodiversity: Implications for river management. *Aquatic Conservation: Maine and Freshwater Ecosystems*, 16(3), 267–288.

Clarkson, B.R., Ausseil, A.E. & Gerbeaux, P. (2013). Wetland ecosystem services. In Dymond, J.R. (ed.), *Ecosystem Services in New Zealand – Conditions and Trends*. Lincoln, New Zealand: Manaaki Whenua Press.

Costanza, R., d'Arge, R., de Groot, R. Farber, S., Grasso, M., Hannon, B., Limburg, K., et al. (1997). The value of the world's ecosystem services and natural capital. *Nature*, 387, 253–260.

Costanza, R., de Groot, R., Farber, S., Grasso, M., Hannon, B., Limburg, K., & Van Den Belt, M. (1998). The value of the world's ecosystem services and natural capital. *Ecological Economics*, 25(1), 3–15.

DeRose, R.C. & Basher, L. R. (2011). Measurement of river bank and cliff erosion from sequential LIDAR and historical aerial photography. *Geomorphology*, 126, 132–147.

DeRose, R.C., Prosser, I.P., Weisse, M. & Hughes A.O. (2003). *Patterns of Erosion and Sediment and Nutrient Transport in the Murray-Darling Basin*. Technical Report 32/03, Canberra: CSIRO Land and Water.

Downs, P.W. & Piégay, H. (2019). Catchment-scale cumulative impact of human activities on river channels in the late Anthropocene: implications, limitations, prospect. *Geomorphology*, 338, 88–104.

Engineers, Washington DC (2010). Wetlands research program technical report WRP-DE-4. In Burt, T. & Allison, R. (eds.), *Sediment Cascades: An Integrated Approach*. Oxford, UK: Wiley-Blackwell, 471 pp.

Everard, M. (2016). Biodiversity in wetlands. In Finlayson C. et al. (eds.), *The Wetland Book*. Dordrecht: Springer.

Evrard, O., Navratil, O., Ayrault, S., Ahmadi, M., Némery, J., Legout, C., Lefévre, I., Poirel, A., Bonté, P. & Esteves, M. (2011). Combining suspended sediment monitoring and fingerprinting to determine the spatial origin of fine sediment in a mountainous river catchment. *Earth Surface Processes and Landforms*, 36 (8), 1072–1089.

Fausch, K. D., Torgersen, C. E., Baxter, C. V. & Li, H.W. (2002). Landscapes to riverscapes: Bridging the gap between research and conservation of stream fishes: a continuous view of the river is needed to understand how processes interacting among scales set the context for stream fishes and their habitat. *BioScience*, 52 (6), 483–498.

Florsheim, J.L., Mount, J.F. & Chin, A. (2008). Bank erosion as a desirable attribute of rivers. *BioScience*, 58, 519–529.

Florsheim, J.L., Mount, J.F. & Constantine, C.R. (2006). A geomorphic monitoring and adaptive assessment framework to assess the effect of lowland floodplain river restoration on channel–floodplain sediment continuity. *River Research and Applications*, 22(3), 353–375.

Fryirs, K. (2013). (Dis)connectivity in catchment sediment cascades: A fresh look at the sediment delivery problem. *Earth Surface Processes and Landforms*, 38 (1), 30–46.

Fryirs, K.A. (2015). Developing and using geomorphic condition assessments for river rehabilitation planning, implementation and monitoring. *WiresWater*, 2(6), 649–667.

Fryirs, K. & Brierley, G. (1998). The character and age structure of valley fills in upper Wolumla Creek catchment, south coast, New South Wales, Australia. *Earth Surface Processes and Landforms*, 23(3), 271–287.

Fryirs, K. & Brierley, G.J. (1999). Slope–channel decoupling in Wolumla catchment, New South Wales, Australia: The changing nature of sediment sources following European settlement. *Catena*, 35(1), 41–63.

Fryirs, K. & Brierley, G.J. (2001). Variability in sediment delivery and storage along river courses in Bega catchment, NSW, Australia: implications for geomorphic river recovery. *Geomorphology*, 38(3–4), 237–265.

Fryirs, K. & Brierley, G. (2005). *Practical application of the River Styles Framework as a tool for catchment-wide river management: a case study from Bega catchment, New South Wales*. Macquarie University, 227 pp.

Fryirs, K.A. & Brierley, G.J. (2016). Assessing the geomorphic recovery potential of rivers: forecasting future trajectories of adjustment for use in management. *WiresWater*, 3(5), 727–748.

Fryirs, K.A., Brierley, G.J., Hancock, F., Cohen, T.J., Brooks, A.P., Reinfelds, I., Cook, N. & Raine, A. (2018). Tracking geomorphic recovery in process-based river management. *Land Degradation and Development*, 29(9), 3221–3244.

Fryirs, K.A., Brierley, G.J., Preston, N.J. & Kasai, M. (2007). Buffers, barriers and blankets: The (dis)connectivity of catchment-scale sediment cascades. *Catena*, 70, 49–67.

Fryirs, K.A., Farebrother, W. & Hose, G.C. (2019). Understanding the spatial distribution and physical attributes of upland swamps in the Sydney Basin as a template for their conservation and management. *Australian Geographer*, 50, 91–110.

Fryirs, K., Spink, A. & Brierley, G. (2009). Post-European settlement response gradients of river sensitivity and recovery across the upper Hunter catchment, Australia. *Earth Surface Processes and Landforms*, 34(7), 897–918.

Fryirs, K., Hancock, F., Healey, M., Mould, S., Dobbs, L., Riches, M., Raine, A. & Brierley, G. (2021). Things we can do now that we could not do before: Developing and using a cross-scalar, state-wide database to support geomorphologically-informed river management. *PloS One*, 16(1), e0244719.

Fuller, I.C. & Death, R.G. (2018). The science of connected ecosystems: What is the role of catchment-scale connectivity for healthy river ecology? *Land Degradation & Development*, 29(5), 1413–1426.

Fuller, I. C. & Marden, M. (2011). Slope–channel coupling in steepland terrain: A field-based conceptual model from the Tarndale gully and fan, Waipaoa catchment, New Zealand. *Geomorphology*, 128, 105–115.

Fuller, I. C., Strohmaier, F., McColl, S. T., Tunnicliffe, J. & Marden, M. (2020). Badass gully morphodynamics and sediment generation in Waipaoa Catchment, New Zealand. *Earth Surface Processes and Landforms*, 45, 3917–3930.

GDC (2020). *Te mahinga arai waipuke o Waipaoa Waipaoa flood control.* Available at: www.gdc.govt.nz/council/major-projects/waipaoa-river-flood-control-scheme

Gomez, B., Banbury, K., Marden, M., Trustrum, N. A., Peacock, D. H., & Hoskin, P. J. (2003). Gully erosion and sediment production: Te Weraroa Stream, New Zealand. *Water Resources Research*, 39(7).

Gomez, B., Coleman, S., Sy, V., Peacock, D. & Kent, M. (2007). Channel change, bankfull and effective discharges on a vertically accreting, meandering, gravel-bed river. *Earth Surface Processes and Landforms*, 32, 770–785.

Gomez, B., Cui, Y., Kettner, A., Peacock, D. & Syvitski, J. (2009). Simulating changes to the sediment transport regime of the Waipaoa River, New Zealand, driven by climate change in the twenty-first century. *Global and Planetary Change*, 67, 153–166.

Gomez, B., Eden, D. N., Peacock, D. H. & Pinkney, E. J. (1998). Floodplain construction by recent, rapid vertical accretion: Waipaoa River, New Zealand. *Earth Surface Processes and Landforms*, 23, 405–413.

Gomez, B. & Livingston, D. M. (2012). The river it goes right on: Post-glacial landscape evolution in the upper Waipaoa River basin, eastern North Island, New Zealand. *Geomorphology*, 159, 73–83.

Gomez, B., Rosser, B. J., Peacock, D. H., Hicks, D. M. & Palmer, J. A. (2001). Downstream fining in a rapidly aggrading gravel bed river. *Water Resources Research*, 37, 1813–1823.

Gore, D., Brierley, G., Pickard, J. & Jansen, J. (2000). Anatomy of a floodout in semi-arid eastern Australia. *Zeitschrift fur Geomorphologie*, 122, 113–139.

Graves, B.P., Ralph, T.J., Hesse, P.P., Westaway, K.E., Kobayashi, T., Gadd, P.S. & Mazumder, D. (2019). Macro-charcoal accumulation in floodplain wetlands: problems and prospects for reconstruction of fire regimes and environmental conditions. *PLoS ONE*, 14, e0224011.

Gurnell, A. (2014). Plants as river system engineers. *Earth Surface Processes and Landforms*, 39(1):4–25.

Hamilton, D. & Kelman, E. (1952). *Soil Conservation Survey of the Waipaoa River Catchment, Poverty Bay-New Zealand.* Soil Conservation, Ministry of Works.

Harvey, A.M. (2001). Coupling between hillslopes and channels in upland fluvial systems: Implications for landscape sensitivity illustrated from the Howgill Fells, northwest England. *Catena*, 42, 225–250.

Harvey, A.M. (2002). Effective timescales of coupling within fluvial systems. *Geomorphology*, 44, 175–201.

Hesse, P.P., Williams, R., Ralph, T.J., Larkin, Z.T., Fryirs, K.A., Westaway, K.E. & Yonge, D. (2018). Dramatic reduction in size of the lowland Macquarie River in response to Late Quaternary climate-driven hydrologic change. *Quaternary Research*, 90, 360–379.

Hillman, M., Brierley, G. & Fryirs, K. (2008). Social and biophysical connectivity of river systems. In Brierley, G.J., & Fryirs, K.A. (eds.), *River Futures: An Integrative Scientific Approach to River Repair.* Washington, DC: Island Press, 125–142.

Hicks, D. M., Gomez, B. & Trustrum, N. A. (2000). Erosion thresholds and suspended sediment yields, Waipaoa River Basin, New Zealand. *Water Resources Research*, 36, 1129–1142. www.doi.org/10.1029/1999WR900340

Hicks, D. M., Shankar, U., McKerchar, A. I., Basher, L., Lynn, I., Page, M. & Jessen, M. (2011). Suspended sediment yields from New Zealand rivers. *Journal of Hydrology (New Zealand)*, 81–142.

Hooke, J.M. (2003). Coarse sediment connectivity in river channel systems: A conceptual framework and methodology. *Geomorphology*, 56, 79–94.

Jain, V. & Tandon, S.K. (2010). Conceptual assessment of (dis) connectivity and its application to the Ganga River dispersal system. *Geomorphology*, 118(3–4), 349–358.

James, L.A. (2010). Secular sediment waves, channel bed waves, and legacy sediment. *Geography Compass*, 4(6), 576–598.

James, L.A. (2013). Legacy sediment: definitions and processes of episodically produced anthropogenic sediment. *Anthropocene*, 2, 16–26.

Jones, K. E. & Preston, N. J. (2012). Spatial and temporal patterns of off-slope sediment delivery for small catchments subject to shallow landslides within the Waipaoa catchment, New Zealand. *Geomorphology*, 141, 150–159.

Kasai, M., Brierley, G.J., Page, M.J., Marutani, T. & Trustrum, N.A. (2005). Impacts of land use change on patterns of sediment flux in Weraamaia catchment, New Zealand. *Catena*, 64(1), 27–60.

Keesstra, S., Nunes, J., Saco, P., Parsons, A., Pöppl, R., Pereira, P., Novara, A., Comino, J.R., Masselink, R. & Cerda, A. (2018). The way forward: can connectivity be useful to design better measuring and modelling schemes for water and sediment dynamics? *Science of the Total Environment*, 644, 1557–1572.

Kingsford, R.T. (2000). Ecological impacts of dams, water diversions and river management on floodplain wetlands in Australia. *Austral Ecology*, 25, 109–127.

Kondolf, G.M. (2011). Setting goals in river restoration: when and where can the river "heal itself". Stream restoration in dynamic fluvial systems: scientific approaches, analyses, and tools. *Geophysical Monograph Series*, 194, 29–43.

Kondolf, G.M. & Micheli, E.R. (1995). Evaluating stream restoration projects. *Environmental Management*, 19(1), 1–15.

Kondolf, G.M., Boulton, A.J., O'Daniel, S., Poole, G.C., Rahel, F.J., Stanley, E.H., Wohl, E., Bång, A., Carlstrom, J., Cristoni, C. & Huber, H. (2006). Process-based ecological river restoration: visualizing three-dimensional connectivity and dynamic vectors to recover lost linkages. *Ecology and Society*, 11(2).

Kondolf, G.M., Smeltzer, M.W. & Railsback, S.F. (2001). Design and performance of a channel reconstruction project in a coastal California gravel-bed stream. *Environmental Management*, 28(6), 761–776.

Kobayashi, T., Ryder, D.S., Ralph, T.J., Mazumder, D., Saintilan, N., Iles, J., Knowles, L., Thomas, R., & Hunter, S. (2011). Longitudinal spatial variation in ecological conditions in an in-channel floodplain river system during flow pulses. *River Research and Applications*, 27, 461–472.

Kobayashi, T., Ralph, T.J., Ryder, D.S., Hunter, S.J., Shiel, R.J., & Segers, H. (2015). Spatial dissimilarities in plankton structure and function during flood pulses in a semi-arid floodplain wetland system. *Hydrobiologia*, 747, 19–31.

Kuehl, S. A., Alexander, C. R., Blair, N. E., Harris, C. K., Marsaglia, K. M., Ogston, A. S., Orpin, A. R., Roering, J. J., Bever, A. J. & Bilderback, E. L. (2016). A source-to-sink perspective of the Waipaoa River margin. *Earth-Science Reviews*, 153, 301–334.

Larkin, Z.T., Ralph, T.J., Tooth, S., Fryirs, K.A. & Carthey, J.R. (2020). Identifying threshold responses of Australian dryland rivers to future hydroclimatic change. *Scientific Reports*, 10, 6653.

Larkin, Z.T., Ralph, T.J., Tooth, S. & McCarthy, T.S. (2017). The interplay between extrinsic and intrinsic controls in determining floodplain wetland characteristics in the South African drylands. *Earth Surface Processes and Landforms*, 42, 1092–1109.

Leenman, A. & Tunnicliffe, J. (2020). Tributary-junction fans as buffers in the sediment cascade: A multi-decadal study. *Earth Surface Processes and Landforms*, 45(2), 265–279.

Leenman, A. & Tunnicliffe, J. (2018). Genesis of a major gully mass-wasting complex, and implications for valley filling, East Cape, New Zealand. *Geological Society of America Bulletin*, 130(7–8), 1121–1130.

Lisenby, P.E., Fryirs, K.A. & Thompson, C.J. (2020). River sensitivity and sediment connectivity as tools for assessing future geomorphic channel behavior. *International Journal of River Basin Management*, 18(3), 279–293.

Lisenby, P.E., Tooth, S. & Ralph, T.J. (2019). Product vs. process? The role of geomorphology in wetland characterization. *Science of the Total Environment*, 663, 980–991.

Mach, K.J. & Siders, A.R. (2021). Reframing strategic, managed retreat for transformative climate adaptation. *Science*, 372(6548), 1294–1299.

Marden, M. (2011). *Sedimentation History of Waipaoa Catchment*. Landcare Research.

Marden, M. (2012). Effectiveness of reforestation in erosion mitigation and implications for future sediment yields, East Coast catchments, New Zealand: A review. *New Zealand Geographer*, 68, 24–35. www.doi.org/10.1111/j.1745-7939.2012.01218.x

Marden, M., Arnold, G., Gomez, B. & Rowan, D. (2005). Pre-and post-reforestation gully development in Mangatu Forest, East Coast, North Island, New Zealand. *River Research and Applications*, 21, 757–771.

Marden, M., Fuller, I. C., Herzig, A. & Betts, H. D. (2018). Badass gullies: Fluvio-mass-movement gully complexes in New Zealand's East Coast region, and potential for remediation. *Geomorphology*, 307, 12–23.

Marden, M., Herzig, A. & Basher, L. (2014). Erosion process contribution to sediment yield before and after the establishment of exotic forest: Waipaoa catchment, New Zealand. *Geomorphology*, 226, 162–174.

Mitsch, W.J. & Gosselink, J.G. (2015). *Wetlands* (5th ed). New York: Wiley, 744 pp.

Olley, J.M. & Wasson, R.J. (2003). Changes in the flux of sediment in the Upper Murrumbidgee catchment, Southeastern Australia, since European settlement. *Hydrological Processes*, 17, 3307–3320.

Oyston, S.M., Ralph, T.J. & Hesse, P.P. (2014). Cutting down, back and out: assessment of channel erosion in a sensitive floodplain wetland. In Vietz, G., Rutherfurd, I.D. & Hughes, R. (eds.), *Proceedings of the 7th Australian Stream Managament Conference, 27–30 July 2014*. Townsville, QLD: River Basin Management Society, pp. 143–149.

Page, M., Trustrum, N., Brackley, H. & Baisden, T. (2004). Erosion-related soil carbon fluxes in a pastoral steepland catchment, New Zealand. *Agriculture, Ecosystems & Environment*, 103, 561–579.

Piegay, H., Darby, S.E., Mosselman, E., & Surian, N. (2005). A review of techniques available for delimiting the erodible river corridor: a sustainable approach to managing bank erosion. *River Research and Applications*, 21, 773–789.

Pöppl, R.E., Keesstra, S.D. & Hein, T. (2015). The geomorphic legacy of small dams – An Austrian study. *Anthropocene*, 10, 43–55.

Pöppl, R.E., Keesstra, S.D. & Maroulis, J.(2017). A conceptual connectivity framework for understanding geomorphic change in human-impacted fluvial systems. *Geomorphology*, 277, 237–250.

Pöppl, R.E., Coulthard, T., Keesstra, S.D. & Keiler, M., J. (2019). Modeling the impact of dam removal on channel evolution and sediment delivery in a multiple dam setting. *International Journal of Sediment Research*, 34(6), 537–549.

Pöppl, R.E., Fryirs, K.A., Tunnicliffe, J. & Brierley, G.J. (2020). Managing sediment (dis) connectivity in fluvial systems. *Science of the Total Environment*, 736, 139627.

Prosser, I.P., Rutherfurd, I.D., Olley, J.M., Young, W.J., Wallbrink, P.J., & Moran, C.J. (2001). Large-scale patterns of erosion and sediment transport in river networks, with examples from Australia. *Marine and Freshwater Research*, 52, 81–99.

Ralph, T.J., & Hesse, P.P. (2010). Downstream hydrogeomorphic changes along the Macquarie River, southeastern Australia, leading to channel breakdown and flood-plain wetlands. *Geomorphology*, 118, 48–64.

Ralph, T.J., Hesse, P.P., & Kobayashi, T. (2016). Wandering wetlands: Spatial patterns of historical channel and floodplain change in the Ramsar-listed Macquarie Marshes, Australia. *Marine and Freshwater Research*, 67, 782–802.

Ralph, T.J., Kobayashi, T., García, A., Hesse, P.P., Yonge, D., Bleakley, N., & Ingleton, T. (2011). Paleoecological responses to avulsion and floodplain evolution in a semiarid Australian freshwater wetland. *Australian Journal of Earth Sciences*, 58, 75–91.

Ralph, T.J., Larkin, Z., Farebrother, W., Ocock, J., Hosking, T., Kobayashi, Y., Hughes, M., Hesse, P. & Fryirs, K. (2021) Exploring the relationship between channel bed control structures and stream power in low-gradient floodplain wetlands. In *Proceedings of the 10th Australian Stream Management Conference*, 2–4 August 2021, Kingscliff, NSW, 74316, River Basin Management Society, Melbourne, VIC.

Ralph, T.J. & Rogers, K. (2011). Floodplain wetlands of the Murray-Darling Basin and their freshwater biota. In Rogers, K. & Ralph, T. J. (eds.) *Floodplain Wetland Biota in the Murray-Darling Basin: Water and Habitat Requirements*, Collingwood, VIC: CSIRO Publishing, pp. 1–16.

Rayburg, S. & Thoms, M. (2009). A coupled hydraulic–hydrologic modelling approach to deriving a water balance model for a complex floodplain wetland system. *Hydrology Research*, 40, 364–379.

Reid, L. & Page, M. (2003). Magnitude and frequency of landsliding in a large New Zealand catchment. *Geomorphology*, 49, 71–88.

Rittenhouse T.A. & Peterman W.E. (2018). Connectivity of wetlands. In Finlayson, C. et al. (eds.), *The Wetland Book*. Dordrecht: Springer.

Rogers, K. & Ralph, T.J. (eds.), (2011). *Floodplain Wetland Biota in the Murray-Darling Basin: Water and Habitat Requirements*. Collingwood, VIC: CSIRO Publishing, 348 pp.

Schmidt, J.C., Webb, R.H., Valdez, R.A., Marzolf, G.R. & Stevens, L.E. (1998). Science and values in river restoration in the Grand Canyon: there is no restoration or reha-bilitation strategy that will improve the status of every riverine resource. *BioScience*, 48(9), 735–747.

Scorpio, V., Aucelli, P. P., Giano, S. I., Pisano, L., Robustelli, G., Rosskopf, C.M. & Schiattarella, M. (2015). River channel adjustments in Southern Italy over the past 150 years and implications for channel recovery. *Geomorphology*, 251, 77–90.

Sear, D.A., Newson, M.D. & Brookes, A. (1995). Sediment-related river maintenance: the role of fluvial geomorphology. *Earth Surface Processes and Landforms*, 20(7), 629–647.

Semeniuk, C. & Semeniuk, V. (1995). A geomorphic approach to global classification for inland wetlands. *Vegetatio*, 118, 103–124.

Simon, A., Doyle, M., Kondolf, M., Shields Jr, F.D., Rhoads, B. & McPhillips, M. (2007). Critical evaluation of how the Rosgen classification and associated "natural channel design" methods fail to integrate and quantify fluvial processes and channel response. *Journal of the American Water Resources Association*, 43(5), 1117–1131.

Singh, M., Sinha, R. & Tandon, S.K. (2021). Geomorphic connectivity and its application for understanding landscape complexities: a focus on the hydro-geomorphic systems of India. *Earth Surface Processes and Landforms*, 46(1), 110–130.

Spink, A., Fryirs, K. & Brierley, G. (2009). The relationship between geomorphic river adjustment and management actions over the last 50 years in the upper Hunter catchment, NSW, Australia. *River Research and Applications*, 25(7), 904–928.

Taylor, R. J., Massey, C., Fuller, I. C., Marden, M., Archibald, G. & Ries, W. (2018). Quantifying sediment connectivity in an actively eroding gully complex, Waipaoa catchment, New Zealand. *Geomorphology*, 307, 24–37.

Thompson, C., Fryirs, K. & Croke, J. (2016). The disconnected sediment conveyor belt: Patterns of longitudinal and lateral erosion and deposition during a catastrophic flood in the Lockyer Valley, southeast Queensland, Australia. *River Research and Applications*, 32, 540–551.

Tooth, S., Ellery, W., Grenfell, M., Thomas, A., Kotze, D. & Ralph, T. (2015). *10 reasons Why the Geomorphology of Wetlands Is Important*. Wetlands in Drylands Research Network.

Tubridy, D., Scott, M. & Lennon, M. (2021). Managed retreat in response to flooding: lessons from the past for contemporary climate change adaptation. *Planning Perspectives*, 1–20.

Tunnicliffe, J., Brierley, G., Fuller, I. C., Leenman, A., Marden, M. & Peacock, D. (2018). Reaction and relaxation in a coarse-grained fluvial system following catchment-wide disturbance. *Geomorphology*, 307, 50–64.

Turley, M., Hassan, M. A. & Slaymaker, O. (2021). Quantifying Sediment Connectivity: Moving Towards a Holistic Assessment Through a Mixed Methods Approach. *Earth Surface Processes and Landforms*.

Wainwright, J., Turnbull, L., Ibrahim, T.G., Lexartza-Artza, I., Thornton, S.F. & Brazier, R.E. (2011). Linking environmental regimes, space and time: Interpretations of structural and functional connectivity. *Geomorphology*, 126(3–4), 387–404.

Walling, D.E. (1983). The sediment delivery problem. *Journal of Hydrology*, 65, 209–237.

Walter, R.C. & Merritts, D.J. (2008). Natural streams and the legacy of water-powered mills. *Science*, 319(5861), 299–304.

Warrick, J.A., Bountry, J.A., East, A.E., Magirl, C.S., Randle, T.J., Gelfenbaum, G., Ritchie, A.C., Pess, G.R., Leung, V. & Duda, J.J. (2015). Large-scale dam removal on the Elwha River, Washington, USA: Source-to-sink sediment budget and synthesis. *Geomorphology*, 246, 729–750.

Wethered, A.S., Ralph, T.J., Smith, H.G., Fryirs, K.A. & Heijnis, H. (2015). Quantifying fluvial (dis)connectivity in an agricultural catchment using a geomorphic approach and sediment source tracing. *Journal of Soils and Sediments*, 15, 2052–2066.

Wheaton, J.M., Bennett, S., Bouwes, N., Maestas, J.D. & Shahverdian, S.M. (eds.), (2019). *Low-Tech Process-Based Restoration of Riverscapes: Design Manual*. Version 1.0. Logan, UT: Utah State University Restoration Consortium. Available at: http://lowtechpbr.restoration.usu.edu/manual

Williams, R.T. & Fryirs, K.A. (2020). The morphology and geomorphic evolution of a large chain-of-ponds river system. *Earth Surface Processes and Landforms*, 45, 1732–1748.

Wohl, E, (2015). Legacy effects on sediments in river corridors. *Earth Science Reviews*, 147, 30–53.

Wohl, E. (2018). Geomorphic context in rivers. *Progress in Physical Geography: Earth and Environment*, 42(6), 841–857.

Wohl, E. & Beckman, N.D. (2014). Leaky rivers: Implications of the loss of longitudinal fluvial disconnectivity in headwater streams. *Geomorphology*, 205, 27–35.

Wohl, E., Bledsoe, B.P., Jacobson, R.B., Poff, N. L., Rathburn, S.L., Walters, D.M. & Wilcox, A.C. (2015). The natural sediment regime in rivers: broadening the foundation for ecosystem management. *BioScience*, 65(4), 358–371.

Wohl, E., Brierley, G., Cadol, D., Coulthard, T., Covino, T., Fryirs, K., Grant, G., Pöpplet al. (2019). Connectivity as an emergent property of geomorphic systems. *Earth Surface Processes Landforms*, 44, 4–26.

Xu, X., Chen, M., Yang, G., Jiang, B. & Zhang, J. (2020). Wetland ecosystem services research: a critical review. *Global Ecology and Conservation*, 22, e01027.

Ziliani, L. & Surian, N. (2012). Evolutionary trajectory of channel morphology and controlling factors in a large gravel-bed river. *Geomorphology*, 173, 104–117.

Ziliani, L. & Surian, N. (2016). Reconstructing temporal changes and prediction of channel evolution in a large Alpine river: The Tagliamento River, Italy. *Aquatic Sciences*, 78(1), 83–94.

14

Drylands

PATRICIA SACO, JOSE RODRIGUEZ, MARIANO MORENO-DE LAS HERAS,
STEVEN SANDI ROJAS AND JUAN QUIJANO-BARON

This chapter describes the application of the connectivity framework to drylands, by establishing the relation between spatial vegetation patterns and connectivity (Section 14.1), identifying the main processes governing the connectivity of water, sediment and wind (Sections 14.2 and 14.3), exploring system feedbacks and evolution leading to degradation thresholds (Section 14.4) and discussing how human activities and climate change can produce system collapse and desertification (Section 14.5). The chapter concludes with an overview of the implications of the connectivity analysis for the management of drylands (Section 14.6).

14.1 Spatial Vegetation Patterns and Connectivity in Drylands

Drylands occupy around 40% of the Earth's land surface and are usually identified as areas with an aridity index (ratio between rainfall and potential evapotranspiration) below 0.65 (Reynolds et al., 2007; Okin et al., 2015). Rainfall in drylands tends to be not only scarce but also highly unpredictable, and temperature and solar radiation are extremely high (Saco et al., 2018). These conditions of low annual precipitation and high potential evapotranspiration rates lead to soil moisture scarcity and strong ecological–hydrological feedbacks and interactions occurring across fine to coarse scales (Wilcox et al., 2003; Ludwig et al., 2005). The key role of water redistribution for the maintenance of vegetation patterns has led to numerous studies to better understand ecosystem stability and landscape processes (Ye et al., 2016; Ratajczak et al., 2017). Generally, the vegetation of these regions consists of sparse cover, organised as a mosaic of patches with high biomass cover interspersed within bare soil component.

The spatial organization of the patchy vegetation in drylands can be either periodic or random (Tongway & Ludwig, 2001). Regular or periodic patterns, with spotted, labyrinthic, gapped, or banded organization, are very common in arid and semiarid ecosystems worldwide (Figure 14.1). Numerous field and modeling

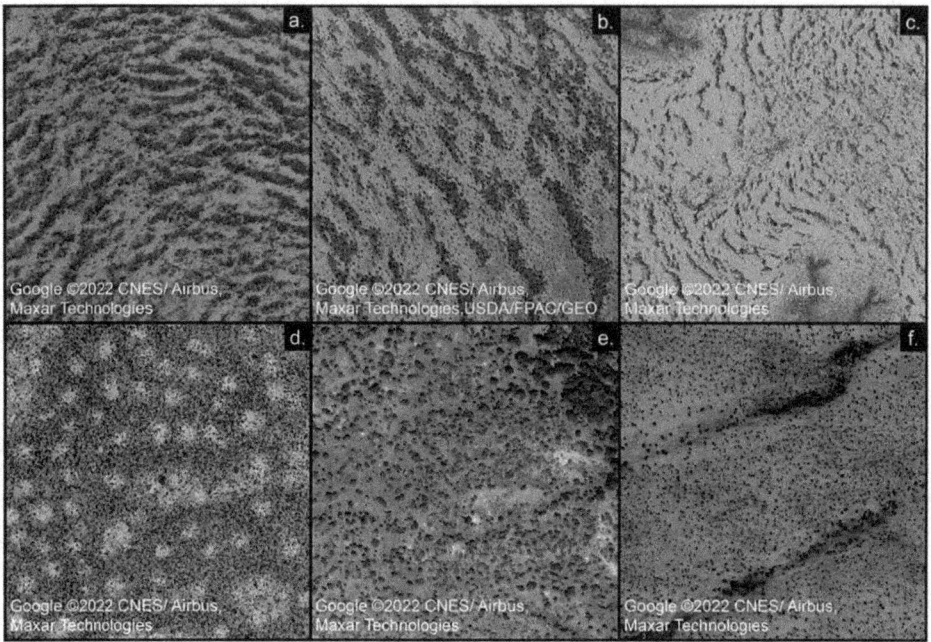

Figure 14.1   Examples of vegetation patterns in drylands around the world. (a) Northern Territory, Australia (Google Earth[TM] 23°29'6.80"S, 133°51'46.02"E), (b) Texas, USA (Google Earth[TM] 30°48'44.78"N, 103°23'28.41"W), (c) Baja California, Mexico (Google Earth[TM] 26°52'43.87"N, 112°51'56.27"W), (d) Kursin, Kenya (Google Earth[TM] 1° 9'3.10"N, 40°22'21.49"E), (e) W Benin Niger National Park, Benin (Google Earth[TM] 12° 0'32.05"N, 2°59'36.25"E), (f) Nara, Mali (Google Earth[TM] 15° 6'17.70"N, 8°10'48.37"W). A black and white version of this figure will appear in some formats. For the colour version, refer to the plate section.

studies have been reported over the last few decades, aimed at understanding the emergence and self-organization of these remarkable patterns, as well as their role in landscape function and resilience (Meron et al., 2004; Rietkerk et al., 2004; Borgogno et al., 2009; Moreno-de las Heras et al., 2012; Saco et al., 2020).

Connectivity can be defined as the degree to which resources (water, sediment, energy, organisms) can be transferred or redistributed across the landscape (Bracken et al., 2013; Okin et al., 2015) (see also Chapter 2). This concept has been extensively used in recent years by earth scientists for understanding landscape water and sediment redistribution in drylands (Okin et al., 2015; Poeppl et al., 2017; Keesstra et al., 2018; Wohl et al., 2019; Saco et al., 2020; Bracken & Croke, 2007; Fryirs et al., 2007; Turnbull et al., 2008; Wainwright et al., 2011). The natural biotic and abiotic structural features that define landscape connectivity, may differ across catchments and geographical regions, but topography, vegetation, soils and geological units tend to have an overarching control (Bracken & Croke, 2007; Lisenby & Fryirs, 2017). Structures such as terraces, channels, hydraulic structures and other man-made features can also significantly affect drainage patterns in drylands, and therefore hydraulic and sediment connectivity (Mekonnen et al., 2017; Rodrigo Comino et al., 2018; Moreno-de-las-Heras et al., 2019).

Recent work on landscape hydrologic and sediment connectivity has emphasised the importance and usefulness of these concepts for the analysis of dryland function, and as a potential tool for management strategies to improve landscape resilience (Okin et al., 2009; Moreno-de las Heras et al., 2012; Okin et al., 2015; Masselink et al., 2016; Wohl et al., 2019). Most of this work has focussed on the fast response of hydrological processes, but the implications of the emergent system connectivity and structures (i.e., vegetation patterns, topography) at longer time scales have been also shown to have critical importance (Saco et al., 2007; Saco & Moreno-de las Heras, 2013; Saco et al., 2018; Wohl et al., 2019; Saco et al., 2020; Bautista & Mayor, 2021; Stewart et al., 2014). For example, little research has focussed on implications of the eco-hydro-geomorphic coevolution of system structures in response to external stressors (induced by, for example, climate change and anthropogenic pressures).

14.2 Water Redistribution and Hydrologic and Sediment Connectivity in Drylands

As mentioned earlier, dryland areas around the world typically display a patchy vegetation cover, and this patchiness emerges from nonlinear feedbacks between biomass growth and water redistribution (Meron et al., 2004; Rietkerk et al., 2004; Saco and Moreno-de las Heras, 2013; Mayor et al., 2019). A key mechanism leading to the development and maintenance of these patterns is the emergence of a spatially variable infiltration field with low infiltration rates in the bare areas and high infiltration rates in the vegetated areas (Tongway & Ludwig, 2001; Ludwig et al., 2005). Spatially variable infiltration rates have been observed in many field sites, with some studies recording infiltration rates up to 10 times higher under perennial vegetation patches than in the bare soil interpatch areas (Bhark & Small, 2003; Ludwig et al., 2005). Enhanced infiltration in vegetated patches is due to improved soil aggregation and macroporosity induced by biological activity (e.g., termites, ants, and earthworms are very active in semi-arid areas) and plant roots (Ludwig et al., 2005). On the other hand, bare soil patches tend to be covered by biological soil crusts, which limit infiltration and are mostly absent in the vegetation patches (Belnap et al., 2003; Tongway & Hindley, 2004). The bare soil areas with lower infiltration become a source of runoff that can be absorbed by vegetation patches with higher infiltration. The amount of water infiltrated into the vegetation patches, and/or received as runon from the bare areas, can be much higher than direct rainfall (Valentin et al., 1999; Dunkerley, 2002; Wilcox et al., 2003).

These mechanisms lead to increased soil moisture in the vegetation patches due to the higher infiltration, decreased evaporation due to the canopy, and run-on capture (Scanlon et al., 2007; Magliano et al., 2019). The runoff–runon mechanism triggers a positive feedback, as increased soil moisture in vegetated patches promotes plant growth reinforcing the pattern (Puigdefabregas et al., 1999; Valentin et al., 1999; Wilcox et al., 2003; Rossi et al., 2018; Saco et al., 2020). The redistribution of water and sediments from source areas (bare patches) to sink areas (vegetation patches) is a fundamental process within drylands that may be altered if the vegetation patch structure is disturbed (Cammeraat & Imeson, 1999; Wilcox et al., 2003; Imeson & Prinsen, 2004). Vegetation patterns are therefore linked to both hydrologic processes through their effect on soil moisture patterns, runoff redistribution and evapotranspiration, and to geomorphologic processes as they determine the spatial distribution of erosion-deposition areas. The downslope routing of water, sediments, nutrients, seeds, and so on, is determined by the spatial connectivity between vegetated and bare patches (Imeson & Prinsen, 2004).

The concepts of structural and functional connectivity (Chapter 2), facilitate the analysis of the spatial and temporal dynamics of runoff and sediment connectivity of drylands at hillslope and catchment scales (Wainwright et al., 2011; Okin et al., 2015). While structural connectivity is linked to the effect of topography and vegetation patterns, as explained in the previous paragraphs, functional (or process-based) connectivity refers to the activation of connections of runoff and/ or sediment pathways (Moreno-de-las-Heras et al., 2020). Functional hydrologic connectivity depends on the dynamics of runoff generation, routing and infiltration, which vary as a function of rainfall intensity, whereas sediment functional connectivity varies depending on rates of soil entrainment, detachment, deposition and remobilization (Bracken et al., 2013; Turnbull & Wainwright, 2019).

14.3 Wind Connectivity in Drylands

The main effect of wind on arid landscapes is as an agent of transport for seeds (Peters et al., 2004) and sediment (Wainwright et al., 1999) and also as a control on evapotranspiration (Dickinson, 1984). Connectivity is affected by wind erosion in a similar way to water erosion (see also Chapter 5). Erosion by wind depends on wind speed and surface roughness (Kawamura, 1951) as well as on soil erodibility (Gillette et al., 1980). Both factors depend on vegetation cover and structure. For example, shrublands have a sparser coverage than grasslands, so they experience more erosion and sediment transport. In addition, vegetation in drylands exhibit patches and discontinuous coverage, and wind erosivity is also a function of the size of the gaps. Gap size and wind velocity determine the extent of the wake behind the vegetation clusters and the degree of penetration of the wind shear stresses within the gaps of bare soil (Okin, 2008; Okin et al., 2015; Webb et al., 2021). Shrublands and other woody species typically display larger gaps than grasses, which results in higher wind erosion and a higher connectivity.

Connectivity changes due to wind occur at different scales. Annuals that make up the shrub understory (Li et al., 2008) develop over weeks or months and reduce the bare soil areas, protect the soil from wind and decrease connectivity (Bergametti & Gillette, 2010). Over longer time scales (years to decades), changes in vegetation cover, for example, from grassland to shrublands, increases erosion opportunities for wind and thus connectivity. Long-term changes can also occur in the topography of the landscape due to wind, for example, dunes can develop in the shrub areas, with important effects on hydrology and connectivity (Rango et al., 2000; Ravi et al., 2007). All these changes together with differences in soil properties (Peters et al., 2015) and historic legacies (Monger et al., 2015) result in a mosaic of grasslands interspersed with woody plants (Iwaniec et al., 2021), which in turn leads to a complex pattern of connectivity.

Wind and water erosion effects on connectivity are similar (increase in erosion results in increased connectivity), but their response to climate variability and change are different. During periods of higher rainfall, connectivity due to water erosion is likely to increase (more runoff), but connectivity due to wind erosion is expected to decrease because of higher vegetative cover. The opposite will occur during dry periods. However, these two opposing responses can potentially add up instead of counteracting each other in the future according to most predictions of climate change. This is because climate variability is expected to increase in arid zones (Power et al., 2013; Cai et al., 2015; Huang et al., 2016; Huang et al., 2017), which will lead to higher intensity precipitation and longer drought conditions. Connectivity will increase due to higher fluvial erosion during the rainy periods and will also increase during the longer drought periods due to wind erosion. In addition, because shrubland is more resistant to droughts than grassland (Gherardi & Sala, 2015), longer droughts will favour the encroachment of shrublands over grasslands. Shrubland is associated with sparser coverage and larger connectivity than grassland, and this increase in connectivity may eventually lead to desertification (Reynolds & Stafford Smith, 2002; Havstad et al., 2007; Yahdjian et al., 2015), as already observed in many dryland areas around the world (IPCC, 2021). Long-term transitions of grasslands to shrublands can result in irreversible transitions, as reductions of the grass species seed-banks limit the potential re-establishment of grasses (Moreno-de las Heras et al., 2016).

Trends of encroachment of woody vegetation on grassland are already occurring worldwide, as evidenced by observations in North America (van Auken, 2009), semi-arid and tropical Australia (Fensham et al., 2005; Saintilan et al., 2021), Europe (Maestre et al., 2009), India (Misra, 1983) and China (Peng et al., 2013).

This encroachment has contributed to an increase in tree canopy cover across the globe (Song et al., 2018) despite tree mortality due to the combination of elevated temperatures and severe drought (Sitch et al., 2015; Birch et al., 2019; Locosselli et al., 2020). Despite this average global greening, climate change has degraded 12.6% (5.43 million km^2) of drylands since the 1980s (Burrell et al., 2020).

Management of erosion by wind to limit connectivity requires monitoring of indicators of vegetation cover and structure. These indicators are different from the indicators of water erosion because typical connectivity measures like fractional ground cover only account for some of the physical features that determine wind connectivity (Webb et al., 2021). Ground cover is an important factor in wind erosion as vegetation protects the soil surface and reduces its erodibility, it traps sediment (Zobell et al., 2020) and absorbs momentum from the wind reducing its shear stress at the surface (Raupach et al., 1993; Okin, 2008). However, ground cover is not capable of capturing the aerodynamic effects associated with wind erosion, such as the wake behind the vegetation clusters and the degree of penetration of the wind shear stresses within the gaps of bare soil. Wind strength is heavily reduced in the downward side of vegetation due to a sheltering effect and only fully recovers at distances between 10 and 20 times the height of the vegetation (Bradley & Mulhearn, 1983; Leenders et al., 2011; Mayaud et al., 2017). As a result, taller vegetation will produce a larger sheltering effect than shorter vegetation, and sites with short vegetation and large bare soil gaps will be the most affected by wind erosion (Gillette et al., 2006). Indicators must include both gap size and arrangement (from ground cover products) and also vegetation height (from ground or high-resolution lidar observations) particularly in sites with mixed plant communities (Chappell et al., 2018; Pi et al., 2020).

14.4 Degradation and Threshold Behaviour in Drylands

Threshold behaviour in landscape systems is usually linked to strong feedbacks between structures and/or system processes (Turnbull et al., 2008; Moreno-de las Heras et al., 2012), which is the case for drylands due to the presence of strong ecohydrological feedbacks (Wainwright, 2009; Turnbull et al., 2012; Okin et al., 2018). Ecohydrological feedbacks can lead to stabilizing (i.e., negative feedback) effects that increase the resilience of the landscape, but sometimes can also lead to amplifying (i.e., positive feedback) effects that reduce the resilience of the landscape (Turnbull et al., 2012; Mayor et al., 2019; Saco et al., 2020). Positive feedbacks that decrease the resilience of the landscape, can drive the system towards a critical threshold leading to an alternative state in response to an abrupt perturbation, or sometimes even in response to small changes in environmental conditions. In some cases, the new alternative ecosystem state can consist of an irreversible degraded state. Using case studies in arid systems, Turnbull et al. (2012) show how external perturbations (i.e., climate shifts, fire, surface compaction, grazing, shrub encroachment, land use changes) can affect soil and vegetation composition leading to salinization, biocrust disturbance, sediment erosion and deposition, runoff amplification, and so on. These processes can result in irreversible degradation due to positive ecohydrological feedbacks. Similarly, Scheffer et al. (2001) analyse sudden ecosystem changes facilitated by a gradual loss of resilience, which in the case of arid systems can be due to loss of vegetation cover. When vegetation cover is lost beyond a certain threshold, water disappears from topsoils due to runoff or infiltration to deeper layers where is not available for plants that could potentially recolonise, so the system is unable to recover. These irreversible changes are often associated with hysteresis in the relationship between ecosystem state and exogenous forces, a situation in which transitions are hard to reverse as the system does not return to the former state even if the exogenous force returns to its former condition (Suding & Hobbs, 2009).

As described in Section 14.2, vegetation patterns can provide stabilizing properties because of their source-sink function (local negative ecohydrological feedbacks). In this context, vegetation patches function as sinks for water and sediments and therefore reduce overland flow and erosion and help ecosystems to resist stressors and recover from disturbances (Cammeraat & Imeson, 1999; Turnbull et al., 2012; Mayor et al., 2019). However, vegetation loss or removal increases hydrological connectivity and erosion potential, and can trigger the development of rills or gullies reinforcing the increase in connectivity (Turnbull et al., 2008; Saco & Moreno-de las Heras, 2013; Baartman et al., 2018). Feedbacks between ecohydrological and erosion processes can generate positive hillslope scale (global) feedbacks, which can lead to irreversible landscape degradation (Moreno-de las Heras et al., 2011). If episodes of widespread landscape erosion occur, this can lead to changes in topography that further reinforce the increase of connectivity because of the effect of rills and gullies on enabling water losses from the landscape. Simulations of Mediterranean-dry reclaimed slopes have been used to define vegetation thresholds for unrilled slope conditions (Moreno-de las Heras et al., 2011). Beyond this threshold, the appearance of rills leads to a decrease in water retained in the landscape and eventually vegetation death, further increasing the potential for fluvial erosion and soil loss, reinforcing the degradation process. The increase in connectivity (i.e., through vegetation death or removal) can reduce the plant patches' ability to capture/obstruct surface run-on critical for vegetation growth (Ludwig et al., 2007; Tongway & Ludwig, 2011). This reduction in ecosystem functionality and productivity can lead to non-linear threshold behaviour. Hence, the integrity of natural vegetation patterns, which is one of the key factors determining the hydrologic and sediment connectivity of the landscape, can be used as an indicator of ecosystem health and resilience (Moreno-de las Heras et al., 2012; Kéfi et al., 2016; Berdugo et al., 2017; Urgeghe et al., 2021). Early detection of degradation signals is therefore of paramount importance in drylands, as degraded landscape conditions may be difficult to revert (Saco et al., 2018).

Recent studies have underlined the value of incorporating measures associated to vegetation patch size distributions and landscape hydrological connectivity for the maintenance of a healthy landscape function (Moreno-de las Heras et al., 2011; Schneider & Kéfi, 2016). To this end, recent studies have investigated the use of hydrological connectivity for identifying transitions and threshold behaviour, and its potential use for the development of early warning indicators of degradation in drylands (Zurlini et al., 2014; Okin et al., 2015; Rodríguez et al., 2018; Saco et al., 2018; Mayor et al., 2019). For example, a study of banded Mulga landscapes in Australia (Saco et al., 2020), based on remote sensing and field observations showed that vegetation cover disturbances had a significant impact on the structural connectivity, leading to a drastic increase in structural connectivity (measured using mean flow length metrics) for landscapes with plant cover below a critical threshold of around 30%. The increase in connectivity was shown to display a threshold-like behaviour (Figure 14.2), triggering degradation and loss of landscape functionality (measured using vegetation rainfall use efficiency). This drastic loss of ecosystem functionality was linked to the existence of underlying non-linear feedbacks (Saco et al., 2020), and could be explained by a change from prevailing negative (local) ecohydrologic source-sink feedbacks for landscapes with low connectivity, to an increasing dominance of global positive feedbacks leading to losses of resources at global (hillslope) scales as hydrologic connectivity and sediment increases. In related work, (Rodríguez et al., 2018), analysed changes in a connectivity metric as a function of landscape cover. Their findings suggested that deviations from the expected value of their connectivity metric could be used as early warning index for imminent transitions to a degraded state.

The importance of ecohydrology and erosion feedbacks described earlier highlights the need of accounting for the coevolving nature of landform-vegetation structures for the understanding of degradation processes (Saco et al., 2020). Ecohydrological processes (including surface runoff redistribution and vegetation growth/death) and landform evolution processes (driving changes in topography, slopes, flow paths) and their feedbacks are key drivers of the continuously changing coevolving landscape system. Connectivity is an emerging property of this landscape system that is changing as well, both in terms of its more temporary persistent structural features (vegetation distribution and patterns, landform features) and its short-response functional counterparts (hydrology and sediment processes). As explained earlier, the strong feedbacks in the system imply that climatic or

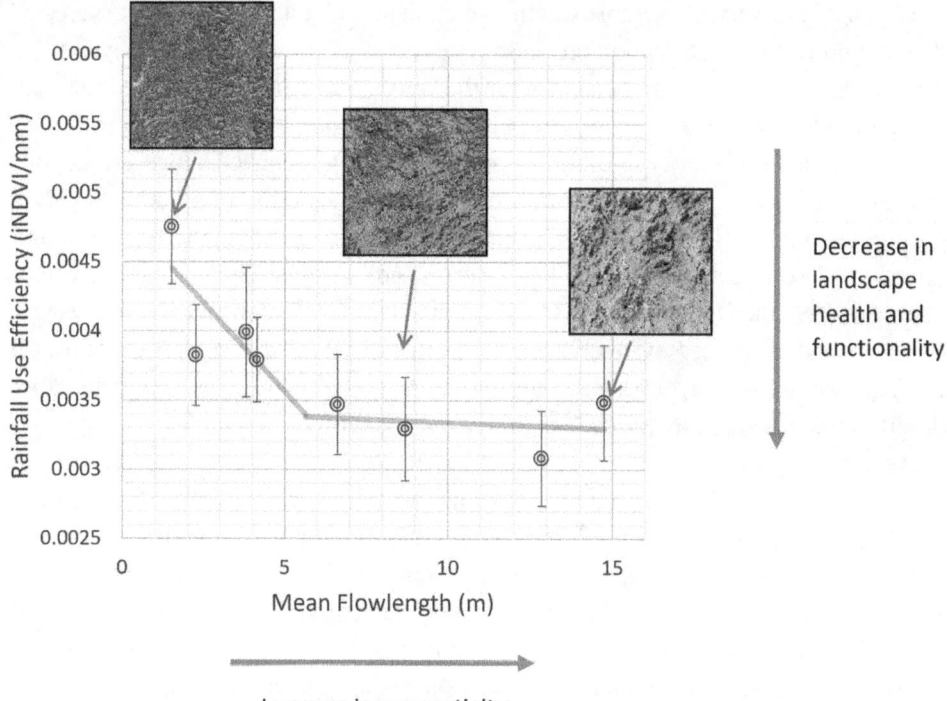

Figure 14.2   Response of landscape health and functionality (quantified using rainfall use efficiency over a period of 13 years) to changes in hydrologic connectivity (quantified using mean flow length) for Mulga dryland sites in Australia with different levels of degradation (adapted from Saco et al., 2020). A black and white version of this figure will appear in some formats. For the colour version, refer to the plate section.

man-made disturbances affect connectivity and can trigger cascade effects in the system, potentially leading to desertification as will be explained in Section 14.5.

14.5 Climate and Anthropogenic Impacts on Connectivity Leading to Desertification

Climate change projections indicate that drylands will receive less rainfall over the next century, which will critically increase the pressure on natural ecosystems as they will compete for water with irrigation and human and livestock consumption (Huang et al., 2017). Increased aridity, longer droughts and a reduced vegetation productivity are expected in drylands (Seager et al., 2014; Huang et al., 2016), which will also be affected by climate variability. Anomalies in weather patterns such as El Niño Southern Oscillation and the interdecadal Pacific oscillation are expected to increase in strength and frequency (Power et al., 2013; Cai et al., 2015), generating more dry years and more intense storm episodes.

Drylands are very vulnerable to climate change and anthropogenic pressures as they are prone to desertification processes (Reynolds et al., 2007; Bestelmeyer et al., 2015). Desertification is associated with the drastic reduction in both productivity and the ability of the landscape to support wild or commercial grazing, crops, and therefore human livelihoods (Bestelmeyer et al., 2015). Climate change projections indicate that between 180 and 280 million people will be affected by desertification during this century (IPCC, 2019). The global importance and potentially devastating effects of desertification has been long recognised, with the United Nations establishing in 1994 the United Nations Convention to Combat Desertification framework to understand the processes involved in desertification and the development of useful management tools to avoid regime shifts to degraded states. Almost 30 years after, desertification is still a major challenge that requires further research.

As a general rule, longer droughts are expected to result in a reduction of arid vegetation, increases in bare-soil areas and increased connectivity (Okin et al., 2009). Increases in rainfall intensity and wind speed are also expected due to climate change (IPCC, 2021), which will increase functional connectivity, intensify erosion, and trigger feedback processes that could lead to degradation and desertification. In addition, longer and more intense droughts can lead to shrubland invasion of grassland as explained in Section 14.3, a phenomenon that is already occurring in many dryland areas around the world (IPCC, 2021).

Human disturbances associated to poor or inefficient land management practices can also result in increases in connectivity, and eventually to degradation and desertification. Overgrazing or logging can alter the structure of vegetation patches reducing their density and/or size which leads to a 'leaky' more connected system (Saco et al., 2007). Because a system with high connectivity is less efficient at trapping runoff and sediments, it loses valuable water and nutrient resources (Ludwig et al., 2005) and triggers a positive-feedback loop that reinforces the degradation process (Lavee et al., 1998). Removal of vegetation leading to increased connectivity can also be associated with the creation of arable fields, which makes them extremely vulnerable during prolonged droughts. For example, increased connectivity in agricultural fields during the 1930s drought in the United States has been linked to the desertification of extensive areas and the creation of the Dust Bowl (Peters et al., 2007). Trails, roads and preferential paths associated with intensive livestock or offroad vehicle traffic can also lead to degradation and desertification, starting as small, connected areas but rapidly expanding due to water and wind erosion.

14.6 Implications for Management of Drylands

Reducing connectivity as a way of managing degraded landscapes can be achieved using simple methodologies. For instance, in their landmark work, Ludwig and Tongway (1996) used simple obstacles made of brush piles laid perpendicular to

Figure 14.3 Experimental restoration in Lake Mere, Australia. (a) pre-experimental conditions, (b) brush piles set in the field with initial vegetation growth, and (c) established vegetation (photo credit David J. Tongway, CSIRO). A black and white version of this figure will appear in some formats. For the colour version, refer to the plate section.

the main downslope flow direction in order to reduce runoff pathways and generate sinks of resources (Figure 14.3). The obstacles captured water and seeds, facilitating the establishment of pioneering grasses and eventually woody vegetation. Later works in other types of vegetation have confirmed the suitability of simple branch and organic matter piles or artificial structures as a low-cost technique to reduce connectivity and restore degraded arid sites (Fick et al., 2016; Kimiti et al., 2017; Peters et al., 2020). Sometimes these techniques are combined with revegetation to increase soil permeability and maximise the incorporation of resources (water and nutrients), which speed up recovery (Bautista & Mayor, 2021).

Reduction of connectivity has also been achieved by promoting grassland over shrublands in areas that have experienced shrub encroachment. Peters et al. (2020), reports on successful manipulation experiments in the Jornada Experimental Range in New Mexico, in which a combination of removal of mesquite shrubs and placement of artificial connectivity modifiers made from cloth (ConMods) promoted the recruitment of perennial grasses, effectively reducing the system connectivity (Figure 14.4). In turn, Johnson et al. (2021) used a herbicide treatment to control shrub encroachment in a grassland in south-eastern Arizona, which resulted in an increase of a perennial non-native grass that reduced soil erodibility and sediment connectivity. However, herbicide treatments must be used with caution as the increase in grass cover will only occur if climatic and soil conditions are consistent with grass-dominated systems (Johnson et al., 2021). If shrub cover is reduced via herbicide treatments without a simultaneous increase in grass cover, the connectivity of the system will increase and result in further degradation.

Figure 14.4   Connectivity modifiers (ConMOd) installed in a degraded grassland plot in Canyonlands National Park (July 26, 2016) (adapted from USGS, public domain). A black and white version of this figure will appear in some formats. For the colour version, refer to the plate section.

Other practices that reduce connectivity include runoff-harvesting techniques (Bainbridge, 2007; Paz-Kagan et al., 2017), which consist of collection areas that intercept flow and resources and are typically situated as contour banks along the hillslopes. Plantation techniques that collect resources and runoff into planting holes have demonstrated potential for introduced seedling survival in degraded sites with high connectivity (Fuentes et al., 2017). Most of these measures to reduce connectivity are applied at a small scale and focus on restoring or maintaining the source sink mechanisms for redistribution of water and nutrients. Larger-scale modifications are also possible (flow control dams) but the small-scale measures have been reported to be more cost-effective (Stavi et al., 2020). A combination of small-scale and cost-effective practices, however, may be used for managing and monitoring hydrological connectivity at catchment or landscape/slope scale. Bautista and Mayor (2021) suggest that encouraging diverse forest types, interspersing natural vegetation in flow concentrated areas, reducing hydrologic connectivity with hedges and vegetated corridors, and establishing riparian vegetation will have connectivity effects across the catchment. In addition, improving sink-source dynamics for capturing runoff (through trenches, branch bundles, mulches, etc.) will have an immediate effect at a landscape or slope scale. These combined effects at different scales will lead to reduced connectivity, less sediment export and more runoff capture.

Another good example of small-scale measures for the use of restoration techniques based on connectivity principles is described in a recent experiment by

Vicente et al. (2022), who made use of suitable sink microsites as key spots for introducing woody vegetation in Mediterranean-dry, restored post-mining hillslopes in northeast Spain. In these environments, moderate erosional activity during the first stages of reclamation can lead to the formation of source-sink systems, where bare areas and vegetation patches are connected by spatially discontinuous rills and other surface micro-topographical depressions that act as preferential pathways for internal transmission of water and sediment resources within the hillslopes (Merino-Martín et al., 2015). Sink areas, covered by perennial grasses, are located in rill discontinuities and micro-topographical splays, where runoff and sediments produced in barely covered source patches during precipitation events are discharged. Vicente et al. (2022) compared between 2012 and 2020 the survival and plant growth rates of seedlings of two tree species (*Quercus ilex* and *Pinus nigra*) introduced in these sink and source patches. Overall, the higher availability of water resources in the sink patches largely improved seedling survival both in the short- and long-term (1–4 years and eight years since plantation, respectively). Seedling growth was also enhanced in the sink patches, particularly during wet years, when higher precipitation allowed sink patches to capture large amounts of overland flow produced in the runoff and sediment generating source patches.

The experiment detailed in Vicente et al. (2022) constitutes a passive framework on the use hillslope connectivity patterns for enhancing tree seedling establishment in restoration projects, where the actual distribution of source and sink patches and their spatial connections are not actively manipulated. Further plantation investigations by Urgeghe and Bautista (2015) for optimizing the management of hillslope connective patterns in the introduction of woody plants (*Olea europaea* and *Pistacea lentiscus*) in a semi-arid landscape (Albaterra catchment, SE Spain) concluded that the most relevant factors for the survival and growth of the introduced seedlings were the interplay between the size and connectivity of upslope bare-soil areas and the relative location of the hillslope where the seedlings were planted. These evidences of the positive effects of the size of the upslope bare-soil micro-catchment area of the planting holes on seedling performance offer new clues for active management of hillslope connectivity patterns in the introduction of woody vegetation. The spatial pattern of the plantations can be designed in terms of an optimal source-to-sink area ratio that maximises the input of surface water resources into planted vegetation without exceeding the overland flow capture capacity of the sink patches where seedling are introduced, minimizing resource leaks from the system (Bautista & Mayor, 2021).

At large spatial scales, the use of structural indexes of connectivity can facilitate establishing priority areas for restoring human-managed landscapes in drylands. For example, Calsamiglia et al. (2017, 2018a) apply a structural sediment connectivity index (IC) developed by Borselli et al. (2008) and modified by Cavalli et al. (2013) for exploring the stability of abandoned agricultural terraces and the organization of traditional agricultural tile drainages in catchments of Mallorca (Spain). These traditional anthropogenic features of Mediterranean agricultural landscapes act generating general disconnectivity between compartments of the catchments, while diverting water and sediment fluxes through controlled pathways. However, their collapse after land abandonment can promote rapid geomorphological readjustment and activation of high erosion processes in the landscape, causing land degradation. This is for example, the case of traditional agricultural terraces in Mediterranean areas, where the degradation of these structures can provoke cascade effects resulting in gully formation, mass movements and general terrace failure (Moreno de las Heras et al., 2019). As demonstrated by Calsamiglia et al. (2017), IC can be applied for detecting preferential flow concentration pathways promoting the collapse of dry-stone walls and the reworking of stored soils and sediments in the terraces, providing a cost-effective strategy for establishing priority areas of restoration in the most vulnerable structures of the catchment. Similar conclusions were obtained by Marchamalo et al. (2016) in a small semi-arid catchment (Cárcavo basin) in SE Spain, where they explored runoff and erosion hotspots by applying repeated connectivity mapping for targeting mitigation and restoration measures using vegetation.

Monitoring and assessment for the management of drylands must also go beyond the typical protocols that document only vegetation cover and composition and incorporate changes in connectivity (Webb et al., 2021). For example, the National Resource Inventory implemented by the National Resource Conservation Service of the United States Department of Agriculture includes an assessment of the size distribution of intercanopy gaps in more than 10,000 points across the country (Herrick et al., 2005). This indicator can be used for management by determining threshold values above which connectivity requires control measures, for example, limit livestock movement or offroad traffic (Okin et al., 2009). However, similar indicators are not available globally.

As explained before, dryland systems are often characterised by the presence of tipping points at which the system transitions abruptly to a degraded state that maybe irreversible. More recent work has focussed on identifying early warning indicators of transition to degraded states that could be used as a preventive management tool (Saco et al., 2018). Some of these indicators look at changes over time of connectivity measures obtained using remote sensing tools, like the size distribution of vegetation patches. Xu et al. (2015) and Alados et al. (2017) used the skewness of the patch-size distribution and plant aggregation to develop early warning signals accounting for the role of plant–plant interactions and local facilitation mechanisms. Berdugo et al. (2017) found a bimodal distribution of an ecosystem functionality index in drylands around the world, suggesting two alternative states (healthy and degraded), which were correlated with patch size distribution, but not captured by vegetation cover. Another measure of water and sediment connectivity, the flow length index (Mayor et al., 2008), has been shown to capture increases in connectivity that trigger degradation. Moreno-de las Heras et al. (2012) related the index to water use efficiency (ratio of normalised difference vegetation index (NDVI) over rainfall) in Mulga landscapes of Australia and found thresholds in flow length above which degradation occurred. Rodríguez et al. (2018) and van den Elsen et al. (2020) derived expressions for expected values of the flow length index for different landscape cover, and associated deviations from the expected value to early warning signals for imminent degradation.

References

Alados, C. L., Saiz, H., Gartzia, M., Nuche, P., Escós, J., Navarro, T., & Pueyo, Y. (2017). Plant–plant interactions scale up to produce vegetation spatial patterns: the influence of long- and short-term process. *Ecosphere, 8* (8), p. e01915. www.doi.org/10.1002/ecs2.1915

Ali, G. A., & Roy, A. G. (2009). Revisiting hydrologic sampling strategies for an accurate assessment of hydrologic connectivity in humid temperate systems. *Geography Compass, 3* (1), pp. 350–374. www.doi.org/10.1111/j.1749-8198.2008.00180.x

Baartman, J. E., Temme, A. J., & Saco, P. M. (2018). The effect of landform variation on vegetation patterning and related sediment dynamics. *Earth Surface Processes and Landforms, 43*(10), pp. 2121–2135.

Baartman, J. E. M., Nunes, J. P., Masselink, R., Darboux, F., Bielders, C., Degré, A., Cantreul, V., et al. (2020). What do models tell us about water and sediment connectivity? *Geomorphology, 367*, p. 107300. www.doi.org/10.1016/j.geomorph.2020.107300

Bainbridge, D. A. (2007). Arid lands research needs next twenty-five years. *Annals of Arid Zone, 46* (3&4), pp. 1–28.

Bautista, S., & Mayor, Á. G. (2021). El papel de la (des)conectividad ecohidrológica en el funcionamiento y el manejo de las zonas áridas. *Ecosistemas, 30* (3), pp. 2265. www.doi.org/10.7818/ECOS.2265

Belnap, J., Büdel, B., & Lange, O. L. (2003). Biological soil crusts: Characteristics and distribution. In J. Belnap & O. L. Lange (eds.), *Biological Soil Crusts: Structure, Function, and Management* (pp. 3–30). Berlin, Heidelberg: Springer.

Berdugo, M., Kéfi, S., Soliveres, S., & Maestre, F. T. (2017). Plant spatial patterns identify alternative ecosystem multifunctionality states in global drylands. *Nature Ecology & Evolution, 1* (2), p. 3. www.doi.org/10.1038/s41559-016-0003

Bergametti, G., & Gillette, D. A. (2010). Aeolian sediment fluxes measured over various plant/soil complexes in the Chihuahuan desert. *Journal of Geophysical Research: Earth Surface, 115* (F3). www.doi.org/10.1029/2009JF001543

Bestelmeyer, B. T., Okin, G. S., Duniway, M. C., Archer, S. R., Sayre, N. F., Williamson, J. C., & Herrick, J. E. (2015). Desertification, land use, and the transformation of global drylands. *Frontiers in Ecology and the Environment, 13* (1), pp. 28–36. www.doi.org/10.1890/140162

Bhark, E. W., & Small, E. E. (2003). Association between plant canopies and the spatial patterns of infiltration in shrubland and grassland of the Chihuahuan desert, New Mexico. *Ecosystems, 6* (2), pp. 0185–0196. www.doi.org/10.1007/s10021-002-0210-9

Birch, J. D., Lutz, J. A., Hogg, E. H., Simard, S. W., Pelletier, R., LaRoi, G. H., & Karst, J. (2019). Decline of an ecotone forest: 50 years of demography in the southern boreal forest. *Ecosphere, 10* (4), pp. e02698. www.doi.org/10.1002/ecs2.2698

Borgogno, F., D'Odorico, P., Laio, F., & Ridolfi, L. (2009). Mathematical models of vegetation pattern formation in ecohydrology. *Reviews of Geophysics, 47* (1). www.doi .org/10.1029/2007RG000256

Borselli, L., Cassi, P., & Torri, D. (2008). Prolegomena to sediment and flow connectivity in the landscape: A GIS and field numerical assessment. *CATENA, 75* (3), pp. 268–277. www.doi.org/10.1016/j.catena.2008.07.006

Bracken, L. J., & Croke, J. (2007). The concept of hydrological connectivity and its contribution to understanding runoff-dominated geomorphic systems. *Hydrological Processes, 21* (13), pp. 1749–1763. www.doi.org/10.1002/hyp.6313

Bracken, L. J., Wainwright, J., Ali, G. A., Tetzlaff, D., Smith, M. W., Reaney, S. M., & Roy, A. G. (2013). Concepts of hydrological connectivity: Research approaches, pathways and future agendas. *Earth-Science Reviews, 119*, pp. 17–34. www.doi .org/10.1016/j.earscirev.2013.02.001

Bradley, E. F., & Mulhearn, P. J. (1983). Development of velocity and shear stress distribution in the wake of a porous shelter fence. *Journal of Wind Engineering and Industrial Aerodynamics, 15* (1), pp. 145–156. www.doi.org/10.1016/0167-6105(83)90185-X

Burrell, A. L., Evans, J. P., & De Kauwe, M. G. (2020). Anthropogenic climate change has driven over 5 million km2 of drylands towards desertification. *Nature Communications, 11* (1), pp. 3853. www.doi.org/10.1038/s41467-020-17710-7

Cai, W., Wang, G., Santoso, A., McPhaden, M. J., Wu, L., Jin, F.-F., Timmermann, A., et al. (2015). Increased frequency of extreme La Niña events under greenhouse warming. *Nature Climate Change, 5* (2), pp. 132–137. www.doi.org/10.1038/ nclimate2492

Calsamiglia, A., Fortesa, J., García-Comendador, J., Lucas-Borja, M. E., Calvo-Cases, A., & Estrany, J. (2018a). Spatial patterns of sediment connectivity in terraced lands: Anthropogenic controls of catchment sensitivity. *Land Degradation & Development, 29* (4), pp. 1198–1210. www.doi.org/10.1002/ldr.2840

Calsamiglia, A., García-Comendador, J., Fortesa, J., López-Tarazón, J. A., Crema, S., Cavalli, M., Calvo-Cases, A., & Estrany, J. (2018b). Effects of agricultural drainage systems on sediment connectivity in a small Mediterranean lowland catchment. *Geomorphology, 318*, pp. 162–171. www.doi.org/10.1016/j.geomorph.2018.06.011

Calsamiglia, A., Lucas-Borja, M. E., Fortesa, J., García-Comendador, J., & Estrany, J. (2017). Changes in soil quality and hydrological connectivity caused by the abandonment of terraces in a Mediterranean burned catchment. *Forests, 8* (9), p. 333

Cammeraat, L. H., & Imeson, A. C. (1999). The evolution and significance of soil–vegetation patterns following land abandonment and fire in Spain. *CATENA, 37* (1), pp. 107–127. www.doi.org/10.1016/S0341-8162(98)00072-1

Cavalli, M., Trevisani, S., Comiti, F., & Marchi, L. (2013). Geomorphometric assessment of spatial sediment connectivity in small Alpine catchments. *Geomorphology, 188*, pp. 31–41. www.doi.org/10.1016/j.geomorph.2012.05.007

Chappell, A., Webb, N. P., Guerschman, J. P., Thomas, D. T., Mata, G., Handcock, R. N., Leys, J. F., & Butler, H. J. (2018). Improving ground cover monitoring for wind erosion assessment using MODIS BRDF parameters. *Remote Sensing of Environment, 204*, pp. 756–768. www.doi.org/10.1016/j.rse.2017.09.026

Dickinson, R. E. (1984). Modeling evapotranspiration for three-dimensional global climate models. In J. E. Hansen & T. Takahashi (Eds.), *Climate Processes and Climate Sensitivity* (Vol. 5, pp. 58–72). Washington, DC: American Geophysical Union.

Dunkerley, D. L. (2002). Infiltration rates and soil moisture in a groved mulga community near Alice Springs, arid central Australia: evidence for complex internal rainwater redistribution in a runoff–runon landscape. *Journal of Arid Environments*, *51* (2), pp. 199–219. www.doi.org/10.1006/jare.2001.0941

Fensham, R. J., Fairfax, R. J., & Archer, S. R. (2005). Rainfall, land use and woody vegetation cover change in semi-arid Australian savanna. *Journal of Ecology*, *93* (3), pp. 596–606. www.doi.org/10.1111/j.1365-2745.2005.00998.x

Fick, S. E., Decker, C., Duniway, M. C., & Miller, M. E. (2016). Small-scale barriers mitigate desertification processes and enhance plant recruitment in a degraded semiarid grassland. *Ecosphere*, *7* (6), pp. e01354. www.doi.org/10.1002/ecs2.1354

Fryirs, K. A., Brierley, G. J., Preston, N. J., & Kasai, M. (2007). Buffers, barriers and blankets: The (dis)connectivity of catchment-scale sediment cascades. *CATENA*, *70* (1), pp. 49–67. www.doi.org/10.1016/j.catena.2006.07.007

Fuentes, D., Smanis, A., & Valdecantos, A. (2017). Recreating sink areas on semi-arid degraded slopes by restoration. *Land Degradation & Development*, *28* (3), pp. 1005–1015. www.doi.org/https://doi.org/10.1002/ldr.2671

Gherardi, L. A., & Sala, O. E. (2015). Enhanced precipitation variability decreases grass- and increases shrub-productivity. *Proceedings of the National Academy of Sciences*, *112* (41), pp. 12735–12740. www.doi.org/10.1073/pnas.1506433112

Gillette, D. A., Adams, J., Endo, A., Smith, D., & Kihl, R. (1980). Threshold velocities for input of soil particles into the air by desert soils. *Journal of Geophysical Research: Oceans*, *85* (C10), pp. 5621–5630. www.doi.org/10.1029/JC085iC10p05621

Gillette, D. A., Herrick, J. E., & Herbert, G. A. (2006). Wind characteristics of mesquite streets in the Northern Chihuahuan Desert, New Mexico, USA. *Environmental Fluid Mechanics*, *6* (3), pp. 241–275. www.doi.org/10.1007/s10652-005-6022-7

Havstad, K. M., Peters, D. P. C., Skaggs, R., Brown, J., Bestelmeyer, B., Fredrickson, E., Herrick, J., & Wright, J. (2007). Ecological services to and from rangelands of the United States. *Ecological Economics*, *64* (2), pp. 261–268. www.doi.org/10.1016/j.ecolecon.2007.08.005

Herrick, J. E., van Zee, J. W., McCord, S. E., Courtright, E. M., Karl, J. W., & Burkett, L. M. (2005). *Monitoring manual for grassland, shrubland and savanna ecosystems: Vol. I: Core Methods*. Las Cruces, New Mexico: USDA-ARS Jornada Experimental Range.

Huang, J., Li, Y., Fu, C., Chen, F., Fu, Q., Dai, A., Shinoda, M., et al. (2017). Dryland climate change: Recent progress and challenges. *Reviews of Geophysics*, *55* (3), pp. 719–778. www.doi.org/10.1002/2016RG000550

Huang, L., He, B., Chen, A., Wang, H., Liu, J., Lü, A., & Chen, Z. (2016). Drought dominates the interannual variability in global terrestrial net primary production by controlling semi-arid ecosystems. *Scientific Reports*, *6* (1), pp. 24639. www.doi.org/10.1038/srep24639

Imeson, A. C., & Prinsen, H. A. M. (2004). Vegetation patterns as biological indicators for identifying runoff and sediment source and sink areas for semi-arid landscapes in Spain. *Agriculture, Ecosystems & Environment*, *104* (2), pp. 333–342. www.doi.org/10.1016/j.agee.2004.01.033

IPCC. (2019). Summary for policymakers. In P. R. Shukla, J. Skea, E. Calvo Buendia, V. Masson-Delmotte, H.-O. Pörtner, D. C. Roberts, P. Zhai, R. Slade, S. Connors, R. van Diemen, M. Ferrat, E. Haughey, S. Luz, S. Neogi, M. Pathak, J. Petzold, J. Portugal Pereira, P. Vyas, E. Huntley, K. Kissick, M. Belkacemi, & J. Malley (eds.), *Climate Change and Land: an IPCC special report on climate change, desertification, land degradation, sustainable land management, food security, and greenhouse gas fluxes in terrestrial ecosystems*: In Press.

IPCC. (2021). Summary for policymakers. In V. Delmotte, P. Zhai, A. Pirani, S. L. Connors, C. Péan, S. Berger, N. Caud, Y. Chen, L. Goldfarb, M. I. Gomis, M. Huang, K. Leitzell, E. Lonnoy, J. B. R. Matthews, T. K. Maycock, T. Waterfield, O. Yelekçi, R. Yu, & B. Zhou (eds.), *Climate Change 2021: The Physical Science Basis. Contribution of Working Group I to the Sixth Assessment Report of the Intergovernmental Panel on Climate Change* (pp. 3–32). Cambridge, UK and New York: Cambridge University Press.

Iwaniec, D. M., Gooseff, M., Suding, K. N., Samuel Johnson, D., Reed, D. C., Peters, D. P. C., Adams, B., et al. (2021). Connectivity: insights from the U.S. Long Term Ecological Research Network. *Ecosphere, 12* (5), pp. e03432. www.doi.org/10.1002/ecs2.3432

Johnson, J. C., Williams, C. J., Guertin, D. P., Archer, S. R., Heilman, P., Pierson, F. B., & Wei, H. (2021). Restoration of a shrub-encroached semi-arid grassland: Implications for structural, hydrologic, and sediment connectivity. *Ecohydrology, 14* (4), pp. e2281. www.doi.org/10.1002/eco.2281

Kawamura, R. (1951). *Study of Sand Movement by Wind*: University of California.

Keessstra, S., Nunes, J. P., Saco, P., Parsons, T., Poeppl, R., Masselink, R., & Cerdà, A. (2018). The way forward: Can connectivity be useful to design better measuring and modelling schemes for water and sediment dynamics? *Science of The Total Environment, 644*, pp. 1557–1572. www.doi.org/10.1016/j.scitotenv.2018.06.342

Kéfi, S., Holmgren, M., & Scheffer, M. (2016). When can positive interactions cause alternative stable states in ecosystems? *Functional Ecology, 30* (1), pp. 88–97. www.doi .org/10.1111/1365-2435.12601

Kimiti, D. W., Riginos, C., & Belnap, J. (2017). Low-cost grass restoration using erosion barriers in a degraded African rangeland. *Restoration Ecology, 25* (3), pp. 376–384. www.doi.org/10.1111/rec.12426

Lavee, H., Imeson, A. C., & Sarah, P. (1998). The impact of climate change on geomorphology and desertification along a mediterranean-arid transect. *Land Degradation & Development, 9* (5), pp. 407–422. www.doi.org/10.1002/(SICI)1099-145X(199809/10)9:5<407::AID-LDR302>3.0.CO;2-6

Leenders, J. K., Sterk, G., & Van Boxel, J. H. (2011). Modelling wind-blown sediment transport around single vegetation elements. *Earth Surface Processes and Landforms, 36* (9), pp. 1218–1229. www.doi.org/10.1002/esp.2147

Li, J., Okin, G. S., Alvarez, L., & Epstein, H. (2008). Effects of wind erosion on the spatial heterogeneity of soil nutrients in two desert grassland communities. *Biogeochemistry, 88* (1), pp. 73–88. www.doi.org/10.1007/s10533-008-9195-6

Lisenby, P. E., & Fryirs, K. A. (2017). 'Out with the Old?' Why coarse spatial datasets are still useful for catchment-scale investigations of sediment (dis)connectivity. *Earth Surface Processes and Landforms, 42* (10), pp. 1588–1596. www.doi.org/10.1002/esp.4131

Locosselli, G. M., Brienen, R. J. W., Leite, M. D. S., Gloor, M., Krottenthaler, S., de Oliveira, A. A., Barichivich, J., et al. (2020). Global tree-ring analysis reveals rapid decrease in tropical tree longevity with temperature. *Proceedings of the National Academy of Sciences, 117* (52), pp. 33358–33364. www.doi.org/10.1073/pnas.2003873117

Ludwig, J. A., Bastin, G. N., Chewings, V. H., Eager, R. W., & Liedloff, A. C. (2007). Leakiness: A new index for monitoring the health of arid and semiarid landscapes using remotely sensed vegetation cover and elevation data. *Ecological Indicators, 7* (2), pp. 442–454. www.doi.org/10.1016/j.ecolind.2006.05.001

Ludwig, J. A., & Tongway, D. J. (1996). Rehabilitation of semiarid landscapes in australia. ii. restoring vegetation patches. *Restoration Ecology, 4* (4), pp. 398–406. www.doi .org/10.1111/j.1526-100X.1996.tb00192.x

Ludwig, J. A., Wilcox, B. P., Breshears, D. D., Tongway, D. J., & Imeson, A. C. (2005). Vegetation patches and runoff-erosion as interacting ecohydrological processes in semiarid landscapes. *Ecology, 86* (2), pp. 288–297. www.doi.org/10.1890/03-0569

Maestre, F. T., Bowker, M. A., Puche, M. D., Belén Hinojosa, M., Martínez, I., García-Palacios, P., Castillo, A. P., et al. (2009). Shrub encroachment can reverse desertification in semi-arid Mediterranean grasslands. *Ecology Letters, 12* (9), pp. 930–941. www.doi.org/10.1111/j.1461-0248.2009.01352.x

Magliano, P. N., Whitworth-Hulse, J. I., & Baldi, G. (2019). Interception, throughfall and stemflow partition in drylands: Global synthesis and meta-analysis. *Journal of Hydrology, 568,* pp. 638–645. www.doi.org/10.1016/j.jhydrol.2018.10.042

Marchamalo, M., Hooke, J. M., & Sandercock, P. J. (2016). Flow and sediment connectivity in semi-arid landscapes in SE Spain: Patterns and controls. *Land Degradation & Development, 27* (4), pp. 1032–1044. www.doi.org/10.1002/ldr.2352

Masselink, R. J. H., Keesstra, S. D., Temme, A. J. A. M., Seeger, M., Giménez, R., & Casalí, J. (2016). Modelling discharge and sediment yield at catchment scale using connectivity components. *Land Degradation & Development, 27* (4), pp. 933–945. www.doi.org/10.1002/ldr.2512

Mayaud, J. R., Wiggs, G. F. S., & Bailey, R. M. (2017). A field-based parameterization of wind flow recovery in the lee of dryland plants. *Earth Surface Processes and Landforms, 42* (2), pp. 378–386. www.doi.org/10.1002/esp.4082

Mayor, A. G., Bautista, S., Rodriguez, F., & Kéfi, S. (2019). Connectivity-Mediated Ecohydrological Feedbacks and Regime Shifts in Drylands. *Ecosystems, 22* (7), pp. 1497–1511. www.doi.org/10.1007/s10021-019-00366-w

Mayor, Á. G., Bautista, S., Small, E. E., Dixon, M., & Bellot, J. (2008). Measurement of the connectivity of runoff source areas as determined by vegetation pattern and topography: A tool for assessing potential water and soil losses in drylands. *Water Resources Research, 44* (10). www.doi.org/10.1029/2007WR006367

Mekonnen, M., Keesstra, S. D., Baartman, J. E. M., Stroosnijder, L., & Maroulis, J. (2017). Reducing sediment connectivity through man-made and natural sediment sinks in the Minizr Catchment, Northwest Ethiopia. *Land Degradation & Development, 28* (2), pp. 708–717. www.doi.org/10.1002/ldr.2629

Merino-Martín, L., Moreno-de las Heras, M., Espigares, T., & Nicolau, J. M. (2015). Overland flow directs soil moisture and ecosystem processes at patch scale in Mediterranean restored hillslopes. *CATENA, 133,* 71–84. www.doi.org/10.1016/j.catena.2015.05.002

Meron, E., Gilad, E., von Hardenberg, J., Shachak, M., & Zarmi, Y. (2004). Vegetation patterns along a rainfall gradient. *Chaos, Solitons & Fractals, 19* (2), pp. 367–376. www.doi.org/10.1016/S0960-0779(03)00049-3

Misra, R. (1983). Indian savannas. In F. Bourliere (ed.), *Ecosystems of the world* (pp. 151–166). New York: Elsevier.

Monger, C., Sala, O. E., Duniway, M. C., Goldfus, H., Meir, I. A., Poch, R. M., Throop, H. L., & Vivoni, E. R. (2015). Legacy effects in linked ecological–soil–geomorphic systems of drylands. *Frontiers in Ecology and the Environment, 13* (1), pp. 13–19. www.doi.org/10.1890/140269

Moreno-de las Heras, M., Díaz-Sierra, R., Nicolau, J. M., & Zavala, M. A. (2011). Evaluating restoration of man-made slopes: a threshold approach balancing vegetation and rill erosion. *Earth Surface Processes and Landforms, 36* (10), pp. 1367–1377. https://doi.org/10.1002/esp.2160

Moreno-de las Heras, M., Lindenberger, F., Latron, J., Lana-Renault, N., Llorens, P., Arnáez, J., Romero-Díaz, A., & Gallart, F. (2019). Hydro-geomorphological consequences of the abandonment of agricultural terraces in the Mediterranean region: Key controlling factors and landscape stability patterns. *Geomorphology*, *333*, pp. 73–91. www.doi.org/10.1016/j.geomorph.2019.02.014

Moreno-de las Heras, M., Merino-Martín, L., Saco, P. M., Espigares, T., Gallart, F., & Nicolau, J. M. (2020). Structural and functional control of surface-patch to hillslope runoff and sediment connectivity in Mediterranean dry reclaimed slope systems. *Hydrology and Earth System Sciences*, *24* (5), pp. 2855–2872. www.doi.org/10.5194/hess-24-2855-2020

Moreno-de las Heras, M., Saco, P. M., Willgoose, G. R., & Tongway, D. J. (2011). Assessing landscape structure and pattern fragmentation in semiarid ecosystems using patch-size distributions. *Ecological Applications*, *21* (7), pp. 2793–2805. www.doi.org/10.1890/10-2113.1

Moreno-de las Heras, M., Saco, P. M., Willgoose, G. R., & Tongway, D. J. (2012). Variations in hydrological connectivity of Australian semiarid landscapes indicate abrupt changes in rainfall-use efficiency of vegetation. *Journal of Geophysical Research: Biogeosciences*, *117* (G3). www.doi.org/10.1029/2011JG001839

Moreno-de las Heras, M., Turnbull, L., & Wainwright, J. (2016). Seed-bank structure and plant-recruitment conditions regulate the dynamics of a grassland-shrubland Chihuahuan ecotone. *Ecology*, *97* (9), 2303–2318. https://doi.org/10.1002/ecy.1446

Okin, G. S. (2008). A new model of wind erosion in the presence of vegetation. *Journal of Geophysical Research: Earth Surface*, *113* (F2). www.doi.org/10.1029/2007JF000758

Okin, G. S., Moreno de las Heras, M., Saco, P. M., Throop, H. L., Vivoni, E. R., Parsons, A. J., Wainwright, J., & Peters, D. P. (2015). Connectivity in dryland landscapes: shifting concepts of spatial interactions. *Frontiers in Ecology and the Environment*, *13* (1), pp. 20–27. www.doi.org/10.1890/140163

Okin, G. S., Parsons, A. J., Wainwright, J., Herrick, J. E., Bestelmeyer, B. T., Peters, D. C., & Fredrickson, E. L. (2009). Do changes in connectivity explain desertification? *BioScience*, *59* (3), pp. 237–244. www.doi.org/10.1525/bio.2009.59.3.8

Okin, G. S., Sala, O. E., Vivoni, E. R., Zhang, J., & Bhattachan, A. (2018). The interactive role of wind and water in functioning of drylands: What does the future hold? *BioScience*, *68* (9), pp. 670–677. www.doi.org/10.1093/biosci/biy067

Paz-Kagan, T., Ohana-Levi, N., Shachak, M., Zaady, E., & Karnieli, A. (2017). Ecosystem effects of integrating human-made runoff-harvesting systems into natural dryland watersheds. *Journal of Arid Environments*, *147*, pp. 133–143. www.doi.org/10.1016/j.jaridenv.2017.07.015

Peng, H.-Y., Li, X.-Y., Li, G.-Y., Zhang, Z.-H., Zhang, S.-Y., Li, L., Zhao, G.-Q., et al. (2013). Shrub encroachment with increasing anthropogenic disturbance in the semi-arid Inner Mongolian grasslands of China. *CATENA*, *109*, pp. 39–48. www.doi.org/10.1016/j.catena.2013.05.008

Peters, D. P., Havstad, K. M., Archer, S. R., & Sala, O. E. (2015). Beyond desertification: new paradigms for dryland landscapes. *Frontiers in Ecology and the Environment*, *13* (1), pp. 4–12. www.doi.org/10.1890/140276

Peters, D. P., Sala, O. E., Allen, C. D., Covich, A., & Brunson, M. (2007). Cascading events in linked ecological and socioeconomic systems. *Frontiers in Ecology and the Environment*, *5* (4), pp. 221–224. www.doi.org/10.1890/1540-9295(2007)5[221:CEILEA]2.0.CO;2

Peters, D. P. C., Okin, G. S., Herrick, J. E., Savoy, H. M., Anderson, J. P., Scroggs, S. L. P., & Zhang, J. (2020). Modifying connectivity to promote state change reversal: the importance of geomorphic context and plant–soil feedbacks. *Ecology*, *101* (9), pp. e03069. www.doi.org/10.1002/ecy.3069

Peters, D. P. C., Yao, J., & Havstad, K. M. (2004). Insights to invasive species dynamics from desertification studies. *Weed Technology*, *18* (sp1), pp. 1221–1225.

Pi, H., Webb, N. P., Huggins, D. R., & Sharratt, B. (2020). Critical standing crop residue amounts for wind erosion control in the inland Pacific Northwest, USA. *CATENA*, *195*, pp. 104742. www.doi.org/10.1016/j.catena.2020.104742

Poeppl, R. E., Keesstra, S. D., & Maroulis, J. (2017). A conceptual connectivity framework for understanding geomorphic change in human-impacted fluvial systems. *Geomorphology*, *277*, pp. 237–250. www.doi.org/10.1016/j.geomorph.2016.07.033

Power, S., Delage, F., Chung, C., Kociuba, G., & Keay, K. (2013). Robust twenty-first-century projections of El Niño and related precipitation variability. *Nature*, *502* (7472), pp. 541–545. www.doi.org/10.1038/nature12580

Puigdefabregas, J., Sole, A., Gutierrez, L., Del Barrio, G., & Boer, M. (1999). Scales and processes of water and sediment redistribution in drylands: Results from the Rambla Honda field site in Southeast Spain. *Earth Science Reviews*, *48* (1–2), pp. 39–70. www.doi.org/10.1016/S0012-8252(99)00046-X

Rango, A., Chopping, M., Ritchie, J., Havstad, K., Kustas, W., & Schmugge, T. (2000). Morphological Characteristics of shrub coppice dunes in desert grasslands of Southern New Mexico derived from scanning LIDAR. *Remote Sensing of Environment*, *74* (1), pp. 26–44. www.doi.org/10.1016/S0034-4257(00)00084-5

Ratajczak, Z., D'Odorico, P., Collins, S. L., Bestelmeyer, B. T., Isbell, F. I., & Nippert, J. B. (2017). The interactive effects of press/pulse intensity and duration on regime shifts at multiple scales. *Ecological Monographs*, *87* (2), pp. 198–218. www.doi.org/10.1002/ecm.1249

Raupach, M. R., Gillette, D. A., & Leys, J. F. (1993). The effect of roughness elements on wind erosion threshold. *Journal of Geophysical Research: Atmospheres*, *98* (D2), pp. 3023–3029. www.doi.org/10.1029/92JD01922

Ravi, S., D'Odorico, P., & Okin, G. S. (2007). Hydrologic and aeolian controls on vegetation patterns in arid landscapes. *Geophysical Research Letters*, *34* (24). www.doi.org/10.1029/2007GL031023

Reynolds, J. F., & Stafford Smith, D. M. (2002). *Global desertification: Do humans cause deserts? Dahlem Workshop Report 88.* Berlin, Germany: Dahlem University Press.

Reynolds, J. F., Stafford Smith, D. M., Lambin, E. F., Turner, B. L., Mortimore, M., Batterbury, S. P. J., Downing, T. E., et al. (2007). Global desertification: Building a science for dryland development. *Science*, *316* (5826), pp. 847–851. www.doi.org/10.1126/science.1131634

Rietkerk, M., Dekker, S. C., de Ruiter, P. C., & van de Koppel, J. (2004). Self-organized patchiness and catastrophic shifts in ecosystems. *Science*, *305* (5692), pp. 1926–1929. www.doi.org/10.1126/science.1101867

Rodrigo Comino, J., Keesstra, S. D., & Cerdà, A. (2018). Connectivity assessment in Mediterranean vineyards using improved stock unearthing method, LiDAR and soil erosion field surveys. *Earth Surface Processes and Landforms*, *43* (10), pp. 2193–2206. www.doi.org/10.1002/esp.4385

Rodríguez, F., Mayor, Á. G., Rietkerk, M., & Bautista, S. (2018). A null model for assessing the cover-independent role of bare soil connectivity as indicator of dryland functioning and dynamics. *Ecological Indicators*, *94*, pp. 512–519. www.doi.org/10.1016/j.ecolind.2017.10.023

Rossi, M. J., Ares, J. O., Jobbágy, E. G., Vivoni, E. R., Vervoort, R. W., Schreiner-McGraw, A. P., & Saco, P. M. (2018). Vegetation and terrain drivers of infiltration depth along a semiarid hillslope. *Science of the Total Environment*, *644*, pp. 1399–1408

Saco, P. M., & Moreno-de las Heras, M. (2013). Ecogeomorphic coevolution of semiarid hillslopes: Emergence of banded and striped vegetation patterns through interaction of biotic and abiotic processes. *Water Resources Research*, *49* (1), pp. 115–126. www.doi.org/10.1029/2012WR012001

Saco, P. M., Moreno-de las Heras, M., Keesstra, S., Baartman, J., Yetemen, O., & Rodríguez, J. F. (2018). Vegetation and soil degradation in drylands: Non linear feedbacks and early warning signals. *Current Opinion in Environmental Science & Health*, *5*, pp. 67–72. www.doi.org/10.1016/j.coesh.2018.06.001

Saco, P. M., Rodríguez, J. F., Moreno-de las Heras, M., Keesstra, S., Azadi, S., Sandi, S., Baartman, J., Rodrigo-Comino, J., et al. (2020). Using hydrological connectivity to detect transitions and degradation thresholds: Applications to dryland systems. *CATENA*, *186*, pp. 104354. www.doi.org/10.1016/j.catena.2019.104354

Saco, P. M., Willgoose, G. R., & Hancock, G. R. (2007). Eco-geomorphology of banded vegetation patterns in arid and semi-arid regions. *Hydrology and Earth System Sciences*, *11* (6), pp. 1717–1730. www.doi.org/10.5194/hess-11-1717-2007

Saintilan, N., Bowen, S., Maguire, O., Karimi, S. S., Wen, L., Powell, M., Colloff, M. J., et al. (2021). Resilience of trees and the vulnerability of grasslands to climate change in temperate Australian wetlands. *Landscape Ecology*, *36* (3), pp. 803–814. www.doi.org/10.1007/s10980-020-01176-5

Scanlon, T. M., Caylor, K. K., Levin, S. A., & Rodriguez-Iturbe, I. (2007). Positive feedbacks promote power-law clustering of Kalahari vegetation. *Nature*, *449* (7159), pp. 209–212. www.doi.org/10.1038/nature06060

Scheffer, M., Carpenter, S., Foley, J. A., Folke, C., & Walker, B. (2001). Catastrophic shifts in ecosystems. *Nature*, *413* (6856), pp. 591–596. www.doi.org/10.1038/35098000

Schneider, F. D., & Kéfi, S. (2016). Spatially heterogeneous pressure raises risk of catastrophic shifts. *Theoretical Ecology*, *9* (2), pp. 207–217. www.doi.org/10.1007/s12080-015-0289-1

Seager, R., Liu, H., Henderson, N., Simpson, I., Kelley, C., Shaw, T., Kushnir, Y., & Ting, M. (2014). Causes of increasing aridification of the mediterranean region in response to rising greenhouse gases. *Journal of Climate*, *27* (12), pp. 4655–4676. www.doi.org/10.1175/JCLI-D-13-00446.1

Sitch, S., Friedlingstein, P., Gruber, N., Jones, S. D., Murray-Tortarolo, G., Ahlström, A., Doney, S. C., et al. (2015). Recent trends and drivers of regional sources and sinks of carbon dioxide. *Biogeosciences*, *12* (3), pp. 653–679. www.doi.org/10.5194/bg-12-653-2015

Song, X. P., Hansen, M. C., Stehman, S. V., Potapov, P. V., Tyukavina, A., Vermote, E. F., & Townshend, J. R. (2018). Global land change from 1982 to 2016. *Nature*, *560* (7720), pp. 639–643.

Stavi, I., Siad, S. M., Kyriazopoulos, A. P., & Halbac-Cotoara-Zamfir, R. (2020). Water runoff harvesting systems for restoration of degraded rangelands: A review of challenges and opportunities. *Journal of Environmental Management*, *255*, pp. 109823. www.doi.org/10.1016/j.jenvman.2019.109823

Stewart, J., Parsons, A. J., Wainwright, J., Okin, G. S., Bestelmeyer, B. T., Fredrickson, E. L., & Schlesinger, W. H. (2014). Modeling emergent patterns of dynamic desert ecosystems. *Ecological Monographs*, *84* (3), pp. 373–410. www.doi.org/10.1890/12-1253.1

Suding, K. N., & Hobbs, R. J. (2009). Threshold models in restoration and conservation: a developing framework. *Trends in Ecology & Evolution*, *24* (5), pp. 271–279. www.doi.org/10.1016/j.tree.2008.11.012

Tongway, D. J., & Hindley, N. (2004). *Landscape function analysis: procedures for monitoring and assessing landscapes*. Canberra, Australia: CSIRO Sustainable Ecosystems.

Tongway, D. J., & Ludwig, J. A. (2001). Theories on the origins, maintenance, dynamics, and functioning of banded landscapes. In D. J. Tongway, C. Valentin, & J. Seghieri (eds.), *Banded Vegetation Patterning in Arid and Semiarid Environments: Ecological Processes and Consequences for Management* (pp. 20–31). New York: Springer.

Tongway, D. J., & Ludwig, J. A. (2011). *Restoring disturbed landscapes: putting principles into practice.* Island Press.

Turnbull, L., & Wainwright, J. (2019). From structure to function: Understanding shrub encroachment in drylands using hydrological and sediment connectivity. *Ecological Indicators*, *98*, pp. 608–618. www.doi.org/10.1016/j.ecolind.2018.11.039

Turnbull, L., Wainwright, J., & Brazier, R. E. (2008). A conceptual framework for understanding semi-arid land degradation: ecohydrological interactions across multiple-space and time scales. *Ecohydrology*, *1* (1), pp. 23–34. www.doi.org/10.1002/eco.4

Turnbull, L., Wilcox, B. P., Belnap, J., Ravi, S., D'Odorico, P., Childers, D., Gwenzi, W., et al. (2012). Understanding the role of ecohydrological feedbacks in ecosystem state change in drylands. *Ecohydrology*, *5* (2), pp. 174–183. www.doi.org/10.1002/eco.265

Urgeghe, A. M., & Bautista, S. (2015). Size and connectivity of upslope runoff-source areas modulate the performance of woody plants in Mediterranean drylands. *Ecohydrology*, *8* (7), pp. 1292–1303. www.doi.org/10.1002/eco.1582

Urgeghe, A.M., Mayor, A.G., Turrión, D., Rodríguez, F., & Bautista, S.(2021).Disentangling the independent effects of vegetation cover and pattern on runoff and sediment yield in dryland systems – Uncovering processes through mimicked plant patches. *Journal of Arid Environment*, *193*, pp. 104585.

Valentin, C., d'Herbès, J. M., & Poesen, J. (1999). Soil and water components of banded vegetation patterns. *CATENA*, *37* (1), pp. 1–24. www.doi.org/10.1016/S0341-8162(99)00053-3

van Auken, O. W. (2009). Causes and consequences of woody plant encroachment into western North American grasslands. *Journal of Environmental Management*, *90* (10), pp. 2931–2942. www.doi.org/10.1016/j.jenvman.2009.04.023

van den Elsen, E., Stringer, L. C., De Ita, C., Hessel, R., Kéfi, S., Schneider, F. D., Bautista, S., et al. (2020). Advances in understanding and managing catastrophic ecosystem shifts in Mediterranean ecosystems. *Frontiers in Ecology and Evolution*, *8*. Original Research. www.doi.org/10.3389/fevo.2020.561101

Vicente, E., Moreno-de las Heras, M., Merino-Martín, L., Nicolau, J. M., & Espigares, T. (2022). Assessing the effects of nurse shrubs, sink patches and plant water-use strategies for the establishment of late-successional tree seedlings in Mediterranean reclaimed mining hillslopes. *Ecological Engineering*, *176*, pp. 106538. www.doi.org/10.1016/j.ecoleng.2021.106538

Wainwright, J. (2009). Desert ecogeomorphology. In A. J. Parsons & A. D. Abrahams (eds.), *Geomorphology of Desert Environments* (pp. 21–66). Dordrecht: Springer Netherlands.

Wainwright, J., Parsons, A. J., & Abrahams, A. D. (1999). Rainfall energy under creosotebush. *Journal of Arid Environments*, *43* (2), pp. 111–120. www.doi.org/10.1006/jare.1999.0540

Wainwright, J., Turnbull, L., Ibrahim, T. G., Lexartza-Artza, I., Thornton, S. F., & Brazier, R. E. (2011). Linking environmental régimes, space and time: Interpretations of structural and functional connectivity. *Geomorphology*, *126* (3), pp. 387–404. www.doi.org/10.1016/j.geomorph.2010.07.027

Webb, N. P., McCord, S. E., Edwards, B. L., Herrick, J. E., Kachergis, E., Okin, G. S., & Van Zee, J. W. (2021). Vegetation canopy gap size and height: critical indicators for wind erosion monitoring and management. *Rangeland Ecology & Management*, *76*, pp. 78–83. www.doi.org/10.1016/j.rama.2021.02.003

Wilcox, B. P., Breshears, D. D., & Allen, C. D. (2003). Ecohydrology of a resource-coserving semiarid woodland: Effects of scale and disturbance. *Ecological Monographs*, **73** (2), pp. 223–239. www.doi.org/10.1890/0012-9615(2003)073[0223:EOARSW]2.0.CO;2

Wohl, E., Brierley, G., Cadol, D., Coulthard, T. J., Covino, T., Fryirs, K. A., Grant, G., et al. (2019). Connectivity as an emergent property of geomorphic systems. *Earth Surface Processes and Landforms*, **44** (1), pp. 4–26. www.doi.org/10.1002/esp.4434

Xu, C., van Nes, E. H., Holmgren, M., Kéfi, S., & Scheffer, M. (2015). Local facilitation may cause tipping points on a landscape level preceded by early-warning indicators. *The American Naturalist*, **186** (4), pp. E81–E90. www.doi.org/10.1086/682674

Yahdjian, L., Sala, O. E., & Havstad, K. M. (2015). Rangeland ecosystem services: shifting focus from supply to reconciling supply and demand. *Frontiers in Ecology and the Environment*, **13** (1), pp. 44–51. www.doi.org/10.1890/140156

Ye, J.-S., Reynolds, J. F., Maestre, F. T., & Li, F.-M. (2016). Hydrological and ecological responses of ecosystems to extreme precipitation regimes: A test of empirical-based hypotheses with an ecosystem model. *Perspectives in Plant Ecology, Evolution and Systematics*, **22**, pp. 36–46. www.doi.org/10.1016/j.ppees.2016.08.001

Zobell, R. A., Cameron, A., Goodrich, S., Huber, A., & Grandy, D. (2020). Ground cover – What are the critical criteria and why does it matter? *Rangeland Ecology & Management*, **73** (4), pp. 569–576. www.doi.org/10.1016/j.rama.2020.02.002

Zurlini, G., Jones, K. B., Riitters, K. H., Li, B.-L., & Petrosillo, I. (2014). Early warning signals of regime shifts from cross-scale connectivity of land-cover patterns. *Ecological Indicators*, **45**, pp. 549–560. www.doi.org/10.1016/j.ecolind.2014.05.018

15

Coasts and Deltas

ROBERT TWILLEY AND PAOLA PASSALACQUA

15.1 Introduction

As introduced in Chapter 8, coastal deltaic floodplains are distinct geomorphic units that form at the mouth of major rivers where jet-plume deposits develop (Wellner et al., 2005; Syvitski et al., 2009; Fagherazzi et al., 2015; Islam, 2016; Twilley et al., 2019). These landforms have important ecosystem services such as trapping sediment (Nardin & Edmonds, 2014), reducing storm surge (Barbier et al., 2013; Wamsley et al., 2010), and processing riverine nutrients (Perez et al., 2003; Mitsch et al., 2005; Lane et al., 2011; Li et al., 2020), making coastal deltaic floodplains significant landscapes at the edge of continents. At the same time, these densely populated landscapes (Edmonds et al., 2020) and their ecosystem services are at high risk due to a combination of factors including increasing rates of sea level rise, accelerated subsidence, extraction of resources from the subsurface and extensive human interventions (e.g., dams and levees). Coastal-zone management strategies, defined as the set of human interventions to preserve, sustain, or restore ecosystem services in the coastal zone, often aim at reestablishing the connectivity of rivers and their floodplains. Dis-connectivity is in fact responsible for degrading many coastal ecosystems as it prevents the natural distribution of water, solids and solutes over the delta plain. Interventions such as the design of river diversions in lower Mississippi and Yellow Rivers, and tidal river management in the Ganges-Brahmaputra-Meghna Delta in Bangladesh, are examples of how reconnecting rivers and their floodplains is used to restore ecosystem functioning. Thus, concepts of connectivity are at the core of coastal-zone management interventions.

With a similar approach to that taken in Chapter 8, in this chapter we will focus on river deltas and, without loss of generality, we will use many examples from the Mississippi River Delta Plain (Figure 15.1). With portions of the coast aggrading (Wax Lake Delta), portions of the coast degrading (eastern bays such as Terrebonne and Barataria Bay), and ongoing large-scale restoration projects (Coastal Protection

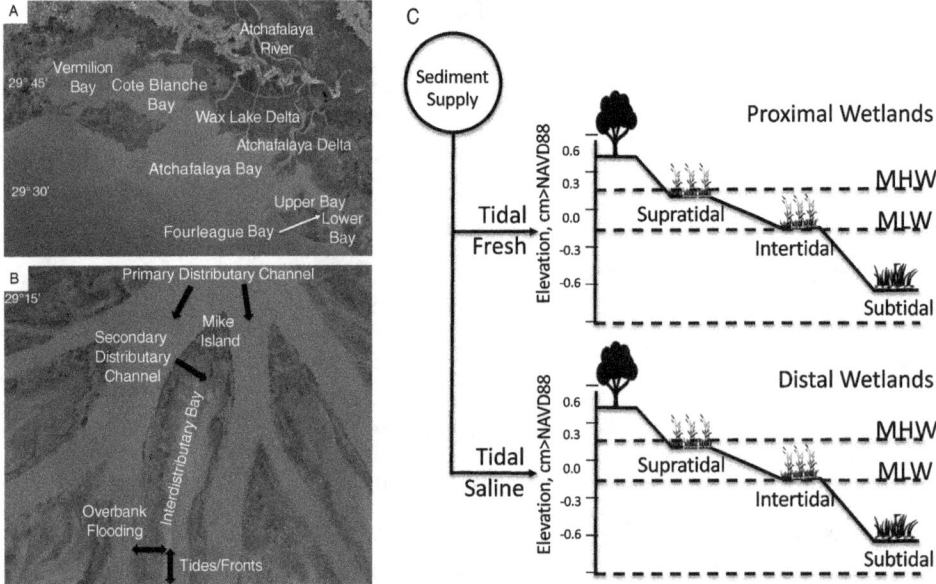

Figure 15.1   (A) Map of the large delta estuaries of the Atchafalaya River that include Fourleague Bay, Atchafalaya Bay, Cote Blanche Bay and Vermilion Bay (Landsat Image, Google Earth, 2021). (B) Morphologic features of coastal deltaic floodplain that defines the connectivity between primary and secondary channels with delta islands that include wetlands defined by hydrogeomorphic zones and interdistributary bay that is coupled to Gulf of Mexico by tides and fronts. (C) Diagrams of marsh platforms illustrating the hydrogeomorphic zones in the proximal and distal wetlands of an active coastal basin, such as described for the Atchafalaya River Delta Estuaries. (Figure from Twilley et al., 2019). A black and white version of this figure will appear in some formats. For the colour version, refer to the plate section.

& Restoration Authority, 2017), coastal Louisiana offers the opportunity to discuss how connectivity concepts are leveraged in coastal-zone management with concrete examples. The design strategy for restoring the Mississippi River Delta embraces the delta cycle concept, which is associated with re-establishing river processes that control the self-organization of coastal morphology and ecosystem dynamics or ecogeomorphology (Day Jr. et al., 2007; Blum & Roberts, 2009; Paola et al., 2011; Ma et al., 2018). This restoration strategy is one of the key examples of how connectivity supports coastal-zone management. The exclusion of other coastal landforms in the coastal zone does not ignore their relevance to ecosystem services but allows us to focus on the landforms that transport water, solids and solutes from major river basins to the oceans. The network structure that characterizes these landforms describes how connectivity concepts capture the functioning of the exchange between channels and wetlands as critical elements of source-to-sink networks, providing excellent examples of how connectivity concepts are relevant to coastal-zone management. The approach and ideas developed in this chapter can be expanded to other coastal landforms.

As defined in Chapter 8 and throughout this book, structural connectivity describes the physical adjacency that allows for the interaction between channels and floodplains, whereas functional connectivity describes the actual interaction, as flow magnitude and direction, across time (Wainwright et al., 2011). Process connectivity, which describes the transfer of information through these interactions and the processes which control them (Passalacqua, 2017; Sendrowski & Passalacqua, 2017), is influenced by structural and functional connectivity and captures the behaviour of the system across spatial and temporal scales.

15.2 What Does it Mean to Manage Coastal Landscapes Effectively?

The effectiveness of restoring an active delta basin depends on process-based ecosystem design approaches to recreate the hydrologic conditions of functional deltaic floodplains (Shaffer et al., 1999; Day et al., 2018; Rutherford et al., 2018; Wiegman et al., 2018). Coastal deltaic floodplains are coupled to downstream estuarine ecosystems, which are also considered part of an active delta and thus also targeted by coastal management strategies (Madden et al., 1988; Perez et al., 2000; Day et al., 2011; Lane et al., 2011). The coastal morphology and ecosystem dynamics, including ecological succession and biogeochemistry, that result from engineered river diversions to reconnect river processes are shaped by the hydrogeomorphology of the flood-pulse (Heiler et al., 1995; Passalacqua, 2017). Therefore, sedimentation processes, as a function of the connectivity between river channels and deltaic floodplains in both the proximal and distal regions of the outfall area of river diversions, control the ecosystem design of active deltaic coastal basins (Figure 15.1).

The key management question relative to ecosystem design for an inactive coastal basin is how to overcome elevation deficits of wetland platforms that have accumulated for decades causing marsh instability. This is particularly problematic in areas where sulfate can produce H_2S which stresses the ability of marsh production to overcome several decades of elevation deficits for marsh platforms. Another way to provide sediment input is via dredged sediments (Mendelssohn & Kuhn, 2003), but this is expensive and energy-intensive and likely unsustainable in the long-term (Wiegman et al., 2018). However, the use of diversions to overcome the accretion deficit requires flood-pulse operations that extend beyond the normal winter-spring flood pulse season. Some combination of marsh creation and beneficial use of dredge materials may offer limited recovery to elevation deficits at the coastal basin scale but using long-term maintenance of elevation platforms with large sediment plumes from river diversions along with beneficial use of sediment may achieve large basin impacts.

A major issue with reintroducing river sediment to make up for decades of elevation deficits in marsh platforms is the consequence that river nutrients, particularly nitrate, have increased concentration fourfold in the last four decades (Rabalais et al., 1996; Rabalais et al., 2002; Broussard & Turner, 2009) raising concerns over eutrophication in distal estuarine waters (Twilley & Rivera-Monroy, 2009; Bargu et al., 2019). As in many major coastal basins globally, nitrogen fertilizer associated with industrial agriculture has increased nitrogen loading, leading to the incidence and severity of harmful algal blooms that may harm oysters, alter food web structure, or cause fish kills (Riekenberg et al., 2014). There is concern that large river-pulses from diversions, that are intended to benefit from large sediment inputs to remedy decades of elevation deficits in wetland platforms, may cause water quality problems in coastal basins (Lane et al., 2011; Riekenberg et al., 2014; Roberts & Doty, 2015; White et al., 2019). There are several small diversions operating mainly to introduce freshwater to control for isohalines in estuarine waters of coastal Louisiana. But a significant sediment diversion project is being designed for the mid-Barataria coastal basin rated at 2124 m^3/s, which will be one of the largest river diversion projects in the USA. The effect of nutrient loading from a river-pulse of this magnitude to the eutrophication of coastal bays and estuaries downstream depends on the operations that define how much nutrient load is delivered during the seasonal cycles of estuarine productivity. This potential impact may be mitigated by the ability of wetlands to remove excess nutrients depending on the diversion (Twilley et al. 2021). The connectivity of river resources to wetland platforms is defined by these diversion operations including when and how much water with sediment will benefit wetland restoration, relative to how solutes in river water may damage the fishery productivity of delta estuaries. In Section 15.3, we will discuss these coastal management issues and trade-offs in terms of the ecosystem services that characterize coastal systems and the role that structural, functional and process connectivity play in each of them.

15.3 Connectivity Concepts

15.3.1 Hydrological Connectivity

Hydrological connectivity is defined in the context of coastal management as the percent of river channel water interacting with floodplain wetlands (Junk et al., 1989; Hiatt & Passalacqua, 2015; Amoros & Bornette, 2002; Covino, 2017). As discussed in Chapter 8, the hydrological exchange between rivers and wetlands in coastal systems is regulated by both structural and functional connectivity and this exchange is fundamental for the distribution of water, solids and solutes across delta plains. Structural connectivity takes the form of secondary channels that connect the main delta network and the island interiors, and levee topography facilitating overbank flow (Figure 15.1). Functional connectivity takes the form of various forcings, such as discharge, wind, tides, and the presence of vegetation, that control the magnitude and direction of the flux exchange between these two sub-environments. Similar to alluvial floodplains, coastal deltaic floodplains are flooded by primary distributary channels when river stage reaches the elevation of natural levees (bankfull stage), transporting sediment along with dissolved and particulate nutrients from channels to floodplain ecosystems (Noe & Hupp, 2005; Amoros & Bornette, 2002; Asselman & Middelkoop, 1995). However, in contrast to alluvial floodplains, river-floodplain exchange is observed also at low discharges (Hiatt & Passalacqua, 2015). Additionally, coastal deltaic floodplains experience inundation also due to coastal forcings such as tides and meteorological fronts (Walker & Hammack, 2000; Geleynse et al., 2015).

Hydrological connectivity controls the extent of ecosystem services such as sediment connectivity and nutrient removal. These ecosystem services, in turn, affect hydrological connectivity; as ecological succession increases vegetation density on delta islands, connectivity between primary and secondary channels as well as overbank flooding is reduced in response to the increased drag coefficient of vegetation (Hiatt & Passalacqua, 2017). The seasonality of vegetation density and river stage may control this dynamic, and together with meteorological fronts, they control sediment deposition patterns and continued successional patterns as delta islands age (Bevington et al., 2017).

Exploring hydrologic connectivity for management purposes often involves the use of hydrodynamic models able to reproduce the complex interaction of seasonal river and tide forcings, hydrogeomorphology and vegetation roughness (Hiatt & Passalacqua, 2015; Christensen et al., 2020). Numerical modelling efforts have demonstrated the role of river (Edmonds & Slingerland, 2010; Verschelling et al., 2017), tides and wind (Bryan et al., 2016; Hanegan & Georgiou, 2014; Horstman et al., 2013), waves (Everett et al., 2019; Nardin et al., 2020), hurricanes (Hu et al., 2015; Liu et al., 2018; Xing et al., 2017), and vegetation (Nardin et al., 2020, 2016; Nardin & Edmonds, 2014; Temmerman et al., 2003) in coastal wetland hydrodynamics and sediment transport, thus providing key information in support of the design and implementation of coastal management strategies. Hydrodynamic models coupled to biogeochemical processes have been developed for wetland treatment systems (Day Jr. et al., 2004; Kadlec & Wallace, 2008), but fewer models exist for the complex coastal zone where deltaic floodplains process nutrients before export to the coastal ocean (Martin & Reddy, 1997; Kelly-Gerreyn et al., 1999; Dettmann, 2001; Kelly-Gerreyn et al., 2001).

15.3.2 Sedimentation Patterns

As for water fluxes, sediment transport and resulting sedimentation patterns are also controlled by structural and functional connectivity. Episodic events, such as hurricanes and cold fronts, force water upstream and downstream in islands, depending on southerly pre-frontal and northerly post-frontal wind conditions, respectively (Roberts & Doty, 2015). These events can play a significant role in the sediment transport of coastal deltaic floodplains (Bevington et al., 2017). For example, at the Wax Lake Delta (WLD), extremely high peak discharge of river floods occurring in 2008 and 2011 resulted in a mean net elevation gain of 4.9–5.4 cm over each flood season, respectively. This result is similar to patterns observed in the Atchafalaya River Delta where delta growth only occurred during floods with mean monthly discharge >14,000 m^3/s (Rejmánek et al., 1987). While large floods add a considerable amount of sediment, lower discharge floods also contributed sediment to the deltaic floodplain wetlands (Bevington et al.,2017) and to the growth of delta networks (Shaw & Mohrig, 2014). The total sediment subsidy from both large and small floods is likely necessary to maintain land building due to a large amount of elevation loss that occurs as a result of annual winter cold front passages (most likely due to erosion, Liu et al., 2018). Hurricanes have also been observed to result in net elevation gain; at WLD, for example, Hurricanes Gustav and Ike resulted in a total net elevation gain of 1.2 cm. However, the long-term contribution of hurricane-derived sediments to deltaic wetlands is estimated to be just 22% of the long-term contribution of large river floods (Bevington et al., 2017). It is likely that sediment subsidies resulting from hurricanes are even less when measured over a larger temporal scale.

The resuspension of sediment occurs as a result of the waves, currents, and storm surge associated with hurricane passage (Walker, 2001). The sediment is then redeposited as the surge moves inland into coastal wetlands, resulting in a measurable elevation gain attributed to hurricanes (Baumann et al., 1984; Rejmanek et al., 1988; Nyman et al., 1995; Cahoon, 2006; Turner et al., 2006; McKee & Cherry, 2009; Morton & Barras, 2011; Tweel & Turner, 2012). However, if erosion is not quantified and gross deposition is reported (i.e., positive elevation change), then landscape-scale values overestimate the total sediment attributable to hurricanes along the northern Gulf of Mexico coast (Turner et al., 2006; Tweel & Turner, 2012). While sediment deposition during hurricanes is still an appreciable sediment subsidy for coastal wetlands, especially in abandoned delta lobes that receive very little riverine sediment input (McKee & Cherry, 2009; Baustian et al., 2012), it only represents a small contribution in coastal wetlands with appreciable riverine sediment delivery (Tornqvist et al., 2007; Bevington et al., 2017).

The seasonality of sedimentation is similar in both the coastal deltaic floodplain platforms and distal estuarine platforms as a function of river stage and meteorological fronts (Twilley et al., 2019). There is evidence that biotic feedbacks on sedimentation reduce connectivity with vegetation density in the proximal wetlands, where primary channels convey sediment past delta islands as marsh platform elevation increases (Hiatt & Passalacqua, 2017; Christensen et al., 2020). Estuarine bays in the distal region serve as reservoirs of fine sediments that are resupplied in the flood-pulse season and redistributed to the wetland platform when water elevations are maximum during a combination of river flood and meteorological fronts (Restreppo et al., 2018). Biotic feedbacks to sedimentation are positive in distal estuarine wetlands, as observed for most salt marshes. During the calm season of late summer and fall, the lower water levels drain wetland platforms, consolidating the sediment deposited during the winter season. This seasonal decrease in water levels relative to marsh elevation reduces the negative feedback of longer durations of flooding with lower marsh platforms.

15.3.3 Chronosequence Zonation

Deltaic floodplains with tidal freshwater and estuarine wetlands can be defined by the elevation of the wetland platform that controls the frequency and duration of flooding (hydroperiod), an example of the feedback between structural (elevation) and functional (frequency and duration of flooding) connectivity elements and reflected in the resulting couplings among system's variables (process connectivity) (Figure 15.2). Distinct hydrogeomorphic zones are defined by soil elevation relative to a tidal datum (Wagner et al., 2017; Bevington & Twilley, 2018). Subtidal zones are those described as below mean low water (MLW), intertidal zones as those between MLW and mean high water (MHW), and supratidal as those above MHW (Figure 15.2). The subtidal zone is generally vegetated by submersed aquatic vegetation (SAV) and floating leaf vegetation. Maximum vegetated depth depends on water turbidity for SAV. Intertidal and supratidal hydrogeomorphic zones occur mainly along the natural levees and may limit inundation from primary distributary channels to island interiors. Wetlands in the proximal and distal sedimentation regimes of an active delta can be classified into salinity zones in addition to hydrogeomorphic zones. Both proximal and distal wetland types are influenced by tides, but the freshwater floodplain wetlands are diverse with several freshwater species differentiated by island topography, since tolerance to salinity is not a stress. In the distal sedimentation region, salinity gradients are due to mixing of river discharge with saline tidal waters, resulting in intermediate, brackish and saltmarsh zones (Visser et al., 1998; Sasser et al., 2008).

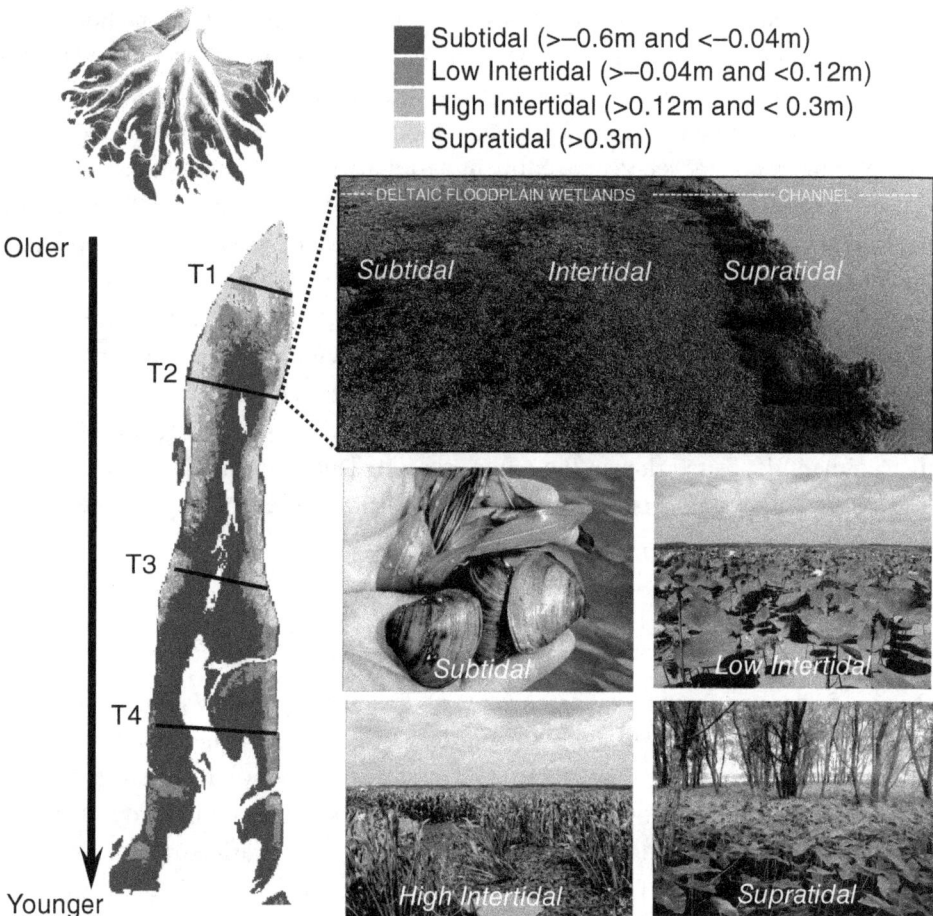

Figure 15.2 Map of Mike Island in WLD, Louisiana, showing the distribution of hydrogeomorphic zones based on elevation records from USGS Atchafalaya 2 project LiDAR Survey 2012 digital elevation model (4 m resolution). The black lines across Mike Island delineate four chronosequence transects (from younger to older: T4–T1) mainly defined by the distance to apex of Mike Island and the characterization of cross-sectional morphology (Figure from Bevington & Twilley, 2018). Three hydrogeomorphic zones (subtidal, intertidal and supratidal) are distinguished by sediment surface elevation relative to MHW and MLW. A black and white version of this figure will appear in some formats. For the colour version, refer to the plate section.

Hydrogeomorphic zones of coastal deltaic floodplains in the proximal sedimentation zone can also be defined by time since subaerial emergence, to account for variability in vegetation community composition and soil successional development. The amount of time that each emergent zone has existed at a given elevation controls the geomorphology and formation of deltaic wetlands and contributes to patterns in vegetation community composition and other ecosystem processes such as nutrient biogeochemistry, organic matter composition and nutrient sequestration rates (Bevington & Twilley, 2018; Twilley et al., 2019). The younger stage of island development in deltaic floodplains is dominated by subtidal hydrogeomorphic zones, and as the islands age there is an infilling of the interdistributary bay resulting in decrease in subtidal zone and increase in intertidal hydrogeomorphic zones (Figure 15.2). Older areas of deltaic floodplains are often supratidal resulting from the collective effects of physical and organic sediment deposition and ecological processes contributing to infilling of these regions (Bevington & Twilley, 2018). In addition, there is no chronosequence applied to defining ecological processes on marsh platforms of estuarine wetlands in the distal sedimentation zone. Most of the estuarine wetlands in the distal region of an active coastal basin are at high intertidal hydrogeomorphic elevation.

15.4 Management of Ecosystem Services

15.4.1 Mitigating Marsh Elevation Deficits

Shifts in vegetation type and net primary production follow the geomorphological displacement of hydrogeomorphic zones as platform elevations change relative to MHW, resulting in the successional patterns that can be predicted using marsh equilibrium models (Morris et al., 2002; Morris, 2006; Cahoon et al., 2011). These shifts in net primary production with elevation and/or in response to stochastic disturbances, such as hurricanes and large river floods, are based on adaptations of specific vegetation to hydrogeomorphic zones that outcompete each other depending on growth relative to percent time inundation, known as geomorphological displacement (Morris, 2006). Therefore, to mitigate elevation deficits, connectivity must account for how sediment input changes elevation capital causing a shift in organic production that in return contributes to increase in soil development and elevation. Percent time inundation will determine both inorganic and organic sedimentation in deltaic floodplains based on the optimum growth range of vegetation at a site (Morris et al., 2002).

Geomorphological displacement in the ecological succession of deltaic floodplains has important implications to operations of river diversions and beneficial use of dredged sediments. Cumulative impacts of elevation deficits can be defined by the amount of time depositional environments have been abandoned from river connectivity together with rates of relative sea level rise due to eustatic change plus subsidence. Decades without river connectivity in delta floodplains with high relative sea level rise will require extensive sediment input on a large scale if Mississippi Delta floodplains are to remain emerged above MLW. River diversions for regional-scale wetland restoration strategies (Boesch et al., 1994; Day Jr. et al., 2007; Day et al., 2009) are being designed to optimize large sediment quantities on coastal deltaic and salt marsh platforms. Extensive river connectivity is essential to maintain an active delta growth (Roberts, 1997; Day Jr. et al., 2007; Paola et al., 2011; Twilley et al., 2019).

15.4.2 Mitigating CO_2 Enrichment

Recent global analyses of carbon dynamics identify river networks and terrestrial-ocean interface ecosystems as critical components of global land-atmosphere feedbacks. Early studies estimated that deltaic floodplains annually bury 70 Tg of the terrestrial organic carbon flux to the oceans, but these estimates do not explicitly account for blue carbon, that is, the contribution of primary production in coastal deltaic wetlands. Blue carbon traditionally refers to the C sequestered in three coastal environments: mangrove forests, seagrass meadows and tidal salt marshes (Mcleod et al., 2012; Siikamäki et al., 2012; Grimsditch et al., 2013). Even though 40–50% of all coastal and marine organic carbon (C) burial occurs in coastal deltas (Hedges & Keil, 1995; Blair & Aller, 2012), research on natural global C sinks is traditionally focused on open ocean or terrestrial forest ecosystems. Recent 'blue carbon' research emphasizes the importance of coastal wetlands on the global C cycle but mostly excludes coastal deltaic floodplains with some exceptions (Shields et al., 2016, 2018, 2019; Holmquist et al., 2018).

Taking as example the WLD, based on ^{137}Cs, marsh accretion is 1.43 cm/yr (DeLaune et al., 2016), which is close to earlier estimates of 1.4 cm/yr (DeLaune et al., 1987). Based on these accretion rates, C sequestration rates range from 131 to 342 g m^{-2} yr^{-1} for marsh surface profiles. These C sequestration rates are comparable to the decadal value reported by Smith et al. (1985) for Louisiana deltaic freshwater marsh (224 g m^{-2} yr^{-1}). For the WLD, representing 60 years of delta formation, the organic carbon accumulation rate was 250 g m^{-2} yr^{-1} (Shields et al., 2017). One of the mechanisms of preferential carbon storage in deltas is the interaction with iron, which is high in this active delta (Shields et al., 2016). Furthermore, coastal deltas are already more effective at C burial than typical blue carbon systems such as mangroves (226 $\pm$ 39 g C m^{-2} yr^{-1}), seagrass beds (138 $\pm$ 38 g C m^{-2} yr^{-1}) and salt marshes (218 $\pm$ 24 g C m^{-2} yr^{-1}) (McLeod et al., 2011). These metrics make the case for greater inclusion of deltaic floodplains in global blue carbon studies.

Based on new estimates of delta areas globally (including many smaller deltas), deltaic floodplains annually sequester an amount of carbon equivalent to nearly 2% of fossil fuel carbon burned in 2017, meaning, every 50 years the world's deltas sequester enough carbon to offset current annual CO_2 emissions from fossil fuels combustion (Twilley et al. submitted). However, a dramatic reduction in carbon sequestration potential in deltas could be expected as sediment delivery to the world's coastal oceans rapidly declines through the Anthropocene. This sediment budget deficit disrupts ecogeomorphic feedbacks (e.g., relative dependent contribution of organic and inorganic matter to soil formation) that maintain sediment accretion rates with sea level and threatens the significance of deltas as net sinks in the carbon-climate feedback.

15.4.3 Mitigating Nitrogen Enrichment

An important issue of coastal management is the dramatically increased nitrogen (N) loading into riverine, estuarine and coastal ecosystems due to agricultural fertilization. The application of N fertilizers to agricultural fields has increased by about 800% over the last several decades, resulting in significant amounts of inorganic N leaching into rivers, causing a two to fourfold increase in nitrate (NO_3^-) concentrations since 1960 in estuarine and coastal ecosystems in North America (Goolsby et al., 2000; Howarth et al., 2002; Rabalais et al., 2002). Present NO_3^- concentration ranges from 54 to 106 µM, which is about 5–10 times higher than historical NO_3^- concentrations (Goolsby et al., 2000; Rabalais et al., 2002). The elevated loads of inorganic N enhance coastal net primary productivity, stimulate harmful algal blooms, decrease water quality, and exacerbate hypoxia (oxygen depletion < 2 mg L^{-1} in bottom water) (Rabalais et al., 2002; Paerl, 2006; Diaz & Rosenberg, 2008). It has been argued that ecosystem processes in large river estuaries that control the fate of material flux from major river basins to the ocean are disproportionate to other types of estuaries on continental margins (Bianchi & Allison, 2009). One of the reasons for the significant processing of nutrients in large river delta estuaries is the extensive deltaic floodplain area connected to waters as they flow from river basins to the sea. This is why the lack of connectivity between rivers and floodplains in coastal systems due to engineering interventions has exacerbated the issues of eutrophication in coastal waters.

The connectivity of channel waters with the wetland platform determines how much of the large nutrient load carried by primary channels is processed by coastal deltaic floodplains (Figure 15.3). Water exchange from primary channels to secondary channels of delta floodplains and/or overbank flooding determines water age and temperature, which in turn control the effectiveness of these floodplains in processing nutrients (Hiatt et al., 2018; Hiatt & Passalacqua, 2015; Twilley et al., 2019; Christensen et al., 2020). Water age is the amount of time elapsed since a parcel of water entered the system from a particular source (Delhez et al., 1999; Deleersnijder et al., 2001; Shen & Haas, 2004; Viero & Defina, 2016) and is commonly used when studying nutrient and sediment transport since it complements water residence time (Kadlec & Reddy, 2001; Kadlec, 2010). Water residence time increases biogeochemical processes such as nitrate update and reduction through

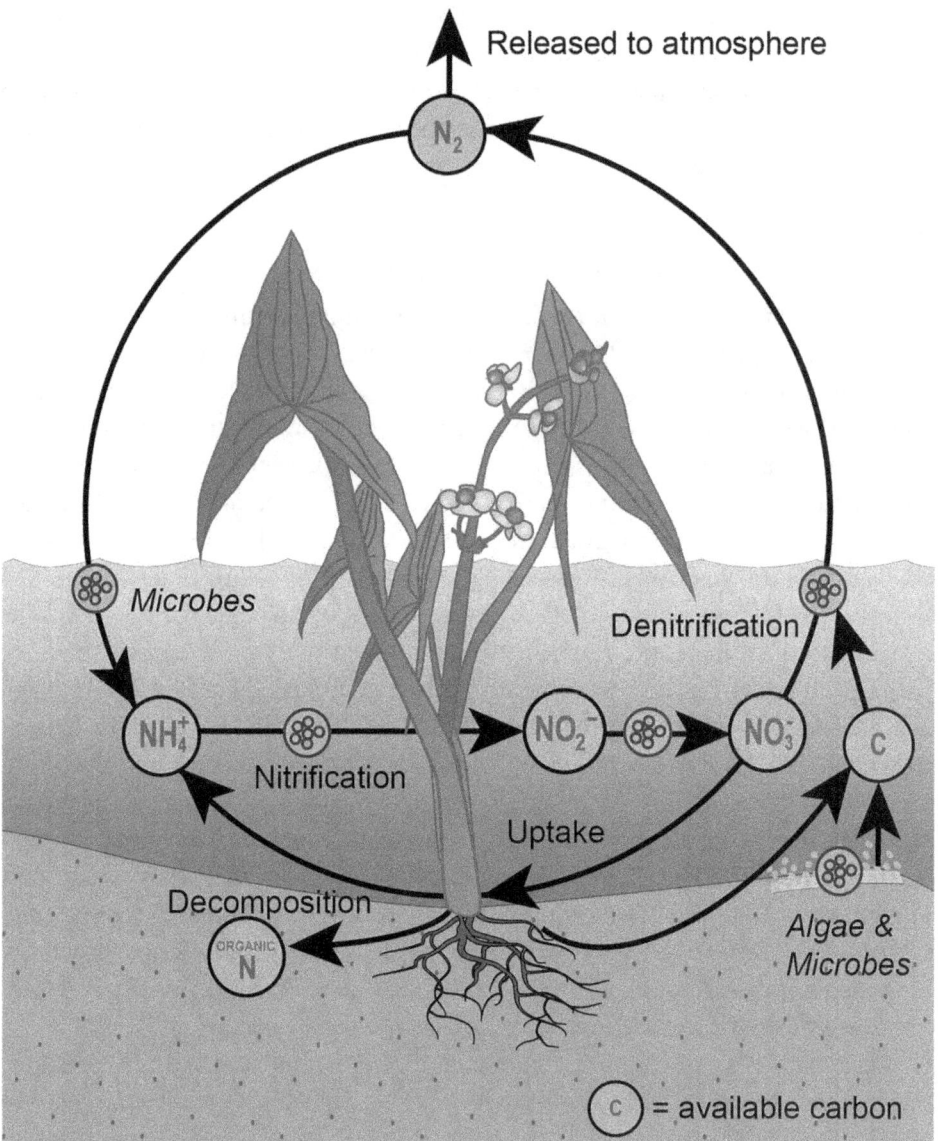

Figure 15.3   A simplified illustration of the nitrogen and phosphorus cycles in a wetland Treatment Wetlands'; images from IAN, University of Maryland (modified from Kadlec & Knight (1996), 'Treatment Wetlands'; images from IAN, University of Maryland). A black and white version of this figure will appear in some formats. For the colour version, refer to the plate section.

denitrification (Dettmann, 2001; Kaushal et al., 2008; Klocker et al., 2009; Li et al., 2020). Similarly, biogeochemical cycling is impacted by water temperature, particularly microbially mediated processes (Stanford et al., 1975; Bachand & Horne, 1999; Kadlec & Reddy, 2001). Water temperatures within floodplains are often warmer than adjacent channels, depending on connectivity, water depth, water age and climate (Twilley et al. 2021). Therefore, connectivity, residence time and water temperature are key factors controlling biogeochemical processes that determine effectiveness of coastal deltaic floodplains in nutrient reduction mitigation.

NO_3^- removal by inundated floodplain sediments is influenced by sediment characteristics (such as bulk density, particle size and organic matter content), together with hydroperiod that establishes complex pathways of N cycling (Ensign et al., 2012; Noe et al., 2013; Li et al., 2020). Denitrification is categorized as 'direct denitrification' that reduces external NO_3^- to N_2 gas and 'coupled nitrification-denitrification' that uses NO_3^- generated by in situ nitrification (Figure 15.2). In contrast, dissimilatory nitrate reduction to ammonium (DNRA) converts NO_3^- to NH_4^+ causing N to be recycled or buried in floodplain wetlands, or exported to downstream ecosystems (Gardner et al., 2006; Bernard et al., 2015). Recent studies on NO_3^- transformations describe how fluxes depend on morphological development of coastal deltaic floodplains (Henry & Twilley, 2014; Li et al., 2020). The shift from mineral to organic sedimentation from younger subtidal to older supratidal hydrogeomorphic zones increases sediment organic matter (SOM) concentrations, which have been shown to control rates of NO_3^- removal at sediment-water interface when hydrogeomorphic zones are inundated (Henry & Twilley, 2014; Li et al., 2020). Studies indicate that denitrification is positively related to NO_3^- loading, particularly under higher SOM concentrations where oxygen is depleted due to higher sediment respiration rates that inhibit nitrification (Cornwell et al., 1999; Eyre & Ferguson, 2009; Scaroni et al., 2011; Henry & Twilley, 2014; Li et al., 2020). Direct denitrification is the major process of NO_3^- removal in coastal deltaic floodplains of the Atchafalaya River compared to coupled nitrification–denitrification and DNRA (Li et al., 2021). Benthic nutrient fluxes vary along deltaic chronosequence and hydrogeomorphic zones, connecting the geomorphology and soil development with how delta ecosystems process nutrients prior to export to coastal ocean environments (Li et al., 2020; Li & Twilley, 2021).

The increased SOM concentrations and decrease in connectivity associated with stages of ecological succession determine the NO_3^- removal capacity of coastal deltaic floodplains (Twilley et al., 2019; Li et al., 2020). The subtidal site with lower SOM has lower elevation and is inundated over 97% of the year (Li et al., 2020). Though the lower-SOM site has lower NO_3- removal rates, its annual N_2 production via denitrification is substantial because the lower-SOM site has more contact with riverine NO_3^- (Li et al., 2020). The inverse interaction between SOM and connectivity as hydrogeomorphic zones develop from subtidal (higher connectivity, lower SOM) to supratidal zones (lower connectivity, higher SOM) controls total NO_3^- reduction mainly via direct denitrification during sediment inundation. The estimated annual NO_3^- removal by the coastal deltaic floodplain at the WLD is about 896 Mg N yr^{-1}, which accounts for 10–27% of total NO_3^- load (Li &

Twilley, 2021; Li et al., 2020). Over 90% of the annual N removal occurs during warmer temperatures ($\geq 17\ ^\circ$C) providing evidence of how residence time and water temperature are important controls of N loss. About 60% of this NO_3^- removal in delta floodplains occurs in the subtidal hydrogeomorphic zones compared to only 40% of NO_3^- removal in the intertidal and supratidal zone, considering the connectivity, land area and denitrification fluxes (Li et al., 2020).

15.5 Connectivity and Ecosystem Design

As discussed in this chapter and in Chapter 8, delta islands form hydrogeomorphic zones that represent marsh platforms (Wagner et al., 2017; Bevington & Twilley, 2018). These landforms change in elevation, resulting in predictive vegetation patterns with increased levels of above and below ground production. This ecological succession results in biotic feedbacks, including increased sediment organic accumulation, that shift the elevation capital from mineral to more organic based soils. The combined effect of sediment type and vegetation has contributed significantly to how we can understand ecosystem design using the concepts of river reoccupation to inactive coastal basins (Figure 15.4; Edmonds & Slingerland, 2010; Paola et al., 2011; Carney et al., 2018). The challenge is to improve ecogeomorphic models to more accurately associate designs of fluvial processes to ecological outcomes. Percent time inundation defines a connectivity threshold that restricts the range of marsh net productivity that can achieve stability of an active delta floodplain. Connectivity provides knowledge on the natural feedback mechanisms that could aid in the engineering design of future restoration strategies using delta floodplains as model for ecosystem design (Ross et al., 2015). Most sediment transport to and deposition on the marsh appear to occur during weather-induced flooding events and flood river stage (Baumann et al., 1984; Rejmanek et al., 1988; Restreppo et al., 2018), but low discharge events and interarrival times between floods also play an important role in sediment redistribution and delta network formation (Shaw & Mohrig, 2014).

The alluvial and coastal deltaic floodplains along with large delta estuaries of the Atchafalaya Coastal Basin demonstrate that self-organization of landscapes can be used to calibrate how longitudinal and lateral connectivity by river reoccupation can restore an active delta (Edmonds et al., 2011; Twilley et al., 2019). High sediment input will be necessary on a large scale if deltaic floodplains in inactive coastal basins of the Mississippi River Delta are to survive high rates of sea level rise in a delta with the significant subsidence rates that are impacting present elevation deficits. The ecosystem design of river diversions is to utilize the flood-pulse season, along with coastal fronts, to distribute sediments on proximal and distal marsh platforms during high water level stands during winter and spring (Boesch et al., 1994; Paola et al., 2011; Blum & Roberts, 2012). This approach supports the Coastal Protection and Restoration Authority (2017) objective of using natural processes of the delta cycle to recover floodplains that have become submerged over the last several decades (Figure 15.4). However, it is argued that optimizing sediment delivery to overcome cumulative impacts of elevation deficits in inactive coastal basins such as the Barataria Bay would require freshwater discharge rates and durations of flood-pulse operations that would exceed the natural flood-pulse cycle of an active deltaic estuary (Turner & Boyer, 1997; Das et al., 2012; Peyronnin et al., 2017; Day et al., 2018). The connectivity of water, solids and solutes in these ecosystem designs are optimized for ecogeomorphic processes that result in emergent wetland floodplains in 50 year horizons (Snedden et al., 2007; Wang et al., 2014; Elsey-Quirk et al., 2019; White et al., 2019).

River diversions are designed to reduce elevation deficits by transforming inactive delta basins to active delta cycle and recover cumulative impacts of wetland loss over several decades (Figure 15.4). The trade-offs of aggressive connectivity designs are defined by potential negative impacts to fishing communities (Soniat et al., 2013; Rose et al., 2014; de Mutsert et al., 2017). Large river diversions operated beyond the flood-pulse season of large river estuaries result in delivery of water, solids, and solutes that change seasonal water quality conditions of salinity, water temperatures and NO_3^- concentrations. Extending low salinities and cooler temperatures into the late spring and early summer can disrupt the biological cycles that define the estuary as a nursery in areas that presently allow for these functions to occur. Spatial distributions in salinity, wetland coverage, and biological productivity of a delta estuary shift inland in inactive deltas over time (Gagliano & Van Beek, 1975; Gosselink et al., 1998; Twilley et al., 2016). There is also a regressive shift in these zones as the coastal basin becomes active and connected with a river (Twilley et al., 2019). Shifts in habitat types and biological productivity in the transgressive and regressive shifts of the delta cycle have been very well described (Figure 15.4). For example, oyster reefs and marine habitats dominate in inactive, disconnected, coastal basins, whereas tidal freshwater and brackish deltaic floodplains dominate in an active, connected, delta region (Melancon Jr et al., 1998; Soniat et al., 2013). The distinction is that the thresholds of when connectivity or dis-connectivity occur are within the permitting process of ecosystem design for river diversions. Governance will determine the relative shifts in these habitat types as inactive basins become active. As described earlier, optimizing to recover from cumulative elevation deficits with connectivity operations will minimize marine habitats and fisheries to achieve the elevation thresholds that are needed to re-establish delta floodplain productivity (Caffey et al., 2014; Peyronnin et al., 2017). This establishes a conflict in connectivity strategies between delta floodplain stability and estuarine fisheries within coastal basins. This conflict is

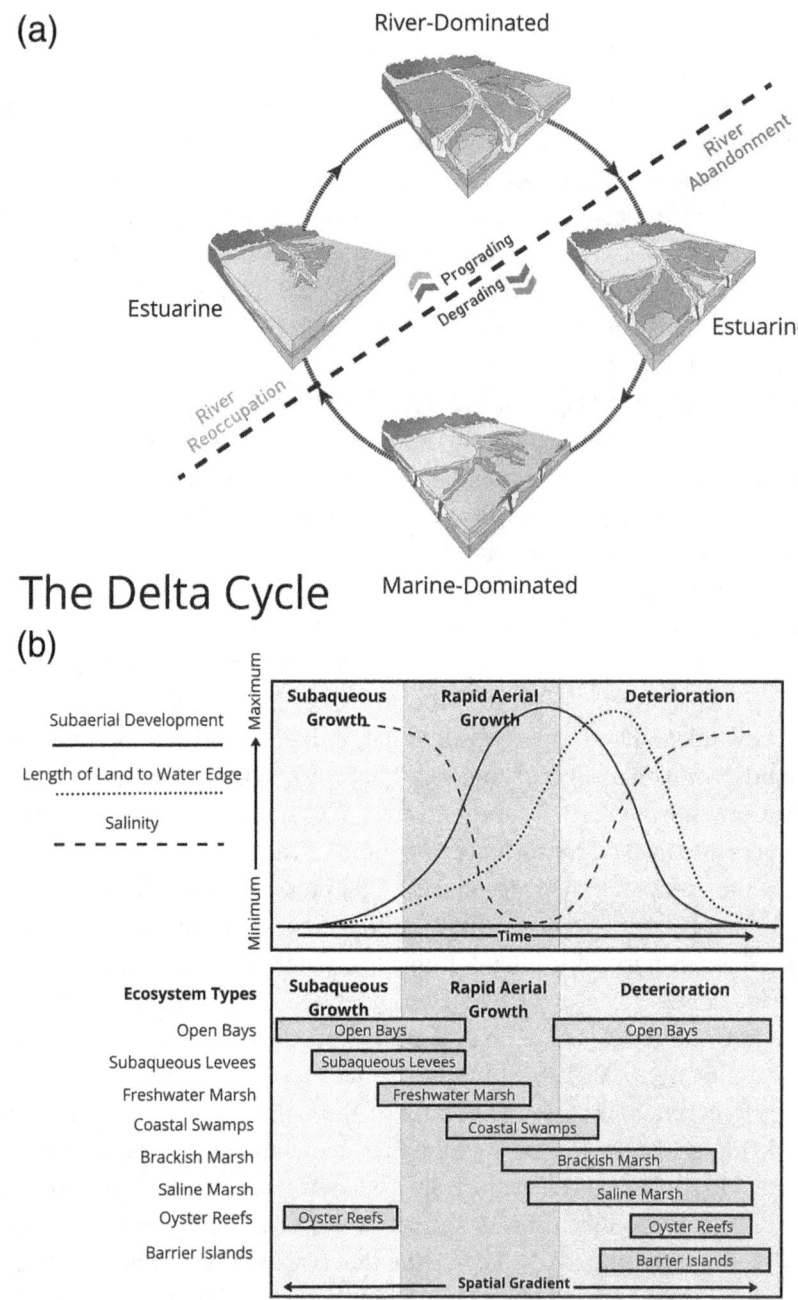

Figure 15.4 (a) Landscape changes of delta cycle associated with river occupation (prograding active delta) compared to stages with river abandonment (degrading inactive delta) modified from Penland et al. (1988). (b) Ecosystem development along the spatial and temporal gradients of delta cycle associated with magnitude of sediment delivery to coastal basins including specific attributes of coastal basins (subaerial development, length of land to water edge, salinity, estuarine secondary productivity) and distribution of ecosystem types in a coastal basin with magnitude of river input (modified from Gagliano & Van Beek, 1975; Gosselink et al., 1998). Numbers on the subaerial development line correspond to delta lobes in Mississippi River Delta. Twilley et al. (2019). A black and white version of this figure will appear in some formats. For the colour version of panel 15.4 (a), refer to the plate section.

not entirely associated with presence and absence of a fishery, as the shifts in habitat along the longitudinal axis of large river delta-front estuaries do not mean the productivity of nekton is lost, but it occurs where the delta front estuary may be extended out into the near shore environments (Rose et al., 2014; de Mutsert et al., 2017). However, benthic fishery, such as oyster harvesting, is regulated by leases of specific bottom areas of a coastal basin. The effects of river diversions on these industries are particularly controversial as shifts in these habitats with connectivity operations result in direct impact on existing fishery economics (Soniat et al., 2013). These complex linkages between deltaic floodplain stability and fisheries associated with connectivity operations demonstrate the importance of these concepts to regional economic development and management of the coastal zone.

The migration of coastal waters inland during the transgressive processes of an inactive delta basin of the Mississippi River Delta are also associated with flooding of local communities in the coastal zone (Twilley et al., 2016; Siverd et al., 2019). The decrease in the relative delta floodplain area since 1932 around Lafitte, Louisiana, is associated with the dis-connectivity of Barataria Bay from the Mississippi River (Siverd et al., 2020). An increase in construction of levees to establish protection from coastal flooding has cost local inhabitants about $19,000 per person per km inland migration of the Gulf of Mexico from 1973 to 2010 (inland migration has been 17 km during this time, Twilley et al., 2016). The indication is that decisions on connectivity and dis-connectivity of deltaic floodplains with river channels may enhance economic development in many sectors of local and national economics, but also have consequences to those that live in the coastal zone that becomes an inactive coastal basin (Barbier, 2015). The consideration of trade-offs in how ecosystem design and connectivity operations are decided, in the process of connecting inactive delta basins with river channels, involves more ecosystem services than just fisheries. The emergent delta floodplain area can be a significant contributor to local economics of flood protection, as well as impact fisheries as described in the previous paragraph. The key is that decisions on the magnitude and seasonal connectivity of water, solids, and solutes are incredibly important as complex design decisions impact how ecosystems and people adapt to changing conditions of deltaic continental margins around the world.

References

Amoros, C., & Bornette, G. (2002). Connectivity and biocomplexity in waterbodies of riverine floodplains. *Freshwater Biology,* 47(4): 761–776.

Asselman, N. E., & Middelkoop, H. (1995). Floodplain sedimentation: Quantities, patterns and processes. *Earth Surface Processes and Landforms,* 20(6), 481–499.

Bachand, P. A., & Horne, A. J. (1999). Denitrification in constructed free-water surface wetlands: II. Effects of vegetation and temperature. *Ecological Engineering,* 14(1–2), 17–32.

Barbier, E. B. (2015). Valuing the storm protection service of estuarine and coastal ecosystems. *Ecosystem Services*, 11, 32–38.

Barbier, E. B., Georgiou, I., Enchelmeyer, B., & Reed, D. J. (2013). The value of wetlands in protecting southeast Louisiana from hurricane storm surges. *Plos One*, 8(3): e58715.

Bargu, S., Justic, D., White, J. R., Lane, R., Day, J., Paerl, H., & Raynie, R. (2019). Mississippi River diversions and phytoplankton dynamics in deltaic Gulf of Mexico estuaries: a review. *Estuarine, Coastal and Shelf Science*, 221, 39–52.

Baumann, R. H., Day, J. W., Jr., & Miller, C. A. (1984). Mississippi deltaic wetland survival: sedimentation versus coastal submergence. *Science*, 224, 1093–1095.

Baustian, J. J., Mendelssohn, I. A., & Hester, M. W. (2012). Vegetation's importance in regulating surface elevation in a coastal salt marsh facing elevated rates of sea level rise. *Global Change Biology*, 18, 3377–3382. doi:10.1111/j.1365-2486.2012.02792.x

Bernard, R. J., Mortazavi, B., & Kleinhuizen, A. A. (2015). Dissimilatory nitrate reduction to ammonium (DNRA) seasonally dominates NO_3^- reduction pathways in an anthropogenically impacted sub-tropical coastal lagoon. *Biogeochemistry*, 125(1), 47–64.

Bevington, A. E., & Twilley, R. R. (2018). Island Edge Morphodynamics along a Chronosequence in a Prograding Deltaic Floodplain Wetland. *Journal of Coastal Research*, 34 (4), 806–817.

Bevington, A. E., Twilley, R. R., Sasser, C. E., & Holm, G. O. (2017). Contribution of river floods, hurricanes, and cold fronts to elevation change in a deltaic floodplain, northern Gulf of Mexico, USA. *Estuarine, Coastal and Shelf Science*, 191, 188–200. doi:10.1016/j.ecss.2017.04.010

Bianchi, T. S., & Allison, M. A. (2009). Large-river delta-front estuaries as natural "recorders" of global environmental change. *Proceedings of the National Academy of Sciences*, 106(20), 8085–8092.

Blair, N. E., & Aller, R. C. (2012). The fate of terrestrial organic carbon in the marine environment. *Annual Review of Marine Science*, 4, 401–423. doi:10.1146/annurev-marine-120709-142717

Blum, M. D., & Roberts, H. H. (2009). Drowning of the Mississippi delta due to insufficient sediment supply and global sea-level rise. *Nature Geoscience*, 2(7), 488–491. doi:10.1038/ngeo553

Blum, M. D., & Roberts, H. H. (2012). The Mississippi delta region: past, present, and future. *Annual Review of Earth and Planetary Sciences*, 40, 655–683.

Boesch, D. F., Josselyn, M. N., Mehta, A. J., Morris, J. T., Nuttle, W. K., Simenstad, C. A., & Swift, D. J. P. (1994). Scientific assessment of coastal wetland loss, restoration and management in Louisiana. *Journal of Coastal Research, Special issue No* 20, 1–103.

Broussard, W., & Turner, R. E. (2009). A century of changing land-use and water-quality relationships in the continental US. *Frontiers in Ecology and the Environment*, 7(6), 302–307.

Bryan, K. R., Nardin, W., Mullarney, J. C., & Fagherazzi, S. (2016). The role of cross-shore tidal dynamics in controlling intertidal sediment exchange in mangroves in Cù Lao Dung, Vietnam. *Continental Shelf Research* 147, 128–143, ISSN 0278-4343, https://doi.org/10.1016/j.csr.2017.06.014

Caffey, R. H., Wang, H., & Petrolia, D. R. (2014). Trajectory economics: Assessing the flow of ecosystem services from coastal restoration. *Ecological Economics*, 100, 74–84.

Cahoon, D. R. (2006). A review of major storm impacts on coastal wetland elevations. *Estuaries and Coasts*, 29(6A), 889–898.

Cahoon, D. R., White, D. A., & Lynch, J. C. (2011). Sediment infilling and wetland formation dynamics in an active crevasse splay of the Mississippi River delta. *Geomorphology*, 131, 57–68. doi:10.1016/j.geomorph.2010.12.002

Carney, J. A., R. R. Twilley, C. Agre, J. Hird, I. Georgiou, & J. Shelden. (2018). The giving delta. In E. Mossop (Ed.), *Sustainable Coastal Design and Planning*, pp. 239–254. CRC Press, Boca Roton, FL.

Christensen, A., Twilley, R. R., Willson, C. S., & Castañeda-Moya, E. (2020). Simulating hydrological connectivity and water age within a coastal deltaic floodplain of the Mississippi river delta. *Estuarine, Coastal and Shelf Science*, 245, 106995.

Coastal Protection and Restoration Authority. (2017). Louisiana's Comprehensive Master Plan for a Sustainable Coast (*Technical Report*). Coastal Protection and Restoration Authority of Louisiana. Baton Rouge, LA.

Cornwell, J. C., Kemp, W. M., & Kana, T. M. (1999). Denitrification in coastal ecosysems: methods, environmental controls, and ecosystem level controls, a review. *Aquatic Ecology*, 33, 41–54.

Covino, T. (2017). Hydrologic connectivity as a framework for understanding biogeochemical flux through watersheds and along fluvial networks. *Geomorphology*, 277, 133–144.

Das, A., Justic, D., Inoue, M., Hoda, A., Huang, H., & Park, D. (2012). Impacts of Mississippi River diversions on salinity gradients in a deltaic Louisiana estuary: Ecological and management implications. *Estuarine, Coastal and Shelf Science*, 111, 17–26. doi:10.1016/j.ecss.2012.06.005

Day Jr., J. W., Boesch, D. F., Clairain, E. J., Kemp, G. P., Laska, S. D., Mitsch, W. J., & Whigham, D. F. (2007). Restoration of the Mississippi delta: Lessons from Hurricanes Katrina and Rita. *Science*, 315, 1679–1684. doi:10.1126/science.1137030

Day Jr., J. W., Ko, J.-Y. Y., Rybczyk, J., Sabins, D., Bean, R., Berthelot, G., Twilley, R. (2004). The use of wetlands in the Mississippi delta for wastewater assimilation: A review. *Ocean & Coastal Management*, 47, 671–691. doi:10.1016/J.Ocecoaman.2004.12.007

Day, J. W., Cable, J. E., Cowan, J. H., DeLaune, R., de Mutsert, K., Fry, B., Wissel, B. (2009). The impacts of pulsed reintroduction of river water on a Mississippi delta coastal basin. *Journal of Coastal Research*, 10054, 225–243. doi:10.2112/SI54-015.1

Day, J. W., Kemp, G. P., Reed, D. J., Cahoon, D. R., Boumans, R. M., Suhayda, J. M., & Gambrell, R. (2011). Vegetation death and rapid loss of surface elevation in two contrasting Mississippi delta salt marshes: The role of sedimentation, autocompaction and sea-level rise. *Ecological Engineering*, 37, 229–240. doi:10.1016/j.ecoleng.2010.11.021

Day, J. W., Lane, R. R., D'Elia, C. F., Wiegman, A. R. H., Rutherford, J. S., Shaffer, G. P., Brantley, C. G., & Kemp, G. P. (2018). Large infrequently operated river diversions for Mississippi delta restoration. *Mississippi Delta Restoration: Pathways to a Sustainable Future*, 2018: 113–133.

de Mutsert, K., Lewis, K., Milroy, S., Buszowski, J., & Steenbeek, J. (2017). Using ecosystem modeling to evaluate trade-offs in coastal management: Effects of large-scale river diversions on fish and fisheries. *Ecological Modelling*, 360, 14–26.

DeLaune, R., Sasser, C., Evers-Hebert, E., White, J., & Roberts, H. (2016). Influence of the Wax Lake Delta sediment diversion on aboveground plant productivity and carbon storage in deltaic island and mainland coastal marshes. *Estuarine, Coastal and Shelf Science*, 177, 83–89.

DeLaune, R. D., Smith, C. J., Patrick, W. H., & Roberts, H. H. (1987). Rejuvenated marsh and bay-bottom accretion on the rapidly subsiding coastal plain of U.S. Gulf Coast: A second-order effect of the emerging Atchafalaya Delta. *Estuarine, Costal and Shelf Science*, 25, 381–389.

Deleersnijder, E., Campin, J.-M., & Delhez, E. J. (2001). The concept of age in marine modelling: I. Theory and preliminary model results. *Journal of Marine Systems*, 28(3–4), 229–267.

Delhez, E. J., Campin, J.-M., Hirst, A. C., & Deleersnijder, E. (1999). Toward a general theory of the age in ocean modelling. *Ocean Modelling*, 1(1), 17–27.

Dettmann, E. H. (2001). Effect of water residence time on annual export and denitrification of Nitrogen in Estuaries: A model analysis. *Estuaries*, 24, 481. doi:10.2307/1353250

Diaz, R. J., & Rosenberg, R. (2008). Spreading dead zones and consequences for marine ecosystems. *Science*, 321(5891), 926–929.

Edmonds, D. A., Paola, C., Hoyal, D. C., & Sheets, B. A. (2011). Quantitative metrics that describe river deltas and their channel networks. *Journal of Geophysical Research: Earth Surface*, 116(F4).

Edmonds, D. A., & Slingerland, R. L. (2010). Significant effect of sediment cohesion on delta morphology. *Nature Geoscience*, 3, 105–109. doi:10.1038/ngeo730

Edmonds, D. A., Caldwell, R. L., Brondizio, E. S. & Mo Siani, S. (2020). Coastal flooding will disproportionately impact people on river deltas. *Nature Geoscience*, 11, 1–8, https://doi.org/10.1038/s41467-020-18531-4

Elsey-Quirk, T., Graham, S. A., Mendelssohn, I. A., Snedden, G., Day, J. W., Twilley, R., Lane, R. (2019). Mississippi river sediment diversions and coastal wetland sustainability: Synthesis of responses to freshwater, sediment, and nutrient inputs. *Estuarine, Coastal and Shelf Science*, 221, 170–183.

Ensign, S., Siporin, K., Piehler, M., Doyle, M., & Leonard, L. (2013). Hydrologic versus biogeochemical controls of denitrification in tidal freshwater wetlands. *Estuaries and Coasts*, 36, 519–532.

Everett, T., Chen, Q., Karimpour, A., & Twilley, R. (2019). Quantification of swell energy and its impact on wetlands in a deltaic estuary. *Estuaries and Coasts*, 42, 68–84. https://doi.org/10.1007/s12237-018-0454-z

Eyre, B. D., & Ferguson, A. J. P. (2009). Denitrification efficiency for defining critical loads of carbon in shallow coastal ecosystems. In *Eutrophication in Coastal Ecosystems: Towards better understanding and management strategies Selected Papers from the Second International Symposium on Research and Management of Eutrophication in Coastal Ecosystems, 20–23 June 2006, Nyborg, Denmark*, pp. 137–146. Springer, Netherlands.

Fagherazzi, S., Edmonds, D. A., Nardin, W., Leonardi, N., Canestrelli, A., Falcini, F., & Slingerland, R. L. (2015). Dynamics of river mouth deposits. *Reviews of Geophysics*, 53(3), 642–672.

Gagliano, S. M., & Van Beek, J. L. (1975). *An approach to multiuse management in the Mississippi Delta system.* Houston Geological Society, Deltas: Models for Exploration.

Gardner, W. S., McCarthy, M. J., An, S., Sobolev, D., Sell, K. S., & Brock, D. (2006). Nitrogen fixation and dissimilatory nitrate reduction to ammonium (DNRA) support nitrogen dynamics in Texas estuaries. *Limnology and Oceanography*, 51(1), 558–568.

Geleynse, N., Hiatt, M., Sangireddy, H., & Passalacqua, P. (2015). Identifying environmental controls on the shoreline of a natural river delta. *Journal of Geophysical Research F: Earth Surface* 120, 877–893.

Goolsby, D. A., Battaglin, W. A., Aulenbach, B. T., & Hooper, R. P. (2000). Nitrogen flux and sources in the Mississippi River Basin. *Science of The Total Environment* 248(2–3), 75–86. doi:10.1016/S0048-9697(99)00532-X

Gosselink, J. G., Coleman, J. M., & Stewart Jr, R. E. (1998). Coastal Louisiana. *Status and Trends of the Nation's Biological Resources*, 2, 385–436.

Grimsditch, G., Alder, J., Nakamura, T., Kenchington, R., & Tamelander, J. (2013). The blue carbon special edition – Introduction and overview. *Ocean & Coastal Management*, 83, 1–4. doi:10.1016/j.ocecoaman.2012.04.020

Hanegan, K. & Georgiou, I. (2015). Tidal modulated flow and sediment flux through Wax Lake Delta distributary channels: Implications for delta development. *Proceedings of IAHS*, 367, 391–398, https://doi.org/10.5194/piahs-367-391-2015

Hedges, J. I., & Keil, R. G. (1995). Sedimentary organic matter preservation: an assessment and speculative synthesis. *Marine Chemistry*, 49, 81–115. doi:10.1016/0304-4203(95)00008-F

Heiler, G., Hein, T., Schiemer, F., & Bornette, G. (1995). Hydrological connectivity and flood pulses as the central aspects for the integrity of a river-floodplain system. *Regulated Rivers: Research & Management*, 11, 351–361. doi:10.1002/rrr.3450110309

Henry, K. M., & Twilley, R. R. (2014). Nutrient biogeochemistry during the early stages of delta development in the Mississippi river deltaic plain. *Ecosystems*, 17, 327–343.

Hiatt, M., & Passalacqua, P. (2015). Hydrological connectivity in river deltas: The first-order importance of channel-island exchange. *Water Resources Research*, 51(4), 2264–2282.

Hiatt, M., & Passalacqua, P. (2017), What controls the transition from confined to unconfined flow? Analysis of hydraulics in a coastal river delta. *Journal of Hydraulic Engineering*, 143, 6, doi: 10.1061/(ASCE)HY.1943-7900.0001309

Hiatt, M., Castañeda-Moya, E., Twilley, R., Hodges, B. R., & Passalacqua, P. (2018). Channel-island connectivity affects water exposure time distributions in a coastal river delta. *Water Resources Research*, 54(3), 2212–2232.

Holmquist, J. R., Windham-Myers, L., Bliss, N., Crooks, S., Morris, J. T., Megonigal, J. P., Drexler, J. (2018). Accuracy and precision of tidal wetland soil carbon mapping in the conterminous United States. *Scientific Reports*, 8(1), 1–16.

Horstman, Erik M., Dohmen-Janssen, C. Marjolein, & Hulscher, Suzanne J. M. H. (2013). Flow routing in mangrove forests: A field study in Trang province, Thailand. *Continental Shelf Research*, 71, 52–67, ISSN 0278-4343, https://doi.org/10.1016/j.csr.2013.10.002

Howarth, R. W., Sharpley, A., & Walker, D. (2002). Sources of nutrient to coastal waters in the United States (implications for achieving coastal water quality goals). *Estuaries*, 25, 656–676. doi:10.1007/BF02804898

Hu, Kelin, Chen, Qin, & Wang, Hongqing. (2015). A numerical study of vegetation impact on reducing storm surge by wetlands in a semi-enclosed estuary. *Coastal Engineering*, 95, 66–76, ISSN 0378-3839, https://doi.org/10.1016/j.coastaleng.2014.09.008

Islam, S. (2016). Deltaic floodplains development and wetland ecosystems management in the Ganges–Brahmaputra–Meghna Rivers Delta in Bangladesh. *Sustainable Water Resources Management*, 2(3): 237–256.

Junk, W. J., Bayley, P., & Sparks, R. (1989). The flood pulse concept in river-floodplain systems. *Canadian Special Publication of Fisheries and Aquatic Sciences*, 106, 110–127. doi:10.1371/journal.pone.0028909

Kadlec, R., & Wallace, S. (2008). *Treatment Wetlands*, Second Edition. doi:10.1201/9781420012514

Kadlec, R. H. (2010). Nitrate dynamics in event-driven wetlands. *Ecological Engineering*, 36(4), 503–516. doi:10.1016/j.ecoleng.2009.11.020

Kadlec, R. H. and Knight, R.L. (1996). *Treatment Wetlands*. CRC Press LLC, Boca Raton, FL. ISBN 0-87371-930-1. 893 pages.

Kadlec, R. H., & Reddy, K. (2001). Temperature effects in treatment wetlands. *Water environment research*, 73(5), 543–557.

Kaushal, S. S., Groffman, P. M., Mayer, P. M., Striz, E., & Gold, A. J. (2008). Effects of stream restoration on denitrification in an urbanizing watershed. *Ecological Applications*, 18(3), 789–804.

Kelly-Gerreyn, B., Hydes, D., Trimmer, M., & Nedwell, D. (1999). Calibration of an early diagenesis model for high nitrate, low reactive sediments in a temperate latitude estuary (Great Ouse, UK). *Marine Ecology Progress Series*, 177, 37–50.

Kelly-Gerreyn, B. A., Trimmer, M., & Hydes, D. J. (2001). A diagenetic model discriminating denitrification and dissimilatory nitrate reduction to ammonium in a temperate estuarine sediment. *Marine Ecology Progress Series*, 220, 33–46. doi:10.3354/meps220033

Klocker, C. A., Kaushal, S. S., Groffman, P. M., Mayer, P. M., & Morgan, R. P. (2009). Nitrogen uptake and denitrification in restored and unrestored streams in urban Maryland, USA. *Aquatic sciences*, 71(4), 411–424.

Lane, R. R., Madden, C. J., Day, J. W., Jr., & Solet, D. J. (2011). Hydrologic and nutrient dynamics of a coastal bay and wetland receiving discharge from the Atchafalaya River. *Hydrobiologia*, 658(1), 55–66. doi:10.1007/s10750-010-0468-4

Li, S., Christensen, A., & Twilley, R. R. (2020). Benthic fluxes of dissolved oxygen and nutrients across hydrogeomorphic zones in a coastal deltaic floodplain within the Mississippi River delta plain. *Biogeochemistry*, 149, 115–140.

Li, S., & Twilley, R. R. (2021). Nitrogen dynamics of inundated sediments in an emerging coastal deltaic floodplain in mississippi river delta using isotope pairing technique to test response to nitrate enrichment and sediment organic matter. *Estuaries and Coasts*, 44:1899–1915.

Li, S., Twilley, R. R., & Hou, A. (2021). Heterotrophic nitrogen fixation in response to nitrate loading and sediment organic matter in an emerging coastal deltaic floodplain within the Mississippi River Delta plain. *Limnology and Oceanography*, 66(5), 1961–1978.

Liu, K., Chen, Q., Hu, K., Xu, K., & Twilley, R. R. (2018). Modeling hurricane-induced wetland-bay and bay-shelf sediment fluxes. *Coastal Engineering* 135: 77–90.

Ma, H., Larsen, L. G., & Wagner, R. W. (2018). Ecogeomorphic feedbacks that grow deltas. *Journal of Geophysical Research: Earth Surface*, 123(12), 3228–3250.

Madden, C. J., Day, J. W., Jr., & Randall, J. M. (1988). Freshwater and marine coupling in estuaries of the Mississippi River deltaic plain. *Limnology and Oceanography*, 33(4), 982–1004.

Martin, J. F., & Reddy, K. R. (1997). Interaction and spatial distribution of wetland nitrogen processes. *Ecological Modelling*, 105, 1–21. doi:10.1016/S0304-3800(97)00122-1

McKee, K. L., & Cherry, J. A. (2009). Hurricane Katrina sediment slowed elevation loss in subsiding brackish marshes of the Mississippi River Delta. *Wetlands*, 29(1), 2–15.

McLeod, E., Chmura, G. L., Bouillon, S., Salm, R., Bjork, M., Duarte, C. M., & Silliman, B. R. (2011). A blueprint for blue carbon: toward an improved understanding of the role of vegetated coastal habitats in sequestering $CO2$. *Frontiers in Ecology and the Environment*, 9(10), 552–560. doi:10.1890/110004

Melancon Jr, E., Soniat, T., Cheramie, V., Dugas, R., Barras, J., & Lagarde, M. (1998). Oyster resource zones of the Barataria and Terrebonne estuaries of Louisiana. *Journal of Shellfish Research*, 17(4), 1143–1148.

Mendelssohn, I. A., & Kuhn, N. L. (2003). Sediment subsidy: effects on soil-plant responses in a rapidly submerging coastal salt marsh. *Ecological Engineering*, 21(2–3), 115–128. doi:10.1016/j.ecoleng.2003.09.006

Mitsch, W. J., Day, J. W., Zhang, L., & Lane, R. R. (2005). Nitrate-nitrogen retention in wetlands in the Mississippi River Basin. *Ecological Engineering*, 24, 267–278.

Morris, J. T. (2006). Competition among marsh macrophytes by means of geomorphological displacement in the intertidal zone. *Estuarine, Coastal and Shelf Science*, 69(3,Äì4), 395–402. http://dx.doi.org/10.1016/j.ecss.2006.05.025

Morris, J. T., Sundareshwar, P. V., Nietch, C. T., Kjerfve, B., & Cahoon, D. R. (2002). Responses of coastal wetlands to rising sea level. *Ecology*, 83(10), 2869–2877. doi:10.2307/3072022

Morton, R. A., & Barras, J. A. (2011). Hurricane Impacts on Coastal Wetlands: A Half-Century Record of Storm-Generated Features from Southern Louisiana. *Journal of Coastal Research*, 275, 27–43. doi:10.2112/JCOASTRES-D-10-00185.1

Nardin, W., & Edmonds, D. A. (2014). Optimum vegetation height and density for inorganic sedimentation in deltaic marshes. *Nature Geoscience*, 7(10), 722.

Nardin, W., Edmonds, D. A., & Fagherazzi, S. (2016). Influence of vegetation on spatial patterns of sediment deposition in deltaic islands during flood. *Advances in Water Resources*, 93, 236–248.

Nardin, W., Lera, S., & Nienhuis, J. (2020) Effect of offshore waves and vegetation on the sediment budget in the Virginia Coast Reserve (VA). *Earth Surface Processes and Landforms*, 45, 3055–3068. https://doi.org/10.1002/esp.4951

Noe, G. B., & Hupp, C. R. (2005). Carbon, nitrogen, and phosphorus accumulation in floodplains of Atlantic Coastal Plain rivers, USA. *Ecological Applications*, 15(4), 1178–1190. doi:10.1890/04-1677

Noe, G. B., Hupp, C. R., & Rybicki, N. B. (2013). Hydrogeomorphology Influences Soil Nitrogen and Phosphorus Mineralization in Floodplain Wetlands. *Ecosystems*, 16(1), 75–94. doi:10.1007/s10021-012-9597-0

Nyman, J. A., Crozier, C. R., & DeLaune, R. D. (1995). Roles and patterns of hurricane sedimentation in an estuarine marsh landscape. *Estuarine, Coastal and Shelf Science*, 40, 665–679.

Paerl, H. W. (2006). Assessing and managing nutrient-enhanced eutrophication in estuarine and coastal waters: Interactive effects of human and climatic perturbations. *Ecological Engineering*, 26, 40–45.

Paola, C., Twilley, R. R., Edmonds, D. A., Kim, W., Mohrig, D., Parker, G., & Voller, V. R. (2011). Natural Processes in Delta Restoration: Application to the Mississippi Delta. *Annual Review of Marine Science*, 3, 67–91. doi:10.1146/annurev-marine-120709-142856

Passalacqua, P. (2017). The Delta Connectome: A network-based framework for studying connectivity in river deltas. *Geomorphology*, 277, 50–62. doi:10.1016/j.geomorph.2016.04.001

Penland, S., Boyd, R., & Suter, J. R. (1988). Transgressive depositional systems of the Mississippi Delta plain; a model for barrier shoreline and shelf sand development. *Journal of Sedimentary Research*, 58(6), 932–949.

Perez, B. C., Day, J. W., Jr., Rouse, L. J., Shaw, R. F., & Wang, M. (2000). Influence of Atchafalaya River discharge and winter frontal passage on suspended sediment concentration and flux in Fourleague Bay, Louisiana. *Estuarine, Coastal and Shelf Science*, 50, 271–290.

Perez, B. C., Day, J. W., Justic, D., & Twilley, R. R. (2003). Nitrogen and phosphorus transport between Fourleague Bay, LA, and the Gulf of Mexico: the role of winter cold fronts and Atchafalaya River discharge. *Estuarine Coastal and Shelf Science*, 57(5–6), 1065–1078. doi:10.1016/S0272-7714(03)00010-6

Peyronnin, N. S., Caffey, R. H., Cowan, J. H., Justic, D., Kolker, A. S., Laska, S. B., Wilkins, J. G. (2017). Optimizing sediment diversion operations: Working group recommendations for integrating complex ecological and social landscape interactions. *Water (Switzerland)*, 9. doi:10.3390/w9060368

Rabalais, N. N., Turner, R. E., Justić, D., Dortch, Q., Wiseman, W. J., & Gupta, B. K. S. (1996). Nutrient changes in the Mississippi River and system responses on the adjacent continental shelf. *Estuaries*, 19(2), 386–407.

Rabalais, N. N., Turner, R. E., & Scavia, D. (2002). Beyond Science into Policy: Gulf of Mexico Hypoxia and the Mississippi River: Nutrient policy development for the Mississippi River watershed reflects the accumulated scientific evidence that the increase in nitrogen loading is the primary factor in the worsening of hypoxia in the northern Gulf of Mexico. *AIBS Bulletin*, 52(2), 129–142.

Rejmánek, M., Sasser, C. E., & Gosselink, J. G. (1987). Modeling of vegetation dynamics in the Mississippi River deltaic plain. *Vegetatio*, 69(1–3), 133–140.

Rejmánek, M., Sasser, C. E., & Peterson, G. W. (1988). Hurricane-induced sediment deposition in a Gulf coast marsh. *Estuarine, Coastal and Shelf Science*, 27(2), 217–222.

Restreppo, G. A., Bentley, S. J., Wang, J., & Xu, K. (2018). Riverine Sediment Contribution to Distal Deltaic Wetlands: Fourleague Bay, LA. *Estuaries and Coasts*, 1–13.

Riekenberg, J., Bargu, S., & Twilley, R. (2014). Phytoplankton Community Shifts and Harmful Algae Presence in a Diversion Influenced Estuary. *Estuaries and Coasts*, 38, 2213–2226. doi:10.1007/s12237-014-9925-z

Roberts, B. J., & Doty, S. M. (2015). Spatial and temporal patterns of benthic respiration and net nutrient fluxes in the Atchafalaya River Delta Estuary. *Estuaries and Coasts*, 38(6), 1918–1936.

Roberts, H. H. (1997). Dynamic changes of the Holocene Mississippi River delta plain: the delta cycle. *Journal of Coastal Research*, 605–627.

Rose, K. A., Huang, H., Justic, D., & de Mutsert, K. (2014). Simulating fish movement responses to and potential salinity stress from large-scale river diversions. *Marine and Coastal Fisheries*, 6, 43–61. doi:10.1080/19425120.2013.866999

Ross, M. R. V., Emily, S., Bernhardt, E. S., Doyle, M. W., & Heffernan, J. B. (2015). Designer ecosystems: Incorporating design approaches into applied ecology. *Annual Review of Environment and Resources* 40, 419–443. doi:10.1146/annurev-environ-121012-100957

Rutherford, J. S., Day, J. W., D'Elia, C. F., Wiegman, A. R., Willson, C. S., Caffey, R. H., Batker, D. (2018). Evaluating trade-offs of a large, infrequent sediment diversion for restoration of a forested wetland in the Mississippi delta. *Estuarine, Coastal and Shelf Science*, 203, 80–89.

Sasser, C. E., Visser, J. M., Mouton, E., Linscombe, J., & Hartley, S. B. (2008). Vegetation types in coastal Louisiana in 2007. *Estuaries*, 21, 818–828.

Scaroni, A. E., Nyman, J. A., & Lindau, C. W. (2011). Comparison of denitrification characteristics among three habitat types of a large river floodplain: Atchafalaya River Basin, Louisiana. *Hydrobiologia*, 658(1), 17–25.

Sendrowski, A. & Passalacqua, P. (2017). Process connectivity in a naturally prograding river delta, *Water Resources Research*, 53(3), 1841–1863, doi:10.1002/2016 WR019768.

Shaffer, P. W., Kentula, M. E., & Gwin, S. E. (1999). Characterization of wetland hydrology using hydrogeomorphic classification. *Wetlands*, 19, 490–504. doi:10.1007/BF03161688

Shaw, J. B., & Mohrig, D. (2014). The importance of erosion in distributary channel network growth, Wax Lake Delta, Louisiana, USA. *Geology*, 42(1), 31–34.

Shen, J., & Haas, L. (2004). Calculating age and residence time in the tidal York River using three-dimensional model experiments. *Estuarine, Coastal and Shelf Science*, 61(3), 449–461.

Shields, M. R., Bianchi, T. S., Gélinas, Y., Allison, M. A., & Twilley, R. R. (2016). Enhanced terrestrial carbon preservation promoted by reactive iron in deltaic sediments. *Geophysical Research Letters*, 43, 1149–1157. doi:10.1002/2015GL067388

Shields, M.R., Bianchi, T.S., Kolker, A.S., Kenney, W.F., Mohrig, D., Osborne, T.Z., & Curtis, J.H. (2019). Factors controlling storage, sources, and diagenetic state of organic carbon in a prograding subaerial delta: Wax Lake Delta, Louisiana: *Journal of Geophysical Research – Biogeosciences,* 124, doi:10.1029/2018JG004683

Shields, M. R., Bianchi, T. S., Mohrig, D., Hutchings, J., Kenney, W. F., Kolker, A. S., & Curtis, J. H. (2017). Carbon storage in the Mississippi River Delta enhanced by ecosystem engineering: *Nature Geoscience,* 10 (11), doi:10.1038/NGEO3044

Siikamäki, J., Sanchirico, J. N., Jardine, S. L., Siikamaki, J., Sanchirico, J. N., & Jardine, S. L. (2012). Global economic potential for reducing carbon dioxide emissions from mangrove loss. *Proceedings of the National Academy of Sciences,* 109, 14369–14374. doi:10.1073/pnas.1200519109

Siverd, C. G., Hagen, S. C., Bilskie, M. V., Braud, D. H., Peele, R. H., Foster-Martinez, M. R., & Twilley, R. R. (2019). Coastal Louisiana landscape and storm surge evolution: 1850–2110. *Climatic Change,* 157(3), 445–468.

Siverd, C. G., Hagen, S. C., Bilskie, M. V., Braud, D. H., & Twilley, R. R. (2020). Quantifying storm surge and risk reduction costs: A case study for Lafitte, Louisiana. *Climatic Change,* 161(1), 201–223.

Smith, C. J., DeLaune, R. D., & Patrick, W. H., Jr. (1985). Fate of riverine nitrate entering an estuary: I. Denitrification and nitrogen burial. *Estuaries,* 8, 15–21.

Snedden, G. A., Cable, J. E., Swarzenski, C., & Swenson, E. (2007). Sediment discharge into a subsiding Louisiana seltaic estuary through a Mississippi River diversion. *Estuarine, Coastal and Shelf Science,* 71, 181–193.

Soniat, T. M., Conzelmann, C. P., Byrd, J. D., Roszell, D. P., Bridevaux, J. L., Suir, K. J., & Colley, S. B. (2013). Predicting the effects of proposed Mississippi River Diversions on Oyster habitat quality; Application of an Oyster habitat suitability index model. *Journal of Shellfish Research,* 32, 629–638. doi:10.2983/035.032.0302

Stanford, G., Dzienia, S., & Vander Pol, R. A. (1975). Effect of temperature on denitrification rate in soils. *Soil Science Society of America Journal,* 39(5), 867–870.

Syvitski, J. P., Kettner, A. J., Overeem, I., Hutton, E. W., Hannon, M. T., Brakenridge, G. R., & Giosan, L. (2009). Sinking deltas due to human activities. *Nature Geoscience,* 2(10), 681–686.

Temmerman, S., Govers, G., Wartel, S., & Meire, P. (2003). Spatial and temporal factors controlling short-term sedimentation in a salt and freshwater tidal marsh, Scheldt estuary, Belgium, SW Netherlands. *Earth Surface Processes and Landforms: The Journal of the British Geomorphological Research Group* 28(7): 739–755.

Tornqvist, T. E., Paola, C., Parker, G., Liu, K., Mohrig, D., Holbrook, J. M., & Twilley, R. R. (2007). Comment on "Wetland sedimentation from Hurricanes Katrina and Rita". *Science,* 316(5822). doi: 10.1126/Science.1136780

Turner, R. E., Baustian, J. J., Swenson, E., & Spicer, J. S. (2006). Wetland sedimentation from hurricanes Katrina and Rita. *Science,* 314, 449–452.

Turner, R. E., & Boyer, M. E. (1997). Mississippi river diversions, coastal wetland restoration/creation and an economy of scale. *Ecological Engineering,* 8(2), 117–128. http://dx.doi.org/10.1016/S0925-8574(97)00258-9

Tweel, A. W., & Turner, R. E. (2012). Landscape-scale analysis of wetland sediment deposition from four tropical cyclone events. *Plos One,* 7(11), e50528.

Twilley, R., Day, J., Bevington, A., Castañeda-Moya, E., Christensen, A., Holm, G., & Aarons, A. (2019). Ecogeomorphology of coastal deltaic floodplains and estuaries in an active delta: Insights from the Atchafalaya coastal basin. *Estuarine, Coastal and Shelf Science,* 106341.

Twilley, R. R., & Rivera-Monroy, V. H. (2009). Sediment and nutrient trade-offs in restoring Mississippi river delta: restoration versus eutrophicaion. *Journal of Contemporary Water Research Education,* 141, 1–6.

Twilley, R. R., Bentley, S. J., Chen, Q., Edmonds, D. A., Hagen, S. C., Lam, N. S., & McCall, A. (2016). Co-evolution of wetland landscapes, flooding, and human settlement in the Mississippi river delta plain. *Sustainability Science*, 11, 711–731. doi:10.1007/s11625-016-0374-4

Twilley, R.R., S. Rick, D. Bond, J. Baker. 2021. Benthic Nutrient Fluxes Across Subtidal and Intertidal Habitats in Breton Sound in Response to River-Pulses of a Diversion in Mississippi River Delta. Water (ISSN 2073–4441).

Verschelling, E., Van der Deijl, E. C., Van der Perk, M., Sloff, K., & Middelkoop, H. (2017). Effects of discharge, wind and tide on sedimentation in a recently restored tidal freshwater wetland. *Hydrological Processes*, 31, 2827–2841.

Viero, D. P., & Defina, A. (2016). Water age, exposure time, and local flushing time in semi-enclosed, tidal basins with negligible freshwater inflow. *Journal of Marine Systems*, 156, 16–29.

Visser, J. M., Sasser, C. E., Chabreck, R. H., & Linscombe, R. (1998). Marsh vegetation types of the Mississippi river deltaic plain. *Estuaries*, 21(4), 818–828.

Wagner, W., Lague, D., Mohrig, D., Passalacqua, P., Shaw, J., & Moffett, K. (2017). Elevation change and stability on a prograding delta. *Geophysical Research Letters*, 44(4), 1786–1794.

Wainwright, J., Turnbull, L., Ibrahim, T. G., Lexartza-Artza, I., Thornton, S. F., & Brazier, R. E. (2011) Linking environmental régimes, space and time: Interpretations of structural and functional connectivity. *Geomorphology*, 126(3–4), 387–404, ISSN 0169-555X, https://doi.org/10.1016/j.geomorph.2010.07.027

Wamsley, T.V., Cialone, M. A., Smith, J. M., Atkinson, J. H. & Rosati, J. D. (2010). The potential of wetlands in reducing storm surge. *Ocean Engineering*, 37(1): 59–68.

Walker, N. D. (2001). Tropical storm and hurricane wind effects on water level, salinity, and sediment transport in the river-influenced Atchafalaya-Vermilion Bay system, Louisiana, USA. *Estuaries*, 24(4), 498–508.

Walker, N. D., & Hammack, A. B. (2000). Impacts of winter storms on circulation and sediment transport: Atchafalaya-Vermilion Bay Region, Louisiana, U.S.A., *Journal of Coastal Research*, 16(4), 996–1010.

Wang, H., Steyer, G. D., Couvillion, B. R., Rybczyk, J. M., Beck, H. J., Sleavin, W. J., & Rivera-Monroy, V. H. (2014). Forecasting landscape effects of Mississippi River diversions on elevation and accretion in Louisiana deltaic wetlands under future environmental uncertainty scenarios. *Estuarine, Coastal and Shelf Science*, 138, 57–68. doi:10.1016/j.ecss.2013.12.020

Wellner, R., Beaubouef, R., Van Wagoner, J., Roberts, H. H., Sun, T., & Wagoner, J. V. (2005). Jet-plume depositional bodies; the primary building blocks of Wax Lake Delta. *Transactions – Gulf Coast Association of Geological Societies*, 55, 867–909.

White, J. R., DeLaune, R. D., Justic, D., Day, J. W., Pahl, J., Lane, R. R., & Twilley, R. R. (2019). Consequences of Mississippi river diversions on nutrient dynamics of coastal wetland soils and estuarine sediments: A review. *Estuarine, Coastal and Shelf Science*, 224, 209–216.

Wiegman, A. R., Day, J. W., D'Elia, C. F., Rutherford, J. S., Morris, J. T., Roy, E. D., & Snyder, B. F. (2018). Modeling impacts of sea-level rise, oil price, and management strategy on the costs of sustaining Mississippi delta marshes with hydraulic dredging. *Science of the Total Environment*, 618, 1547–1559.

Xing, Fei, Syvitski, J. P. M., Kettner, A. J., Meselhe, E. A., Atkinson, J. H., & Khadka, A. K. (2017). Morphological responses of the Wax Lake Delta, Louisiana, to Hurricanes Rita. *Elementa: Science of the Anthropocene*, 5, 80. https://doi.org/10.1525/elementa.125

Index